Edited by
Sarah Jane George and
Jason Johnson

Atherosclerosis

Related Titles

Kim, J. S., Caplan, L. R., Wong, K. S. L.

Intracranial Atherosclerosis

2009
ISBN: 978-1-4051-7822-8

Bader, M. (ed.)

Cardiovascular Hormone Systems
From Molecular Mechanisms to Novel Therapeutics

2008
ISBN: 978-3-527-31920-6

Xu, Q. (ed.)

A Handbook of Mouse Models of Cardiovascular Disease

2006
ISBN: 978-0-470-01610-7

Edited by Sarah Jane George and Jason Johnson

Atherosclerosis

Molecular and Cellular Mechanisms

WILEY-VCH

WILEY-VCH Verlag GmbH & Co. KGaA

The Editors

Dr. Sarah Jane George
Bristol Royal Infirmary
Upper Maudlin Street
Bristol BS2 8HW
United Kingdom

Dr. Jason Johnson
Bristol Royal Infirmary
Upper Maudlin Street
Bristol BS2 8HW
United Kingdom

Cover
Cover figure shows Picrosirius red staining of a brachiocephalic artery plaque taken from an Apolipoprotein E deficient mice after eight weeks of high fat feeding. Inset shows a schematic diagram of an unstable atherosclerotic plaque. Pictures provided with kind permission of Sarah J. George.

Library of Congress Card No.: applied for

British Library Cataloguing-in-Publication Data
A catalogue record for this book is available from the British Library.

Bibliographic information published by the Deutsche Nationalbibliothek
The Deutsche Nationalbibliothek lists this publication in the Deutsche Nationalbibliografie; detailed bibliographic data are available on the Internet at http://dnb.d-nb.de.

© 2010 WILEY-VCH Verlag GmbH & Co. KGaA, Weinheim

Cover Design   Adam-Design, Weinheim
Typesetting   Toppan Best-set Premedia Limited
Printing and Binding   Strauss GmbH, Mörlenbach

Printed in the Federal Republic of Germany
Printed on acid-free paper

ISBN: 978-3-527-32448-4

Contents

Atherosclerosis: Molecular and Cellular Mechanisms. Edited by Sarah Jane George and Jason Johnson
Copyright © 2010 WILEY-VCH Verlag GmbH & Co. KGaA, Weinheim
ISBN: 978-3-527-32448-4

Preface

Cardiovascular disease, which causes coronary artery and peripheral artery disease, is the largest single cause of death and disability in the industrialised world. Atherosclerosis is the pathology which underlies cardiovascular disease. Consequently, a greater understanding of the mechanisms underlying atherosclerosis is essential for the development of new therapeutic approaches for the treatment of cardiovascular disease.

Atherosclerosis: Molecular and Cellular Mechanisms is designed to present evidence for the mechanisms underlying atherosclerosis. Atherosclerosis is multifactorial and involves several cell types and therefore this book is aimed to reflect this. This book begins with an introductory chapter, which provides an overview to the pathogenesis of atherosclerosis and sets the scene for the remainder of the volume. The subsequent seventeen chapters are written by world experts and focus on specific cellular and molecular mechanisms to provide detailed up-to-date evidence on one specific area. The book is divided into five sections to reflect key and contemporary areas which modulate atherosclerosis; pro-inflammatory factors, proteases, hyperlipidemia, oxidative stress, and cell growth and phenotype. Although the full picture is not resolved, this book presents our current understanding of some of the key pieces of the jigsaw, which we hope will help to develop better treatments and survival of patients in the future.

We hope that *Atherosclerosis: Molecular and Cellular Mechanisms* will be a key volume which is useful for both basic science and clinical investigators beginning in the field of atherosclerosis such as PhD, MD or undergraduate students or for those with more experience which will benefit from this summary of the latest findings in this field. We are extremely grateful to our authors who have provided excellent authoritative chapters for this book.

Bristol, December 2009   *Sarah Jane George and*
                                                                     Jason Lee Johnson

List of Contributors

Hafid Ait-Oufella
Georges-Pompidou European Hospital
Paris-Cardiovascular Research Center
Inserm U970
56 rue Leblanc
75015 Paris
France

Faisal Ali
Imperial College London
Hammersmith Hospital
Bywaters Centre for Vascular
Inflammation
Cardiovascular Sciences
National Heart and Lung Institute
Du Cane Road
London W12 0HS
UK

Lili Bai
Cardiovascular Research Institute
Maastricht (CARIM)
Experimental Vascular Pathology
Division (EVP)
Department of Pathology
Maastricht University Medical Center
P. Debyelaan 25
6229 HX Maastricht
The Netherlands

Marie-Luce Bochaton-Piallat
University of Geneva
Department of Pathology and
Immunology
Faculty of Medicine
1, rue Michel Servet
Geneva 1211
Switzerland

Barbara Bottazzi
Istituto Clinico Humanitas
via Manzoni 56
Rozzano
Milan 20089
Italy

Kevin G.S. Carson
Bristol Heart Institute
Research Floor Level 7
Bristol Royal Infirmary
Upper Maudlin Street
Bristol BS2 8HW
UK

Keith M. Channon
University of Oxford
John Radcliffe Hospital
Department of Cardiovascular
Medicine
Oxford OX3 9DU
UK

Atherosclerosis: Molecular and Cellular Mechanisms. Edited by Sarah Jane George and Jason Johnson
Copyright © 2010 WILEY-VCH Verlag GmbH & Co. KGaA, Weinheim
ISBN: 978-3-527-32448-4

Matteo Coen
University of Geneva
Department of Pathology and
Immunology
Faculty of Medicine
1, rue Michel Servet
Geneva 1211
Switzerland
St Anna University-Hospital
Operative Unit of Vascular and
Endovascular Surgery
Ferrara 44100
Italy

Colin Cunnington
University of Oxford
John Radcliffe Hospital
Department of Cardiovascular
Medicine
Oxford OX3 9DU
UK

Kristina Edfeldt
University of Copenhagen
Faculty of Health Sciences
The Novo Nordisk Foundation Center
for Protein Research
Copenhagen N 2200
Denmark

Cecilia Garlanda
Istituto Clinico Humanitas
via Manzoni 56
Rozzano
Milan 20089
Italy

Sarah Jane George
Bristol Heart Institute
Research Floor Level 7
Bristol Royal Infirmary
Upper Maudlin Street
Bristol BS2 8HW
UK

Florence Gizard
University of Kentucky
Division of Endocrinology and Molecu-
lar Medicine
Department of Medicine
Lexington, KY 40503
USA

David R. Greaves
University of Oxford
Sir William Dunn School of Pathology
South Parks Road
Oxford OX1 3RE
UK

Matthew T. Harper
University of Bristol
Department of Physiology and
Pharmacology
University Walk
Bristol BS8 1TD
UK

Sylvia Heeneman
Cardiovascular Research Institute
Maastricht (CARIM)
Experimental Vascular Pathology
Division (EVP)
Department of Pathology
Maastricht University Medical Center
P. Debyelaan 25
6229 HX Maastricht
The Netherlands

Yuqing Huo
University of Minnesota
Cardiovascular Division & Vascular
Biology Center
Department of Medicine
420 Delaware St SE
MMC508
Minneapolis, MN 55455
USA

Christopher L. Jackson
Bristol Heart Institute
Research Floor Level 7
Bristol Royal Infirmary
Upper Maudlin Street
Bristol BS2 8HW
UK

Nicholas P. Jenkins
Bristol Heart Institute
Research Floor Level 7
Bristol Royal Infirmary
Upper Maudlin Street
Bristol BS2 8HW
UK

Jamie Y. Jeremy
University of Bristol
Bristol Heart Institute
Research Floor Level 7
Bristol Royal Infirmary
Upper Maudlin Street
Bristol BS2 8HW
UK

Jason L. Johnson
Bristol Heart Institute
Research Floor Level 7
Bristol Royal Infirmary
Upper Maudlin Street
Bristol BS2 8HW
UK

Matthew L. Jones
University of Bristol
Department of Physiology and
Pharmacology
University Walk
Bristol BS8 1TD
UK

Olga Konopatskaya
University of Bristol
Department of Physiology and
Pharmacology
University Walk
Bristol BS8 1TD
UK

Petri T. Kovanen
Wihuri Research Institute
Kalliolinnantie 4
Helsinki 00140
Finland

Roberto Latini
Mario Negri Institute for Pharmaco-
logical Research
Department of Cardiovascular Research
via La Masa 19
Milano 20156
Italy
New York Medical College
Department of Medicine
Valhalla, NY 10595
USA

Anna M. Lundberg
Center for Molecular Medicine
Karolinska Institute
Stockholm 17176
Sweden

Esther Lutgens
Cardiovascular Research Institute
Maastricht (CARIM)
Experimental Vascular Pathology
Division (EVP)
Department of Pathology
Maastricht University Medical Center
P. Debyelaan 25
6229 HX Maastricht
The Netherlands

Cressida Lyon
Bristol Heart Institute
Research Floor Level 7
Bristol Royal Infirmary
Upper Maudlin Street
Bristol BS2 8HW
UK

Lucy MacCarthy-Morrogh
University of Bristol
Department of Physiology and
Pharmacology
University Walk
Bristol BS8 1TD
UK

Ziad Mallat
Georges-Pompidou European Hospital
Paris-Cardiovascular Research Center
Inserm U970
56 rue Leblanc
75015 Paris
France

Alberto Mantovani
Istituto Clinico Humanitas
via Manzoni 56
Rozzano
Milan 20089
Italy
University of Milan
Institute of General Pathology
Faculty of Medicine
Milan 20100
Italy

Justin C. Mason
Imperial College London
Hammersmith Hospital
Bywaters Centre for Vascular
Inflammation
Cardiovascular Sciences
National Heart and Lung Institute
Du Cane Road
London W12 0HS
UK

Carina Mill
University of Bristol
Bristol Heart Institute
Research Floor Level 7
Bristol Royal Infirmary
Upper Maudlin Street
Bristol BS2 8HW
UK

Fabiola Molla
Mario Negri Institute for Pharmaco-
logical Research
Department of Cardiovascular Research
via La Masa 19
Milano 20156
Italy

Kathryn J. Moore
Harvard Medical School
Lipid Metabolism Unit
Department of Medicine
Massachusetts General Hospital
Boston, MA 02114
USA

Saima Muzaffar
University of Bristol
Bristol Heart Institute
Research Floor Level 7
Bristol Royal Infirmary
Upper Maudlin Street
Bristol BS2 8HW
UK

Andrew C. Newby
University of Bristol
Bristol Heart Institute
Bristol BS2 8HW
UK

Inés Pineda-Torra
University College London
Centre for Clinical Pharmacology and
Therapeutics
Division of Medicine
London WC1E 6JJ
UK

Alastair W. Poole
University of Bristol
Department of Physiology and
Pharmacology
University Walk
Bristol BS8 1TD
UK

Katey Rayner
Harvard Medical School
Lipid Metabolism Unit
Department of Medicine
Massachusetts General Hospital
Boston, MA 02114
USA

Nilima Shukla
University of Bristol
Bristol Heart Institute
Research Floor Level 7
Bristol Royal Infirmary
Upper Maudlin Street
Bristol BS2 8HW
UK

Alain Tedgui
Georges-Pompidou European Hospital
Paris-Cardiovascular Research Center
Inserm U970
56 rue Leblanc
75015 Paris
France

Huan Wang
University of Minnesota
Cardiovascular Division & Vascular
Biology Center
Department of Medicine
420 Delaware St SE
MMC508
Minneapolis, MN 55455
USA

Gemma E. White
University of Oxford
Sir William Dunn School of Pathology
South Parks Road
Oxford OX1 3RE
UK

Stephen J. White
University of Bristol
Bristol Heart Institute
Bristol BS2 8HW
UK

Zhong-qun Yan
Center for Molecular Medicine
Karolinska Institute
Stockholm 17176
Sweden

PART I
Introduction

Atherosclerosis: Molecular and Cellular Mechanisms. Edited by Sarah Jane George and Jason Johnson
Copyright © 2010 WILEY-VCH Verlag GmbH & Co. KGaA, Weinheim
ISBN: 978-3-527-32448-4

1
Pathogenesis of Atherosclerosis

Sarah Jane George and Cressida Lyon

1.1
Epidemiology of Coronary Heart Disease

Cardiovascular disease (CVD) is the most frequent cause of premature death in modern industrialized countries, accounting for 4.35 million deaths each year in Europe, and 35% of deaths in the United Kingdom [1, 2]. In addition, it is an increasing cause of death in developing countries. The main forms of CVD are coronary heart disease (CHD) and stroke. CHD alone is the most common cause of death in Europe (1.95 million and 94 000 deaths per year in Europe and the United Kingdom, respectively) [1, 2]. A wide geographical variation in CHD mortality rates exists, the highest in Eastern and Central Europe and the lowest rates in Italy, France, and Japan [2]. Mortality rates are also generally much higher in men than women and increase with age. In addition to geographical variation and gender, CHD mortality rates are also influenced by ethnic origin and social class [2]. For example, within the United Kingdom the people from the South Asian community have a higher than average rate, while those of an Afro-Caribbean origin have a lower than average rate [2]. CHD mortality rates are greater in the most deprived social groups than those observed in affluent groups [2].

1.2
Risk Factors

The term "risk factor" is widely used to describe characteristics found in individuals that have been shown in observational epidemiological studies, autopsy studies, metabolic studies, and genetic studies to relate to the subsequent occurrence of a disease. In global terms, risk factor categories cover personal, lifestyle, biochemical, and physiological characteristics, some of which are modifiable, while others are not. A summary of accepted risk factors for CHD are shown in Table 1.1. Smoking increases the risk of CHD. The long-term risk of smoking to individuals has been quantified in a 50-year cohort of British doctors. The study found that mortality from CHD was around 60% higher in smokers (and 80% in heavy

Atherosclerosis: Molecular and Cellular Mechanisms. Edited by Sarah Jane George and Jason Johnson
Copyright © 2010 WILEY-VCH Verlag GmbH & Co. KGaA, Weinheim
ISBN: 978-3-527-32448-4

Table 1.1 Coronary heart disease risk factors.

Lifestyle	• Tobacco smoking • Diet high in saturated fat and calories, low in fruit and vegetables, high in sugar • Physical inactivity • Stress • Excess alcohol • Obesity
Biochemical or physiological	• High plasma cholesterol • High blood pressure • Low plasma HDL-cholesterol • High plasma triglyceride • Diabetes mellitus • Thrombogenic factors
Personal	• Age • Gender • Family history • Personal history

smokers) than in non-smokers, and about half of all regular smokers will eventually suffer a fatal myocardial infarction (MI) [3]. It is now universally recognized that a poor diet increases the risk of chronic diseases, particularly CVD and cancer. The World Health Organization (WHO) estimated in 2002 that just under 30% of CHD and almost 20% of stroke in developed countries was related to levels of fruit and vegetable consumption below 600 g/day [4]. The WHO has yet to calculate the precise proportion of the disease burden caused by other dietary factors, however, such as high sodium intake or high saturated fat intake.

People who are physically active (aerobic activity) have a lower risk of CHD. It has been estimated that over 20% of CHD and 10% of stroke in developed countries is related to physical inactivity [4]. While moderate alcohol consumption (one or two drinks a day) reduces the risk of CVD, at high levels of intake, particularly in "binges," the risk of CVD is increased. The WHO estimates that 2% of CHD and almost 5% of stroke in men in developed countries is related to alcohol [4]. Overweight and obesity increase the risk of CHD. As well as being an independent risk factor, obesity is also a major risk factor for high blood pressure, raised blood cholesterol, diabetes, and impaired glucose tolerance [5].

The adverse effects of excess weight are more pronounced when fat is concentrated in the abdomen. Around a third of CHD and ischemic stroke in developed countries is related to body mass index (BMI) levels in excess of the theoretical minimum (21 kg/m^2) [4]. Risk of CHD is also directly related to blood cholesterol levels. The WHO estimates that over 60% of CHD and around 40% of ischemic stroke in developed countries is due to total blood cholesterol levels in excess of the theoretical minimum (3.8 mmol/l) [4].

Risk of CHD is directly related to both systolic and diastolic blood pressure levels. The INTERHEART study estimated that 22% of heart attacks in Western Europe were associated with a history of high blood pressure, and that those with a history of hypertension were at just under twice the risk of a heart attack compared with those with no history of hypertension [6]. Diabetes substantially increases the risk of CHD. Men with non-insulin-dependent (type 2) diabetes have a two- to fourfold greater annual risk of CHD, with an even higher (three- to five-fold) risk in women with type 2 diabetes [7]. Diabetes not only increases the risk of CHD but also magnifies the effect of other risk factors for CHD, such as raised cholesterol levels, raised blood pressure, smoking, and obesity. The INTERHEART case–control study estimated that 15% of heart attacks in Western Europe are related to diagnosed diabetes, and that people with diagnosed diabetes are at three times the risk of a heart attack compared with those without [6].

In summary, there are many risk factors for CHD and stroke, some of which cannot be modified, such as age and gender. The severity of each risk factor is associated with a greater risk. It is of note that many of these risk factors are associated, such as diet, physical activity, and obesity. Moreover, multiple risk factors increase the likelihood of CHD and stroke.

1.3
Atherosclerosis

Atherosclerosis is the pathology that underlies CHD. It is defined as a focal, inflammatory fibro-proliferative response to multiple forms of endothelial injury. The response-to-injury hypothesis was proposed by Russell Ross and colleagues over 30 years ago [8], and has been refined and developed since.

The normal artery wall is composed of two organized layers: intima and media (Figure 1.1a). The intima is made up of a single layer of endothelial cells that are seated on basement membrane and then the internal elastic lamina (IEL). Beneath the IEL is the medial layer, comprising vascular smooth muscle cells (VSMCs) surrounded by basement membrane and embedded in interstitial extracellular matrix. The boundary of the media is marked by the external elastic lamina (EEL). All infants have focal thickening of the coronary artery intima due to VSMC pro-liferation [9]. Although focal thickening is an important hallmark of the developing atherosclerotic plaque, this is considered to be an adaptive response to turbulent blood flow rather than pathological.

Endothelial dysfunction initiated by the risk factors already described permits the entry of lipids and inflammatory cells into the artery wall (Figure 1.1b). Once in the artery, monocytes differentiate into macrophages which take up the lipid and become foam cell macrophages. This results in the formation of lesions termed "fatty streaks," recognized as the onset of atherosclerosis (Figure 1.1c). Fatty streaks are small, slightly raised lesions caused by focal collections of foam cell macrophages in the intima. They may be precursors of larger atherosclerotic plaques, but may also regress. Progression of the fatty streak to a more complex

A: Normal Artery

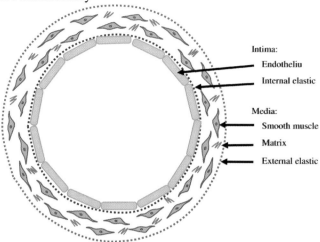

Intima:
  Endotheliu
  Internal elastic

Media:
  Smooth muscle
  Matrix
  External elastic

B: Endothelial Dysfunction

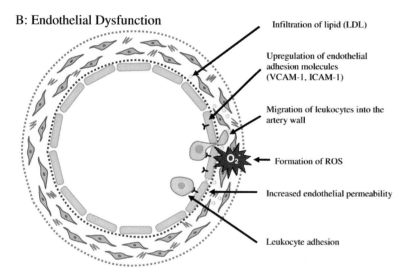

Infiltration of lipid (LDL)

Upregulation of endothelial
adhesion molecules
(VCAM-1, ICAM-1)

Migration of leukocytes into the
artery wall

Formation of ROS

Increased endothelial permeability

Leukocyte adhesion

Figure 1.1 Atherosclerotic plaque formation. A: Normal artery wall is composed of the intima – a layer of endothelial cells seated on the internal elastic lamina – and the media, where vascular smooth muscle cells (VSMCs) are embedded in extracellular matrix and surrounded by the external elastic lamina. B: Endothelial dysfunction. Damage to the endothelium results due to exposure to risk factors and leads to the presence of reactive oxygen species (ROS), upregulation of adhesion molecules, and increased permeability. Consequently, leukocytes adhere to the artery wall. Leukocytes invade into the wall. In addition, low-density lipoprotein (LDL) infiltrates into the artery wall.

C: Fatty Streak Formation

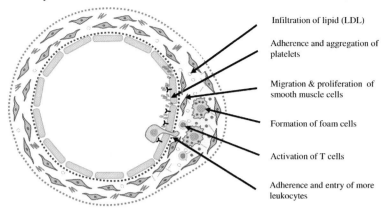

Infiltration of lipid (LDL)

Adherence and aggregation of platelets

Migration & proliferation of smooth muscle cells

Formation of foam cells

Activation of T cells

Adherence and entry of more leukocytes

D: Stable Atherosclerotic Plaque

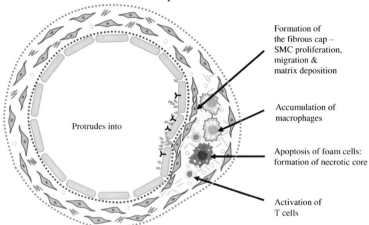

Formation of the fibrous cap – SMC proliferation, migration & matrix deposition

Accumulation of macrophages

Protrudes into

Apoptosis of foam cells: formation of necrotic core

Activation of T cells

Figure 1.1 *Continued* C: Fatty streak formation. Platelets adhere to the damaged endothelial layer and degranulate. Once in the artery wall LDL is modified and then is more readily taken up by macrophages which develop into foam cells. Release of chemokines and growth factors leads to further recruitment of inflammatory cells, activation of T cells and VSMC migration and proliferation. D: Stable atherosclerotic plaque formation. VSMCs migration and proliferation lead to the formation of a fibrous cap. Apoptosis of foam cells causes the formation of a necrotic core. Further accumulation of activated T cells and macrophages occurs.

E: Unstable Atherosclerotic Plaque

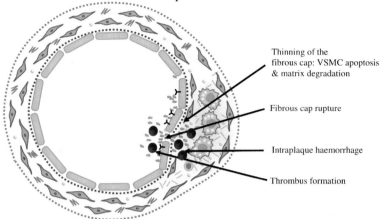

Thinning of the
fibrous cap: VSMC apoptosis
& matrix degradation

Fibrous cap rupture

Intraplaque haemorrhage

Thrombus formation

F: Ruptured Atherosclerotic Plaque

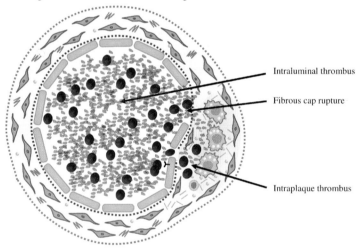

Intraluminal thrombus

Fibrous cap rupture

Intraplaque thrombus

Figure 1.1 *Continued* E: Unstable atherosclerotic plaque formation. Thinning of the fibrous cap occurs as the result of VSMC apoptosis and matrix degradation. Intraplaque hemorrhage occurs. Both these factors render the plaque more unstable and likely to rupture. F: Atherosclerotic plaque rupture. Rupture of the fibrous cap or erosion of the luminal surface precipitates thrombus formation, which can be occlusive.

lesion occurs due to the formation of a necrotic core and a fibrous cap (Figure 1.1d).

Foam cell macrophages, engorged with lipid, begin to die and release their contents, which contributes to the formation of a necrotic core. The release of the cytoplasmic contents of the foam cells leads to the accumulation of extracellular lipids and growth factors which induce inflammation. The occurrence of VSMC migration and proliferation results in the formation of a fibrous cap. VSMCs migrate into the intima where they proliferate and deposit extracellular matrix. The increase in cell number and presence of matrix causes augmentation of the bulk of the plaque, which now protrudes into the lumen. This is termed a stable advanced plaque (Figure 1.1d). The size and composition of the plaque determine its outcome.

Classification schemes have been devised to categorize the various plaque types [10–13]. A plaque with a large necrotic core, high content of inflammatory cells, and thin fibrous cap is termed an "unstable plaque" (Figure 1.1e), and is more prone to rupture than a plaque with a smaller necrotic core, lower content of inflammatory cells, and thick fibrous cap, termed a "stable plaque" (Figure 1.1d). Rupture of the plaque leads to thrombus formation, which can occlude the lumen and cause the symptoms of MI or stroke (Figure 1.1f). However, plaque rupture does not always lead to occlusion of the artery and the plaque may restabilize and heal over. This is at a cost since the "healed plaque" is larger [14] and repeated episodes of plaque rupture and healing is associated with a greater incidence of a fatal event [15].

The aim of this book is to discuss the key molecules that drive the processes underlying the pathogenesis of atherosclerosis. A greater understanding of the molecules involved in atherosclerosis will no doubt lead to potential clinical targets and design of new therapies. The subsequent chapters focus on particular key molecules, however to serve as an introduction to these chapters we will briefly outline the key cellular processes in atherosclerosis.

1.4
Initiation of Plaque Formation – Endothelial Dysfunction

Atherosclerosis is thought to be initiated by damage to the endothelium, resulting in altered endothelial function (Figure 1.1b). The risk factors outlined above are generally the cause of this damage through one or more of the following pathways: high levels of oxidized low-density lipoprotein (LDL); free radicals such as reactive oxygen species (ROS); genetic variations; elevated plasma homocysteine concentrations; infectious microorganisms such as herpes virus or *Chlamydia pneumoniae*; shear stress in areas of turbulent blood flow or endogenous inflammatory signals such as cytokines [16].

Loss of normal nitric oxide (NO) production from the endothelium of the vessel wall occurs during endothelial dysfunction. NO regulates many aspects of cardiovascular homeostasis, including blood pressure and flow, smooth muscle contraction, inflammation, and platelet activation, and is a well-known atheroprotective factor. The molecular mechanisms behind this loss of NO remain unclear, however recent studies suggest that changes in the activity and regulation of endothelial NO synthase by its cofactor tetrahydrobiopterin (BH4) is an important contributor [17]. See Chapter 14 by Cunnington and Channon for more detail.

The anti-inflammatory enzyme heme oxygenase-1 (HO-1) has also been reported to have cytoprotective effects during endothelial dysfunction. HO-1 is the inducible form of the heme oxygenase system and its importance in vasculoprotection is demonstrated by the severe and persistent endothelial damage observed in human HO-1 deficiency and *Hmox1*$^{-/-}$ mice (reviewed in [18]). HO-1 exerts a potent protective effect against atherogenesis, cardiac ischemia/reperfusion injury, and both graft rejection and accelerated arteriosclerosis post-transplantation. HO-1 protects tissues during inflammatory stress through degradation of pro-oxidative heme, production of bilirubin and carbon monoxide, and regulation of cellular iron [19].

It has been proposed that these vasculoprotective effects are through multiple mechanisms, including regulation of the cell cycle and angiogenesis, redox homeostasis, and the modulation of complement system. More detail on this topic is provided in Chapter 15 by Mason and Ali.

ROS are produced by enzyme systems present in the vascular wall, including NADPH oxidases (NOX), xanthine oxidase, and NO synthase, which are increased in association with risk factors. The NOX family, which is expressed by neutrophils, monocytes, VSMCs, endothelial cells (ECs), fibroblasts, macrophages, and mast cells, is a particularly important source of ROS in the vessel wall. Abnormal NOX activity is seen during endothelial dysfunction and results in the overproduction of ROS, including superoxide (O_2^-), which promotes further endothelial dysfunction through a number of effects including endothelial cell apoptosis, vasoconstriction, lipid peroxidation, cell proliferation, and isoprostane formation. O_2^- also reacts with NO, negating its cardioprotective effects and further ROS production in the form of peroxynitrite. ROS modulate many processes involved in atherosclerosis, including inflammation, apoptosis, VSMC replication, angiogenesis, and matrix turnover and therefore increased expression and/or activity of NADPH oxidases has been proposed to play an important role in atherosclerosis (see review by [20]). For further detail see Chapter 13 by Jeremy and colleagues.

1.5
Inflammation

Platelets adhere to dysfunctional endothelial cells at lesion-prone sites such as the carotid artery bifurcation prior to leukocyte invasion and atherosclerotic plaque formation [21], indicating their involvement in atherogenesis (see Chapter 17 by Harper and colleagues for more details). Platelet adherence and aggregation is stimulated by integrins, P-selectin, fibrin, thromboxane A$_2$, and tissue factor [16]. Once adherent the platelets become activated and release, or expose on their cell surface, multiple inflammatory factors, proteases, and vasoactive substances, which further promote endothelial cell dysfunction and inflammation [22]. This leads to recruitment and adhesion of neutrophils and monocytes to the area. In addition, platelets directly bind leukocytes and therefore act as bridging cells between the endothelium and plaque-forming cells [23]. Platelets also release platelet factor 4 (PF4) in atherosclerotic plaques [24] and promote monocyte differentiation into macrophages and uptake of oxidized LDL by macrophages [25].

Endothelial cell dysfunction also directly results in leukocyte invasion, leading to the formation of a fatty streak (Figure 1.1b). Leukocyte recruitment occurs through a series of events involving adhesion molecules. First, leukocytes roll along the endothelial surface of the vessel wall, which is mediated by endothelial expression of selectins, including P-selectin. Second, leukocytes adhere tightly to the endothelium and transmigrate through by extravasation, and these processes are mediated by integrins and immunoglobulin superfamily members, such as vascular cell adhesion molecule 1 (VCAM-1) and intercellular adhesion molecule 1 (ICAM-1) (see review by [26]). More detail is provided in Chapter 3 by Wang and Huo.

Endothelial damage increases endothelium permeability and therefore once bound, leukocytes can migrate through the endothelium and into the intima in response to the various chemokines released by intimal cells and damaged endothelium [27]. Chemokines are presented on the endothelial surface by glycosaminoglycan (GAG) binding or as a membrane-bound molecules, for example CX_3CL1 (fractalkine) [28]. This permits strong binding of the monocyte via the cognate chemokine receptor. In the intima, the monocytes and T lymphocytes mature and release chemoattractant cytokines (or chemokines), for example Chemokin (C-C motif) ligand 2 (CCL-2 also known as MCP-1), which establishes a chemokine gradient for migration and further amplifies the inflammatory response. Interestingly, chemokines also regulate retention of macrophages in the plaque [29].

T cells also accumulate in the atherosclerotic plaque in response to chemokines chemokine (C-X-C motif) ligand 9, 10 and 11 (CXCL9, CXCL10, and CXCL11). Direct evidence for the involvement of CCL2 and its receptor CCR2 in atherogenesis come from $ApoE^{-/-}$ mice that lack either CCL2 or CCR2 expression, as these mice have dramatically less macrophage recruitment and atherosclerotic plaque formation [30, 31]. Chapter 2 by White, Channon, and Greaves provides further details regarding the role of chemokines in atherosclerosis. Inflammation continues throughout atherosclerotic plaque development, and mediates a myriad of proinflammatory cytokines, including interferon-γ (IFN-γ), tumor necrosis factor-α (TNF-α), interleukin-1 (IL-1), IL-2, and IL-18 (see review by [32]). Chapter 4 by Tedgui, Ait-Oufella, and Mallat provides further details. It is of note that some cytokines, such as IL-10 and transforming growth factor-β (TGF-β) are atheroprotective, through the induction of T_{reg} cells (see review by [33]).

The importance of immunity in atherosclerosis is not restricted to the adaptive immune system: innate immunity is also important. Toll-like receptors (TLRs) are pathogen-associated molecular pattern receptors which respond to invading microorganisms, as well as modified LDL and "danger signals" released from damaged tissues, by initiating an inflammatory response. For example, TLR4 is responsible for cellular activation after recognition of bacterial lipopolysaccharides (LPS), as well as microbial and eukaryotic heat shock protein-60 (HSP-60). TLR4 is expressed in human atherosclerotic plaques and interestingly, polymorphisms in the TLR4 gene have varying effects on risk of acute coronary syndromes [34] and TLR4 deletion in $ApoE^{-/-}$ mice reduced atherosclerosis [35]. Further evidence implicating TLRs in atherosclerosis comes from studies sing $TLR2^{-/-}$ mice crossed with LDLR$^{-/-}$ mice, which developed smaller atherosclerotic lesions [36, 37]. See Chapter 5 by Yan, Edfeldt, and Lundberg for additional information.

Furthermore, the classic short pentraxin, C-reactive protein (CRP), which is used as a prognostic marker for ischemic heart disease, is a component of the innate immune system. The mechanisms behind the involvement of CRP in heart disease are unknown, as experimentation has been made difficult due to a genetic difference between mouse and human CRP. However, a recent study has shown that the long pentraxin PTX3, which is evolutionally conserved, provides a cardioprotective effect in mice via modulation of the complement pathway. Levels of PTX3 are elevated in myocardial ischemia in both mouse and human, therefore PTX3 could also have potential as a prognostic marker [38]. See Chapter 7 by Mantovani and colleagues for additional details.

Peroxisome proliferator–activated receptors (PPARs) expressed by endothelial cells in the vessel wall are an important family of nuclear receptors that regulate monocyte recruitment, adhesion, and transmigration as well as having anti-inflammatory properties and the ability to reduce oxidative stress (see [39] and Chapter 6 by Gizard and Pineda-Torra for more detailed information). PPARs are also expressed by monocytes and macrophage and T cells and are associated with atheroprotective processes including inhibition of inflammatory molecules and increased cholesterol efflux.

1.6
Foam Cell Formation

Increased permeability of the endothelium also results in infiltration of LDL into the blood vessel wall. Once in the blood vessel wall LDL is modified and is more readily taken up by cells [40]. In the intima, growth factors (monocyte colony-stimulating factor, M-CSF) and cytokines (TNF-α and IFN-γ) are released and cause monocytes to differentiate into macrophages and mature into active macrophages. The maturation process causes macrophages to increase their expression of various scavenger receptors, including scavenger receptors A and B1, and CD36, which can bind and internalize modified LDL (reviewed in [41]). The relative importance of scavenger receptors in atherosclerotic plaque development is controversial. Currently, it is thought that while scavenger receptor A is pro-atherogenic (primarily as a consequence of uptake of minimally oxidized LDL by macrophages in the arterial wall), scavenger receptor B is protective against atherosclerosis at later timepoints (primarily because of its function in the removal of high-density lipoprotein (HDL)-cholesterol esters in the liver) [42]. Lipid ingestion, which is likely to be beneficial at the onset, results in the accumulation of fatty droplets within the macrophages, which take on a foamy appearance under microscopy and are therefore termed foam cells [43]. Foam cells release growth factors and cytokines, which are involved in lesion progression, as well as matrix-degrading metalloproteinases (MMPs), which stimulate matrix degradation. See Chapter 12 by Moore and Rayne for further information. Together with T lymphocytes, foam cells form the fatty streak.

The proteolytic enzymes chymase and tryptase derived from the mast cells have been shown to profoundly modify the composition and function of HDL particles and chymase proteolytically modifies LDL [44]. Stimulated mast cells therefore can accelerate foam cell formation in the intimal areas in which mast cells and macrophages coexist, and are surrounded with intimal fluid enriched with plasma-derived LDL and HDL particles. Further details are available in Chapter 11 by Kovanen.

As well as being taken up by macrophages, cholesterol can also be effluxed via transporters, such as ABCA-1 and ABCG-1, to extracellular HDL-based acceptors, such as apolipoprotein E. This process is known as reverse cholesterol transport (RCT). After efflux to HDL, cholesterol may be esterified in the plasma by the enzyme lecithin:cholesterol acyltransferase and is ultimately transported from HDL to the liver, either directly via the scavenger receptor BI or after transfer to apolipoprotein B-containing lipoproteins by the cholesteryl ester transfer protein. Cholesterol efflux is thought to be atheroprotective by removing the cholesterol accumulation and therefore reducing foam cell formation [45]. Promotion of macrophage RCT is a potential therapeutic approach to preventing or regressing atherosclerotic vascular disease, but robust measures of RCT in humans will be needed in order to confidently advance RCT-promoting therapies in clinical development (reviewed in [46]).

Monocytes and macrophages display a high level of heterogeneity to allow them to specialize in particular functions. Many macrophage phenotypes have been identified and at least two of these (M1 and M2) have been shown to be present in atherosclerotic plaques. The phenotype expressed by a macrophage depends on the chemical signals used to induce macrophage differentiation. Th1 cytokines, such as IFN-γ, IL-1β and lipopolysaccharide, induce a "classical" activation profile called M1. On the other hand, Th2 cytokines, such as IL-4 and IL-13, induce an "alternative" activation program called M2. The difference in the activity of these two subtypes is extreme, while M1 macrophages are proinflammatory, M2 are anti-inflammatory [47]. Macrophages can change from one phenotype to another and therefore the modulation of this phenotypic change is an interesting therapeutic strategy. Macrophage phenotypes are discussed in detail in Chapter 18 by Johnson and Jenkins.

1.7
Vascular Smooth Muscle Cell Migration and Proliferation

As fatty streaks develop into more complex, advanced atherosclerotic plaques, they develop a fibrous cap, consisting of VSMCs, which form a protective layer between the lesion and the lumen of the vessel. In the development from a fatty streak to an advanced plaque, VSMCs migrate from the media into the intima, where they undergo proliferation and lay down extracellular matrix (Figure 1.1d). During these processes, VSMCs undergo phenotypic modulation from contractile to a synthetic phenotype resulting in VSMC de-differentiation. See Chapter 16 by Coen and Bochaton-Piallat for further detail. Fibrous cap formation is thought to be stimulated by growth factors and cytokines, including platelet-derived growth factor (PDGF), fibroblast growth factor 2 (FGF-2), and transforming growth factor β (TGF-β) released from degranulated platelets, endothelial cells, macrophages, foam cells, and VSMCs. In addition, degradation of the extracellular matrix and cell–cell contacts mediated by proteases including MMPs, permits the migration of VSMCs from the media into the intima [48–50].

Plasminogen-deficient mice have significantly reduced intimal thickening, suggesting that plasmin is required for VSMC migration into the intima, by increasing extracellular matrix degradation either directly, or indirectly through activation of MMPs [51]. In the intima, VSMCs proliferate and synthesize extracellular matrix components. The cells and matrix components form the fibrous cap (sclerosis) of the atherosclerotic plaque by arranging themselves as a layer above the lipid core. The VSMCs of the fibrous cap are synthetic, and have high levels of migration and proliferation.

As the lesion becomes larger and more complex, inflammatory cytokines, such as IFN-γ, cause macrophages to undergo apoptosis. This leads to the formation of a necrotic core, which is characteristic of an advanced lesion.

1.8
Plaque Rupture and Thrombus Formation

After the formation of the early lesion, it is thought that plaque growth is discontinuous and occurs in bursts [43]. These "bursts" of growth may occur due to the rupture and subsequent repair of plaques. The composition of a plaque is thought to determine the risk of rupture. Plaques with thick fibrous plaques are considered to be clinically stable, while plaques with a high lipid content and little or no fibrous cap are considered vulnerable and "rupture-prone" [52].

There are three main precipitating factors of plaque rupture: (i) thinning of the fibrous cap, (ii) matrix degradation, and (iii) formation and thrombosis of microvessels within the plaque:

i) Thinning of the fibrous cap can occur following various processes. First, VSMC apoptosis results in reduced matrix synthesis, reducing the tensile strength of the fibrous cap [53]. A reduction in the number of VSMCs within the cap also reduces the potential for extracellular matrix production. Second, macrophage apoptosis increases the size of the necrotic core, exerting physical pressure on the fibrous cap and leading to further VSMC apoptosis [52]. Third, inflammation occurring within the plaque can result in protease activation (see point ii below), leading to degradation of extracellular proteins, including collagen, and thinning of the cap.

ii) Matrix degradation reduces the amount of collagen in the plaque, resulting in reduced tensile strength. Macrophages secrete several classes of extracellular proteases, including serine proteases, cathepsins, and MMPs. Plasmin, a serine protease which degrades fibrin and also activates MMPs, modulates plaque development and stability, its exact role, however, remains unclear, due to the lack of specific inhibitory drugs and various human and mouse studies which have yielded contradictory results [51]. See Chapter 10 by Jackson and Carson for further details. Cathepsins are a versatile family of extracellular matrix proteases that are implicated in plaque stability, as well as other processes involved in atherosclerosis such as monocyte invasion, foam cell formation, apoptosis, and thrombosis. Cathepsin expression in macrophages colocalizes with areas of elastin fragmentation, and cathepsin S and K knockout mice, show a smaller plaque area, a reduction in plaque development, and reduced elastin breaks [54]. See Chapter 9 by Bai, Lutgens and Heeneman for a detailed description. MMPs colocalize with areas of degraded extracellular matrix in human plaques, and increased plasma MMP levels are associated with acute coronary syndromes. MMP knockout and TIMP (inhibitor of MMPs) transgenic mice show differential effects of MMPs on atherosclerotic plaque progression and stability (see reviews [55, 56] and Chapter 8 by White and Newby). Mast cells release a serine protease called chymase, which causes VSMC apoptosis through degradation of the extracellular matrix protein fibronectin and focal adhesion kinase, of particular importance to VSMC survival [57]. In addition, mast cell proteases (chymase, tryptase, and cathepsin G) cause MMP activation [58–60]. A detailed description of the role of mast cells is provided in Chapter 11 by Kovanen.

iii) The formation of microvessels within atherosclerotic plaques (neoangiogenesis) is more prominent in unstable plaques than in stable fibrous plaques, suggesting that microvessels and subsequent intraplaque hemorrhage due to microvessel incompetence are associated with plaque instability [61]. Mast cells located near the neovessels contain basic fibroblast growth factor (bFGF), a potent pro-angiogenic factor, and therefore mast cells are thought to contribute to the growth of the neovascular sprouts [62]. In addition, mast cells contain other pro-angiogenic factors, such as histamine, heparin, tryptase, chymase, vascular endothelial growth factor (VEGF), and nerve growth factor (NGF), rendering the mast cell potentially a powerful pro-angiogenic effector cell (see review by [63]). Focal collections of T cell- and macrophage-derived angiogenic factors contribute to: (1) the arborization of vasa vasorum around the necrotic core; (2) the formation of immature vessels; and (3) loss of basement membrane around functional capillaries. These processes initiate leakage of red blood cells (RBCs) into the plaque and induce a cycle of inflammation and neovascularization. By contributing to the deposition of free cholesterol, macrophage infiltration, and enlargement of the necrotic core, the accumulation of RBC membranes within an atherosclerotic plaque may represent a potent atherogenic stimulus and may increase the risk of plaque destabilization [64] (see review by [65]). Excessive proteolytic activity produced by various cell types, including macrophages and mast cells, at the inflamed tissue site may damage the neovessels [66].

Plaque rupture exposes highly thrombogenic plaque constituents to the bloodstream, leading to platelet aggregation and thrombus formation (Figure 1.1e and f). The thrombus can completely occlude the blood vessel, resulting in clinical symptoms such as heart attack or stroke. Therefore a greater understanding of the mechanisms regulating platelets and thrombus formation is vital. Protein kinase C appears to be an important mediator of this process, as different isoforms play distinct roles in the platelet activation process (see review by [67]). Further details are provided in Chapter 17 by Harper and colleagues.

The plaque rupture may heal either by regeneration or scarring [52]. Both of these processes lead to an increase in plaque size. TGF-β is released during the natural repair process, and stimulates VSMCs to produce connective tissue and extracellular matrix components, essential for fibrous cap reticulation [68]. Thrombin, which is released during thrombus formation, is a stimulator of VSMC growth, and is also fundamental for fibrous cap formation. Thus both TGF-β and thrombin appear important contributors to rupture repair.

In addition to plaque rupture, superficial erosion can lead to the formation of a platelet-rich thrombus formation, as collagen and other factors are uncovered leading to platelet adhesion and activation [43]. The resultant thrombus may lead to clinical symptoms and accounts for 30–40% of acute thrombotic events [69]. However, the precise mechanisms underlying plaque erosion are poorly defined. Mast cell proteases including chymase, tryptase, and cathepsin G may contribute to weakening and leakage of the microvessels due to cleavage of VE-cadherin, fibronectin, and activation of MMPs, which cause apoptosis or loosening of the attachment of endothelial cells and thereby erosion [70].

In summary, atherosclerosis is a complex pathology that develops over decades and is modulated by various risk factors. Atherosclerosis involves several cell types and factors as outlined in more detail in subsequent chapters. Despite a wealth of research for more than 30 years which has highlighted many underlying mechanisms, there are still many unknowns. Future research will no doubt increase our knowledge and help with the development of new therapies for atherosclerosis.

References

1 Peterson, S., Peto, V., and Rayner, M. (2005) *European Cardiovascular Disease Statistics*, British Heart Foundation, London.

2 Allender, S., Peto, V., Scarborough, P., Kaur, A., and Rayner, M. (2008) *Coronary Heart Disease Statistics*, British Heart Foundation, London.

3 Coll, R., Peto, R., Boreham, J., and Sutherland, I. (2005) Mortality from cancer in relation to smoking: 50 years observations on British doctors. *Br. J. Cancer*, **92** (3), 426–429.

4 World Health Organization (2002) *The World Health Report 2002. Reducing Risks, Promoting Healthy Life*, World Health Organization, Geneva.

5 World Health Organization (2000) *Obesity – Preventing and Managing the Global Epidemic. Report of a WHO Consultation on Obesity*, World Health Organization, Geneva.

6 Yusuf, S., Hawken, S., Ounpuu, S., Dans, T., Avezum, A., Lanas, F., *et al.* (2004) Effect of potentially modifiable risk factors associated with myocardial

infarction in 52 countries (the INTERHEART study): case-control study. *Lancet*, **364** (9438), 937–952.

7 Garcia, M., McNamara, P., Gordon, T., and Kannell, W. (1974) Morbidity and mortality in the Framingham population. Sixteen year follow-up. *Diabetes*, **23**, 105–111.

8 Ross, R., Glomset, J., and Harker, L. (1977) Response to injury and atherogenesis. *Am. J. Pathol.*, **86** (3), 675–684.

9 Milei, J., Ottaviani, G., Lavezzi, A.M., Grana, D.R., Stella, I., and Matturri, L. (2008) Perinatal and infant early atherosclerotic coronary lesions. *Can. J. Cardiol.*, **24** (2), 137–141.

10 Virmani, R., Kolodgie, F.D., Burke, A.P., Farb, A., and Schwartz, S.M. (2000) Lessons from sudden coronary death: a comprehensive morphological classification scheme for atherosclerotic lesions. *Arterioscler. Thromb. Vasc. Biol.*, **20** (5), 1262–1275.

11 Stary, H.C., Chandler, A.B., Glagov, S., Guyton, J.R., Insull, W., Rosenfeld, M.E., *et al.* (1994) A definition of initial, fatty streak, and intermediate lesions of atherosclerosis – a report from the committee on vascular-lesions of the council on arteriosclerosis, American Heart Association. *Arterioscler. Thromb.*, **14** (5), 840–856.

12 Stary, H.C., Blankenhorn, D.H., Chandler, A.B., Glagov, S., Insull, W., Jr., Richardson M., *et al.* (1992) A definition of the intima of human arteries and of its atherosclerosis-prone regions. A report from the committee on vascular lesions of the council on arteriosclerosis, American Heart Association. *Arterioscler. Thromb.*, **12** (1), 120–134.

13 Stary, H.C., Chandler, A.B., Dinsmore, R.E., Fuster, V., Glagov, S., Insull W., Jr., *et al.* (1995) A definition of advanced types of atherosclerotic lesions and a histological classification of atherosclerosis: a report from the committee on vascular lesions of the council on arteriosclerosis, American Heart Association. *Arterioscler. Thromb. Vasc. Biol.*, **15** (9), 1512–1531.

14 Fuster, V., Badimon, L., Badimon, J.J., and Chesebro, J.H. (1992) The pathogenesis of coronary artery disease and the acute coronary syndromes (first of two parts). *N. Engl. J. Med.*, **326** (4), 242–250.

15 Burke, A.P., Kolodgie, F.D., Farb, A., Weber, D.K., Malcom, G.T., Smialek, J., *et al.* (2001) Healed plaque ruptures and sudden coronary death. Evidence that subclinical rupture has a role in plaque progression. *Circulation*, **103**, 934–940.

16 Ross, R. (1999) Atherosclerosis – an inflammatory disease. *N. Engl. J. Med.*, **340** (2), 115–123.

17 Channon, K. (2004) Tetrahydrobiopterin: regulator of endothelial nitric oxide synthase in vascular disease. *Trends Cardiovasc. Med.*, **14**, 323–327.

18 Kinderlerer, A.R., Gregoire, I.P., Hamdulay, S.S., Ali, F., Steinberg, R., Silva, G., *et al.* (2009) Heme oxygenase-1 expression enhances vascular endothelial resistance to complement-mediated injury through induction of decay-accelerating factor: a role for increased bilirubin and ferritin. *Blood*, **113** (7), 1598–1607.

19 True, A.L., Olive, M., Boehm, M., San, H., Westrick, R.J., Raghavachari, N., *et al.* (2007) Heme oxygenase-1 deficiency accelerates formation of arterial thrombosis through oxidative damage to the endothelium, which is rescued by inhaled carbon monoxide. *Circ. Res.*, **101** (9), 893–901.

20 Muzaffar, S., Shukla, N., and Jeremy, J. (2005) Nicotinamide adenine dinucleotide phosphate oxidase: a promiscuous therapeutic target for cardiovascular drugs? *Trends Cardiovasc. Med.*, **15**, 278–282.

21 Massberg, S., Brand, K., Gruner, S., Page, S., Muller, E., Muller, I., *et al.* (2002) A critical role of platelet adhesion in the initiation of atherosclerotic lesion formation. *J. Exp. Med.*, **196** (7), 887–896.

22 Langer, H.F. and Gawaz, M. (2008) Platelet-vessel wall interactions in atherosclerotic disease. *Thromb. Haemost.*, **99** (3), 480–486.

23 Seizer, P., Gawaz, M., and May, A.E. (2008) Platelet-monocyte interactions – a dangerous liaison linking thrombosis,

inflammation and atherosclerosis. *Curr. Med. Chem.*, **15** (20), 1976–1980.

24 Pitsilos, S., Hunt, J., Mohler, E.R., Prabhakar, A.M., Poncz, M., Dawicki, J., *et al.* (2003) Platelet factor 4 localization in carotid atherosclerotic plaques: correlation with clinical parameters. *Thromb. Haemost.*, **90** (6), 1112–1120.

25 Nassar, T., Sachais, B.S., Se, A., Kowalska, M.A., Bdeir, K., Leitersdorf, E., *et al.* (2003) Platelet factor 4 enhances the binding of oxidized low-density lipoprotein to vascular wall cells. *J. Biol. Chem.*, **278** (8), 6187–6193.

26 Huo, Y. and Ley, K. (2001) Adhesion molecules and atherogenesis. *Acta Physiol. Scand.*, **173**, 35–43.

27 Libby, P. (2002) Atherosclerosis: the new view. *Sci. Am.*, **May**, 47–55.

28 Proudfoot, A.E.I. (2006) The biological relevance of chemokine-proteoglycan interactions. *Biochem. Soc. Trans.*, **34** (Pt 3), 422–426.

29 Barlic, J., Zhang, Y., Foley, J.F., and Murphy, P.M. (2006) Oxidized lipid-driven chemokine receptor switch, CCR2 to CX3CR1, mediates adhesion of human macrophages to coronary artery smooth muscle cells Through a peroxisome proliferator-activated receptor γ-dependent pathway. *Circulation*, **114** (8), 807–819.

30 Gosling, J., Slaymaker, S., Gu, L., Tseng, S., Zlot, C.H., Young, S.G., *et al.* (1999) MCP-1 deficiency reduces susceptibility to atherosclerosis in mice that overexpress human apolipoprotein B. *J. Clin. Invest.*, **103** (6), 773–778.

31 Boring, L., Gosling, J., Cleary, M., and Charo, I.F. (1998) Decreased lesion formation in CCR2$^{-/-}$ mice reveals a role for chemokines in the initiation of atherosclerosis. *Nature*, **394** (6696), 894–897.

32 Tedgui, A. and Mallat, Z. (2006) Cytokines in atherosclerosis: pathogenic and regulatory pathways. *Physiol. Rev.*, **86**, 515–581.

33 Askenasy, N., Kaminitz, A., and Yarkoni, S. (2008) Mechanisms of T regulatory cell function. *Autoimmun. Rev.*, **7** (5), 370.

34 Edfeldt, K., Bennet, A.M., Eriksson, P., Frostegard, J., Wiman, B., Hamsten, A.,

et al. (2004) Association of hypo-responsive toll-like receptor 4 variants with risk of myocardial infarction. *Eur. Heart J.*, **25** (16), 1447–1453.

35 Michelsen, K.S., Wong, M.H., Shah, P.K., Zhang, W., Yano, J., Doherty, T.M., *et al.* (2004) Lack of Toll-like receptor 4 or myeloid differentiation factor 88 reduces atherosclerosis and alters plaque phenotype in mice deficient in apolipoprotein E. *Proc. Natl Acad. Sci. U. S. A.*, **101** (29), 10679–10684.

36 Mullick, A.E., Soldau, K., Kiosses, W.B., Bell, T.A., III, Tobias, P.S., and Curtiss, L.K. (2008) Increased endothelial expression of Toll-like receptor 2 at sites of disturbed blood flow exacerbates early atherogenic events. *J. Exp. Med.*, **205** (2), 373–383.

37 Mullick, A.E., Tobias, P.S., and Curtiss, L.K. (2005) Modulation of atherosclerosis in mice by Toll-like receptor 2. *J. Clin. Invest.*, **115** (11), 3149–3156.

38 Salio, M., Chimenti, S., De Angelis, N., Molla, F., Maina, V., Nebuloni, M., *et al.* (2008) Cardioprotective function of the long pentraxin PTX3 in acute myocardial infarction. *Circulation*, **117** (8), 1055–1064.

39 Flavell, D.M., Jamshidi, Y., Hawe, E., Torra, L.P., Taskinen, M.R., Frick, M.H., *et al.* (2002) Peroxisome proliferator-activated receptor alpha gene variants influence progression of coronary atherosclerosis and risk of coronary artery disease. *Circulation*, **105** (12), 1440–1445.

40 Steinberg, D. (2009) The LDL modification hypothesis of atherogenesis: an update. *J. Lipid Res.*, **50** (Suppl.), S376–S381.

41 Li, A.C. and Glass, C.K. (2002) The macrophage foam cell as a target for therapeutic intervention. *Nat. Med.*, **8** (11), 1235–1242.

42 van Berkel, T.J.C., Out, R., Hoekstra, M., Kuiper, J., Biessen, E., and van Eck, M. (2005) Scavenger receptors: friend or foe in atherosclerosis? *Curr. Opin. Lipidol.*, **16** (5), 525–535.

43 Libby, P. (2002) Inflammation in atherosclerosis. *Nature*, **420**, 868–874.

44 Kovanen, P.T. (1996) Mast cells in human fatty streaks and atheromas:

implications for intimal lipid accumulation. *Curr. Opin. Lipidol.*, **7** (5), 281–286.

45 Ouimet, M., Wang, M.D., Cadotte, N., Ho, K., and Marcel, Y.L. (2008) Epoxycholesterol impairs cholesteryl ester hydrolysis in macrophage foam cells, resulting in decreased cholesterol efflux. *Arterioscler. Thromb. Vasc. Biol.*, **28** (6), 1144–1150.

46 Rader, D.J., Alexander, E.T., Weibel, G.L., Billheimer, J., and Rothblat, G.H. (2009) The role of reverse cholesterol transport in animals and humans and relationship to atherosclerosis. *J. Lipid Res*, **50** (Suppl.), S189–S194.

47 Bouhlel, M.A., Derudas, B., Rigamonti, E., Dievart, R., Brozek, J., Haulon, S., *et al.* (2007) PPARgamma activation primes human monocytes into alternative M2 macrophages with anti-inflammatory properties. *Cell Metab.*, **6**, 137–143.

48 Newby, A.C. (2006) Matrix metalloproteinases regulate migration, proliferation, and death of vascular smooth muscle cells by degrading matrix and non-matrix substrates. *Cardiovasc. Res.*, **69** (3), 614.

49 George, S.J. and Beeching, C.A. (2006) Cadherin:catenin complex. A novel regulator of smooth muscle cell behaviour. *Cardiovasc. Res.*, **188** (1), 1–11.

50 George, S.J. and Dwivedi, A. (2004) Cadherins MMPs and proliferation. *Trends Cardiovasc. Med.*, **14**, 100–105.

51 Fay, W.P., Garg, N., and Sunkar, M. (2007) Vascular functions of the plasminogen activation system. *Arterioscler. Thromb. Vasc. Biol.*, **27** (6), 1231–1237.

52 van der Wal, A.C. and Becker, A.E. (1999) Atherosclerotic plaque rupture – pathologic basis of plaque stability and instability. *Cardiovasc. Res.*, **41** (2), 334–344.

53 Clarke, M. and Bennett, M. (2006) The emerging role of vascular smooth muscle cell apoptosis in atherosclerosis and plaque stability. *Am. J. Nephrol.*, **26** (6), 531–535.

54 Lutgens, S.P.M., Cleutjens, K., Daemen, M., and Heeneman, S. (2007) Cathepsin cysteine proteases in cardiovascular disease. *FASEB J.*, **21**, 3029–3041.

55 Johnson, J.L. (2007) Matrix metalloproteinases: influence on smooth muscle cells and atherosclerotic plaque stability. *Exp. Rev. Cardiovasc. Ther.*, **5** (2), 265–282.

56 Newby, A.C. (2008) Metalloproteinase expression in monocytes and macrophages and its relationship to atherosclerotic plaque instability. *Arterioscler. Thromb. Vasc. Biol.*, **28** (12), 2108–2114.

57 Leskinen, M.J., Lindstedt, K.A., Wang, Y.F., and Kovanen, P.T. (2003) Mast cell chymase induces smooth muscle cell apoptosis by a mechanism involving fibronectin degradation and disruption of focal adhesions. *Arterioscler. Thromb. Vasc. Biol.*, **23** (2), 238–243.

58 Johnson, J.L., Jackson, C.L., Angelini, G.D., and George, S.J. (1998) Activation of matrix-degrading metalloproteinases by mast cell proteases in atherosclerotic plaques. *Arterioscler. Thromb. Vasc. Biol.*, **18**, 1707–1715.

59 Saarinen, J., Kalkkinen, N., Welgus, H.G., and Kovanen, P.T. (1994) Activation of human interstitial procollagenase through direct cleavage of the Leu(83)-Thr(84) bond by mast-cell chymase. *J. Biol. Chem.*, **269** (27), 18134–18140.

60 Shamamian, P., Schwartz, J.D., Pocock, B.J.Z., Monea, S., Whiting, D., Marcus, S.G., *et al.* (2001) Activation of progelatinase A (MMP-2) by neutrophil elastase, cathepsin G, and proteinase-3: a role for inflammatory cells in tumor invasion and angiogenesis. *J. Cell. Physiol.*, **189** (2), 197–206.

61 de Boer, O.J., van der Wal, A.C., Teeling, P., and Becker, A.E. (1999) Leukocyte recruitment in rupture prone regions of lipid-rich plaques: a prominent role for neovascularization? *Cardiovasc. Res.*, **41** (2), 443–449.

62 Lappalainen, H., Laine, P., Pentikainen, M.O., Sajantila, A., and Kovanen, P.T. (2004) Mast cells in neovascularized human coronary plaques store and secrete basic fibroblast growth factor, a potent angiogenic mediator. *Arterioscler. Thromb. Vasc. Biol.*, **24** (10), 1880–1885.

63 Norrby, K. (2002) Mast cells and angiogenesis. *Apmis*, **110** (5), 355–371.

64 Kolodgie, F.D., Gold, H.K., Burke, A.P., Fowler, D.R., Kruth, H.S., Weber, D.K., *et al.* (2003) Intraplaque hemorrhage and progression of coronary atheroma. *N. Engl. J. Med.*, **349** (24), 2316–2325.

65 Virmani, R., Kolodgie, F.D., Burke, A.P., Finn, A.V., Gold, H.K., Tulenko, T.N., *et al.* (2005) Atherosclerotic plaque progression and vulnerability to rupture: angiogenesis as a source of intraplaque hemorrhage. *Arterioscler. Thromb. Vasc. Biol.*, **25** (10), 2054–2061.

66 Ribatti, D., Levi-Schaffer, F., and Kovanen, P.T. (2008) Inflammatory angiogenesis in atherogenesis a double-edged sword. *Ann. Med.*, **40** (8), 606–621.

67 Harper, M.T. and Poole, A.W. (2007) Isoform-specific functions of protein kinase C: the platelet paradigm. *Biochem. Soc. Trans.*, **35**, 1005–1008.

68 Amento, E., Ehsani, N., Palmer, H., and Libby, P. (1991) Cytokines and growth factors positively and negatively regulate interstitial collagen gene expression in human vascular smooth muscle cells. *Arterioscler. Thromb. Vasc. Biol.*, **11**, 1223–1230.

69 Kolodgie, F.D., Burke, A.P., Wight, T.N., and Virmani, R. (2004) The accumulation of specific types of proteoglycans in eroded plaques: a role in coronary thrombosis in the absence of rupture. *Curr. Opin. Lipidol.*, **15** (5), 575–582.

70 Mayranpaa, M.I., Heikkila, H.M., Lindstedt, K.A., Walls, A.F., and Kovanen, P.T. (2006) Desquamation of human coronary artery endothelium by human mast cell proteases: implications for plaque erosion. *Coron. Artery Dis.*, **17** (7), 611–621.

PART II
PRO-Inflammatory Factors

Atherosclerosis: Molecular and Cellular Mechanisms. Edited by Sarah Jane George and Jason Johnson
Copyright © 2010 WILEY-VCH Verlag GmbH & Co. KGaA, Weinheim
ISBN: 978-3-527-32448-4

2
Chemokines and Atherosclerosis: A Critical Assessment of Therapeutic Targets

Gemma E. White, Keith M. Channon, and David R. Greaves

2.1
Introduction

Chemokines are a family of low molecular weight proteins that function in homeo-static and pathological conditions to induce the directed migration of leukocytes. Atherosclerosis is a chronic inflammatory disease occurring in the arterial wall over many decades, often leading to acute cardiovascular events including angina, myocardial infarction, and ischemic stroke. Over the last decade we have come to understand that plaque formation is a process of ongoing inflammation character-ized by endothelial damage, recruitment of mononuclear cells, and smooth muscle cell (SMC) proliferation [1]. Many of these processes are driven by chemokines. This chapter will highlight the potential functions of chemokines in atherogenesis, describe the evidence supporting their involvement in atherosclerosis, and discuss possible avenues for chemokine-directed therapy in cardiovascular disease.

2.1.1
Chemokines

Chemokines, or chemoattractant cytokines, are a family of low molecular weight proteins (around 8–10 kDa) that have a wide range of functions in both tissue homeostasis and pathological conditions [2]. They were initially divided into two families, α and β but have since been reclassified on the basis of structure and given a systematic nomenclature [3, 4]. There are currently around 50 known chemokines, divided into four families: C, CC, CXC, and CX_3C by the location of key structural cysteine residues. The CC family (previously the β chemokines) contain two adjacent cysteine residues in the N-terminus, forming two disulfide bonds. The CXC family (or α family) has a single amino acid dividing these two cysteine residues, and can be further divided into two groups (ELR positive or negative) depending on the presence or absence of a conserved glutamic acid–leucine–arginine (ELR) sequence immediately before the cysteine motif. The C

Atherosclerosis: Molecular and Cellular Mechanisms. Edited by Sarah Jane George and Jason Johnson
Copyright © 2010 WILEY-VCH Verlag GmbH & Co. KGaA, Weinheim
ISBN: 978-3-527-32448-4

and CX_3C group have been more recently discovered, and each family contains only a single member. The chemokine family share 20–90% sequence homology and have a highly conserved tertiary structure [5].

In vivo, chemokines can be presented on the cell surface via binding to glycosaminoglycans (GAGs). GAGs are long chains of highly charged polysaccharides attached to proteoglycans on the cell surface and are thought to allow the formation of an immobilized chemokine gradient, especially under flow conditions [6]. The requirement for GAG binding *in vivo* has been functionally demonstrated using various chemokines with impaired GAG-binding activity [7, 8]. For example, a mutant form of CCL7 (MCP-3) that is unable to bind to GAGs was shown to block the chemoattractant effects of wild-type CCL7 and other CC chemokines in the murine air pouch model [9].

2.1.2
Chemokine Receptors

All chemokines signal through seven-transmembrane G protein–coupled receptors (GPCRs), usually $G_{\alpha i}$-coupled receptors. These are also named systematically according to the type of chemokine they bind; for example, CC chemokine receptor 1 is denoted CCR1, CXC chemokine receptor 2 is denoted CXCR2 [4]. To date, around 20 chemokine receptors have been cloned – a much smaller number than their chemokine ligands. Thus, most chemokine receptors have multiple high-affinity chemokine ligands and most chemokines are promiscuous in receptor binding. There are a few notable exceptions, including CX_3CL1/CX_3CR1, CXCL13/CXCR5 and CXCL16/CXCR6, which have a single ligand–receptor pair. Functional chemokine receptors, like other GPCRs, contain a conserved aspartic acid–arginine–tyrosine (DRY) motif in the third intracellular loop of the receptor, which is essential for signaling and believed to be involved in stabilizing the GPCR in either an active or inactive conformation [10]. Two chemokine receptors that do not have a DRY motif are the decoy receptors DARC (Duffy antigen receptor for chemokines) and D6, which are able to bind chemokines but do not signal and therefore may act as a molecular "sink" for chemokines *in vivo* [11, 12].

2.1.3
Chemokine Functions

Chemokines function in both innate and adaptive immune responses and were originally identified by their role in directing leukocytes to sites of inflammation or injury [13, 14]. The CXC family of chemokines (mostly ELR$^+$, e.g., CXCL8) are primarily involved in recruiting polymorphonuclear leukocytes to sites of acute inflammation, and are released rapidly in response to bacterial products such as lipopolysaccharide (LPS) or inflammatory mediators at a site of injury [2]. CC chemokines are mainly involved in recruitment of mononuclear cells to sites of chronic inflammation. Thus chemokines produced at a site of injury of infection establish a chemoattractant gradient and are presented on the endothelium of postcapillary venules. Chemokine presentation plays a critical role in leukocyte recruitment by triggering rapid activation of rolling leukocytes, leading to firm adhesion and diapedesis [15].

More recently, we have come to appreciate that certain chemokines have homeostatic functions. Chemokines acting via CXCR2, for example, have a crucial role in angiogenesis, which is often dysregulated in malignant cells, facilitating neovascularization of tumors [16]. Homeostatic chemokines are also essential for the appropriate trafficking of mature lymphocytes through the secondary lymphoid organs in response to the chemokines CCL19 and CCL21 via the chemokine receptor CCR7 [17, 18].

2.2
Key Chemokine Functions in Atherogenesis

Experimental evidence derived from animal models, primary cell culture systems, immunohistochemistry of atherosclerotic plaques, and genetic association studies have all shown a role for numerous chemokines in atherosclerosis. Following endothelial damage and activation, chemokine expression is upregulated and these are presented on the endothelial surface by GAG binding or, in the case of CX_3CL1 (fractalkine), expressed as membrane-bound molecules [19]. This enables firm adhesion of circulating monocytes expressing the cognate chemokine receptor (e.g., CX_3CR1) to the endothelial surface [20, 21]. Monocytes then enter the subendothelial space of the vessel following a chemokine gradient whereupon they differentiate to become macrophages. CCL2 and CX_3CL1, among others, have a role in monocyte recruitment into the plaque and their relative contribution was the subject of two recent studies [22, 23]. Using $Cx_3cl1^{-/-}/Ccr2^{-/-}$/apolipoprotein E $(Apoe)^{-/-}$ triple knockout mice, Saederup et $al.$ [23] demonstrated that these chemokines have independent and additive roles in macrophage recruitment to plaques, which may result from recruitment of different monocyte subsets expressing variable levels of receptor [22, 24]. Figure 2.1 illustrates the role of CCL2 and CX_3CL1 in monocyte recruitment into early atherosclerotic plaques. For a more detailed discussion of monocyte subsets see recent papers and reviews from the laboratories of Frederic Geissman and Gwen Randolph [24–26] and Chapter 18 by Johnson and Jenkins.

It is not only the recruitment of monocytes into the plaque, but also the retention of macrophages within the plaque, which contributes to plaque progression [27]. Emigration of monocytes from the plaque was detectable under experimental conditions of plaque regression but not in plaques undergoing progression – indicating that reduced monocyte trafficking may be as important as monocyte recruitment in the development of atherosclerosis. A subsequent study using the same aortic transplantation model of plaque regression indicated that the expression and function of CCR7 were essential for emigration from the plaque [28]. Chemokines also have an established role in monocyte retention in the progressive plaque. Barlic et $al.$ have suggested a model whereby monocyte exposure to oxi-

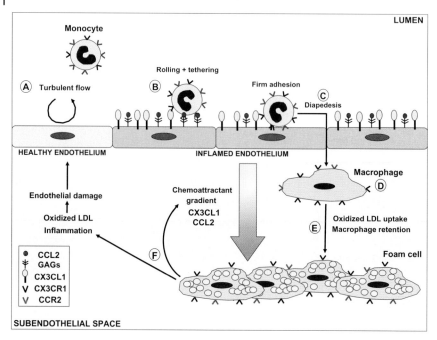

Figure 2.1 Functional effects of chemokines in early plaque development: the role of CCL2 and CX₃CL1 in monocyte recruitment. Endothelial damage/dysfunction occurs in response to circulating oxidized LDL and systemic inflammation, particularly in regions of turbulent blood flow and low shear stress (A). Activated endothelial cells express a variety of adhesion molecules and present chemokines on their surface, either bound to glycosaminoglycans (GAGs) (e.g., CCL2) or as membrane-bound molecules (e.g., CX₃CL1). Monocytes expressing the cognate chemokine receptor (e.g., CCR2) roll along the endothelial surface (B), before undergoing firm adhesion, diapedesis, and migration into the plaque, following a chemoattractant gradient (C). Monocyte chemotaxis into the subendothelial space is driven by many chemokines, including CX₃CL1 and CCL2. Recent evidence suggests that CCL2 and CX₃CL1 have independent roles in atherogenesis and may recruit different subsets of monocytes into the plaque. Once in the vessel wall, monocytes differentiate to become macrophages, with changes in their chemokine receptor expression that act to retain macrophages in the plaque (D). Activated macrophages take up oxidized LDL via scavenger receptors to become foam cells (E), which in turn secrete chemokines and other inflammatory cytokines to drive further monocyte recruitment and exacerbate endothelial damage (F).

dized lipids leads to loss of CCR2 (which is promigratory) and gain of CX₃CR1 (which is pro-adhesive), thus facilitating plaque progression [29]. Indeed, they show that macrophages exposed to oxidized LDL express high levels of CX₃CR1, which enables adhesion to SMCs, providing a retention mechanism.

T cells, particularly those of the CD4⁺ Th1 subset, also accumulate in the plaque in response to chemokines such as CXCL9, CXCL10, and CXCL11 (MIG, IP-10, and I-TAC respectively) via expression of the receptor CXCR3 [30]. Th1 cells secrete interferon-γ (IFN-γ), leading to classical macrophage activation, LDL

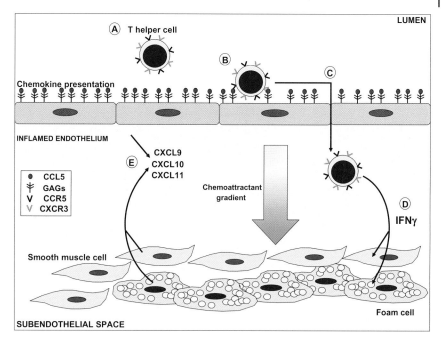

Figure 2.2 Functional effects of chemokines in later plaque development: T cell recruitment and the role of interferon-γ. Circulating CD4+ T cells express the chemokine receptors CCR5 and CXCR3 (A). T cells adhere to activated endothelial cells expressing chemokines that include CCL5, CXCL9, CXCL10, and CXCL11 (B). T cells enter the plaque (C) and secrete interferon-γ in response to antigen presented by macrophages (D). Interferon-γ induces local expression of the chemokines CXCL9 and CXCL11 by endothelial cells and macrophages, and CXCL10 by smooth muscle cells, endothelial cells, and macrophages (E). This facilitates continued T cell recruitment and atherosclerotic plaque progression.

uptake, and foam cell formation [31, 32]. These are the hallmarks of the early fatty streak lesion. Figure 2.2 illustrates the role of IFN-γ and the chemokines CXCL9–11 in T cell recruitment into more advanced atherosclerotic plaques.

Chemokines have been assigned roles in almost every aspect of plaque progression including SMC apoptosis and proliferation (GW and DRG, unpublished data), monocyte anti-apoptosis [33], platelet activation [34], matrix metalloprotease (MMP) expression [35], and neovascularization in advanced plaques [36, 37].

2.3
Experimental Evidence for the Role of Chemokines in Atherosclerosis

In this section we will focus on a few chemokine ligand–receptor pairs that have been most extensively studies in the context of atherogenesis.

2.3.1
CCL2 Acting via CCR2

CCL2 (MCP-1) was one of the first chemokines shown to have a role in atherosclerosis and it remains the best-characterized chemokine in atherogenesis. CCL2 is a potent monocyte chemoattractant, is highly expressed in human plaques, and is produced by endothelial cells, monocytes/macrophages, and SMCs [38, 39]. Gene deletion studies have confirmed a key role for CCL2–CCR2 interactions in atherogenesis. Knockout of CCR2 in $Apoe^{-/-}$ animals led to a reduction of total lesion area and plaque macrophage content [40, 41]. Furthermore, $Ccr2^{+/-}$ $Apoe^{-/-}$ animals showed an intermediate reduction in lesion size, indicating a gene dosage effect [40]. $Ccl2^{-/-}/Ldlr^{-/-}$ mice had an 83% reduction in total lesion area compared with $Ldlr^{-/-}$ controls after 12–14 weeks Western-type diet [42]. Bone marrow transplantation of CCR2-deficient leukocytes into $Apoe3$-Leiden mice led to a reduction in atherosclerosis compared with wild-type bone marrow transplantation [43].

Interestingly, in a model of intimal hyperplasia following wire injury, CCR2 deficiency resulted in a 64% reduction in intimal hyperplasia without any significant effect on macrophage infiltration [44]. CCL2 also has a potential role in SMC proliferation, suggesting the possibility of multiple roles for CCL2 in atherogenesis [45]. Recent work from the authors' laboratories showed a reduction in vein graft accelerated atherosclerosis in $Apoe^{-/-}$ mice overexpressing endothelial GTP cyclohydrolase – the rate-limiting enzyme in tetrahydrobiopterin synthesis and an essential cofactor for endothelial nitric oxide generation [46]. This effect was shown to be mediated via reduced expression of CCL2 in the aorta and a decrease in CCR2 ligand bioactivity. This suggests CCL2 may have numerous roles depending on the type of vessel injury.

CCL2 has also been shown to be a direct target for statin therapy [47]. *In vitro*, simvastatin treatment led to a reduction in CCR2 expression on THP-1 monocytes. Treatment of normocholesterolemic patients for two weeks with simvastatin also led to a reduction in CCR2 expression on circulating monocytes without a reduction in plasma LDL during this time period. Furthermore, using an adoptive transfer technique, leukocytes from healthy rats were transferred to hypercholesterolemic littermates. Leukocytes from donor animals treated with simvastatin showed reduced recruitment to the vessel wall compared with untreated animals [47]. In a study involving over 2000 patients with a high risk of cardiovascular events, atorvastatin treatment resulted in a small (4%) but significant drop in plasma CCL2 levels [48].

2.3.2
CCL3, 4, 5 Acting via CCR1 and CCR5

The CC chemokine receptors CCR1 and CCR5 have been implicated in the development of atherosclerosis and share a number of ligands that can be detected in atherosclerotic plaques including CCL5 (RANTES), CCL3 (MIP-1α) and CCL4

(MIP-1β) [49, 50]. CCR1 and CCR5 are differentially expressed on circulating leukocyte populations: CCR1 is highly expressed on monocytes, while levels of CCR5 expression are highest on Th1-like cells [51]. In addition, CCR1 and CCR5 have independent roles in leukocyte entry into the vessel wall: CCR1 mediates leukocyte capture, while CCR5 is involved in leukocyte spreading and transmigration under flow [51]. Furthermore, oligomerization of CCL5 was shown to be crucial for T cell and monocyte arrest under flow mediated by CCR1 [52].

CCL5 (RANTES) is released upon platelet activation and degranulation, whereupon it is immobilized on activated endothelium and can trigger arrest of circulating monocytes [53]. Furthermore, this process was shown to require platelet P-selectin expression and contribute to the development of intimal hyperplasia following wire injury [54].

Several studies have presented conflicting evidence for the role of each receptor in atherosclerosis. CCR5 deficiency, for example, was not protective against early atherosclerosis in $Apoe^{-/-}$ mice [55], while the absence of bone marrow CCR5 had no effect on lesion size but did reduce plaque macrophage content in $Ldlr^{-/-}$ mice [56]. In contrast, bone marrow deficiency of CCR1 had a pro-atherogenic effect in $Ldlr^{-/-}$ mice [56]. A study comparing the two receptors in the development of diet-induced atherosclerosis in $Apoe^{-/-}$ mice showed that only deletion of CCR5 and not CCR1 could protect against plaque development [57].

2.3.3
CXCL9,10, 11 Acting via CXCR3

Atherosclerotic plaques in humans and mice contain both CD4$^+$ T-helper cells and IFN-γ, indicating a Th1 environment [58]. The chemokines CXCL9 (MIG), CXCL10 (IP-10), and CXCL11 (I-TAC) are all induced by IFN-γ and are critically involved in the recruitment of Th1 cells into the plaque via their receptor CXCR3 – indeed virtually all CD4$^+$ T cells in the plaque express CXCR3 [30]. Immunofluorescent staining shows that CXCL10 is expressed in the plaque by SMCs, endothelial cells (ECs), and macrophages, whereas CXCL9 and CXCL11 are found mainly in ECs and macrophages. The role of T cells in mouse models of atherosclerosis is limited to specific timepoints (reviewed in [59]). Specifically, deletion of dividing T cells in $Apoe^{-/-}$ animals had no effect on lesion initiation, but did attenuate lesion progression over an eight-week period [60].

$Cxcr3^{-/-}/Apoe^{-/-}$ mice showed a ~50% reduction in lesion area in the thoracoabdominal aorta but not the aortic root, and lesions contained significantly fewer T cells but a similar number of macrophages compared with $Apoe^{-/-}$ controls [61]. Interestingly, plaques from CXCR3 knockout mice showed a higher level of CD4$^+$ CD25$^+$ regulatory T cells (T_{reg}), which may have a protective effect in early lesion development. In a $Cxcl10^{-/-}/Apoe^{-/-}$ mouse model, plaque size was reduced in the aortic arch and thoracic and abdominal aortas, and there was a 44% reduction in CD4$^+$ T cells [62]. The authors also saw an increase in FoxP3 mRNA (a T_{reg} marker) and an increase in interleukin-10 (IL-10) and transforming growth factor β1 (TGFβ1) mRNA (two key anti-inflammatory T_{reg} cytokines) in CXCL10 knockout animals [63, 64]. Thus, CXCL10 may have a role in regulating T_{reg}/effector T cell ratios in early lesion development.

In a recent study, CXCR3 ligands were shown to be regulated both *in vitro* and *in vivo* by adiponectin, suggesting a novel pathway for the anti-inflammatory function of this adipocytokine [65]. CXCR3 has also recently been implicated in the inflammatory response to arterial wire injury [66]. CXCL9 and CXCL10 were induced locally and systemically in response to wire injury, resulting in enhanced recruitment of CXCR3$^+$ cells, activation of mTORC1, generation of reactive oxygen species and induction of apoptosis. Furthermore, blockade of CXCR3 signaling led to a significant reduction in intimal hyperplasia, suggesting a novel link between T cell stimulation and intimal SMC proliferation. A very recent report has documented an increase in blood pressure in CXCR3 knockout mice [67]. The mechanism whereby deficiency of a single chemokine receptor causes hypertension remains to be determined, but this paper reminds us that chemokines and their receptors have many functions beside recruitment of inflammatory cells into atherosclerotic lesions.

2.3.4
CXCL1/CXCL8 Acting via CXCR2

CXCL8 (IL-8) was identified in humans as a chemoattractant for human neutrophils generated by LPS-stimulated monocytes [13]. CXCL8 was subsequently shown to signal through two chemokine receptors CXCR1 and CXCR2 [68, 69]. Mice do not have a direct homolog of CXCL8 but the equivalent chemokine CXCL1 (KC and in humans GROα) acts via CXCR2 to induce murine neutrophil chemotaxis [70]. In human atherosclerotic lesions, neutrophils are rare or absent except in ruptured plaques where their influx may be associated with acute coronary events [71]. However, CXCL8 can also mediate monocyte adhesion under flow to activated endothelial cells, suggesting a function for this chemokine in atherosclerosis [72]. A recent study suggests that in *Apoe*$^{-/-}$ mice, neutrophils are detectable in the plaque and their depletion attenuates plaque progression [73].

Using lethally irradiated *Ldlr*$^{-/-}$ mice reconstituted with bone marrow from mice, which lacked the murine CXCR2 homolog, the absence of this receptor was shown to induce smaller lesions and seemed to have a direct role in macrophage accumulation in the plaque [74]. In a similar study using *Cxcl1*$^{-/-}$ knockout bone marrow, the authors demonstrated that reconstituted *Apoe*$^{-/-}$ animals had smaller lesions – though the reduction was only about 50% of that seen in the animals with *Cxcr2*$^{-/-}$ leukocytes [75]. In conclusion, CXCL1 and CXCR2 seem to be more important for macrophage accumulation in advanced rather than early plaques.

2.3.5
CX₃CL1 Acting via CX₃CR1

CX_3CL1 is the only member of the CX_3C family of chemokines and is unusual in being synthesized as a membrane-bound chemokine, which can be cleaved from the cell surface under inflammatory conditions [76, 77]. CX_3CL1 is expressed by SMCs, ECs, and macrophages, while CX_3CR1 is found on peripheral monocytes and T cells and on SMCs in the plaque [20, 78]. CX_3CL1 also induces SMC chemotaxis [78] and proliferation *in vitro* (GW and DRG, unpublished observation).

Knockout studies have confirmed a pro-atherogenic function for CX_3CL1. $Cx_3cl1^{-/-}/Apoe^{-/-}$ animals fed a Western-type diet, showed no difference in lesion size at the aortic root but a significant reduction was seen at the brachiocephalic artery (BCA) in both homozygote and heterozygote animals [79]. In $Cx_3cl1^{-/-}/Ldlr^{-/-}$ female animals, lesions were smaller at both the aortic arch and BCA. $Cx_3cr1^{-/-}/Apoe^{-/-}$ mice were generated by two groups [80, 81]. Combadiere *et al.* showed that knockout animals fed on a chow diet had significantly reduced lesion coverage in both the thoracic aorta and aortic arch and a 50% reduction in macrophage content in lesions at the aortic sinus [80]. They also demonstrated a similar SMC and collagen content in knockout compared to wild-type plaques. Very similar results were found by Lesnik *et al.*, with a 36% reduction in plaque area in the thoracic aorta and a 40% reduction in macrophage content in the aortic sinus using animals fed on a Western-type diet [81]. As described for CCR2, $Cx3cr1^{+/-}/Apoe^{-/-}$ animals also show a proportional reduction in lesion size, indicating a gene dosage effect [80].

In addition to animal studies, epidemiologic evidence has shown an altered risk of cardiovascular disease associated with two non-synonymous single nucleotide polymorphisms of CX_3CR1: V249I and T280M [82]. In one study, heterozygosity of the I249 allele was associated with a significantly reduced number of cardiovascular events and was associated with a smaller number of CX_3CL1-binding sites on peripheral blood mononuclear cells [83]. In another study, the M280 allele, which has also been associated with reduced atherosclerosis, showed reduced adhesion to CX_3CL1 [84]. A number of subsequent genetic association studies have shown conflicting effects of the CX_3CR1 polymorphisms on cardiovascular risk [85, 86].

2.4
Strategies for Targeting Chemokines in Atherosclerosis

Many researchers have attempted to target the chemokine network with a variety of strategies. Some of these may provide useful avenues for therapeutic targeting of chemokines in atherosclerosis and other inflammatory diseases.

2.4.1
Neutralizing Antibodies

Evidence for the role of chemokines in inflammatory disease comes from the use of neutralizing antibodies targeted against a single chemokine or receptor. A gene expression study of plaques from $Apoe^{-/-}$ mice showed a time-dependent increase in transcripts for CCL2 (MCP-1), CCL3 (MIP-1α), CCL4 (MIP-1β), and CCL12

(MCP-5) [87]. Administration of a neutralizing monoclonal antibody against CCL2 and CCL12 (named 11K2), reduced plaque size when administered at both an early timepoint (5–17 weeks) and a delayed timepoint (17–29 weeks) indicating a reduction in lesion progression. Treatment reduced total leukocyte and macrophage content in the plaque, while increasing SMC and collagen content, indicating an improvement in plaque stability.

In a randomized, double-blind, placebo-controlled phase II clinical trial, an anti-CCR2 monoclonal antibody (MLN1202) was administered to patients with risk factors for acute coronary disease (see http://clinicaltrials.gov/ct2/show/NCT00715169). A single dose of the antibody was given, and resulted in significantly reduced C-reactive protein levels (a cardiovascular biomarker) for several months after administration of the anti-CCR2 antibody.

There are many examples of therapeutic antibodies used successfully in the clinic, though the chronic nature of atherosclerosis and the possibility of an immune response targeted to the therapeutic antibody may limit the success of this approach. However, local delivery of anti-chemokine antibodies, for example in drug-eluting stents, may be a potential application.

2.4.2
Small Molecule Inhibitors

Small molecule inhibitors of chemokine receptors have been the subject of multiple screening programs and offer a number of clear advantages over other therapies. Many are orally active allowing easy administration, and their shorter half-life (compared with biologicals for example) means the therapy can be controlled and ceased if necessary. Furthermore, potential immunologic responses against biologicals can be avoided. Many small molecule inhibitors of chemokine receptors exist but only a small number have been tested in models of atherosclerosis and vascular injury. It is currently unclear which animal model most closely resembles human disease and is therefore most suitable to test these compounds.

TAK-779 is a small molecule CCR5 antagonist developed as an inhibitor of HIV entry into cells via CCR5, with significant activity against the CXCR3 receptor [88]. In a collar-induced carotid artery atherosclerosis model in $Ldlr^{-/-}$ mice, TAK-779 reduced lesion size by 68%, without affecting macrophage numbers in the plaque [89]. In a second experiment using diet-induced atherosclerosis in $Ldlr^{-/-}$ mice, TAK-779 reduced plaque size in the aortic root by 40% without altering plaque macrophage content [89]. In the collar-induced vascular injury model, T cell numbers in the plaque were dramatically reduced – consistent with the expression of CCR5 and CXCR3 receptors on Th1-type T cells. Indeed, there was a 98% reduction in staining of the Th1 cytokine IFN-γ in the plaque. A particular attraction of this therapy is that treatment of HIV with combination therapy including protease inhibitors leads to increased serum cholesterol and increased mortality from atherosclerosis and subsequent acute coronary events [90]. Thus, treatment with anti-CCR5 therapy may have dual benefits in HIV treatment and prevention of associated atherosclerosis.

Using a specific antagonist of CXCR3 (NBI-74330) and a diet-induced athero-sclerosis model in $Ldlr^{-/-}$ mice, these authors also recently demonstrated that CXCR3 antagonism reduced plaque size (53%) but not plaque macrophage content [91]. Furthermore, the mice had smaller lymph nodes draining from the aortic arch, and the T cells within these lymph nodes were of a more regulatory pheno-type: suggesting a favorable modulation of the adaptive immune response.

2.4.3
Mutant Chemokines

Delivery of mutant chemokines has been shown to be a successful strategy to block atherosclerosis in a number of studies using mouse models of atherogenesis.

2.4.3.1 N-Terminal Chemokine Mutants

The N-terminus of most chemokines is critical for the activation but not binding of their cognate receptors. Thus, N-terminal modification allows generation of chemokine receptor antagonists which are able to bind to their cognate chemokine receptors but not signal. This appears to be a physiological method of regulating chemokine activity: many low-affinity ligands for CCR1 (including CCL6, CCL15, and CCL23) are N-terminally processed by proteases found in inflammatory exu-dates to generate highly active chemokines [92]. When CCL5 (RANTES) is expressed and purified from bacteria the initiating methionine is retained (Met-RANTES) generating a "decoy chemokine," which binds both CCR1 and CCR5 with high affinity but is unable to induce calcium signaling or chemotaxis [93]. When administered to $Ldlr^{-/-}$ mice fed a high-fat diet, Met-RANTES treatment led to a 43% reduction in plaque size at the aortic root and a 58% reduction in the thoracoabdominal aorta [94]. Furthermore, Met-RANTES reduced plaque macro-phage and $CD4^+$ T cell content and increased SMC and collagen content in the plaque – indicating a more stable plaque phenotype.

N-Terminal deletion mutants, administered via gene therapy, have also been used to target CCL2 in atherosclerosis. Inoue *et al.* created an N-terminal mutant of CCL2 (7 amino acid deletion – 7ND), which acted as a dominant negative chem-okine and transfected this into skeletal muscle of mice with established athero-sclerosis (after 20 weeks of fat feeding) [95]. After eight weeks of gene therapy, 7ND-treated mice showed little evidence of lesion progression (compared to the 20-week baseline), while plaques in animals transfected with empty plasmid had more than doubled in size. The authors also showed that 7ND therapy stabilized the plaque by preserving SMC and collagen content, and preventing increases in MMP content of the plaque.

Using the same construct in a different delivery method, Egashira *et al.* admin-istered 7ND in a gene-eluting stent in rabbits and monkeys [96]. 7ND significantly reduced in-stent restenosis via inhibition of monocyte infiltration and SMC pro-liferation. The authors detected no adverse effects in either rabbits or monkeys, and suggest this may be a viable therapeutic strategy for prevention of in stent stenosis.

2.4.3.2 Disruption of Oligomerization

Many chemokines, including CCL5 (RANTES), are known to require oligomerization in order to be presented on the endothelial surface and function *in vivo* [52]. Braunersreuther *et al.* administered the oligomerization mutant [^{44}AANA^{47}]-RANTES to *Ldlr*$^{-/-}$ mice with established lesions as a result of high-fat diet for 11 weeks [97]. [^{44}AANA^{47}]-RANTES significantly attenuated lesion progression in the aortic root and thoracoabdominal aorta. Furthermore, plaques from treated mice contained fewer T cells, macrophages, and lower MMP-9 expression, but had increased numbers of SMCs and more collagen – features consistent with a more stable plaque phenotype. [^{44}AANA^{47}]-RANTES reduced expression of CCR2, CCR5, and CCL2 in the aorta, and leukocyte recruitment to lesions.

Similarly, a monomeric mutant of CCL2 (P8A-CCL2) is unable to oligomerize on GAGs and blocks leukocyte recruitment in a model of sterile peritonitis induced by thioglycollate [98]. Thus, mutant chemokines may provide a possible therapeutic strategy in many inflammatory diseases. A drawback is the requirement for daily injection in mouse disease models, a dosing regime which may limit their usefulness in the clinic.

As well as oligomerizing with itself, CCL5 has been shown to form heteromers with CXCL4 (platelet factor-4), which enhance monocyte arrest on the endothelium [99]. In a very recent study, Koenen *et al.* generated stable peptide antagonists, which block interaction of CCL5 and CXCL4 and can dissociate preformed oligomers [100]. Using intravital microscopy, the peptide antagonist MKEY reduced leukocyte arrest on early atherosclerotic endothelium. MKEY also inhibited lesion formation when administered to *Apoe*$^{-/-}$ mice fed a high-fat diet. This strategy had no effect on the systemic immune response, and may offer a significant advantage over total chemokine blockade, which may result in immunosuppression as seen in *Ccl5*$^{-/-}$ mice that show impaired T cell responses [101].

2.4.4
Broad-Spectrum Chemokine Blockade

As discussed previously, the chemokine network shows a significant level of redundancy: most chemokines bind "promiscuously" to more than one chemokine receptor and many key chemokine receptors have multiple high-affinity chemokine ligands. In general, genetic deletion of a single chemokine or receptor has shown only ~50% reduction in lesion size. Two recent papers have highlighted the need to target multiple chemokines in order to further reduce atherosclerosis in mouse models. Combadiere *et al.* demonstrated that combined inhibition of CCL2, CCR5, and CX$_3$CR1 led to an additive and dramatic 90% reduction of atherosclerosis in *Apoe*$^{-/-}$ mice [22, 23]. Thus, therapeutic strategies aimed at multiple chemokines may prove most successful, particularly in short-term interventions.

Many viruses encode proteins that interfere with the chemokine network as a means of evading the anti-viral arm of the immune system. The specificity of these proteins varies; some target just one family of chemokines, while others target those from multiple families. All poxviruses produce a 35 kDa chemokine-binding protein, known as viral CC chemokine inhibitor (vCCI or 35K), which blocks CC chemokine activity by binding with very high affinity and preventing interaction with their cognate GPCRs [102–104].

Using an adenovirus construct to express 35K derived from vaccinia virus Lister strain, Bursill *et al.* demonstrated that *in vitro* chemotaxis towards CCL5 could be blocked [104]. This adenoviral construct was subsequently injected into *Apoe*$^{-/-}$ mice following a period of six weeks on high-fat diet [105]. After a two-week gene expression window, plaque size was analyzed and mice treated with 35K showed significantly reduced lesion size (55%) and a dramatic reduction in plaque macrophage content (85%) compared to controls. In other studies, including vein graft models of accelerated atherosclerosis in mice and rabbits, 35K dramatically reduces lesion progression and intimal hyperplasia [106, 107]. A recent study also demonstrated that long-term CC chemokine inhibition using a lentiviral-35K construct similarly reduced atherosclerosis throughout a much longer time frame – three months [108].

Herpes viruses also express a chemokine-binding protein (hvCKBP or M3), which has broad specificity: binding chemokines of the C, CC, CXC, and CX₃C subfamilies [109, 110]. This protein has not been extensively tested in models of atherosclerosis or other inflammatory diseases, but has been used in a model of wire injury [111]. Mice were generated which conditionally expressed the M3 transgene in response to doxycycline in a number of tissues. Mice administered doxycycline and therefore expressing M3 in tissues showed significantly reduced intimal hyperplasia (67% smaller intima) compared with animals not expressing the transgene. The authors did not observe any overt toxicity, but blockade of such a wide range of chemokines could result in significant immunosuppression, limiting the usefulness of M3 as a therapy.

2.5
Conclusions and Future Clinical Prospects for Chemokine Targeting

Chemokines have been identified as key players in inflammatory cell recruitment in a wide range of chronic inflammatory diseases. Since chemokines and their receptors represent interesting therapeutic targets, a wide range of tools and reagents have been developed to study the role of different chemokine classes and different chemokine receptors *in vivo*. In this brief book chapter we have sought to illustrate how chemokine receptor knockout mice, anti-chemokine antibodies, mutant chemokines, and viral chemokine-binding proteins have all been used to probe the role of chemokine biology in the pathogenesis of atherosclerosis and other forms of vascular injury. An obvious question that arises from this impressive body of experimental work is – will any of these therapeutic approaches find clinical application in the treatment of cardiovascular disease?

Long-term inhibition of parts of the chemokine network may seem like an unattractive strategy because of the risks associated with disabling leukocyte recruitment to homeostatic and inflammatory cues. Several arguments suggest that these fears may be misplaced. First, chemokine inhibition by any of the modalities discussed is likely to be less than 100% effective and in the case of small molecule inhibitors of chemokine receptors the dose of drug can be adjusted to give less than complete inhibition. Second, in the case of the CCR5 receptor, a significant proportion (2–5%) of the Caucasian population are heterozygous for a loss-of-function mutation of the CCR5 gene, CCR5delta32 [112]. The recent introduction of anti-CCR5 drugs as an adjunct to highly active antiretroviral therapy (HAART) for HIV⁺ individuals will show us if long-term depression of CCR5 activity leads to significant changes in immune responses to pathogens or the development of autoimmune disease. Third, there have been reports of increased susceptibility of some chemokine receptor knockout animals to infectious diseases, for example the increased susceptibility of $Ccr2^{-/-}$ mice to experimental *Mycobacterium tuberculosis* infection [113]. However, there have been no reports of increased susceptibility of chemokine receptor deficient animals to opportunistic pathogens, which infect animals through more natural routes of infection and in more realistic numbers.

Rather than thinking about life-long anti-chemokine therapy, perhaps we should consider if short-term, high-intensity blockade of chemokine activity could offer a significant benefit to patients undergoing percutaneous angioplasty or coronary artery bypass surgery. Experimental animal models of arterial injury strongly suggest that short-term intervention to prevent inflammatory cell recruitment to sites of vascular injury have beneficial effects on the resultant vascular remodeling, providing that they do not prevent the recruitment of endothelial progenitor cells [114]. Mutant chemokines and viral chemokine-binding proteins seem ideally suited for these types of application.

An important question to address is will anti-chemokine drugs and treatments offer a significant increased benefit to patients taking statins? Undoubtedly the success of statins has made the development and introduction of new drugs difficult in the cardiovascular disease arena, but there are significant numbers of patients with accelerated atherosclerosis (e.g., people with diabetes, patients with rheumatoid arthritis and South Asians) who could benefit from novel anti-inflammatory therapies that work in the presence of statins.

A final consideration is whether there exists a subset of patients in which elevated levels of specific chemokines play a key role in the development of atherosclerosis. Several studies looking for novel biomarkers of increased risk of cardiovascular disease have been performed and a number of CC chemokines have been identified as promising candidates for further studies [115, 116]. If safe, small-molecule inhibitors of chemokine receptors are developed for other inflammatory diseases it will be possible to test the ability of these drugs to act synergistically with statins in clinical trials in subgroups of patients at enhanced risk of developing cardiovascular disease complications.

References

1 Libby, P. (2002) Inflammation in atherosclerosis. *Nature,* **420**, 868–874.

2 Charo, I.F. and Ransohoff, R.M. (2006) The many roles of chemokines and chemokine receptors in inflammation. *N. Engl. J. Med.,* **354**, 610–621.

3 Luster, A.D. (1998) Chemokines – chemotactic cytokines that mediate inflammation. *N. Engl. J. Med.,* **338**, 436–445.

4 Zlotnik, A. and Yoshie, O. (2000) Chemokines: a new classification system and their role in immunity. *Immunity,* **12**, 121–127.

5 Allen, S.J., *et al.* (2007) Chemokine: receptor structure, interactions, and antagonism. *Annu. Rev. Immunol.,* **25**, 787–820.

6 Proudfoot, A.E.I. (2006) The biological relevance of chemokine-proteoglycan interactions. *Biochem. Soc. Trans.,* **34**, 422–426.

7 Proudfoot, A.E., *et al.* (2003) Gly-cosaminoglycan binding and oligomeri-zation are essential for the *in vivo* activity of certain chemokines. *Proc. Natl Acad. Sci. U. S. A.,* **100**, 1885–1890.

8 Johnson, Z., *et al.* (2004) Interference with heparin binding and oligomeriza-tion creates a novel anti-inflammatory strategy targeting the chemokine system. *J. Immunol.,* **173**, 5776–5785.

9 Ali, S., *et al.* (2005) A non-gly-cosaminoglycan-binding variant of CC chemokine ligand 7 (monocyte chemoattractant protein-3) antagonizes chemokine-mediated inflammation. *J. Immunol.,* **175**, 1257–1266.

10 Flanagan, C.A. (2005) A GPCR that is not "DRY". *Mol. Pharmacol.,* **68**, 1–3.

11 Horuk, R., *et al.* (1993) A receptor for the malarial parasite *Plasmodium vivax*: the erythrocyte chemokine receptor. *Science,* **261**, 1182–1184.

12 Nibbs, R.J., *et al.* (1997) Cloning and characterization of a novel promiscuous human beta-chemokine receptor D6. *J. Biol. Chem.,* **272**, 32078–32083.

13 Yoshimura, T., *et al.* (1987) Purification of a human monocyte-derived neutrophil chemotactic factor that has peptide sequence similarity to other host defense cytokines. *Proc. Natl Acad. Sci. U. S. A.,* **84**, 9233–9237.

14 Peveri, P., *et al.* (1988) A novel neutrophil-activating factor produced by human mononuclear phagocytes. *J. Exp. Med.,* **167**, 1547–1559.

15 Luster, A.D., *et al.* (2005) Immune cell migration in inflammation: present and future therapeutic targets. *Nat. Immunol.,* **6**, 1182.

16 Strieter, R.M., *et al.* (2005) CXC chemokines in angiogenesis. *Cytokine Growth Factor Rev.,* **16**, 593–609.

17 Cyster, J.G. (1999) Chemokines and cell migration in secondary lymphoid organs. *Science,* **286**, 2098–2102.

18 Ohl, L., *et al.* (2003) Cooperating mechanisms of CXCR5 and CCR7 in development and organization of secondary lymphoid organs. *J. Exp. Med.,* **197**, 1199–1204.

19 Bazan, J.F., *et al.* (1997) A new class of membrane-bound chemokine with a CX3C motif. *Nature,* **385**, 640–644.

20 Imai, T., *et al.* (1997) Identification and molecular characterization of fractalkine receptor CX3CR1, which mediates both leukocyte migration and adhesion. *Cell,* **91**, 521–530.

21 Fong, A.M., *et al.* (1998) Fractalkine and CX3CR1 mediate a novel mechanism of leukocyte capture, firm adhesion, and activation under physiologic flow. *J. Exp. Med.,* **188**, 1413–1419.

22 Combadiere, C., *et al.* (2008) Combined inhibition of CCL2, CX3CR1, and CCR5 abrogates Ly6C(hi) and Ly6C(lo) monocytosis and almost abolishes atherosclerosis in hypercholesterolemic mice. *Circulation,* **117**, 1649–1657.

23 Saederup, N., *et al.* (2008) Fractalkine deficiency markedly reduces macro-phage accumulation and atherosclerotic lesion formation in CCR2-/- mice: evidence for independent chemokine functions in atherogenesis. *Circulation,* **117**, 1642–1648.

24 Tacke, F., *et al.* (2007) Monocyte subsets differentially employ CCR2, CCR5, and

CX3CR1 to accumulate within atherosclerotic plaques. *J. Clin. Invest.*, **117**, 185–194.

25 Randolph, G.J. (2008) Emigration of monocyte-derived cells to lymph nodes during resolution of inflammation and its failure in atherosclerosis. *Curr. Opin. Lipidol.*, **19**, 462–468.

26 Auffray, C., *et al.* (2009) Blood monocytes: development, heterogeneity, and relationship with dendritic cells. *Annu. Rev. Immunol.* **27**.

27 Llodra, J., *et al.* (2004) Emigration of monocyte-derived cells from atherosclerotic lesions characterizes regressive, but not progressive, plaques. *Proc. Natl Acad. Sci. U. S. A.*, **101**, 11779–11784.

28 Trogan, E., *et al.* (2006) Gene expression changes in foam cells and the role of chemokine receptor CCR7 during atherosclerosis regression in ApoE-deficient mice. *Proc. Natl Acad. Sci. U. S. A.*, **103**, 3781–3786.

29 Barlic, J., *et al.* (2006) Oxidized lipid-driven chemokine receptor switch, CCR2 to CX3CR1, mediates adhesion of human macrophages to coronary artery smooth muscle cells through a peroxisome proliferator-activated receptor gamma-dependent pathway. *Circulation*, **114**, 807–819.

30 Mach, F., *et al.* (1999) Differential expression of three T lymphocyte-activating CXC chemokines by human atheroma-associated cells. *J. Clin. Invest.*, **104**, 1041–1050.

31 Yla-Herttuala, S., *et al.* (1989) Evidence for the presence of oxidatively modified low density lipoprotein in atherosclerotic lesions of rabbit and man. *J. Clin. Invest.*, **84**, 1086–1095.

32 Munn, D.H., *et al.* (1995) Activation-induced apoptosis in human macrophages: developmental regulation of a novel cell death pathway by macrophage colony-stimulating factor and interferon gamma. *J. Exp. Med.*, **181**, 127–136.

33 Landsman, L., *et al.* (2008) CX3CR1 is required for monocyte homeostasis and atherogenesis by promoting cell survival. *Blood*, **113**, 963–972.

34 von Hundelshausen, P. and Weber, C. (2007) Platelets as immune cells: bridging inflammation and cardiovascular disease. *Circ. Res*, **100**, 27–40.

35 Kodali, R., *et al.* (2006) Chemokines induce matrix metalloproteinase-2 through activation of epidermal growth factor receptor in arterial smooth muscle cells. *Cardiovasc. Res.*, **69**, 706–715.

36 Moulton, K.S. (2006) Angiogenesis in atherosclerosis: gathering evidence beyond speculation. *Curr. Opin. Lipidol.*, **17**, 548–555.

37 Gelati, M., *et al.* (2008) The angiogenic response of the aorta to injury and inflammatory cytokines requires macrophages. *J. Immunol.*, **181**, 5711–5719.

38 Cushing, S.D., *et al.* (1990) Minimally modified low density lipoprotein induces monocyte chemotactic protein 1 in human endothelial cells and smooth muscle cells. *Proc. Natl Acad. Sci. U. S. A.*, **87**, 5134–5138.

39 Nelken, N.A., *et al.* (1991) Monocyte chemoattractant protein-1 in human atheromatous plaques. *J. Clin. Invest.*, **88**, 1121–1127.

40 Boring, L., *et al.* (1998) Decreased lesion formation in CCR2-/- mice reveals a role for chemokines in the initiation of atherosclerosis. *Nature*, **394**, 894–897.

41 Dawson, T.C., *et al.* (1999) Absence of CC chemokine receptor-2 reduces atherosclerosis in apolipoprotein E-deficient mice. *Atherosclerosis*, **143**, 205–211.

42 Gu, L., *et al.* (1998) Absence of monocyte chemoattractant protein-1 reduces atherosclerosis in low density lipoprotein receptor-deficient mice. *Mol. Cell.*, **2**, 275.

43 Guo, J., *et al.* (2003) Transplantation of monocyte CC-chemokine receptor 2-deficient bone marrow into apoE3-Leiden mice inhibits atherogenesis. *Arterioscler. Thromb. Vasc. Biol.*, **23**, 447–453.

44 Roque, M., *et al.* (2002) CCR2 deficiency decreases intimal hyperplasia after arterial injury. *Arterioscler. Thromb. Vasc. Biol.*, **22**, 554–559.

45 Porreca, E., *et al.* (1997) Monocyte chemotactic protein 1 (MCP-1) is a mitogen for cultured rat vascular

smooth muscle cells. *J. Vasc. Res.*, **34**, 58–65.

46 Ali, Z.A., *et al.* (2008) CCR2-mediated antiinflammatory effects of endothelial tetrahydrobiopterin inhibit vascular injury-induced accelerated atherosclerosis. *Circulation*, **118**, S71–77.

47 Han, K.H., *et al.* (2005) HMG-CoA reductase inhibition reduces monocyte CC chemokine receptor 2 expression and monocyte chemoattractant protein-1-mediated monocyte recruitment *in vivo*. *Circulation*, **111**, 1439–1447.

48 Blanco-Colio, L.M., *et al.* (2007) Elevated ICAM-1 and MCP-1 plasma levels in subjects at high cardiovascular risk are diminished by atorvastatin treatment. Atorvastatin on Inflammatory Markers study: a substudy of Achieve Cholesterol Targets Fast with Atorvastatin Stratified Titration. *Am. Heart J.*, **153**, 881–888.

49 Wilcox, J.N., *et al.* (1994) Local expression of inflammatory cytokines in human atherosclerotic plaques. *J. Atheroscler. Thromb.* **1** (Suppl. 1): S10–S13.

50 Veillard, N.R., *et al.* (2004) Antagonism of RANTES receptors reduces atherosclerotic plaque formation in mice. *Circ. Res.*, **94**, 253–261.

51 Weber, C., *et al.* (2001) Specialized roles of the chemokine receptors CCR1 and CCR5 in the recruitment of monocytes and TH1-like/CD45RO+ T cells. *Blood*, **97**, 1144–1146.

52 Baltus, T., *et al.* (2003) Oligomerization of RANTES is required for CCR1-mediated arrest but not CCR5-mediated transmigration of leukocytes on inflamed endothelium. *Blood*, **102**, 1985–1988.

53 von Hundelshausen, P., *et al.* (2001) RANTES deposition by platelets triggers monocyte arrest on inflamed and atherosclerotic endothelium. *Circulation*, **103**, 1772–1777.

54 Schober, A., *et al.* (2002) Deposition of platelet RANTES triggering monocyte recruitment requires P-selectin and is involved in neointima formation after arterial injury. *Circulation*, **106**, 1523–1529.

55 Kuziel, W.A., *et al.* (2003) CCR5 deficiency is not protective in the early stages of atherogenesis in apoE knockout mice. *Atherosclerosis*, **167**, 25–32.

56 Potteaux, S., *et al.* (2006) Role of bone marrow-cerived CC-chemokine receptor 5 in the development of atherosclerosis of low-density lipoprotein receptor knockout mice. *Arterioscler. Thromb. Vasc. Biol.*, **26**, 1858–1863.

57 Braunersreuther, V., *et al.* (2007) Ccr5 but not Ccr1 deficiency reduces development of diet-induced atherosclerosis in mice. *Arterioscler. Thromb. Vasc. Biol.*, **27**, 373–379.

58 Zhou, X., *et al.* (1998) Hypercholesterolemia is associated with a T helper (Th) 1/Th2 switch of the autoimmune response in atherosclerotic apo E-knockout mice. *J. Clin. Invest.*, **101**, 1717–1725.

59 Robertson, A.-K.L. and Hansson, G.K. (2006) T Cells in atherogenesis: for better or for worse? *Arterioscler. Thromb. Vasc. Biol.*, **26**, 2421–2432.

60 Khallou-Laschet, J., *et al.* (2006) The proatherogenic role of T cells requires cell division and is dependent on the stage of the disease. *Arterioscler. Thromb. Vasc. Biol.*, **26**, 353–358.

61 Veillard, N.R., *et al.* (2005) Differential influence of chemokine receptors CCR2 and CXCR3 in development of atherosclerosis *in vivo*. *Circulation*, **112**, 870–878.

62 Heller, E.A., *et al.* (2006) Chemokine CXCL10 promotes atherogenesis by modulating the local balance of effector and regulatory T cells. *Circulation*, **113**, 2301–2312.

63 Mallat, Z., *et al.* (1999) Protective role of interleukin-10 in atherosclerosis. *Circ. Res.*, **85**, e17–e24.

64 Mallat, Z., *et al.* (2001) Inhibition of transforming growth factor-{beta} signaling accelerates atherosclerosis and induces an unstable plaque phenotype in mice. *Circ. Res.*, **89**, 930–934.

65 Okamoto, Y., *et al.* (2008) Adiponectin inhibits the production of CXC receptor 3 chemokine ligands in macrophages and reduces T-lymphocyte recruitment in atherogenesis. *Circ. Res.*, **102**, 218–225.

66 Schwarz, J.B.K., *et al.* (2009) Novel role of the CXC chemokine receptor 3 in inflammatory response to arterial injury. Involvement of mTORC1. *Circ. Res.*, **104**, 189–200.

67 Hsu, H.-H., *et al.* (2009) Hypertension in mice lacking the CXCR3 chemokine receptor. *Am. J. Physiol. Renal. Physiol.*, **296**, F780–F789.

68 Holmes, W.E., *et al.* (1991) Structure and functional expression of a human interleukin-8 receptor. *Science*, **253**, 1278–1280.

69 Murphy, P.M. and Tiffany, H.L. (1991) Cloning of complementary DNA encoding a functional human inter-leukin-8 receptor. *Science*, **253**, 1280–1283.

70 Bozic, C.R., *et al.* (1994) The murine interleukin 8 type B receptor homo-logue and its ligands. Expression and biological characterization. *J. Biol. Chem.*, **269**, 29355–29358.

71 Naruko, T., *et al.* (2002) Neutrophil infiltration of culprit lesions in acute coronary syndromes. *Circulation*, **106**, 2894–2900.

72 Gerszten, R.E., *et al.* (1999) MCP-1 and IL-8 trigger firm adhesion of monocytes to vascular endothelium under flow conditions. *Nature*, **398**, 718–723.

73 Zernecke, A., *et al.* (2008) Protective role of CXC receptor 4/CXC ligand 12 unveils the importance of neutrophils in atherosclerosis. *Circ. Res.*, **102**, 209–217.

74 Boisvert, W.A., *et al.* (1998) A leukocyte homologue of the IL-8 receptor CXCR-2 mediates the accumulation of macro-phages in atherosclerotic lesions of LDL receptor-deficient mice. *J. Clin. Invest.*, **101**, 353–363.

75 Boisvert, W.A., *et al.* (2006) Up-regulated expression of the CXCR2 ligand KC/GRO-α in atherosclerotic lesions plays a central role in macro-phage accumulation and lesion progression. *Am. J. Pathol.*, **168**, 1385–1395.

76 Garton, K.J., *et al.* (2001) Tumor necrosis factor-alpha-converting enzyme (ADAM17) mediates the cleavage and

shedding of fractalkine (CX3CL1). *J. Biol. Chem.*, **276**, 37993–38001.

77 Tsou, C.L., *et al.* (2001) Tumor necrosis factor-alpha-converting enzyme mediates the inducible cleavage of fractalkine. *J. Biol. Chem.*, **276**, 44622–44626.

78 Lucas, A.D., *et al.* (2003) Smooth muscle cells in human atherosclerotic plaques express the fractalkine receptor CX3CR1 and undergo chemotaxis to the CX3C chemokine fractalkine (CX3CL1). *Circulation*, **108**, 2498–2504.

79 Teupser, D., *et al.* (2004) Major reduction of atherosclerosis in fractalkine (CX3CL1)-deficient mice is at the brachiocephalic artery, not the aortic root. *Proc. Natl Acad. Sci. U. S. A.*, **101**, 17795–17800.

80 Combadiere, C., *et al.* (2003) Decreased atherosclerotic lesion formation in CX3CR1/apolipoprotein E double knockout mice. *Circulation*, **107**, 1009–1016.

81 Lesnik, P., *et al.* (2003) Decreased atherosclerosis in CX3CR1-/- mice reveals a role for fractalkine in atherogenesis. *J. Clin. Invest.*, **111**, 333–340.

82 Faure, S., *et al.* (2000) Rapid progres-sion to AIDS in HIV+ individuals with a structural variant of the chemokine receptor CX3CR1. *Science*, **287**, 2274–2277.

83 Moatti, D., *et al.* (2001) Polymorphism in the fractalkine receptor CX3CR1 as a genetic risk factor for coronary artery disease. *Blood*, **97**, 1925–1928.

84 McDermott, D.H., *et al.* (2003) Chemokine receptor mutant CX3CR1-M280 has impaired adhesive function and correlates with protection from cardiovascular disease in humans. *J. Clin. Invest.*, **111**, 1241–1250.

85 Lavergne, E., *et al.* (2005) Adverse associations between CX3CR1 polymorphisms and risk of cardiovascu-lar or cerebrovascular disease. *Arterioscler. Thromb. Vasc. Biol.*, **25**, 847–853.

86 Niessner, A., *et al.* (2005) Opposite effects of CX3CR1 receptor polymor-phisms V249I and T280M on the development of acute coronary syndrome. A possible implication of fractalkine in inflammatory activation. *Thromb. Haemost.*, **93**, 949–954.

87 Lutgens, E., *et al.* (2005) Gene profiling in atherosclerosis reveals a key role for small inducible cytokines: validation using a novel monocyte chemoattractant protein monoclonal antibody. *Circulation*, **111**, 3443–3452.

88 Gao, P., *et al.* (2003) The unique target specificity of a nonpeptide chemokine receptor antagonist: selective blockade of two Th1 chemokine receptors CCR5 and CXCR3. *J. Leukoc. Biol.*, **73**, 273–280.

89 van Wanrooij, E.J.A., *et al.* (2005) HIV entry inhibitor TAK-779 attenuates atherogenesis in low-density lipoprotein receptor-deficient mice. *Arterioscler. Thromb. Vasc. Biol.*, **25**, 2642–2647.

90 Barbaro, G. (2002) Cardiovascular manifestations of HIV infection. *Circulation*, **106**, 1420–1425.

91 van Wanrooij, E.J.A., *et al.* (2008) CXCR3 antagonist NBI-74330 attenuates atherosclerotic plaque formation in LDL receptor-deficient mice. *Arterioscler. Thromb. Vasc. Biol.*, **28**, 251–257.

92 Berahovich, R.D., *et al.* (2005) Proteolytic activation of alternative CCR1 ligands in inflammation. *J. Immunol.*, **174**, 7341–7351.

93 Proudfoot, A.E.I., *et al.* (1996) Extension of recombinant human RANTES by the retention of the initiating methionine produces a potent antagonist. *J. Biol. Chem.*, **271**, 2599–2603.

94 Veillard, N.R., *et al.* (2004) Differential expression patterns of proinflammatory and anti-inflammatory mediators during atherogenesis in mice. *Arterioscler. Thromb. Vasc. Biol.*, **24**, 2339–2344.

95 Inoue, S., *et al.* (2002) Anti-monocyte chemoattractant protein-1 gene therapy limits progression and destabilization of established atherosclerosis in apolipoprotein E-knockout mice. *Circulation*, **106**, 2700–2706.

96 Egashira, K., *et al.* (2007) Local delivery of anti-monocyte chemoattractant protein-1 by gene-eluting stents attenuates in-stent stenosis in rabbits

and monkeys. *Arterioscler. Thromb. Vasc. Biol.*, **27**, 2563–2568.

97 Braunersreuther, V., *et al.* (2008) A novel RANTES antagonist prevents progression of established atherosclerotic lesions in mice. *Arterioscler. Thromb. Vasc. Biol.*, **28**, 1090–1096.

98 Handel, T.M., *et al.* (2008) An engineered monomer of CCL2 has anti-inflammatory properties emphasizing the importance of oligomerization for chemokine activity *in vivo*. *J. Leukoc. Biol.*

99 von Hundelshausen, P., *et al.* (2005) Heterophilic interactions of platelet factor 4 and RANTES promote monocyte arrest on endothelium. *Blood*, **105**, 924–930.

100 Koenen, R.R., *et al.* (2009) Disrupting functional interactions between platelet chemokines inhibits atherosclerosis in hyperlipidemic mice. *Nat. Med.*, **15**, 97.

101 Makino, Y., *et al.* (2002) Impaired T Cell function in RANTES-deficient mice. *Clin. Immunol.*, **102**, 302.

102 Alcami, A., *et al.* (1998) Blockade of chemokine activity by a soluble chemokine binding protein from vaccinia virus. *J. Immunol.*, **160**, 624–633.

103 Burns, J.M., *et al.* (2002) Comprehensive mapping of poxvirus vCCI chemokine-binding protein. Expanded range of ligand interactions and unusual dissociation kinetics. *J. Biol. Chem.*, **277**, 2785–2789.

104 Bursill, C.A., *et al.* (2003) Adenoviral-mediated delivery of a viral chemokine binding protein blocks CC-chemokine activity *in vitro* and *in vivo*. *Immunobiology*, **207**, 187–196.

105 Bursill, C.A., *et al.* (2004) Broad-spectrum CC-chemokine blockade by gene transfer inhibits macrophage recruitment and atherosclerotic plaque formation in apolipoprotein E-knockout mice. *Circulation*, **110**, 2460–2466.

106 Ali, Z.A., *et al.* (2005) Gene transfer of a broad spectrum CC-chemokine inhibitor reduces vein graft atherosclerosis in apolipoprotein E-knockout mice. *Circulation*, **112**, I-235–I-241.

107 Puhakka, H.L., *et al.* (2005) Effects of vaccinia virus anti-inflammatory protein

35K and TIMP-1 gene transfers on vein graft stenosis in rabbits. *In Vivo*, **19**, 515–521.

108 Bursill, C.A., *et al.* (2008) Lentiviral gene transfer to reduce atherosclerosis progression by long-term CC-chemokine inhibition. *Gene Ther.*, **16**, 93–102.

109 Parry, C.M., *et al.* (2000) A broad spectrum secreted chemokine binding protein encoded by a herpesvirus. *J. Exp. Med.*, **191**, 573–578.

110 Alexander-Brett, J.M. and Fremont, D.H. (2007) Dual GPCR and GAG mimicry by the M3 chemokine decoy receptor. *J. Exp. Med.*, **204**, 3157–3172.

111 Pyo, R., *et al.* (2004) Inhibition of intimal hyperplasia in transgenic mice conditionally expressing the chemokine-binding protein M3. *Am. J. Pathol.*, **164**, 2289–2297.

112 Martinson, J.J., *et al.* (1997) Global distribution of the CCR5 gene 32-basepair deletion. *Nat. Genet.*, **16**, 100.

113 Peters, W., *et al.* (2001) Chemokine receptor 2 serves an early and essential role in resistance to *Mycobacterium tuberculosis. Proc. Natl Acad. Sci. U. S. A.*, **98**, 7958–7963.

114 Hristov, M., *et al.* (2007) Regulation of endothelial progenitor cell homing after arterial injury. *Thromb. Haemost.*, **98**, 274–277.

115 Kraaijeveld, A.O., *et al.* (2007) CC Chemokine ligand-5 (CCL5/RANTES) and CC chemokine ligand-18 (CCL18/PARC) are specific markers of refractory unstable angina pectoris and are transiently raised during severe ischemic symptoms. *Circulation*, **116**, 1931–1941.

116 de Jager, S.C.A., *et al.* (2008) CCL3 (MIP-1α levels are elevated during acute coronary syndromes and show strong prognostic power for future ischemic events. *J. Mol. Cell. Cardiol.*, **45**, 446.

3
Adhesion Molecules and Atherosclerosis

Huan Wang and Yuqing Huo

3.1
The Leukocyte Adhesion Cascade in Atherosclerosis

Atherosclerosis is a chronic inflammatory disease of the arterial vessel wall. Recruitment of leukocytes to the vessel wall, including monocytes, lymphocytes, mast cells, and neutrophils, is the first step in the initiation of atherosclerosis [1]. Leukocyte infiltration into the vessel wall is completed via at least five steps of the adhesion cascade: capture, rolling, slow rolling, firm adhesion, and transmigration. Capture or tethering is the initial contact of leukocytes with the activated endothelium, which results in the movement of leukocytes toward the endothelium, away from the central bloodstream. After leukocytes are captured, they may transiently adhere to the endothelium and begin to roll. The rolling occurs at or below the velocity of freely flowing cells, such as erythrocytes in the same vessel, and at the same radial position. In the presence of activated endothelium, leukocytes roll at a dramatically lower velocity, termed "slow rolling," which is distinct from the much faster rolling described above. Following slow rolling, leukocytes stop on the endothelium, become adherent, and then migrate across the endothelium. For more details on the adhesion cascade, see the review by Ley *et al.* [2].

Each of these five adhesion cascade steps is mediated by various adhesion molecules, which are divided into four major families: (i) selectins, (ii) selectin ligands, (iii) members of the immunoglobulin family, and (iv) integrins. As shown in Figure 3.1, binding of selectins to selectin ligands mediates leukocyte tethering and rolling, whereas interactions between members of the immunoglobulin family and their integrins contribute to leukocyte adhesion and migration [3, 4]. In addition, chemokines and chemokine receptors are important in leukocyte adhesion and migration, as described elsewhere in this book.

Atherosclerosis: Molecular and Cellular Mechanisms. Edited by Sarah Jane George and Jason Johnson
Copyright © 2010 WILEY-VCH Verlag GmbH & Co. KGaA, Weinheim
ISBN: 978-3-527-32448-4

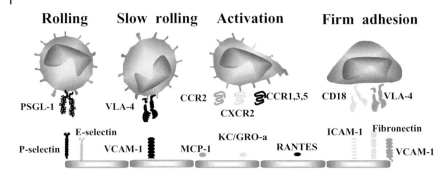

Endothelium at lesion-prone site

Figure 3.1   Monocyte adhesion on atherosclerotic endothelium.

3.2
Background on Adhesion Molecules

3.2.1
Selectins and Their Ligands

The selectin family includes P-selectin, E-selectin, and L-selectin. These molecules are characterized by an N-terminal calcium-dependent lectin domain, an epidermal growth factor–like domain, a series of repeats with similarities to complement binding proteins, a transmembrane segment, and a cytoplasmic region. Selectins are C-type lectins that bind sialylated and fucosylated carbohydrate ligands, and they mediate initial capture, tethering, and rolling along the endothelium [5]. P-selectin is the largest of the known selectins at 140 kDa. It contains nine consensus repeats and extends approximately 40 nm from the endothelial surface. Other names for P-selectin include CD62P, granule membrane protein 140 (GMP-140), and platelet activation–dependent granule to external membrane protein (PADGEM). P-selectin is stored in granules of activated platelets and endothelial cells. It is presented on the cell surface within minutes after stimulation with inflammatory mediators, such as histamine, thrombin, phorbol esters, or surgical trauma. The presentation of P-selectin is short-lived, peaking only for 10 min. Additional synthesis of P-selectin is induced within 2 h by cytokines such as interleukin-1 (IL-1) or tumor necrosis factor-α (TNF-α) [6]. A soluble isoform of P-selectin, resulting from an alternatively spliced mRNA lacking the transmembrane domain, is also observed [7]. Shedding, although not confirmed, might also contribute to the presence of a shorter soluble form of P-selectin. These soluble molecules are not only degradation products, they can have distinct functions, potentially acting as competitors that interact with their normal counterreceptors without triggering cell recruitment, as inhibitors or as agonists [8].

E-Selectin is present on endothelial cells. It is almost absent on the resting endothelium and it is transcriptionally induced by inflammatory cytokines in a process that takes a few hours to peak and then decreases rapidly. The NF-κB and AP-1 transcription factors seem to be crucial in the activation of this gene [9]. L-Selectin is constitutively expressed on the majority of leukocytes. Soluble isoforms of E- and L-selectins, which are measurable in the blood, result from shedding/proteolytic cleavage [10].

Selectin ligands are transmembrane glycoproteins that present oligosaccharide structures to the selectins. Transient bond formation between the selectins and their ligands mediate the early steps of the adhesion cascade. All three selectins can recognize glycoproteins and/or glycolipids containing the tetrasaccharide sialyl-Lewis[x] (sialyl-CD15). This tetrasaccharide is found on all circulating myeloid cells and is composed of sialic acid, galactose, fucose, and N-acetylgalactosamine [11, 12]. PSGL-1 (P-selectin glycoprotein ligand) has been characterized as a ligand for P-selectin. It is a glycoprotein expressed on blood cells and contains the sialyl-Lewis[x] tetrasaccharide. Another P-selectin ligand, CD24, appears to be important for tumor cell binding to P-selectin [13]. For L-selectin, four possible ligands have been identified: GlyCAM-1 (glycosylation-dependent cell adhesion molecule), CD34, MAdCAM-1 (mucosal addressin cell adhesion molecule), and PSGL-1. Evidence suggests that both sulfate and sialic acid in an $\alpha(2,3)$ linkage are essential for L-selectin ligand activity. A specific E-selectin ligand (ESL-1) has been identified as constitutively present on all leukocytes, but its properties are not yet completely understood [14].

3.2.2
Immunoglobulin Superfamily and its Counterreceptors

Leukocyte adhesion to and transmigration through the endothelium are mediated by members of the immunoglobulin superfamily present on endothelial cells and their counterreceptor integrins present on leukocytes [4]. The immunoglobulin superfamily includes intercellular adhesion molecule 1 (ICAM-1), ICAM-2, ICAM-3, platelet endothelial cell adhesion molecule 1 (PECAM-1), and vascular cell adhesion molecule 1 (VCAM-1). These molecules are membrane glycoprotein receptors that possess a variable number of extracellular Ig domains. These 70–100 amino acid domains are composed of two β sheets and give rise to Ig folds that participate in adhesion sites. The genes of this family frequently undergo alternative splicing, allowing the production of multiple isoforms. ICAM-1 (CD54) is widely expressed at a basal level on leukocytes and endothelial cells and it can be upregulated several hours after the initial exposure to proinflammatory cytokines [9]. Binding sites for NF-κB and AP-1 have been identified in the promoter regions of the ICAM-1 gene [15, 16]. ICAM-2 is present on leukocytes, platelets, and endothelial cells, but it is downregulated by inflammatory mediators. ICAM-3 is detected on endothelial cells and leukocytes, and is the only ICAM molecule present on neutrophils [17, 18]. Soluble forms of ICAM-1 and ICAM-3 have been demonstrated to occur as a result of shedding [9, 18–20].

PECAM-1 is constitutively expressed on resting endothelial cells, platelets and endothelial cells. PECAM-1 molecules are particularly dense at the junctions between endothelial cells [21]. Alternative splicing gives rise to soluble PECAM-1 molecules lacking the transmembrane domain [22]. VCAM-1 (CD106) is transcriptionally induced on endothelial cells but it can also be expressed by other cell types such as macrophages, myoblasts, and dendritic cells. VCAM-1 is expressed on the luminal surface of both large and small vessels only after the endothelial cells are stimulated by proinflammatory cytokines. Maximal expression of this molecule is observed approximately 6 h after initial exposure to the inflammatory stimulus. The sustained expression of VCAM-1 lasts over 24 h [23]. A binding site for NF-κB has been localized to the promoter region of the VCAM-1 gene [15, 16]. Proteolytic cleavage involving an unidentified metalloprotease gives rise to a soluble form of VCAM-1 [24]. Inhibitors of NF-κB and AP-1 activation or nuclear translocation have been shown to attenuate VCAM-1 expression induced by cytokines, lipopolysaccharide (LPS), or oxidant [15].

The counterreceptors of the immunoglobulin superfamily are integrins that are expressed on almost all leukocytes. Integrins are heterodimeric transmembrane glycoproteins resulting from the noncovalent association of an α chain and a β chain. In resting cells, they are usually nonadhesive; they are present on the cell surface but in a low-affinity conformation for their ligands. Changes in cell activity trigger signals that modify the conformation of integrins to higher affinity receptors. This represents the "inside-out" signaling, whereas "outside-in" signals are delivered within the cell after ligation between an integrin and its ligand [25]. No soluble form of either integrin has been identified.

ICAM-1, -2, and -3 bind to leukocyte β_2-integrins CD11a/CD18 (LFA-1) and ICAM-1 and -2 also bind to CD11b/CD18 (Mac-1) [26], although in the context of transmigration, CD11a seems to predominantly bind to ICAM-1, whereas CD11b is more promiscuous [27]. VCAM-1 primarily binds to VLA-4 (very late antigen-4 or $\alpha_4\beta_1$) of the β_1 subfamily of integrins, as well as to integrin $\alpha_4\beta_7$ [28]. ICAM-1 and VCAM-1 promote the adhesion of lymphocytes, monocytes, eosinophils, and basophils to the activated endothelium by establishing strong bonds with integrins and inducing firm adhesion of leukocytes to the vascular surface, as well as by participating in leukocyte migration. Inhibition or loss of one or more of the members of the immunoglobulin superfamily decreases leukocyte adhesion and migration and, in some cases, also affects leukocyte rolling [29–31]. PECAM-1 engages in homotypic adhesion to PECAM-1 molecules on other cells and may also bind to aVb3 integrin. PECAM-1 mainly participates in homophilic binding between adjacent cells, and therefore it is involved in endothelial integrity and emigration of cells from the blood compartment into the vessel and underlying tissue [21].

3.3
Adhesion Molecules and Risk Factors of Atherosclerosis

High levels of blood cholesterol and/or blood pressure, diabetes, obesity, and cigarette smoking are known as major risk factors for the development of atherosclerosis. All of these risk factors are able to upregulate the expression of adhesion molecules on the endothelium.

Atherosclerotic lipids, including low-density lipoprotein (LDL) and its modified derivates (oxidized LDL (oxLDL) or minimally modified LDL (MM-LDL)) are significant stimuli for adhesion molecule expression on endothelial cells *in vitro*. LDL induces VCAM-1 expression on human coronary arteries and pig aortic endothelial cells and VCAM-1 and E-selectin expression on human aortic endothelial cells (HAECs) [32]. Treatment of HAECs with MM-LDL causes an increase of P-selectin expression on endothelial cells. Upon histamine treatment, P-selectin surface presentation on endothelial cells previously treated with MM-LDL is elevated [33]. Similarly, pretreatment of HAECs or human umbilical vein endothelial cells (HUVECs) with oxLDL enhances cytokine-induced VCAM-1 and ICAM-1 expression, indicating that signaling regulation by modified lipids augments the ability of vascular endothelial cells to express adhesion molecules [34].

Hypercholesterolemia occurring in animal models of atherosclerosis such as rabbits, LDL-receptor-deficient (LDLR$^{-/-}$) and apolipoprotein E–deficient (ApoE$^{-/-}$) mice also induces the expression of adhesion molecules. During the early phases of high-fat diet–induced atherogenesis in rabbits, the adhesion molecules P-selectin and VCAM-1 are detected focally on the endothelium of the ascending aorta or lesion-prone areas about one week after a high-fat diet [35, 36]. In ApoE$^{-/-}$ mice fed a Western-type diet, hypercholesterolemia induces the expression of VCAM-1 and ICAM-1 on the endothelium of lesion-prone arteries. Unlike VCAM-1, ICAM-1 is also expressed in the uninvolved arteries. Therefore, the expression profile of ICAM-1 is not highly associated with lesion-prone sites in these mice [37]. The PECAM-1, mainly localized on the periphery of endothelial cells, appears not to be regulated by hypercholesterolemia [38].

Hemodynamic condition is another crucial factor in the development of atherosclerosis. It has been observed that lesions are first and mainly located at the bifurcations and branches of arteries. Low and oscillatory shear stresses are major features of the hemodynamic conditions at sites predisposed to atherosclerosis [39]. In an *in vitro* condition, compared with a static one, prolonged oscillatory shear stress increases the expression of E-selectin, VCAM-1, and ICAM-1 on endothelial monolayers. In contrast, laminar flow significantly reduces the expression of VCAM-1 on the endothelium, though it causes a significant increase in the surface expression of ICAM-1 [40–42]. In the aortas of normal chow-fed mice and rabbits, VCAM-1 and ICAM-1 are expressed by endothelial cells of the ascending aorta and proximal arch in which low and oscillatory flow occurs [43]. In rabbit and mouse models in which shear stress has been surgically or equipmentally manipulated in carotid arteries, VCAM-1 expression is greatly increased under low shear stress, and the increase of VCAM-1 expression stays at lesser extent with increased shear stress [43]. ICAM-1 expression is less related to special shear stress [43].

In addition to altered shear stress, increased mechanical force may also participate in modulation of endothelial cell adhesion molecule expression. In a rat model, chronic hypertension increases the expression of P-selectin, ICAM-1, and

VCAM-1 on the endothelium of aortic arteries [44]. Patients with hypertension also exhibit an increase in adhesion molecule expression [45–47].

Upregulation of adhesion molecules is also associated with other atherosclerotic risk factors, such as diabetes mellitus [43], obesity, and smoking [48].

3.4
Adhesion Molecule Expression and Function in Atherosclerosis

Expression of several adhesion molecules was detected on established atherosclerotic lesions (Table 3.1). In human atherosclerotic lesions, strong expression of P-selectin is demonstrated on the endothelium overlying active atherosclerotic plaques, but not on the normal arterial endothelium or on the endothelium overlying inactive fibrous plaques [49]. Spotty E-selectin expression is confined to endothelial cells and occurs on the surface of fibrous and lipid-containing plaques [50]. ICAM-1 expression has been detected on endothelial cells, macrophages and smooth muscle cells of plaques, and it increases after vascular injury [51]. Normal arterial endothelial cells and intimal smooth muscle cells outside plaques display weaker or negative reactions [49, 52]. Unlike E-selectin and ICAM-1, VCAM-1 staining of surface endothelium only occurs in fibrous and lipid-containing plaques [53]. In addition, subsets of plaque smooth muscle cells and macrophages also express VCAM-1 [50, 53], and this expression significantly increases after injury [51]. Adhesion molecules are upregulated in atherosclerotic plaques, suggesting that they may play a critical role in atherogenesis.

Monocytes are recruited to the arterial vessel wall by a variety of adhesion molecules, differentiate into macrophages and become a major component of the atherosclerotic plaque. In an *in vitro* flow model system, freely flowing monocytes first attach to TNF-α-activated HUVECs under flow via L- and P-selectin [54]. Once initially attached, a small number of monocytes begin rolling via L-selectin, as well as endothelial cell ICAM-1 and VCAM-1, while the remaining monocytes become firmly adherent or detached. Monocyte stable arrest and subsequent transendothelial migration occur rapidly and efficiently via either ICAM-1 and/or VCAM-1. In addition to ICAM-1 and VCAM-1, PECAM-1 also seems to play an important role in the transendothelial passage of monocytes [54]. In a similar system utilizing IL-4- activated HUVECs, which selectively expresses VCAM-1 and L-selectin ligand, L-selectin mediates monocyte rolling and facilitates VCAM-1-dependent arrest, whereas VCAM-1 is required for the spreading of firmly attached monocytes on the endothelial cell surface but not for their arrest [55]. To specifically determine the role of VCAM-1 in monocyte interactions with endothelial cells, monocytes were perfused through endothelial cellsinfected with adenovirus carrying VCAM-1. Monocytes were able to roll, firmly adhere, and transmigrate through endothelial cells, indicating a crucial role of VCAM in monocyte recruitment to the arterial wall in atherosclerosis [56, 57].

The above *in vitro* studies examining the roles of adhesion molecules in monocyte interactions with the endothelium only incompletely mimic the pathophysi-

Table 3.1 Expression of adhesion molecules in atherosclerotic lesion.

Adhesion molecules	Atherosclerotic lesions	The expression and location of adhesion molecules
P-selectin	Human atherosclerotic arteries	P-selectin expression is on the endothelium overlying active atherosclerotic plaques, not fibrous plaques [49]
	Aortas of rabbits with hypercholesterolemia	P-selectin expression is on endothelium prior to monocyte and T lymphocyte recruitment to the vessel wall [36]
E-selectin	Human atherosclerotic arteries	E-selectin expression is confined to endothelial cells on the surface of fibrous and lipid-containing plaques [50]
	Aortas of rabbits with hypercholesterolemia	E-selectin expression is confined to very few endothelial cells covering foam cell lesions [36]
	Atherosclerotic aortas from mice	E-selectin mRNA or protein is not detectable on arteries prone to develop or with developed atherosclerotic lesions [37]
ICAM-1	Human atherosclerotic arteries	Expression of ICAM-1 is increased on endothelial layer of all lesion subtypes, on macrophages and smooth muscle cells of the plaques [49, 52]
	Atherosclerotic lesion-prone sites on mouse aortas	ICAM-1 expression is found on the endothelial cell surface in lesion-prone areas. Predominant expression of ICAM-1 is found on endothelium in early lesions and in intimal cells of more advanced lesions; more intense staining on endothelial cells at and adjacent to lesion borders [37, 38]
VCAM-1	Human atherosclerotic arteries	VCAM-1 is expressed on the endothelium in fibrous and lipid-containing plaques, on vascular smooth muscle cells and macrophages in lesion [50, 53]
	Aortas of rabbits with hypercholesterolemia	VCAM-1 is expressed prior to macrophage infiltration and sustained until the end of experiments [36]
	Atherosclerotic lesion-prone sites on mouse aortas	VCAM-1 expression is increased on the endothelial cell surface in lesion-prone sites of mouse arteries [35, 37, 38]
	Mouse atherosclerotic arteries	VCAM-1 expression is predominant on endothelium of early lesions and in intimal cells of mouse and rabbit advanced lesions; more intense staining on endothelial cells at and adjacent to lesion borders [37]

ological process of endothelial activation in atherosclerosis *in vivo*. To investigate the function of adhesion molecules in a model directly relevant to monocyte interactions with the endothelium in atherogenesis, an *ex vivo* model has been developed. In the common carotid arteries of ApoE$^{-/-}$ mice on a Western-type diet for five weeks, VCAM-1, ICAM-1, and P-selectin were expressed on the endothelium near the carotid bifurcation area, where atherosclerotic lesions are prone to develop [58, 59]. Carotid arteries from ApoE$^{-/-}$ mice were cannulated and continuously perfused with fluorescently labeled mononuclear cells under physiological conditions. Monocytes tethered to and rolled on the early atherosclerotic endothelium with endothelial P-selectin and monocyte PSGL-1 but not with L-selectin [59]. Monocyte adhesion to the early atherosclerotic endothelium occurred via monocyte VLA-4. Ligands for VLA-4, including VCAM-1 and fibronectin on the endothelium, both contributed to VLA-4-mediated monocyte adhesion [58, 59].

An *in vivo* model has also been employed to determine the role of adhesion molecules in recruitment of monocytes/macrophages to the arterial wall. Mouse peritoneal macrophages loaded with fluorescent microspheres were injected intravenously into 40-week-old ApoE$^{-/-}$ mice. Labeled macrophages were observed adhering to all stages of atherosclerotic plaques from the early fatty streak to mature calcified lesions. Pretreatment of ApoE$^{-/-}$ mice with monoclonal antibodies directed against ICAM-1 inhibited macrophage homing by 65% compared with isotype-matched controls [60].

3.5
Atherogenesis of Adhesion Molecules in Animal Models

To determine the extent of adhesion molecule involvement in atherosclerosis, blocking antibodies and/or adhesion molecule knockout mice were used to study atherosclerotic lesion formation (Table 3.2). The size of atherosclerotic lesions usually was measured at the end point of these studies.

P-Selectin has been demonstrated to be important in the formation of atherosclerotic lesions in mice. C57BL/6 mice lacking P-selectin are fed a high-fat diet to induce atherosclerotic lesions. The size of lesions is only 37% of that observed in C57BL/6 mice administrated the same treatment. In an LDLR-deficient background, deletion of P-selectin also significantly decreases fatty streaks in the cusp region of the aorta after 8–20 weeks on an atherogenic diet rich in saturated fat and cholesterol. Interestingly, following a much longer period of atherogenic diet, the lesions in P-selectin-deficient mice also advance to the fibrous plaque stage and spread throughout the entire aorta, with no difference regarding the size or distribution of atherosclerotic lesions [61]. In the ApoE$^{-/-}$ gene background, P-selectin-deficient mice exhibit much smaller aortic sinus lesions than control mice after four months on a regular chow diet. When fed a chow diet for a much longer period of time, atherosclerotic lesions advance into the fibrous plaque stage and are distributed throughout the aorta of both ApoE$^{-/-}$ and ApoE$^{-/-}$/P-selectin$^{-/-}$ mice, but this process is delayed in ApoE$^{-/-}$/P-selectin$^{-/-}$ mice. In the aortic sinus,

Table 3.2 Atherogenesis of adhesion molecules.

Adhesion molecules	Model and diet	Influence on atherosclerotic lesions
P-selectin	P-selectin-deficient C57BL/6 mice; high fat, high cholesterol	Aorta lesion size was decreased by 63% at 20 wks [68]
	P-selectin-deficient LDLR$^{-/-}$ mice; high fat, high cholesterol	Lesions were small in males at 8 and 20 wks of diet in males, no difference in female, no difference at 37 wks [61]
	P-selectin-deficient ApoE$^{-/-}$ mice; chow 4 or 15 mo	Aortic sinus lesion size was reduced by 72% at 4 mo; macrophage infiltration was reduced by 50%. At 15 mo, aortic sinus lesion was reduced by 62% [62]
	P-selectin deficient ApoE$^{-/-}$ mice; chow 20 wks	Aortic lesion size in aortas was reduced by 40–50% [63]
	E-selectin-deficient ApoE$^{-/-}$ mouse; chow 20 wks	Aortic lesion size was reduced by 25% [63]
E-selectin	E- and P-selectin-deficient LDLR$^{-/-}$ mouse; high fat, high cholesterol	Aortic lesion size was decreased by 80% at 8 wks, after 37 wks, by 37% and less calcification [64]
P-, E-selectin	PSGL-1 deficient ApoE$^{-/-}$ mice, high fat, high cholesterol	Aortic lesion size was decreased by 50% at 12 wks [65]
PSGL-1	ICAM-1 deficient C57BL/6 mouse; high fat, high cholesterol	Aortic lesion size was decreased by 63% at 20 wks [68]
ICAM-1	ICAM-1-deficient ApoE$^{-/-}$ mouse; chow for 20 wks	Aortic lesion size was reduced by 31% [63]
	CD18 deficient C57BL/6 mouse; high fat, high cholesterol	Aortic lesion size was decreased by 47% [68]
CD18	ICAM-1 and CD18-deficient C57BL/6 mouse; high fat, high cholesterol	Aortic lesion size was decreased by 76% at 20 wks [68]
ICAM-1, CD18	VCAM-1-deficient LDLR$^{-/-}$ mouse; high fat	Aortic arch lesion was reduced by 38% in VCAM-1 hypomorphic mutant, whole aorta lesion reduced by 48% [72]
VCAM-1	FN CS-1 peptide is used in C57BL/6 mouse; high fat, high-cholesterol	FN CS-1 peptide treatment reduced lesion size in aortic sinus by 68%, leukocyte infiltration reduced 48%, lipid accumulation reduced 67% [71]
VLA-4		Lipid accumulation in the aortic sinus reduced by 66% [71]

the lesions of ApoE$^{-/-}$/P-selectin$^{-/-}$ mice were 2.6-fold smaller and less calcified [62]. Collins *et al.* also reported that deletion of P-selectin reduced lesion area by 40% in males, 49% in females in ApoE$^{-/-}$ background at 20 weeks of age on a regular chow diet [63].

In addition, E-selectin is also involved in the formation of atherosclerotic lesions. In ApoE$^{-/-}$ mice at 20 weeks of age on a normal chow diet, E-selectin deficiency reduces lesion size in the aorta by 25% in males and by 26% in females. When combined with P-selectin deficiency, a much more dramatic decrease in lesion formation is observed [63]. In the aortic sinus, LDLR$^{-/-}$ mice lacking both P- and E-selectin (LDLR$^{-/-}$/P/E$^{-/-}$) display fatty streaks five times smaller than those of control mice. However, there is no apparent difference in the density of macrophages in the fatty streak lesions of aortic sinus between two groups of mice. After 22 weeks on the diet, lesions spread throughout the aorta, but this process is delayed in LDLR$^{-/-}$/P/E$^{-/-}$ mice. After 37 weeks on the diet, lesions advance to the fibrous plaque stage in both genotypes, with lesion size in the aortic sinus of LDLR$^{-/-}$/P/E$^{-/-}$ mice demonstrating a decrease of 40%. Compared to lesions in LDLR$^{-/-}$/P/E$^{+/+}$ mice, lesions in LDLR$^{-/-}$/P/E$^{-/-}$ mice appear to be less calcified [64]. So far, there are no data regarding the role of L-selectin in the formation of atherosclerotic lesions.

Selectin receptors are also crucial in the formation of atherosclerotic lesions. In ApoE$^{-/-}$ mice lacking PSGL-1, the size of atherosclerotic lesions on both the aortic sinus and whole aorta is reduced by 40–50% compared with that in control mice fed a Western-type diet for 12 weeks [65]. PSGL-1 contains sialylated and fucosylated oligosaccharides (*O*-glycans). Attachment of the *O*-glycan to PSGL-1 requires core 2 1-6-*N*-glucosaminyltransferase-I (C2GlcNAcT-I), and this modification is crucial for optimal binding of PSGL-1 to selectins. In ApoE$^{-/-}$ mice lacking C2GlcNAcT-I, aortic lesions are reduced by 32% in mice fed a chow diet for six months and by 70% in mice fed a Western-type diet for three months. More importantly, C2GlcNAcT-I deficiency leads to the formation of low-inflammatory, macrophage-poor, and collagen-rich atherosclerotic lesions [66]. Fucosylation of *O*-glycans, which is mediated by alpha(1,3)fucosyltransferases (FucT) IV and VII is also vital for the function of PSGL-1 and other selectin receptors. In ApoE$^{-/-}$ mice, FucT-IV deficiency only yields a subtle decrease in atherosclerosis, whereas FucT-VII deletion reduces the size of atherosclerotic lesion by 30–40% [67].

Endothelial ICAM-1 and its ligand CD18 are also important for the formation of atherosclerotic lesions. In mice lacking functional ICAM-1, CD18, or both, on a high-cholesterol diet for 20 weeks, aortic lesion size is reduced by 63%, 47%, and 76%, respectively [68]. In ApoE$^{-/-}$ mice lacking ICAM-1, the size of atherosclerotic lesions in aortas is decreased by 31% after mice are maintained on a chow diet for 20 weeks [63].

Deletion of VCAM-1 or its counterreceptor α4 causes embryonic lethality [69, 70]. Therefore, no adult mice are available to determine the role of VCAM-1 and α4 in the formation of atherosclerotic lesions. Peptides or antibodies have been used as an alternative. Recruitment of leukocytes to the aortic sinus of C57BL/6 mice is inhibited by 48% in response to chronic infusion of VLA-4-blocking peptide beginning 24–36 h prior to the onset of the atherogenic diet and continuing for four weeks. Mice receiving VLA-4-blocking peptide display a significantly reduced lesion size (68%) compared with mice given control peptide. Using frozen sections stained with Oil Red O, LDLR$^{-/-}$ mice receiving VLA-4-blocking peptide display a significantly reduced area (66%) of lipid accumulation in the aortic sinus [71]. Pretreatment of ApoE$^{-/-}$ mice with VLA-4-blocking antibody reduces macrophage homing to atherosclerotic lesions by 75% [60]. In addition to VCAM-1, VLA-4 also binds fibronectin. Therefore blocking VLA-4 is more effective than blocking VCAM-1. The important role of VCAM-1 was also established in VCAM-1 hypomorphic mice showing 48% reduction in aortic lesions on the LDLR$^{-/-}$ background [72].

3.6
Adhesion Molecules as Clinical Predictors of Atherosclerotic Diseases

Levels of soluble adhesion molecules have been shown to correlate with various cardiovascular risk factors, including hypercholesterolemia and/or hypertriglyceridemia [73–75], low HDL-cholesterol [76], hypertension [45, 77–79], diabetes [80–85], and smoking [86–89]. Soluble ICAM-1 (sICAM-1) is a significant predictor of future coronary artery disease (CAD) events, even after controlling for all other potential factors, including the inflammatory variables fibrinogen, von Willebrand factor, and white blood cell count. However, soluble VCAM-1 (sVCAM-1), E-selectin, and P-selectin are not significant predictors of future CAD events [90–93]. Soluble ICAM-1 is also useful in predicting peripheral arterial disease (PAD). In the initially healthy population of the Physician's Health Study, an increase of sICAM-l but not sVCAM-1 is predictive of future PAD development [94].

Soluble VCAM-1 is a strong independent predictor of future fatal cardiovascular events in patients with established CAD, when compared to most other potential confounders [95]. Soluble E-selectin and ICAM-1 are also associated with prognosis, but the significance of this association was lost after accounting for other inflammatory variables. In acute coronary syndrome, sICAM-1, sVCAM-1, sP-selectin, and/or sE-selectin are all elevated in unstable patients or patients with myocardial infarction [96–103]. In patients with unstable angina or non-Q-wave infarction, the levels of all four soluble adhesion molecules remain elevated for weeks to months after the resolution of the clinical symptoms [104]. An increased concentration of sVCAM-1 is predictive of an increased risk of major adverse cardiovascular events during a six-month follow-up period [105]. Additionally, in two prospective studies of type 2 diabetic patients, sVCAM-1 appeared to be a strong predictor of cardiovascular mortality.

Levels of soluble adhesion molecules also correlate with the extent and severity of atherosclerosis. Although based on small studies, evidence has shown that levels of sP-selectin and sL-selectin are increased in patients with PAD [106, 107]. In addition, increased levels of sE-selectin usually predict arterial restenosis after percutaneous transluminal angiography [108]. The levels of sVCAM-1 and sICAM-1 are significantly associated with carotid intima–media thickness, an index of early atherosclerosis [109]. Furthermore, sVCAM-1 appears to be an indicator of atherosclerotic severity. A few studies have indicated that sVCAM-1 significantly correlates with the degree of peripheral atherosclerosis [110, 111].

The mechanisms responsible for the production and metabolism of these soluble adhesion molecules remain unclear. In a few large prospective studies, sICAM-l, but not sVCAM-1, appears to be associated with incident CAD [90–93, 112]. ICAM-1 is constitutively expressed by a variety of cell types, including cells of the hematopoietic lineage and fibroblasts, and is further upregulated under inflammatory conditions. Therefore, as a general marker of proinflammatory status, sICAM-l correlates with the acute phase of atherosclerosis. VCAM-1, however, is not expressed under baseline conditions but is rapidly induced by pro-atherosclerotic factors in animal models and in humans [53, 113]. Hence, sVCAM-1 does not appear to be a risk factor for healthy individuals without endothelial dysfunction, but it is a strong risk predictor in patients suffering from established atherosclerotic disease, where its expression is expected to be enhanced. Therefore, many studies have recommended sVCAM-1 as the strongest predictor of future cardiovascular events in CAD patients [93, 95], diabetic patients [84, 114], or patients with unstable angina [105]. In conclusion, sICAM-1 predicts symptomatic disease, while sVCAM-1 is a better marker of the extent and severity of atherosclerosis in patients with established disease.

3.7
Adhesion Molecules as Therapeutic Targets

The efficacy of adhesion molecule blocking or deficiency in the suppression of atherosclerotic diseases indicates that most adhesion molecules are promising targets for the prevention and treatment of atherosclerosis. Small chemical molecules against these adhesion molecules, including peptides and antibodies, have been tested in animal models, and some have progressed to clinical trials. VCAM-1 has been considered as the most suitable therapeutic target. First, VCAM-1 is only expressed on activated endothelial cells covering atherosclerotic lesions, and blocking of VCAM-1 may not affect the function of resting endothelial cells. Second, blocking of endothelial cell VCAM-1 dramatically inhibits monocyte recruitment to atherosclerotic arteries. Third, deletion of VCAM-1 leads to significant suppression of atherosclerosis. In addition, inhibition of VCAM-1 or its ligand VLA-4 reduces neointimal formation in mouse carotid arteries and intimal hyperplasia in endarterectomized carotid arteries of primate animals. Most of these characteristics are also associated with other molecules, especially P-selectin and E-selectin.

Nevertheless, many more characteristics need to be assessed to define a certain adhesion molecule as a therapeutic target. First, it must be determined whether inhibition of the adhesion molecule can suppress the growth of advanced atherosclerotic lesions, as this is the stage at which most patients are diagnosed with atherosclerosis. Such a study is necessary because many experimental studies on mice begin at a very early age when atherosclerosis does not start to develop. The efficiency of adhesion molecule inhibition in suppressing the development of atherosclerotic lesions does not necessarily correlated with its effects on the suppression of the growth of advanced lesions. For example, CCR2 plays a robust role in the development of atherosclerotic lesions [115]. However, CCR2 deficiency in bone marrow–derived cells does not stop the growth of established atherosclerotic lesions [116]. Second, more studies are required to clarify whether inhibition of monocyte or endothelial adhesion molecules may interfere with macrophage functions essential for lesion shrinkage. In addition, as shown by many studies, approaches aimed at blocking monocyte recruitment also affect the homing of T cells and NK cells. Thus, patients treated with prospective inhibitors of adhesion molecules for a long period could become immunocompromised. Clearly, these issues must be addressed before considering adhesion molecules as therapeutic targets in the treatment of atherosclerosis.

References

1 Ross, R. (1999) Atherosclerosis – an inflammatory disease. *N. Engl. J. Med.*, **340** (2), 115–126.

2 Ley, K., Laudanna, C., Cybulsky, M.I., and Nourshargh, S. (2007) Getting to the site of inflammation: the leukocyte adhesion cascade updated. *Nat. Rev. Immunol.*, **7** (9), 678–689.

3 Carlos, T.M. and Harlan, J.M. (1994) Leukocyte-endothelial adhesion molecules. *Blood*, **84** (7), 2068–2101.

4 Springer, T.A. (1994) Traffic signals for lymphocyte recirculation and leukocyte emigration: the multistep paradigm. *Cell*, **76** (2), 301–314.

5 McEver, R.P. (2002) Selectins: lectins that initiate cell adhesion under flow. *Curr. Opin. Cell Biol.*, **14** (5), 581–586.

6 McEver, R.P. (1991) GMP-140: a receptor for neutrophils and monocytes on activated platelets and endothelium. *J. Cell. Biochem.*, **45** (2), 156–161.

7 Ishiwata, N., Takio, K., Katayama, M., Watanabe, K., Titani, K., Ikeda, Y., and Handa, M. (1994) Alternatively spliced isoform of P-selectin is present *in vivo* as a soluble molecule. *J. Cell. Biochem.*, **269** (38), 23708–23715.

8 Berger, G., Hartwell, D.W., and Wagner, D.D. (1998) P-selectin and platelet clearance. *Blood*, **92** (11), 4446–4452.

9 Leeuwenberg, J.F., Smeets, E.F., Neefjes, J.J., Shaffer, M.A., Cinek, T., Jeunhomme, T.M., Ahern, T.J., and Buurman, W.A. (1992) E-selectin and intercellular adhesion molecule-1 are released by activated human endothelial cells *in vitro*. *Immunology*, **77** (4), 543–549.

10 Hafezi-Moghadam, A., Thomas, K.L., Prorock, A.J., Huo, Y., and Ley, K. (2001) L-selectin shedding regulates leukocyte recruitment. *J. Exp. Med.*, **193** (7), 863–872.

11 Li, F., Erickson, H.P., James, J.A., Moore, K.L., Cummings, R.D., and McEver, R.P. (1996) Visualization of P-selectin glycoprotein ligand-1 as a highly extended molecule and mapping of protein epitopes for monoclonal antibodies. *J. Biol. Chem.*, **271** (11), 6342–6348.

12 Kansas, G.S. (1996) Selectins and their ligands: current concepts and controversies. *Blood*, **88** (9), 3259–3287.

13 Davenpeck, K.L., Brummet, M.E., Hudson, S.A., Mayer, R.J., and Bochner, B.S. (2000) Activation of human leukocytes reduces surface

P-selectin glycoprotein ligand-1 (PSGL-1, CD162) and adhesion to P-selectin *in vitro*. *J. Immunol.*, **165** (5), 2764–2772.

14 Steegmaier, M., Levinovitz, A., Isenmann, S., Borges, E., Lenter, M., Kocher, H.P., Kleuser, B., and Vestweber, D. (1995) The E-selectin-ligand ESL-1 is a variant of a receptor for fibroblast growth factor. *Nature*, **373** (6515), 615–620.

15 Collins, T., Read, M.A., Neish, A.S., Whitley, M.Z., Thanos, D., and Maniatis, T. (1995) Transcriptional regulation of endothelial cell adhesion molecules: NF-kappa B and cytokine-inducible enhancers. *FASEB J.*, **9** (10), 899–909.

16 Baeuerle, P.A. and Henkel, T. (1994) Function and activation of NF-kappa B in the immune system. *Annu. Rev. Immunol.*, **12**, 141–179.

17 de Fougerolles, A.R., Stacker, S.A., Schwarting, R., and Springer, T.A. (1991) Characterization of ICAM-2 and evidence for a third counter-receptor for LFA-1. *J. Exp. Med.*, **174** (1), 253–267.

18 del Pozo, M.A., Pulido, R., Munoz, C., Alvarez, V., Humbria, A., Campanero, M.R., and Sanchez-Madrid, F. (1994) Regulation of ICAM-3 (CD50) membrane expression on human neutrophils through a proteolytic shedding mechanism. *Eur. J. Immunol.*, **24** (11), 2586–2594.

19 Champagne, B., Tremblay, P., Cantin, A., and St Pierre, Y. (1998) Proteolytic cleavage of ICAM-1 by human neutrophil elastase. *J. Immunol.*, **161** (11), 6398–6405.

20 Fiore, E., Fusco, C., Romero, P., and Stamenkovic, I. (2002) Matrix metallo-proteinase 9 (MMP-9/gelatinase B) proteolytically cleaves ICAM-1 and participates in tumor cell resistance to natural killer cell-mediated cytotoxicity. *Oncogene*, **21** (34), 5213–5223.

21 Newton, J.P., Buckley, C.D., Jones, E.Y., and Simmons, D.L. (1997) Residues on both faces of the first immunoglobulin fold contribute to homophilic binding sites of PECAM-1/CD31. *J. Biol. Chem.*, **272** (33), 20555–20563.

22 Goldberger, A., Middleton, K.A., Oliver, J.A., Paddock, C., Yan, H.C., DeLisser, H.M., Albelda, S.M., and Newman, P.J. (1994) Biosynthesis and processing of the cell adhesion molecule PECAM-1 includes production of a soluble form. *J. Biol. Chem.*, **269** (25), 17183–17191.

23 Ramana, K.V., Bhatnagar, A., and Srivastava, S.K. (2004) Inhibition of aldose reductase attenuates TNF-alpha-induced expression of adhesion molecules in endothelial cells. *FASEB J.*, **18** (11), 1209–1218.

24 Leca, G., Mansur, S.E., and Bensussan, A. (1995) Expression of VCAM-1 (CD106) by a subset of TCR gamma delta-bearing lymphocyte clones. Involvement of a metalloprotease in the specific hydrolytic release of the soluble isoform. *J. Immunol.*, **154** (3), 1069–1077.

25 Hughes, P.E., and Pfaff, M. (1998) Integrin affinity modulation. *Trends Cell Biol.*, **8** (9), 359–364.

26 Diamond, M.S., Staunton, D.E., Marlin, S.D., and Springer, T.A. (1991) Binding of the integrin Mac-1 (CD11b/CD18) to the third immunoglobulin-like domain of ICAM-1 (CD54) and its regulation by glycosylation. *Cell*, **65** (6), 961–971.

27 Shang, X.Z. and Issekutz, A.C. (1998) Contribution of CD11a/CD18, CD11b/CD18, ICAM-1 (CD54) and -2 (CD102) to human monocyte migration through endothelium and connective tissue fibroblast barriers. *Eur. J. Immunol.*, **28** (6), 1970–1979.

28 Elices, M.J., Osborn, L., Takada, Y., Crouse, C., Luhowskyj, S., Hemler, M.E., and Lobb, R.R. (1990) VCAM-1 on activated endothelium interacts with the leukocyte integrin VLA-4 at a site distinct from the VLA-4/fibronectin binding site. *Cell*, **60** (4), 577–584.

29 Lynam, E., Sklar, L.A., Taylor, A.D., Neelamegham, S., Edwards, B.S., Smith, C.W., and Simon, S.I. (1998) Beta2-integrins mediate stable adhesion in collisional interactions between neutrophils and ICAM-1-expressing cells. *J. Leukoc. Biol.*, **64** (5), 622–630.

30 Lefer, A.M. (1999) Role of the beta2-integrins and immunoglobulin superfamily members in myocardial

ischemia-reperfusion. *Ann. Thorac. Surg.*, **68** (5), 1920–1923.

31 Rose, D.M., Grabovsky, V., Alon, R., and Ginsberg, M.H. (2001) The affinity of integrin alpha(4)beta(1) governs lymphocyte migration. *J. Immunol.*, **167** (5), 2824–2830.

32 Allen, S., Khan, S., Al-Mohanna, F., Batten, P., and Yacoub, M. (1998) Native low density lipoprotein-induced calcium transients trigger VCAM-1 and E-selectin expression in cultured human vascular endothelial cells. *J. Clin. Invest.*, **101** (5), 1064–1075.

33 Vora, D.K., Fang, Z.T., Liva, S.M., Tyner, T.R., Parhami, F., Watson, A.D., Drake, T.A., Territo, M.C., and Berliner, J.A. (1997) Induction of P-selectin by oxidized lipoproteins. Separate effects on synthesis and surface expression. *Circ. Res.*, **80** (6), 810–818.

34 Khan, B.V., Parthasarathy, S.S., Alexander, R.W., and Medford, R.M. (1995) Modified low density lipoprotein and its constituents augment cytokine-activated vascular cell adhesion molecule-1 gene expression in human vascular endothelial cells. *J. Clin. Invest.*, **95** (3), 1262–1270.

35 Li, H., Cybulsky, M.I., Gimbrone, M.A., Jr., and Libby P. (1993) An atherogenic diet rapidly induces VCAM-1, a cytokine-regulatable mononuclear leukocyte adhesion molecule, in rabbit aortic endothelium. *Arterioscler. Thromb.*, **13** (2), 197–204.

36 Sakai, A., Kume, N., Nishi, E., Tanoue, K., Miyasaka, M., and Kita, T. (1997) P-selectin and vascular cell adhesion molecule-1 are focally expressed in aortas of hypercholesterolemic rabbits before intimal accumulation of macrophages and T lymphocytes. *Arterioscler. Thromb. Vasc. Biol.*, **17** (2), 310–316.

37 Iiyama, K., Hajra, L., Iiyama, M., Li, H., DiChiara, M., Medoff, B.D., and Cybulsky, M.I. (1999) Patterns of vascular cell adhesion molecule-1 and intercellular adhesion molecule-1 expression in rabbit and mouse atherosclerotic lesions and at sites predisposed to lesion formation. *Circ. Res.*, **85** (2), 199–207.

38 Nakashima, Y., Raines, E.W., Plump, A.S., Breslow, J.L., and Ross, R. (1998) Upregulation of VCAM-1 and ICAM-1 at atherosclerosis-prone sites on the endothelium in the ApoE-deficient mouse. *Arterioscler. Thromb. Vasc. Biol.*, **18** (5), 842–851.

39 Cozzi, P.J., Lyon, R.T., Davis, H.R., Sylora, J., Glagov, S., and Zarins, C.K. (1988) Aortic wall metabolism in relation to susceptibility and resistance to experimental atherosclerosis. *J. Vasc. Surg.*, **7** (5), 706–714.

40 Morigi, M., Zoja, C., Figliuzzi, M., Foppolo, M., Micheletti, G., Bontempelli, M., Saronni, M., Remuzzi, G., and Remuzzi, A. (1995) Fluid shear stress modulates surface expression of adhesion molecules by endothelial cells. *Blood*, **85** (7), 1696–1703.

41 Ando, J., Tsuboi, H., Korenaga, R., Takada, Y., Toyama-Sorimachi, N., Miyasaka, M., and Kamiya, A. (1995) Down-regulation of vascular adhesion molecule-1 by fluid shear stress in cultured mouse endothelial cells. *Ann. N. Y. Acad. Sci.*, **748**, 148–156; discussion 156–147.

42 Tsuboi, H., Ando, J., Korenaga, R., Takada, Y., and Kamiya, A. (1995) Flow stimulates ICAM-1 expression time and shear stress dependently in cultured human endothelial cells. *Biochem. Biophys. Res. Commun.*, **206** (3), 988–996.

43 Walpola, P.L., Gotlieb, A.I., Cybulsky, M.I., and Langille, B.L. (1995) Expression of ICAM-1 and VCAM-1 and monocyte adherence in arteries exposed to altered shear stress. *Arterioscler. Thromb. Vasc. Biol.*, **15** (1), 2–10.

44 Wang, H., Nawata, J., Kakudo, N., Sugimura, K., Suzuki, J., Sakuma, M., Ikeda, J., and Shirato, K. (2004) The upregulation of ICAM-1 and P-selectin requires high blood pressure but not circulating renin-angiotensin system *in vivo. J. Hypertens.*, **22** (7), 1323–1332.

45 DeSouza, C.A., Dengel, D.R., Macko, R.F., Cox, K., and Seals, D.R. (1997) Elevated levels of circulating cell adhesion molecules in uncomplicated essential hypertension. *Am. J. Hypertens.*, **10** (12 Pt 1), 1335–1341.

46 Verhaar, M.C., Beutler, J.J., Gaillard, C.A., Koomans, H.A., Fijnheer, R., and Rabelink, T.J. (1998) Progressive vascular damage in hypertension is associated with increased levels of circulating P-selectin. *J. Hypertens.*, **16** (1), 45–50.

47 De Caterina, R., Ghiadoni, L., Taddei, S., Virdis, A., Almerigogna, F., Basta, G., Lazzerini, G., Bernini, W., and Salvetti, A. (2001) Soluble E-selectin in essential hypertension: a correlate of vascular structural changes. *Am. J. Hypertens.*, **14** (3), 259–266.

48 Shen, Y., Rattan, V., Sultana, C., and Kalra, V.K. (1996) Cigarette smoke condensate-induced adhesion molecule expression and transendothelial migration of monocytes. *Am. J. Physiol.*, **270** (5 Pt 2), H1624–1633.

49 Johnson-Tidey, R.R., McGregor, J.L., Taylor, P.R., and Poston, R.N. (1994) Increase in the adhesion molecule P-selectin in endothelium overlying atherosclerotic plaques. Coexpression with intercellular adhesion molecule-1. *Am. J. Pathol.*, **144** (5), 952–961.

50 Davies, M.J., Gordon, J.L., Gearing, A.J., Pigott, R., Woolf, N., Katz, D., and Kyriakopoulos, A. (1993) The expression of the adhesion molecules ICAM-1, VCAM-1, PECAM, and E-selectin in human atherosclerosis. *J. Pathol.*, **171** (3), 223–229.

51 Manka, D.R., Wiegman, P., Din, S., Sanders, J.M., Green, S.A., Gimple, L.W., Ragosta, M., Powers, E.R., Ley, K., and Sarembock, I.J. (1999) Arterial injury increases expression of inflammatory adhesion molecules in the carotid arteries of apolipoprotein-E-deficient mice. *J. Vasc. Res.*, **36** (5), 372–378.

52 Poston, R.N., Haskard, D.O., Coucher, J.R., Gall, N.P., and Johnson-Tidey, R.R. (1992) Expression of intercellular adhesion molecule-1 in atherosclerotic plaques. *Am. J. Pathol.*, **140** (3), 665–673.

53 O'Brien, K.D., Allen, M.D., McDonald, T.O., Chait, A., Harlan, J.M., Fishbein, D., McCarty, J., Ferguson, M., Hudkins, K., Benjamin, C.D., *et al.* (1993)

Vascular cell adhesion molecule-1 is expressed in human coronary atherosclerotic plaques. Implications for the mode of progression of advanced coronary atherosclerosis. *J. Clin. Invest.*, **92** (2), 945–951.

54 Luscinskas, F.W., Ding, H., Tan, P., Cumming, D., Tedder, T.F., and Gerritsen, M.E. (1996) L- and P-selectins, but not CD49d (VLA-4) integrins, mediate monocyte initial attachment to TNF-alpha-activated vascular endothelium under flow *in vitro*. *J. Immunol.*, **157** (1), 326–335.

55 Luscinskas, F.W., Kansas, G.S., Ding, H., Pizcueta, P., Schleiffenbaum, B.E., Tedder, T.F., and Gimbrone M.A., Jr. (1994) Monocyte rolling, arrest and spreading on IL-4-activated vascular endothelium under flow is mediated via sequential action of L-selectin, beta 1-integrins, and beta 2-integrins. *J. Cell Biol.*, **125** (6), 1417–1427.

56 Gerszten, R.E., Luscinskas, F.W., Ding, H.T., Dichek, D.A., Stoolman, L.M., Gimbrone, M.A., Jr., and Rosenzweig, A. (1996) Adhesion of memory lymphocytes to vascular cell adhesion molecule-1-transduced human vascular endothelial cells under simulated physiological flow conditions *in vitro*. *Circ. Res.*, **79** (6), 1205–1215.

57 Gerszten, R.E., Lim, Y.C., Ding, H.T., Snapp, K., Kansas, G., Dichek, D.A., Cabanas, C., Sanchez-Madrid, F., Gimbrone, M.A., Jr., Rosenzweig A., and Luscinskas, F.W. (1998) Adhesion of monocytes to vascular cell adhesion molecule-1-transduced human endothelial cells: implications for atherogenesis. *Circ. Res.*, **82** (8), 871–878.

58 Huo, Y., Hafezi-Moghadam, A., and Ley, K. (2000) Role of vascular cell adhesion molecule-1 and fibronectin connecting segment-1 in monocyte rolling and adhesion on early atherosclerotic lesions. *Circ. Res.*, **87** (2), 153–159.

59 Ramos, C.L., Huo, Y., Jung, U., Ghosh, S., Manka, D.R., Sarembock, I.J., and Ley, K. (1999) Direct demonstration of P-selectin- and VCAM-1-dependent mononuclear cell rolling in early atherosclerotic lesions of apolipoprotein E-deficient mice. *Circ. Res.*, **84** (11), 1237–1244.

60 Patel, S.S., Thiagarajan, R., Willerson, J.T., and Yeh, E.T. (1998) Inhibition of alpha4 integrin and ICAM-1 markedly attenuate macrophage homing to atherosclerotic plaques in ApoE-deficient mice. *Circulation*, **97** (1), 75–81.

61 Johnson, R.C., Chapman, S.M., Dong, Z.M., Ordovas, J.M., Mayadas, T.N., Herz, J., Hynes, R.O., Schaefer, E.J., and Wagner, D.D. (1997) Absence of P-selectin delays fatty streak formation in mice. *J. Clin. Invest.*, **99** (5), 1037–1043.

62 Dong, Z.M., Brown, A.A., and Wagner, D.D. (2000) Prominent role of P-selectin in the development of advanced atherosclerosis in ApoE-deficient mice. *Circulation*, **101** (19), 2290–2295.

63 Collins, R.G., Velji, R., Guevara, N.V., Hicks, M.J., Chan, L., and Beaudet, A.L. (2000) P-selectin or intercellular adhesion molecule (ICAM)-1 deficiency substantially protects against atherosclerosis in apolipoprotein E-deficient mice. *J. Exp. Med.*, **191** (1), 189–194.

64 Dong, Z.M., Chapman, S.M., Brown, A.A., Frenette, P.S., Hynes, R.O., and Wagner, D.D. (1998) The combined role of P- and E-selectins in atherosclerosis. *J. Clin. Invest.*, **102** (1), 145–152.

65 An, G., Wang, H., Tang, R., Yago, T., McDaniel, J.M., McGee, S., Huo, Y., and Xia, L. (2008) P-selectin glycoprotein ligand-1 is highly expressed on Ly-6Chi monocytes and a major determinant for Ly-6Chi monocyte recruitment to sites of atherosclerosis in mice. *Circulation*, **117** (25), 3227–3237.

66 Wang, H., Tang, R., Zhang, W., Amirikian, K., Geng, Z., Geng, J., Hebbel, R.P., Xia, L., Marth, J.D., Fukuda, M., Katoh, S., and Huo, Y. (2008) Core2 1-6-N-glucosaminyltransferase-I is crucial for the formation of atherosclerotic lesions in apolipoprotein E-deficient mice. *Arterioscler. Thromb. Vasc. Biol.*, **29** (2), 180–187.

67 Homeister, J.W., Daugherty, A., and Lowe, J.B. (2004) Alpha(1,3)fucosyl-transferases FucT-IV and FucT-VII control susceptibility to atherosclerosis in apolipoprotein $E^{-/-}$ mice. *Arterioscler. Thromb. Vasc. Biol.*, **24** (10), 1897–1903.

68 Nageh, M.F., Sandberg, E.T., Marotti, K.R., Lin, A.H., Melchior, E.P., Bullard, D.C., and Beaudet, A.L. (1997) Deficiency of inflammatory cell adhesion molecules protects against atherosclerosis in mice. *Arterioscler. Thromb. Vasc. Biol.*, **17** (8), 1517–1520.

69 Gurtner, G.C., Davis, V., Li, H., McCoy, M.J., Sharpe, A., and Cybulsky, M.I. (1995) Targeted disruption of the murine VCAM1 gene: essential role of VCAM-1 in chorioallantoic fusion and placentation. *Genes Dev.*, **9** (1), 1–14.

70 Arroyo, A.G., Yang, J.T., Rayburn, H., and Hynes, R.O. (1996) Differential requirements for alpha4 integrins during fetal and adult hematopoiesis. *Cell*, **85** (7), 997–1008.

71 Shih, P.T., Brennan, M.L., Vora, D.K., Territo, M.C., Strahl, D., Elices, M.J., Lusis, A.J., and Berliner, J.A. (1999) Blocking very late antigen-4 integrin decreases leukocyte entry and fatty streak formation in mice fed an atherogenic diet. *Circ. Res.*, **84** (3), 345–351.

72 Cybulsky, M.I., Iiyama, K., Li, H., Zhu, S., Chen, M., Iiyama, M., Davis, V., Gutierrez-Ramos, J.C., Connelly, P.W., and Milstone, D.S. (2001) A major role for VCAM-1, but not ICAM-1, in early atherosclerosis. *J. Clin. Invest.*, **107** (10), 1255–1262.

73 Hackman, A., Abe, Y., Insull, W., Jr., Pownall H., Smith, L., Dunn, K., Gotto, A.M., Jr., and Ballantyne C.M. (1996) Levels of soluble cell adhesion molecules in patients with dyslipidemia. *Circulation*, **93** (7), 1334–1338.

74 Abe, Y., El-Masri, B., Kimball, K.T., Pownall, H., Reilly, C.F., Osmundsen, K., Smith, C.W., and Ballantyne, C.M. (1998) Soluble cell adhesion molecules in hypertriglyceridemia and potential significance on monocyte adhesion. *Arterioscler. Thromb. Vasc. Biol.*, **18** (5), 723–731.

75 Lupattelli, G., Lombardini, R., Schillaci, G., Ciuffetti, G., Marchesi, S., Siepi, D., and Mannarino, E. (2000) Flow-mediated vasoactivity and circulating adhesion molecules in hypertriglyceridemia: association with small, dense LDL cholesterol particles. *Am. Heart J.*, **140** (3), 521–526.

76 Calabresi, L., Gomaraschi, M., Villa, B., Omoboni, L., Dmitrieff, C., and Franceschini, G. (2002) Elevated soluble cellular adhesion molecules in subjects with low HDL-cholesterol. *Arterioscler. Thromb. Vasc. Biol.*, **22** (4), 656–661.

77 Blann, A.D., Tse, W., Maxwell, S.J., and Waite, M.A. (1994) Increased levels of the soluble adhesion molecule E-selectin in essential hypertension. *J. Hypertens.*, **12** (8), 925–928.

78 Parissis, J.T., Venetsanou, K.F., Mentzikof, D.G., Kalantzi, M.V., Georgopoulou, M.V., Chrisopoulos, N., and Karas, S.M. (2001) Plasma levels of soluble cellular adhesion molecules in patients with arterial hypertension. Correlations with plasma endothelin-1. *Eur. J. Intern. Med.*, **12** (4), 350–356.

79 Preston, R.A., Ledford, M., Materson, B.J., Baltodano, N.M., Memon, A., and Alonso, A. (2002) Effects of severe, uncontrolled hypertension on endothelial activation: soluble vascular cell adhesion molecule-1, soluble intercellular adhesion molecule-1 and von Willebrand factor. *J. Hypertens.*, **20** (5), 871–877.

80 Cominacini, L., Fratta Pasini, A., Garbin, U., Davoli, A., De Santis, A., Campagnola, M., Rigoni, A., Zenti, M.G., Moghetti, P., and Lo Cascio, V. (1995) Elevated levels of soluble E-selectin in patients with IDDM and NIDDM: relation to metabolic control. *Diabetologia*, **38** (9), 1122–1124.

81 Jilma, B., Fasching, P., Ruthner, C., Rumplmayr, A., Ruzicka, S., Kapiotis, S., Wagner, O.F., and Eichler, H.G. (1996) Elevated circulating P-selectin in insulin dependent diabetes mellitus. *Thromb. Haemost.*, **76** (3), 328–332.

82 Smulders, R.A., Stehouwer, C.D., Schalkwijk, C.G., Donker, A.J., van Hinsbergh, V.W., and TeKoppele, J.M. (1998) Distinct associations of HbA1c and the urinary excretion of pentosidine, an advanced glycosylation end-product, with markers of endothelial function in insulin-dependent diabetes mellitus. *Thromb. Haemost.*, **80** (1), 52–57.

83 Clausen, P., Jacobsen, P., Rossing, K., Jensen, J.S., Parving, H.H., and Feldt-Rasmussen, B. (2000) Plasma concentrations of VCAM-1 and ICAM-1 are elevated in patients with Type 1 diabetes mellitus with microalbuminuria and overt nephropathy. *Diabet. Med.*, **17** (9), 644–649.

84 Jager, A., van Hinsbergh, V.W., Kostense, P.J., Emeis, J.J., Nijpels, G., Dekker, J.M., Heine, R.J., Bouter, L.M., and Stehouwer, C.D. (2000) Increased levels of soluble vascular cell adhesion molecule 1 are associated with risk of cardiovascular mortality in type 2 diabetes: the Hoorn study. *Diabetes*, **49** (3), 485–491.

85 Becker, A., van Hinsbergh, V.W., Jager, A., Kostense, P.J., Dekker, J.M., Nijpels, G., Heine, R.J., Bouter, L.M., and Stehouwer, C.D. (2002) Why is soluble intercellular adhesion molecule-1 related to cardiovascular mortality?. *Eur. J. Clin. Invest.*, **32** (1), 1–8.

86 Blann, A.D., Steele, C., and McCollum, C.N. (1997) The influence of smoking on soluble adhesion molecules and endothelial cell markers. *Thromb. Res.*, **85** (5), 433–438.

87 Barbaux, S.C., Blankenberg, S., Rupprecht, H.J., Francomme, C., Bickel, C., Hafner, G., Nicaud, V., Meyer, J., Cambien, F., and Tiret, L. (2001) Association between P-selectin gene polymorphisms and soluble P-selectin levels and their relation to coronary artery disease. *Arterioscler. Thromb. Vasc. Biol.*, **21** (10), 1668–1673.

88 Mazzone, A., Cusa, C., Mazzucchelli, I., Vezzoli, M., Ottini, E., Ghio, S., Tossini, G., Pacifici, R., and Zuccaro, P. (2001) Cigarette smoking and hypertension influence nitric oxide release and plasma levels of adhesion molecules. *Clin. Chem. Lab. Med.*, **39** (9), 822–826.

89 Takeuchi, N., Kawamura, T., Kanai, A., Nakamura, N., Uno, T., Hara, T., Sano,

T., Sakamoto, N., Hamada, Y., Nakamura, J., and Hotta, N. (2002) The effect of cigarette smoking on soluble adhesion molecules in middle-aged patients with Type 2 diabetes mellitus. *Diabet. Med.*, **19** (1), 57–64.

90 Hwang, S.J., Ballantyne, C.M., Sharrett, A.R., Smith, L.C., Davis, C.E., Gotto, A.M., Jr., and Boerwinkle E. (1997) Circulating adhesion molecules VCAM-1, ICAM-1, and E-selectin in carotid atherosclerosis and incident coronary heart disease cases: the Atherosclerosis Risk In Communities (ARIC) study. *Circulation*, **96** (12), 4219–4225.

91 Ridker, P.M., Hennekens, C.H., Roitman-Johnson, B., Stampfer, M.J., and Allen, J. (1998) Plasma concentration of soluble intercellular adhesion molecule 1 and risks of future myocardial infarction in apparently healthy men. *Lancet*, **351** (9096), 88–92.

92 de Lemos, J.A., Hennekens, C.H., and Ridker, P.M. (2000) Plasma concentration of soluble vascular cell adhesion molecule-1 and subsequent cardiovascular risk. *J. Am. Coll. Cardiol.*, **36** (2), 423–426.

93 Malik, I., Danesh, J., Whincup, P., Bhatia, V., Papacosta, O., Walker, M., Lennon, L., Thomson, A., and Haskard, D. (2001) Soluble adhesion molecules and prediction of coronary heart disease: a prospective study and meta-analysis. *Lancet*, **358** (9286), 971–976.

94 Pradhan, A.D., Rifai, N., and Ridker, P.M. (2002) Soluble intercellular adhesion molecule-1, soluble vascular adhesion molecule-1, and the development of symptomatic peripheral arterial disease in men. *Circulation*, **106** (7), 820–825.

95 Blankenberg, S., Rupprecht, H.J., Bickel, C., Peetz, D., Hafner, G., Tiret, L., and Meyer, J. (2001) Circulating cell adhesion molecules and death in patients with coronary artery disease. *Circulation*, **104** (12), 1336–1342.

96 Ikeda, H., Nakayama, H., Oda, T., Kuwano, K., Muraishi, A., Sugi, K., Koga, Y., and Toshima, H. (1994) Soluble form of P-selectin in patients with acute myocardial infarction. *Coron. Artery Dis.*, **5** (6), 515–518.

97 Ikeda, H., Takajo, Y., Ichiki, K., Ueno, T., Maki, S., Noda, T., Sugi, K., and Imaizumi, T. (1995) Increased soluble form of P-selectin in patients with unstable angina. *Circulation*, **92** (7), 1693–1696.

98 Shyu, K.G., Chang, H., Lin, C.C., and Kuan, P. (1996) Circulating intercellular adhesion molecule-1 and E-selectin in patients with acute coronary syndrome. *Chest*, **109** (6), 1627–1630.

99 Ghaisas, N.K., Shahi, C.N., Foley, B., Goggins, M., Crean, P., Kelly, A., Kelleher, D., and Walsh, M. (1997) Elevated levels of circulating soluble adhesion molecules in peripheral blood of patients with unstable angina. *Am. J. Cardiol.*, **80** (5), 617–619.

100 Miwa, K., Igawa, A., and Inoue, H. (1997) Soluble E-selectin, ICAM-1 and VCAM-1 levels in systemic and coronary circulation in patients with variant angina. *Cardiovasc. Res.*, **36** (1), 37–44.

101 Mulvihill, N.T., Foley, J.B., Murphy, R.T., Pate, G., Crean, P.A., and Walsh, M. (2001) Enhanced endothelial activation in diabetic patients with unstable angina and non-Q-wave myocardial infarction. *Diabet. Med.*, **18** (12), 979–983.

102 Parker, C., 3rd, Vita, J.A., and Freedman, J.E. (2001) Soluble adhesion molecules and unstable coronary artery disease. *Atherosclerosis*, **156** (2), 417–424.

103 Mizia-Stec, K., Zahorska-Markiewicz, B., Mandecki, T., Janowska, J., Szulc, A., and Jastrzebska-Maj, E. (2002) Serum levels of selected adhesion molecules in patients with coronary artery disease. *Int. J. Cardiol.*, **83** (2), 143–150.

104 Mulvihill, N.T., Foley, J.B., Murphy, R., Crean, P., and Walsh, M. (2000) Evidence of prolonged inflammation in unstable angina and non-Q wave myocardial infarction. *J. Am. Coll. Cardiol.*, **36** (4), 1210–1216.

105 Mulvihill, N.T., Foley, J.B., Murphy, R.T., Curtin, R., Crean, P.A., and Walsh, M. (2001) Risk stratification in unstable angina and non-Q wave myocardial infarction using soluble cell adhesion molecules. *Heart*, **85** (6), 623–627.

106 Blann, A., Morris, J., and McCollum, C. (1996) Soluble L-selectin in peripheral arterial disease: relationship with soluble E-selectin and soluble P-selectin. *Atherosclerosis*, **126** (2), 227–231.

107 Blann, A.D., Seigneur, M., Boisseau, M.R., Taberner, D.A., and McCollum, C.N. (1996) Soluble P selectin in peripheral vascular disease: relationship to the location and extent of atherosclerotic disease and its risk factors. *Blood Coagul. Fibrinolysis*, **7** (8), 789–793.

108 Belch, J.J., Shaw, J.W., Kirk, G., McLaren, M., Robb, R., Maple, C., and Morse, P. (1997) The white blood cell adhesion molecule E-selectin predicts restenosis in patients with intermittent claudication undergoing percutaneous transluminal angioplasty. *Circulation*, **95** (8), 2027–2031.

109 Rohde, L.E., Lee, R.T., Rivero, J., Jamacochian, M., Arroyo, L.H., Briggs, W., Rifai, N., Libby, P., Creager, M.A., and Ridker, P.M. (1998) Circulating cell adhesion molecules are correlated with ultrasound-based assessment of carotid atherosclerosis. *Arterioscler. Thromb. Vasc. Biol.*, **18** (11), 1765–1770.

110 De Caterina, R., Basta, G., Lazzerini, G., Dell'Omo, G., Petrucci, R., Morale, M., Carmassi, F., and Pedrinelli, R. (1997) Soluble vascular cell adhesion molecule-1 as a biohumoral correlate of atherosclerosis. *Arterioscler. Thromb. Vasc. Biol.*, **17** (11), 2646–2654.

111 Peter, K., Nawroth, P., Conradt, C., Nordt, T., Weiss, T., Boehme, M., Wunsch, A., Allenberg, J., Kubler, W., and Bode, C. (1997) Circulating vascular cell adhesion molecule-1 correlates with the extent of human atherosclerosis in contrast to circulating intercellular adhesion molecule-1, E-selectin, P-selectin, and thrombomodulin. *Arterioscler. Thromb. Vasc. Biol.*, **17** (3), 505–512.

112 Ridker, P.M., Hennekens, C.H., Buring, J.E., and Rifai, N. (2000) C-reactive protein and other markers of inflammation in the prediction of cardiovascular disease in women. *N. Engl. J. Med.*, **342** (12), 836–843.

113 Cybulsky, M.I., and Gimbrone M.A., Jr. (1991) Endothelial expression of a mononuclear leukocyte adhesion molecule during atherogenesis. *Science*, **251** (4995), 788–791.

114 Stehouwer, C.D., Gall, M.A., Twisk, J.W., Knudsen, E., Emeis, J.J., and Parving, H.H. (2002) Increased urinary albumin excretion, endothelial dysfunction, and chronic low-grade inflammation in type 2 diabetes: progressive, interrelated, and independently associated with risk of death. *Diabetes*, **51** (4), 1157–1165.

115 Dawson, T.C., Kuziel, W.A., Osahar, T.A., and Maeda, N. (1999) Absence of CC chemokine receptor-2 reduces atherosclerosis in apolipoprotein E-deficient mice. *Atherosclerosis*, **143** (1), 205–211.,

116 Guo, J., de Waard, V., Van Eck, M., Hildebrand, R.B., van Wanrooij, E.J., Kuiper, J., Maeda, N., Benson, G.M., Groot, P.H., and Van Berkel, T.J. (2005) Repopulation of apolipoprotein E knockout mice with CCR2-deficient bone marrow progenitor cells does not inhibit ongoing atherosclerotic lesion development. *Arterioscler. Thromb. Vasc. Biol.*, **25** (5), 1014–1019.

4
Cytokines and Atherosclerosis

Alain Tedgui, Hafid Ait-Oufella, and Ziad Mallat

4.1
Introduction

Animal studies using murine models of human atherosclerosis, as well as recent large-scale intervention trials with statin treatment, have enabled us to build up a precise picture of the mechanisms of atherosclerosis, in which both cholesterol and inflammation play a central role. Findings obtained in mice on apolipoprotein E–deficient (ApoE$^{-/-}$) or low-density lipoprotein receptor–deficient (LDLR$^{-/-}$) background deficient for genes encoding pro- or anti-inflammatory cytokines have demonstrated that these mediators act as essential etiologic factors in the development and progression of atherosclerosis.

In humans, a number of clinical studies have shown that inflammatory biomarkers predict future vascular events independently of cholesterol levels. Moreover, the primary or secondary prevention data from controlled trials (e.g., PROVE IT-TIMI 22, JUPITER) provide further evidence for the concomitant importance of lipid profiles and inflammatory status in determining the risk of atherosclerosis in humans. The current view on the mechanisms of the disease is that atherosclerosis is initiated in response to subendothelial cholesterol accumulation and oxidation, which stimulate both innate and adaptive immune responses, leading to chronic inflammation within the arterial wall [1, 2] (Figure 4.1). Atherosclerotic plaques contain small populations of activated B cells, mast cells, and dendritic cells, together with activated T cells and macrophages, trafficking between the blood, the atherosclerotic lesion, and the regional lymph nodes. It is also recognized that vascular cells, endothelial cells (ECs), and smooth muscle cells (SMCs) are involved in the development of the disease, mediating leukocyte recruitment, and extracellular matrix destruction, as well as feeding back to promote perpetuation of inflammation through the release of proinflammatory cytokines and chemokines.

Cytokines play a dual role in atherosclerosis: proinflammatory and Th1-related cytokines promote the development and progression of the disease, whereas anti-inflammatory and regulatory T cell-related cytokines exert clear anti-atherogenic activities. This chapter is an update on our previous review on the pathogenic and regulatory roles of cytokines in atherosclerosis [2].

Atherosclerosis: Molecular and Cellular Mechanisms. Edited by Sarah Jane George and Jason Johnson
Copyright © 2010 WILEY-VCH Verlag GmbH & Co. KGaA, Weinheim
ISBN: 978-3-527-32448-4

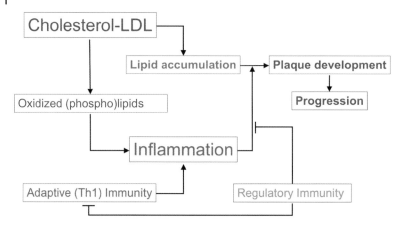

Figure 4.1 Cholesterol is a prerequisite to the development of atherosclerosis. Oxidized (phospho)lipids trigger an inflammatory response that is exacerbated by Th1-driven adaptive immunity, which is counterbalanced by immunosuppressive regulatory T cell response.

4.2
The Cytokine Network

Cytokines are low molecular weight protein mediators that usually act at short range between neighboring cells. The known cytokines consist of more than 100 secreted factors, clustered into several classes: interleukins (35 have been identified to date), tumor necrosis factors (TNF), interferons (IFN), colony-stimulating factors (CSF), transforming growth factors (TGF), and chemokines. They are involved in intercellular communication, which regulates fundamental processes in essentially every important biological pathway, including cell proliferation, inflammation, immunity, migration, fibrosis, repair, body growth, lactation, adiposity, hematopoiesis, and angiogenesis [3, 4].

Cytokines are especially important for regulating inflammatory and immune responses, and have crucial functions in controlling both innate and adaptive immunity. All cells involved in atherosclerosis are capable of producing and responding to cytokines (Figure 4.2).

Cytokines are categorized according to the structural homology of their receptors as class I or class II cytokines [5]. Class I cytokines encompass the interleukin-6 (IL-6) family (IL-6, IL-11, IL-31, oncostatin M (OSM), and leukemia inhibitory factor (LIF)) and γc-family (IL-2, IL-4, IL-7, IL-9, IL-15, and IL-21) based on their sharing the common cytokine receptor γ chain (γc), which is mutated in X-linked severe combined immunodeficiency (SCID). Class II cytokines encompass the IFN family (IFN-α, β, γ) and IL-10-related families (IL-10, IL-19, IL-20, IL-22, IL-24, IL-26). Most ILs, CSFs, and IFNs belong to one of these two classes of cytokines, which mediate their effects through the Janus kinase-signal transducers and activators of transcription (JAK-STAT) pathway.

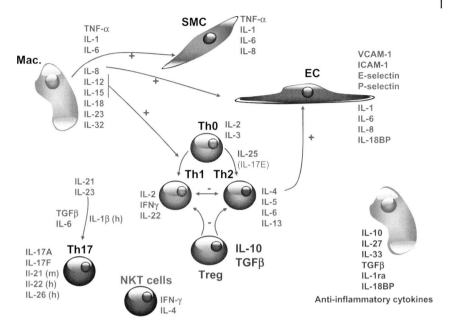

Figure 4.2 Cytokines are produced by several cell types, including inflammatory and vascular cells, as well as adipocytes. IL-12, and IL-18 produced by macrophages are potent inducers of IFN-γ and promote the differentiation of naïve T cells into pro-atherogenic Th1 cells. Macrophage-derived cytokines activate smooth muscle cells (SMCs) and endothelial cells (ECs) to produce an array of proinflammatory mediators. On the other hand, the anti-inflammatory cytokines IL-10 and TGF-β, also produced by macrophages, promote anti-atherogenic T_{reg} cell differentiation. Other anti-inflammatory mediators with potent anti-atherogenic properties include IL-33, IL-1 receptor antagonist (IL-1ra), and IL-18 binding protein (IL-18BP). The effect of IL-27 on atherosclerosis is unknown. In naïve CD4$^+$ T cells from mice (m), Th17 cells are induced in response to a combination of IL-6 or IL-21 and TGF-β. A combination of TGF-β, IL-1β and IL-6, IL-21, or IL-23 is necessary to induce Th17 cells in naïve human (h) CD4$^+$ T cells.

Three other major cytokine families are the IL-1 family (IL-1α, IL-1β, IL-1ra, IL-18, and IL-33), TNF superfamily, and TGF-β superfamily. IL-1 and TNF family members activate the nuclear factor-κB (NF-κB) and mitogen-activated protein (MAP) kinase signaling pathways, while TGF-β superfamily members activate signaling proteins of the Smad family.

4.3
Biological Effects of Cytokines in Atherosclerosis

According to the classical view of inflammation, cytokines are produced by cells of the innate immune system (monocytes, neutrophils, and NK T cells) in response to microbial infection, toxic reagents, trauma, antibodies, or

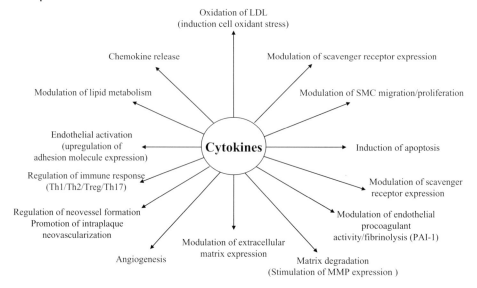

Figure 4.3 Biological effects of cytokines in atherosclerosis.

immune complexes [6]. In the host, Toll-like receptors (TLRs) and the intra-cellular proteins nucleotide-binding oligomerization domain-1 and -2 (NOD1 and NOD2) act as sensors of the conserved molecular motifs present on a wide range of different microbes, the pathogen-associated molecular patterns (PAMPs). Hence, cytokines are secondary mediators of inflammation and not the primary triggers. It is believed that a "sterile inflammation," initiated by oxidized lipo-proteins or phospholipids that can act as endogenous TLR ligands [7] is the primary event leading to a cytokine cascade and subsequent development of atherosclerosis.

The biological effects of proinflammatory cytokines that may cause their pro-atherogenic activities are multiple (Figure 4.3). In the early stages of atherosclero-sis, cytokines can alter the endothelial permeability, promoting subendothelial lipid accumulation and favoring leukocyte transmigration. A number of proin-flammatory cytokines, such as TNF-α and IFN-γ, have been shown to alter the distribution of adhesion receptors involved in cell–cell adhesion (e.g., vascular endothelial (VE)–cadherin–catenin complexes) and prevent the formation of F-actin stress fibers [8], resulting in restructuring of the intercellular junction leading to loss of endothelial permeability. Cytokines play a major role in the activation of adhesion molecule and chemokine expression involved in lym-phocyte/monocyte recruitment, endothelial adhesion, and migration into the inflamed vessel wall [9]. Once in the intima, leukocytes can be permanently acti-vated by locally generated cytokines, which may also accelerate the transformation of macrophages into foam cells by stimulating the expression of scavenger recep-tors and enhancing cell-mediated oxidation.

Lectin-like oxidized LDL receptor-1 (LOX-1), which is detectable in endothelial cells, intimal macrophages, and SMCs of advanced atherosclerotic plaques, can be induced by TNF-α, in addition to its induction by oxidized LDL [10]. IFN-γ induces foam cell formation through upregulation of SR-PSOX [11], the scavenger receptor for phosphatidylserine and ox-LDL, which has been involved in ox-LDL uptake and subsequent foam cell transformation in macrophages [12]. IFN-γ also inhibits ApoE [13] and the ATP-binding cassette transporter A1 (ABCA1) [14], resulting in decreased cholesterol efflux from macrophages [14]. Moreover, TNF-α [15], IL-4, and IL-13 [16] enhance the ability of cell-mediated oxidation.

At a more advanced stage of the disease, proinflammatory cytokines destabilize atherosclerotic plaques by promoting cell apoptosis and matrix degradation. Macrophage apoptosis may contribute to enlargement of the lipid core whereas plaque SMC apoptosis may induce a thinning in the fibrous cap, favoring its rupture [17, 18]. A number of proinflammatory cytokines have been shown to induce SMC and macrophage apoptosis, particularly the association of IL-1, TNF-α, and IFN-γ and promotion of Fas–FasL killing [19].

Membranes of apoptotic cells are particularly rich in proinflammatory oxidized phospholipids, suggesting that accumulation of apoptotic cells and debris within the plaque might increase the risk of plaque progression and complications. Phagocytic clearance of apoptotic cells by professional scavengers is the most important determinant of the steady state and the physiological consequences of apoptosis *in vivo*, suppressing proinflammatory signaling and activating anti-inflammatory pathways [20]. TGF-β and IL-10 production by apoptotic cells or upon ingestion of these cells by macrophages leads to cell deactivation and inhibition of self-reactive T cells [21]. Apoptotic cells may also induce the generation of tolerogenic dendritic cells [20, 22], potentially leading to the development of defined populations of regulatory T cells. However, this mechanism is likely to be impaired in atherosclerosis. Defective clearance of apoptotic cells has been described in atherosclerosis [18]. In a recent study, we have shown that expression of milk fat globule-EGF factor 8 (Mfge8, also known as lactadherin) in macrophages is required for an efficient clearance of apoptotic cell corpses in atherosclerosis and maintains a protective immune response, leading to a limitation of plaque development in LDLR$^{-/-}$ mice [23, 24].

A continuous efficient phagocytosis of apoptotic debris is critical to the maintenance of an anti-inflammatory milieu, counterregulating the proinflammatory response and limiting lesion progression. Pro- and anti-inflammatory cytokines significantly affect the expression of matrix-degrading metalloproteinases (MMPs) and their inhibitors, tissue inhibitors of MMPs (TIMPs), acting synergistically with other cytokines, growth factors, or oxidized lipids to induce substantial remodeling of many components of the extracellular matrix (see [25] for review). For example, IFN-γ inhibits collagen synthesis [26], while IL-1, IL-4, and TNF-α induce a broad range of MMPs in vascular cells, including MMPs-1, -3, -8, and -9. Of note, the Th2-type cytokine IL-4 induces the elastolytic MMP-12 [27]. Diminished synthesis of extracellular matrix proteins together with increased protease activity contribute to the thinning of the fibrous cap, which is a hallmark of vulnerable plaques [28].

Finally, the antithrombotic properties of ECs are deeply affected by cytokines [29]. IL-1 and TNF-α can increase the tissue procoagulant activity and suppress the anticoagulant activity mediated by the thrombomodulin–protein C system, by decreasing gene transcription of thrombomodulin and protein C receptor. Down-regulation of anticoagulant mediators may in turn affect inflammation. Thrombomodulin has direct anti-inflammatory activities on the endothelium, inhibiting MAPK and the NF-κB pathway [30], and activated protein C has been shown to inhibit NF-κB in monocytes. In contrast, proinflammatory cytokines modify the fibrinolytic properties of endothelial cells. They decrease the production of tissue plasminogen activator (tPA) and increase the production of an inhibitor of the tissue plasminogen activator (PAI-1) [31].

4.4
Cytokine Expression in Human Atherosclerotic Plaques

The first evidence that cytokines are expressed in the atherosclerotic plaque stems from the observation in the mid 1980s by Hansson and coworkers that most of the cells present in the plaque express the MHC class II antigen HLA-DR [32], indicating that IFN-γ must be produced in the vicinity of these cells, which was demonstrated later by the same group [33, 34]. By the late 1980s, immunohisto-chemistry, *in situ* hybridization, and RT-PCR techniques were used to identify a number of cytokines in human atherosclerotic plaques from carotid endarterec-tomy specimens, including TNF-α, IL-1, MCP-1 (CCL2), and M-CSF (reviewed in [2]). Hansson and colleagues determined the expression profiling of Th1 and Th2 cytokines in advanced human atherosclerotic plaques [34]. They found that IL-2 was present in 50% of plaques and IFN-γ was detected in some but not all of the IL-2-positive plaques. By contrast, the expression of IL-4 and IL-5 (Th2 cytokines), and TNF-β (lymphotoxin-α), expressed by both Th1 and Th2 cells, was very rare. In addition, IL-10 is produced in atherosclerotic lesions [35, 36] and correlates with diminished expression of inflammatory mediators, while TGF-β is expressed abundantly in plaques [37]. Among the isoforms, TGF-β2 was detected at high frequency and exhibited stronger intensity of staining than TGF-β1 or TGF-β3. The distribution of TGF-β overlapped with that of its transport protein, LTBP, suggesting that TGF-β is actively secreted.

A group of non-collagenous matrix proteins originally identified as important in bone mineralization, including osteopontin (OPN), osteoprotegerin (OPG), and receptor activator of NF-κB ligand (RANKL), have also been detected in macro-phages, ECs, and SMCs in plaques [38–41]. They have pleiotropic effects that influence matrix turnover, cell migration, and inflammation [42, 43]. OPN and OPG expression is greater in symptomatic than in asymptomatic carotid athero-sclerotic plaques, whereas RANKL expression is similar [41]. A better understand-ing of the time-course of cytokine gene expression is important for successful prevention of plaque development and progression. However, this information cannot be easily obtained in humans. In ApoE$^{-/-}$ mice fed a cholesterol-rich diet for four weeks, the expression of the proinflammatory cytokines was much more pronounced than anti-inflammatory cytokines [44]. This imbalance between pro- and anti-inflammatory cytokines might account for the progression of atherosclerosis.

4.5
Effect of Cytokines on Experimental Atherosclerosis

4.5.1
TNF-α and Structurally Related Cytokines

TNF-α was first identified as the factor in serum isolated from endotoxin-treated mice that induced necrosis of a methylcholanthrene-induced murine sarcoma. However, it soon became apparent that TNF-α is the prototype of the proinflammatory cytokines leading to activation of multiple arms of the innate immune system with potent proinflammatory effects on vascular endothelium. Experimental studies using TNF-deficient ApoE$^{-/-}$ mice showed that TNF-α played a key role in atherosclerosis, since atherosclerotic lesion size in the aortic sinus of TNF-α$^{-/-}$/ApoE$^{-/-}$ mice was significantly smaller than that of ApoE$^{-/-}$ mice, and was associated with decreased expression of ICAM-1, VCAM-1, and MCP-1 [45]. Other members of the TNF superfamily may also be involved in the development of atherosclerosis. TNF-related apoptosis-inducing ligand (TRAIL) has been detected in atherosclerotic plaques [46]. The net effect of interactions between TRAIL and its receptors on atherosclerosis, however, is still unknown since they could have multiple consequences on the disease process, which are conflicting [47]. TRAIL can induce both apoptosis or proliferation and migration of ECs and SMCs. It predominantly functions, however, to regulate an immune response or to induce macrophage apoptosis, which should promote atherosclerosis [24].

4.5.2
IL-1 Cytokine Family

Because of their powerful proinflammatory potential and the observation that they are highly expressed by almost all cell types actively implicated in atherosclerosis, members of the IL-1 cytokine family were the first to be investigated in the field of vessel wall inflammation [48]. The IL-1 family comprises five proteins that share considerable sequence homology: IL-1α, IL-1β, IL-1 receptor antagonist (IL-1ra), IL-18, and the newly discovered ligand of the ST2L receptor, IL-33. Like IL-1β and IL-18, IL-33 was found to have strong immunomodulatory functions [49]. However, unlike these two cytokines, which mainly promote Th1-associated responses, IL-33 predominantly induces the production of Th2 cytokines (IL-5 and IL-13) and increases levels of serum immunoglobulins. The IL-33 receptor, ST2L is preferentially expressed on Th2 cells, but not Th1 cells, and can profoundly suppress innate and adaptive immunity [50].

IL-1 Expression of IL-1 family members and their receptors has been demonstrated in atherosclerotic plaques. *In vivo* studies in murine models of atherosclerosis have confirmed the pro-atherogenic properties of IL-1α, IL-1β, and IL-18 associated with upregulation of endothelial adhesion molecules, activation of macrophages, and Th1 immune response [51]. In contrast with this, IL-1ra, a natural antagonist of IL-1, possesses anti-inflammatory properties, mainly through the endogenous inhibition of IL-1 signaling. Administration of recombinant human IL-1ra to ApoE$^{-/-}$ mice [52] or IL-1ra overexpression in LDLR$^{-/-}$ mice [53] or ApoE$^{-/-}$ mice [54] markedly decreased the size of atherosclerotic lesions. In contrast, IL-1ra knockout C57BL/6J mice fed a cholesterol/cholate diet had a threefold decrease in non-HDL-cholesterol and a trend toward increased foam-cell lesion area compared with wild-type littermate controls [53].

IL-18 IL-18 administration increased lesion size in ApoE$^{-/-}$ mice [55], and overexpression of its endogenous inhibitor IL-18 binding protein (IL-18BP) reduced atherosclerosis with profound changes in plaque composition leading to a more stable plaque phenotype [56]. Furthermore, IL-18-deficient ApoE$^{-/-}$ mice reproduced findings observed in ApoE$^{-/-}$ mice in which IL-18 signaling was blocked by overexpression of IL-18BP, with smaller and more stable lesions compared with ApoE$^{-/-}$ mice [57]. It has been suggested that the pro-atherogenic effect of IL-18 is in fact mediated by IFN-γ because the promotion of atherosclerosis by exogenous IL-18 administration was ablated in IFN-γ-deficient ApoE$^{-/-}$ mice [55]. However, the pro-atherogenic effect of IL-18 can occur in the absence of T cells [58]. Intraperitoneal injection of IL-18 in SCID/ApoE$^{-/-}$ mice led to larger lesions and increased circulating IFN-γ compared with mice injected with saline solution. NK cells were the most likely source of IFN-γ.

IL-33 IL-33 was originally described as a modulator of inflammation, tipping the balance towards CD4$^+$ Th2-mediated immune responses. IL-33 may serve as a chemotactic factor for Th2 cells [52] and induce the production of the Th2-associated interleukins IL-4, IL-5, and IL13 [59]. In contrast with IL-1 or IL-18, IL-33 has been shown to reduce atherosclerosis development in ApoE$^{-/-}$ mice on a high-fat diet [60]. IL-33 treatment increased levels of IL-4, IL5, and IL-13, but decreased levels of IFN-γ in serum and lymph node cells. It also enhanced serum levels of IgA, IgE, and IgG1, but decreased IgG2a, indicating a Th1-to-Th2 switch. Conversely, mice treated with soluble ST2, a decoy receptor that neutralizes IL-33, developed significantly larger atherosclerotic plaques [60]. Furthermore, blockade of IL-5 activities with an anti-IL-5 antibody prevented the IL-33-induced reduction in plaque size and reduced the amount of ox-LDL antibodies induced by IL-33. The authors suggested that IL-33 may be atheroprotective via the induction of IL-5 and ox-LDL antibodies.

In agreement with these findings in experimental atherosclerosis, in patients with acute myocardial infarction it has been found that serum ST2 levels were elevated one day post event and declined thereafter. ST2 levels correlated with serum creatine kinase, a standard marker of myocardial injury, and inversely correlated with left ventricular function. Moreover ST2 levels predicted mortality and clinical outcome in these patients [61, 62]. Interestingly, IL-33/ST2L signaling has been reported recently to be mechanically activated in myocardial overload with paracrine cardioprotective effects [63].

4.5.3
Inflammasome

Caspases are a family of cysteine proteases that fulfill a critical role in the execution of apoptosis. Moreover, a subfamily of caspases known as inflammatory caspases are involved in innate immunity, caspase-1 being the prototypic member. Other members include human caspase-4 and -5, and mouse caspase-11 and -12, all of which contain an N-terminal caspase recruitment domain (CARD). Activation of the inflammatory caspases requires the assembly of a unique intracellular complex, designated the inflammasome, that proceeds to cleave and activate IL-1β, IL-18, and IL-33 [64]. Of note, caspase-1$^{-/-}$ mice have defects in the production of IL-1β and IL-18 but only subtle defects in apoptotic pathways [65, 66]. Evidence is accumulating that members of the CATERPILLER (CARD, transcription enhancer, R (purine)-binding, pyrin, lots of leucine repeats) gene family, and, in particular, of the NALP subfamily, are important players in this signaling process. When NALP-1 is activated by factors that are as yet unknown, it interacts with an adapter protein ASC (apoptosis-associated speck-like protein containing a CARD) through homologous pyrin domain (PYD) to induce the assembly of a complex composed of NALP-1, ASC, caspase-1, and caspase-5. This brings the caspases in close proximity to each other, thereby inducing their activation [67]. Upon activation of caspase-1, the 31 kDa IL-1β precursor and the active caspase-1 colocalize to the inner surface of the cell membrane and caspase-1 cleaves the precursor [68]. The active 17 kDa IL-1β is then released into the extracellular compartment. Caspase-1-dependent processing of the 24 kDa IL-18 precursor is believed to be similar to that of IL-1β [69].

Interestingly, a single amino acid mutation in the NALP-3 gene has been reported in humans with Muckle–Wells syndrome, a rare autosomal dominant disease characterized by recurrent fevers, neutrophilia, elevated acute phase proteins, and arthritis [70]. This mutation in NALP-3 results in a high state of activation of caspase-1 in LPS-stimulated monocytes and increased release of IL-1β compared with cells from patients without the mutation. Inflammasome-related proteins might represent novel pharmacological targets to prevent exaggerated production of IL-1 and/or IL-18, and thereby combat inflammatory diseases, including atherosclerosis.

4.5.4
IL-6 Family

IL-6 signaling involves both a specific IL-6 receptor and a ubiquitous signal-transducing protein, gp130, that is also utilized by other members of the IL-6 family, but not by IL-31. IL-6 has been shown to enhance fatty lesion development in mice [71]. IL-6 treatment of C57Bl/6 mice at supraphysiological concentrations increased fatty streak size by approximately fivefold, while treatment of ApoE$^{-/-}$

mice on low- or high-fat diets increased plaque size by approximately twofold [71], suggesting that IL-6 is a pro-atherogenic cytokine. However, one-year-old IL-6$^{-/-}$/ApoE$^{-/-}$ mice showed enhanced plaque formation [72, 73]. Serum cholesterol levels were increased in one study [73], but not in the other [72]. Increased atherosclerosis in IL-6$^{-/-}$/ApoE$^{-/-}$ mice was associated with reduced collagen content in the plaques, blunted synthesis and release of IL-10, and diminished recruitment of inflammatory cells into the atherosclerotic plaque [73] and at one year of age, mice showed more calcified lesions [72]. In younger, 16-week-old IL-6$^{-/-}$/apo-E$^{-/-}$ mice no significant difference in fatty streaks were detected compared with IL-6$^{+/+}$/apo-E$^{-/-}$ mice. Therefore, the role of IL-6 in atherosclerosis appears ambivalent.

Similarly, IL-6 can be viewed as a proinflammatory cytokine, but may also be regarded as an anti-inflammatory cytokine as it induces the synthesis of IL-1ra and release of soluble TNFR, leading to reduced activity of proinflammatory cytokines [74–76]. It also inhibits macrophage SR-A [77]. Like IFN-α, IL-6 activates both STAT1 and STAT3. The relative balance of STAT1 and STAT3 activation determines its proinflammatory versus anti-inflammatory functional profile.

gp130 In a recent report it has been shown that hepatocyte-specific gp130-deficient ApoE$^{-/-}$ mice exhibit less atherosclerosis with decreased macrophage infiltration [78]. In line with these experimental findings, genetic variation within the human gp130 homolog IL-6 signal transducer (IL6ST) was significantly associated with coronary artery disease.

4.5.5
IL-12 Family

IL-12 p35/p40 is a heterodimeric cytokine that plays an important role in Th1 differentiation. Recent findings have shown that both p35 and p40 can form other cytokines with different proteins (IL-23: p19/p40; IL-35: p35/EBI3). Similar to IL-12 and IL-23, IL-27 is a heterodimeric cytokine consisting of EBI3 (an IL-12 p40 homolog originally described to be secreted by Epstein–Barr virus-transformed B cells) and p28, an IL-6 and p35 homolog [79] (Figure 4.4). IL-27 employs a unique receptor subunit, IL-27ra (also known as WSX-1 or TCCR), paired with gp130 for signaling.

IL-12 IL-12 production by dendritic cells and monocytes/macrophages plays a critical role in Th1 differentiation. IL-12 activates the transcription factor STAT4 and a unique Th1 transcription factor, T-box expressed in T cells (T-bet), leading to upregulation of IFN-γ and downregulation of IL-4 and IL-5 expression in T cells. T-bet deficiency clearly reduces lesion development [80]. IL-12 is pro-atherogenic and appears to intervene in the atherosclerotic process during the early phase of the disease in ApoE$^{-/-}$ mice [81]. Thirty-week-old IL-12$^{-/-}$/ApoE$^{-/-}$ mice showed increased lesions, while 40-week-old mice had lesions of equivalent size compared with wild-type ApoE$^{-/-}$ mice [81]. Also, a selective defect of IL-12 synthesis by macrophages due to 12/15-lipoxygenase deficiency reduced plaque formation in

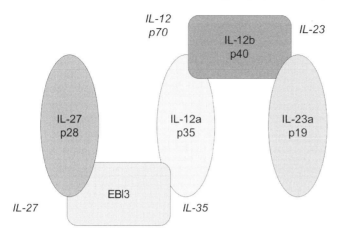

Figure 4.4 The IL-12 family. IL-12 is a heterodimeric cytokine composed of p35 (IL-12a) and p40 (IL-12b) subunits. p40 and p35 form other cytokines with different proteins: IL-23 composed of p40 and p19 (IL-23a), and IL-35 composed of p35 and EBI3 (Epstein–Barr virus–transformed B cells). IL-27 is a heterodimeric cytokine consisting of EBI3 and p28.

Apobec1$^{-/-}$/LDLR$^{-/-}$ mice [82], and injection of IL-12 in ApoE$^{-/-}$ mice promoted lesion development [83].

IL-23 and Th17-Related Cytokines IL-12-deficient mice lack the p40 subunit that is common to IL-23. Therefore, the findings obtained in IL-12$^{-/-}$ mice might also be attributed to the deficiency in IL-23. Further studies are required to elucidate the specific contribution of IL-12 and IL-23 in atherosclerosis. IL-23 has been recently recognized as a key mediator in the differentiation of the newly discovered effector Th lymphocyte subset producing IL-17, Th17 cells (reviewed in [84]). These cells produce IL-17 (also referred to as IL-17A), IL-17F, IL-21, and IL-22. In mice, differentiation of Th17 cells requires the expression of the transcription factors RORγt, STAT3, and IRF4. Remarkably, mice deficient for the p19 subunit of IL-23 are resistant to induction of experimental autoimmune encephalomyelitis due to the lack of Th17 cells [85, 86]. IL-23 and Th17 cells are also required for induction of other autoimmune diseases in mice.

Recently, our group identified novel SOCS3-controlled IL-17 regulatory pathways in atherosclerosis [85] that may have important implication for the understanding of the increased susceptibility to vascular inflammation in patients with dominant-negative STAT3 mutations and defective Th17 cell differentiation [123]. Strikingly, polymorphisms in the IL23R gene are associated with Crohn's disease [87]. However, unlike other chronic inflammatory diseases, myocardial infarction is not associated with IL-23R polymorphisms [88].

IL-27 Nothing is known regarding the role of IL-27 in atherosclerosis. Yet, given its major effects in the regulation of innate and adaptive immunity, it is most likely involved in the control of the disease. Initial studies of IL-27 function revealed a role in promoting early stages of activation of naïve CD4$^+$ cells and differentiation to the Th1 lineage [89, 90]. It was proposed that IL-27 amplifies early responses to weak infectious stimuli to achieve a robust response. More recent work has highlighted a regulatory role for IL-27 in suppressing adaptive immunity. IL-27 suppresses the development of Th1, Th2, and Th17 T cell subsets, and mice deficient in the IL-27 receptor exhibit enhanced inflammation and related toxicity in several infectious disease models [91]. Strikingly, IL-27 has recently been shown to induce STAT3-mediated IL-10 production in activated T cells [92, 93] and to suppress autoimmune inflammation of the central nervous system [94], thus identifying IL-10 as the effector mechanism for the suppressive effects of IL-27.

The role of IL-27 in the regulation of innate immunity seems to be quite different from that of adaptive immunity. IL-27 can induce proinflammatory cytokines in human monocytes and mast cells [95]. Follow-up studies in murine systems have suggested a moderately suppressive function of IL-27 on activated myeloid cells. IL-27 inhibited the function of murine mast cells [96], partially suppressed the production of proinflammatory cytokines by activated murine macrophages [97], and attenuated the antigen-presenting capacity and Th1-promoting function of activated murine dendritic cells *in vitro* [98]. More recent studies indicate that resting murine macrophages are minimally responsive to IL-27, but human monocytes are strongly activated by IL-27 in a STAT1-dependent manner [99]. IL-27 primes human monocytes to enhance inflammatory responses to TLR ligation, strongly suppresses TLR-induced IL-10 production, and alters IL-10 function by increasing IL-10-induced STAT1 activation and attenuating IL-10-induced gene expression.

4.5.6
IL-19/IL-20

IL-19 and IL-20 are two cytokines that were discovered in 2000 and 2001, respectively. Based on the structure and location of their genes, their primary and secondary protein structures and the used receptor complexes, they were classified in the IL-10 family of cytokines. IL-19 and IL-20 are produced by monocytes as well as non-immune tissue cells under inflammatory conditions. They act via a receptor complex that consists of the IL-20R1 and IL-20R2 chains. IL-20 is also able to signal via a second receptor complex (IL-22R1/IL-20R2). Results from animal experiments and massively increased expression of these mediators in human inflamed tissues support the assumption that they play an important role in the pathogenesis of a few inflammatory diseases.

IL-19, IL-20, and IL-24 do not induce activation of signal transducer and activator of transcription (STAT) molecules in immune cells. Instead, several tissues, particularly the skin, tissues from the reproductive and respiratory systems, and various glands appear to be the main targets of these mediators. IL-20 and its receptors, IL-20R1/IL-20R2, are expressed in human and experimental atherosclerotic plaques [100]. Moreover, systemic delivery of IL-20 accelerates atherosclerosis in ApoE$^{-/-}$ mice [100].

4.5.7
IFN-γ and Th1-Related Cytokines

A large body of evidence indicates that the adaptive immune system is involved in the development of atherosclerosis. Deficiency in both T and B cells significantly inhibits lesion development (reviewed in [2, 101]). The protective effect is seen in the early stages of lesion development and in the absence of severe hypercholesterolemia, but varies according to the site of lesion development, being more important in the aortic root and less visible in the thoracic and abdominal aorta or in the brachiocephalic trunk. The majority of pathogenic T cells in atherosclerosis are of the Th1 profile, producing high levels of IFN-γ. Th1-driven responses are detrimental to the atherosclerotic process. IFN-γ activates monocytes/macrophages and dendritic cells, leading to the perpetuation of the pathogenic Th1 response. In addition, IFN-γ acts on vascular cells, inhibiting SMC proliferation and reducing their collagen production while upregulating the expression of MMPs, thereby contributing to the thinning of the fibrous cap.

Deficiency in IFN-γR or IFN-γ significantly reduces lesion development and enhances plaque collagen content [102, 103], whereas exogenous administration of IFN-γ enhances lesion development [103]. Surprisingly, one study reported reduced lesion formation in LDLR$^{-/-}$ mice transplanted with IFN-γ-deficient bone marrow [104]. The reasons for this discrepancy remain to be elucidated.

4.5.8
IL-4/IL-5 and Th2-Related Cytokines

Th2 cells secrete IL-4, IL-5, IL-10, and IL-13 and provide help for antibody production by B cells. Th2 cells are rarely detected within atherosclerotic lesions. However, their induction is promoted in a severe hyperlipidemic context. IL-4 drives Th2 cell differentiation through STAT6, which activates the transcription factor Gata3, leading to upregulation of IL-4 and IL-5 and downregulation of IFN-γ. Thus, Th2-biased responses were proposed to antagonize pro-atherogenic Th1 effects and thereby confer atheroprotection. However, the role of the Th2 pathway in the development of atherosclerosis remains controversial depending on the stage and/or site of the lesion, as well as on the experimental model. In mouse models that are relatively resistant to atherosclerosis, a Th2 bias has been shown to protect against early fatty streak development [105]. However, in more permissive models using LDLR$^{-/-}$ mice, deficiency in IL-4, the prototypic Th2-related cytokine, had no substantial effect on lesion development in one study [106], but was associated with a decrease in atherosclerotic lesion formation in a previous work by the same authors [107], suggesting a potentially pro-atherogenic role.

Prolonged hypercholesterolemia in animal models of atherosclerosis is associated with enhanced IL-4 production, which most probably contributes to plaque progression, since deficiency in IL-4 at these advanced stages greatly hampers plaque progression [81]. However, other Th2-related cytokines, IL-5 and IL-33, appear to exhibit anti-atherogenic properties. Induction of humoral immunity by immunization of hypercholesterolemic ApoE$^{-/-}$ mice with ox-LDL reduced lesion size in association with the production of high levels of IgM-type anti-ox-LDL antibodies, probably from B1 cells (reviewed in [80]). These cells appeared to be stimulated by IL-5 produced by MDA-LDL-specific Th2 cells, since antibody generation and atheroprotection were significantly reduced in mice with genetic deletion of IL-5 in bone marrow cells. More recently, IL-33 has been shown to exhibit atheroprotective effects (see above).

4.5.9
TGFβ/IL-10 and T$_{reg}$-Related Cytokines

Natural T$_{reg}$ cells (nT$_{reg}$) develop in the thymus and recognize specific self-antigen. They are characterized by the expression of CD4, high levels of CD25, and the transcriptional factor Foxp3 (reviewed in [108]). They home to peripheral tissues to maintain self-tolerance and prevent autoimmunity by inhibiting pathogenic lymphocytes. Costimulation, particularly through CD28–CD80/CD86 pathway, is required for their maintenance. Subsets of T$_{reg}$ cells, induced T$_{reg}$ cells (iT$_{reg}$), are also generated in the periphery during active immune responses. Naïve CD4$^+$CD25$^-$ in the periphery can be converted, in the presence of TGF-β, IL-10, or low dose of antigenic peptide into CD4$^+$CD25$^+$FOXP3$^+$ cells (see below). The iT$_{reg}$ cells induced by IL-10 are called Tr1 cells whereas iT$_{reg}$ cells induced by TGF-β are called Th3. These cells mediate suppressor function through the production of IL-10 and TGF-β, respectively (see below).

TGF-β TGF-β inhibits the proliferation, activation, and differentiation of T cells towards Th1 and Th2. In addition, TGF-β1 has been shown to maintain T$_{reg}$ cells in the periphery by acting as a costimulatory factor for expression of Foxp3. Dendritic cells have the capacity to induce T$_{reg}$ cell formation depending on TGF-β. Previous studies have shown that TGF-β has anti-inflammatory and atheroprotective effects. Systemic TGF-β neutralization [109, 110] or genetic deficiency in TGF-β [111] increased lesion development in ApoE$^{-/-}$ mice. Accelerated atherosclerosis was associated with increased infiltration of inflammatory macrophages and T cells within lesions, together with reduced collagen content. Interestingly, specific deletion of TGF-β signaling in T cells was sufficient to induce increased atherosclerosis [112, 113] associated with increased differentiation of T cells towards both Th1 and Th2 phenotypes [112], suggesting T$_{reg}$ cell dysregulation. Indeed, LDLR$^{-/-}$ mice transplanted with bone marrow-derived cells from CD28 or CD80/CD86-deficient mice, known to display a marked reduction in peripheral T$_{reg}$ cell pool, showed accelerated atherosclerosis and enhanced lesional inflammation [114]. Similarly, transfer of CD28-deficient splenocytes to immune-deficient ApoE$^{-/-}$/Rag2$^{-/-}$ mice led to a marked acceleration of atherosclerosis compared with the transfer of wild-type splenocytes.

Similar studies using ICOS-deficient mice clearly showed acceleration of atherosclerosis in association with a reduction in T$_{reg}$ cell number and function [115]. Furthermore, strategies using CD25-neutralizing antibodies in young ApoE$^{-/-}$

mice clearly demonstrated a protective role of T_{reg} cells against atherosclerosis [114]. T_{reg} depletion did not influence lesion size or inflammatory phenotype in mice with specific deletion of TGF-β signaling in T cells, highlighting the role of TGF-β in the atheroprotective effect of T_{reg} cells.

IL-10   IL-10 plays an important role in the control of both innate and adaptive immunity. In lymphocytes, the production of IL-10 has been associated with Th2 subset, T_{reg} cells, and, more recently, with some IFN-γ-producing Th1 cells. Among the regulatory T cells, both nT_{reg} and induced Tr1 cells have the capacity to produce IL-10. IL-10 deficiency promotes atherosclerotic lesion formation, characterized by increased infiltration of inflammatory cells, particularly activated T cells, and by increased production of proinflammatory cytokines [116–118]. Leukocyte-derived IL-10 appears to be instrumental in the prevention of atherosclerotic lesion development and in the modulation of cellular and collagen plaque composition [119].

Consistent with a protective role of IL-10 in atherosclerosis, systemic or local overexpression of IL-10 by adenoviral gene transfer in collar-induced carotid atherosclerosis of LDLR$^{-/-}$ mice was found to be highly efficient in preventing atherosclerosis [120]. Of note, overexpression of IL-10 by activated T lymphocytes reduced atherosclerosis in LDLR$^{-/-}$ mice [121], which, we believe, suggests a protective effect for Tr1-like cells in atherosclerosis. This is consistent with a study from our group showing that transfer of clones of Tr1 cells reduced lesion development and that promotion of endogenous adaptive Tr1 cell response plays a significant role in limiting disease development [122].

4.6
Conclusion

Inflammation plays a major role at all stages of the atherosclerotic process from the early events whereby leukocytes are recruited at sites of subendothelial LDL-cholesterol accumulation to the late events, when plaque rupture occurs, leading to thrombus formation and adverse vascular outcomes. The chronic inflammatory disease of the arterial wall is promoted by both innate and adaptive Th1-driven immunity. Many of immune cells present in the plaque show signs of activation and a whole plethora of cytokines can be found in human or murine atherosclerotic plaques. Murine experimental models of atherosclerosis provide clear evidence that blockade of proinflammatory cytokines; including TNF, IFN-γ, IL-1, IL-12, or IL-18, results in the limitation of plaque development and progression. However, all these cytokines are central for successful host defense against microbial pathogens. Therefore, although therapeutic targeting of these cytokines may be envisioned for a short period of time following acute coronary events, this cannot be a long-term therapeutic strategy to combat atherosclerosis.

As more is discovered about the complex role of adaptive immunity, especially the atheroprotective effects of T_{reg} cells mediated by IL-10 and/or TGF-β, more subtle therapeutic approaches adapted to specific long-term treatment aimed at limiting lesion development and atherosclerosis-related inflammation will be developed. Vaccination-like strategies able to promote antigen-specific T_{reg} cell response should allow targeting of anti-inflammatory/anti-atherogenic cytokine delivery in the plaque, resulting in reduced potential side-effects.

References

1 Hansson, G.K. (2005) Inflammation, atherosclerosis, and coronary artery disease. *N. Engl. J. Med.*, **352** (16), 1685–1695.

2 Tedgui, A. and Mallat, Z. (2006) Cytokines in atherosclerosis: pathogenic and regulatory pathways. *Physiol. Rev.*, **86** (2), 515–581.

3 Boulay, J.L., O'Shea, J.J., and Paul, W.E. (2003) Molecular phylogeny within type I cytokines and their cognate receptors. *Immunity*, **19** (2), 159–163.

4 Vilcek, J. and Feldmann, M. (2004) Historical review: cytokines as therapeutics and targets of therapeutics. *Trends Pharmacol. Sci.*, **25** (4), 201–209.

5 Langer, J.A., Cutrone, E.C., and Kotenko, S. (2004) The Class II cytokine receptor (CRF2) family: overview and patterns of receptor-ligand interactions. *Cytokine Growth Factor Rev.*, **15** (1), 33–48.

6 Nathan, C. (2002) Points of control in inflammation. *Nature*, **420** (6917), 846–852.

7 Tobias, P.S. and Curtiss, L.K. (2007) Toll-like receptors in atherosclerosis. *Biochem. Soc. Trans.*, **35** (Pt 6), 1453–1455.

8 Wojciak-Stothard, B., Entwistle, A., Garg, R., and Ridley, A.J. (1998) Regulation of TNF-alpha-induced reorganization of the actin cytoskeleton and cell-cell junctions by Rho, Rac, and Cdc42 in human endothelial cells. *J. Cell. Physiol.*, **176** (1), 150–165.

9 Zernecke, A., Shagdarsuren, E., and Weber, C. (2008) Chemokines in atherosclerosis: an update. *Arterioscler. Thromb. Vasc. Biol.*, **28** (11), 1897–1908.

10 Kume, N. and Kita, T. (2001) Roles of lectin-like oxidized LDL receptor-1 and its soluble forms in atherogenesis. *Curr. Opin. Lipidol.*, **12** (4), 419–423.

11 Wuttge, D.M., Zhou, X., Sheikine, Y., Wagsater, D., Stemme, V., Hedin, U., et al. (2004) CXCL16/SRPSOX is an interferon-gamma-regulated chemokine and scavenger receptor expressed in atherosclerotic lesions. *Arterioscler. Thromb. Vasc. Biol.*, **24** (4), 750–755.

12 Minami, M., Kume, N., Shimaoka, T., Kataoka, H., Hayashida, K., Akiyama, Y., et al. (2001) Expression of SR-PSOX, a novel cell-surface scavenger receptor for phosphatidylserine and oxidized LDL in human atherosclerotic lesions. *Arterioscler. Thromb. Vasc. Biol.*, **21** (11), 1796–1800.

13 Brand, K., Mackman, N., and Curtiss, L.K. (1993) Interferon-gamma inhibits macrophage apolipoprotein E production by posttranslational mechanisms. *J. Clin. Invest.*, **91** (5), 2031–2039.

14 Panousis, C.G., Evans, G., and TG, Zuckerman, S.H. (2001) F-beta increases cholesterol efflux and ABC-1 expression in macrophage-derived foam cells: opposing the effects of IFN-gamma. *J. Lipid Res.*, **42** (5), 856–863.

15 Maziere, C., Auclair, M., and Maziere, J.C. (1994) Tumor necrosis factor enhances low density lipoprotein oxidative modification by monocytes and endothelial cells. *FEBS Lett.*, **338** (1), 43–46.

16 Folcik, V.A., Aamir, R., and Cathcart, M.K. (1997) Cytokine modulation of LDL oxidation by activated human monocytes. *Arterioscler. Thromb. Vasc. Biol.*, **17** (10), 1954–1961.

17 Mallat, Z. and Tedgui, A. (2001) Current perspective on the role of apoptosis in atherothrombotic disease. *Circ. Res.*, **88** (10), 998–1003.

18 Tabas, I. (2005) Consequences and therapeutic implications of macrophage apoptosis in atherosclerosis: the importance of lesion stage and phagocytic efficiency. *Arterioscler. Thromb. Vasc. Biol.*, **25** (11), 2255–2264.

19 Geng, Y.J. and Libby, P. (2002) Progression of atheroma: a struggle between death and procreation. *Arterioscler. Thromb. Vasc. Biol.*, **22** (9), 1370–1380.

20 Savill, J., Dransfield, I., Gregory, C., and Haslett, C. (2002) A blast from the past: clearance of apoptotic cells regulates immune responses. *Nat. Rev. Immunol.*, **2** (12), 965–975.

21 Steinman, R.M., Turley, S., Mellman, I., and Inaba, K. (2000) The induction of tolerance by dendritic cells that have captured apoptotic cells. *J. Exp. Med.*, **191** (3), 411–416.

22 Albert, M.L. (2004) Death-defying immunity: do apoptotic cells influence antigen processing and presentation?. *Nat. Rev. Immunol.*, **4** (3), 223–231.

23 Ait-Oufella, H., Kinugawa, K., Zoll, J., Simon, T., Boddaert, J., Heeneman, S., *et al.* (2007) Lactadherin deficiency leads to apoptotic cell accumulation and accelerated atherosclerosis in mice. *Circulation*, **115** (16), 2168–2177.

24 Ait-Oufella, H., Pouresmail, V., Simon, T., Blanc-Brude, O., Kinugawa, K., Merval, R., *et al.* (2008) Defective mer receptor tyrosine kinase signaling in bone marrow cells promotes apoptotic cell accumulation and accelerates atherosclerosis. *Arterioscler. Thromb. Vasc. Biol.*, **28** (8), 1429–1431.

25 Newby, A.C. (2005) Dual role of matrix metalloproteinases (matrixins) in intimal thickening and atherosclerotic plaque rupture. *Physiol. Rev.*, **85** (1), 1–31.

26 Amento, E.P., Ehsani, N., Palmer, H., and Libby, P. (1991) Cytokines and growth factors positively and negatively regulate interstitial collagen gene expression in human vascular smooth muscle cells. *Arterioscler. Thromb.*, **11**, 1223–1230.

27 Shimizu, K., Shichiri, M., Libby, P., Lee, R.T., and Mitchell, R.N. (2004) Th2-predominant inflammation and blockade of IFN-gamma signaling induce aneurysms in allografted aortas. *J. Clin. Invest.*, **114** (2), 300–308.

28 Virmani, R., Burke, A.P., Farb, A., and Kolodgie, F.D. (2006) Pathology of the vulnerable plaque. *J. Am. Coll. Cardiol.*, **47** (8 Suppl.), C13–C18.

29 Esmon, C.T. (2004) The impact of the inflammatory response on coagulation. *Thromb. Res.*, **114** (5–6), 321–327.

30 Conway, E.M., Van de Wouwer, M., Pollefeyt, S., Jurk, K., Van Aken, H., De Vriese, A., *et al.* (2002) The lectin-like domain of thrombomodulin confers protection from neutrophil-mediated tissue damage by suppressing adhesion molecule expression via nuclear factor kappaB and mitogen-activated protein kinase pathways. *J. Exp. Med.*, **196** (5), 565–577.

31 Scarpati, E.M. and Sadler, J.E. (1989) Regulation of endothelial cell coagulant properties. Modulation of tissue factor, plasminogen activator inhibitors, and thrombomodulin by phorbol 12myristate 13-acetate and tumor necrosis factor. *J. Biol. Chem.*, **264** (34), 20705–20713.

32 Jonasson, L., Holm, J., Skalli, O., Gabbiani, G., and Hansson, G.K. (1985) Expression of class II transplantation antigen on vascular smooth muscle cells in human atherosclerosis. *J. Clin. Invest.*, **76** (1), 125–131.

33 Hansson, G.K., Holm, J., and Jonasson, L. (1989) Detection of activated T lymphocytes in the human atherosclerotic plaque. *Am. J. Pathol.*, **135**, 169–175.

34 Frostegard, J., Ulfgren, A.K., Nyberg, P., Hedin, U., Swedenborg, J., Andersson, U., *et al.* (1999) Cytokine expression in advanced human atherosclerotic plaques: dominance of pro-inflammatory (Th1) and macrophage-stimulating cytokines. *Atherosclerosis*, **145** (1), 33–43.

35 Uyemura, K., Demer, L.L., Castle, S.C., Jullien, D., Berliner, J.A., Gately, M.K., *et al.* (1996) Cross-regulatory roles of interleukin (IL)-12 and IL-10 in atherosclerosis. *J. Clin. Invest.*, **97** (9), 2130–2138.

36 Mallat, Z., Heymes, C., Ohan, J., Faggin, E., Lesèche, G., and Tedgui, A. (1999) Expression of interleukin10 in human atherosclerotic plaques. Relation to inducible nitric oxide synthase

expression and cell death. *Arterioscler. Thromb. Vasc. Biol.*, **19**, 611–616.

37 Nikol, S., Isner, J.M., Pickering, J.G., Kearney, M., Leclerc, G., and Weir, L. (1992) Expression of transforming growth factor-beta 1 is increased in human vascular restenosis lesions. *J. Clin. Invest.*, **90** (4), 1582–1592.

38 Giachelli, C.M., Bae, N., Almeida, M., Denhardt, D.T., Alpers, C.E., and Schwartz, S.M. (1993) Osteopontin is elevated during neointima formation in rat arteries and is a novel component of human atherosclerotic plaques. *J. Clin. Invest.*, **92**, 1686–1696.

39 O'Brien, E.R., Garvin, M.R., Stewart, D.K., Hinohara, T., Simpson, J.B., Schwartz, S.M., *et al.* (1994) Osteopontin is synthesized by macrophage, smooth muscle, and endothelial cells in primary and restenotic human coronary atherosclerotic plaques. *Arterioscler. Thromb.*, **14** (10), 1648–1656.

40 Dhore, C.R., Cleutjens, J.P., Lutgens, E., Cleutjens, K.B., Geusens, P.P., Kitslaar, P.J., *et al.* (2001) Differential expression of bone matrix regulatory proteins in human atherosclerotic plaques. *Arterioscler. Thromb. Vasc. Biol.*, **21** (12), 1998–2003.

41 Golledge, J., McCann, M., Mangan, S., Lam, A., and Karan, M. (2004) Osteoprotegerin and osteopontin are expressed at high concentrations within symptomatic carotid atherosclerosis. *Stroke*, **35** (7), 1636–1641.

42 Gravallese, E.M. (2003) Osteopontin: a bridge between bone and the immune system. *J. Clin. Invest.*, **112** (2), 147–149.

43 Schoppet, M., Preissner, K.T., and Hofbauer, L.C. (2002) RANK ligand and osteoprotegerin: paracrine regulators of bone metabolism and vascular function. *Arterioscler. Thromb. Vasc. Biol.*, **22** (4), 549–553.

44 Veillard, N.R., Kwak, B., Pelli, G., Mulhaupt, F., James, R.W., Proudfoot, A.E., *et al.* (2004) Antagonism of RANTES receptors reduces atherosclerotic plaque formation in mice. *Circ. Res.*, **94** (2), 253–261.

45 Ohta, H., Wada, H., Niwa, T., Kirii, H., Iwamoto, N., Fujii, H., *et al.* (2005) Disruption of tumor necrosis factor-

alpha gene diminishes the development of atherosclerosis in ApoE-deficient mice. *Atherosclerosis*, **180** (1), 11–17.

46 Michowitz, Y., Goldstein, E., Roth, A., Afek, A., Abashidze, A., Ben Gal, Y., *et al.* (2005) The involvement of tumor necrosis factor-related apoptosis-inducing ligand (TRAIL) in atherosclerosis. *J. Am. Coll. Cardiol.*, **45** (7), 1018–1024.

47 Kavurma, M.M. and Bennett, M.R. (2008) Expression, regulation and function of trail in atherosclerosis. *Biochem. Pharmacol.*, **75** (7), 1441–1450.

48 Loppnow, H. and Libby, P. (1990) Proliferating or interleukin 1-activated human vascular smooth muscle cells secrete copious interleukin 6. *J. Clin. Invest.*, **85** (3), 731–738.

49 Schmitz, J., Owyang, A., Oldham, E., Song, Y., Murphy, E., McClanahan, T.K., *et al.* (2005) IL-33, an interleukin-1-like cytokine that signals via the IL-1 receptor-related protein ST2 and induces T helper type 2-associated cytokines. *Immunity*, **23** (5), 479–490.

50 Xu, D., Chan, W.L., Leung, B.P., Huang, F., Wheeler, R., Piedrafita, D., *et al.* (1998) Selective expression of a stable cell surface molecule on type 2 but not type 1 helper T cells. *J. Exp. Med.*, **187** (5), 787–794.

51 Kirii, H., Niwa, T., Yamada, Y., Wada, H., Saito, K., Iwakura, Y., *et al.* (2003) Lack of interleukin-1beta decreases the severity of atherosclerosis in ApoE-deficient mice. *Arterioscler. Thromb. Vasc. Biol.*, **23** (4), 656–660.

52 Elhage, R., Maret, A., Pieraggi, M.T., Thiers, J.C., Arnal, J.F., and Bayard, F. (1998) Differential effects of interleukin-1 receptor antagonist and tumor necrosis factor binding protein on fatty-streak formation in apolipoprotein E-deficient mice. *Circulation*, **97** (3), 242–244.

53 Devlin, C.M., Kuriakose, G., Hirsch, E., and Tabas, I. (2002) Genetic alterations of IL-1 receptor antagonist in mice affect plasma cholesterol level and foam cell lesion size. *Proc. Natl Acad. Sci. U. S. A.*, **99** (9), 6280–6285.

54 Merhi-Soussi, F., Kwak, B.R., Magne, D., Chadjichristos, C., Berti, M., Pelli, G., *et al.* (2005) Interleukin-1 plays a major role in vascular inflammation and atherosclerosis in male apolipoprotein E-knockout mice. *Cardiovasc. Res.*, **66** (3), 583–593.

55 Whitman, S.C., Ravisankar, P., and Daugherty, A. (2002) Interleukin-18 enhances atherosclerosis in apolipoprotein E$^{-/-}$ mice through release of interferon-γ. *Circ. Res.*, **90**, e17–e22.

56 Mallat, Z., Corbaz, A., Scoazec, A., Graber, P., Alouani, S., Esposito, B., *et al.* (2001) Interleukin-18/interleukin-18 binding protein signaling modulates atherosclerotic lesion development and stability. *Circ. Res.*, **89** (7), E41–5.

57 Elhage, R., Jawien, J., Rudling, M., Ljunggren, H.G., Takeda, K., Akira, S., *et al.* (2003) Reduced atherosclerosis in interleukin-18 deficient apolipoprotein E-knockout mice. *Cardiovasc. Res.*, **59** (1), 234–240.

58 Tenger, C., Sundborger, A., Jawien, J., and Zhou, X. (2005) IL-18 accelerates atherosclerosis accompanied by elevation of IFN-gamma and CXCL16 expression independently of T cells. *Arterioscler. Thromb. Vasc. Biol.*, **25** (4), 791–796.

59 Kakkar, R. and Lee, R.T. (2008) The IL-33/ST2 pathway: therapeutic target and novel biomarker. *Nat. Rev. Drug. Discov.*, **7** (10), 827–840.

60 Miller, A.M., Xu, D., Asquith, D.L., Denby, L., Li, Y., Sattar, N., *et al.* (2008) IL-33 reduces the development of atherosclerosis. *J. Exp. Med.*, **205** (2), 339–346.

61 Shimpo, M., Morrow, D.A., Weinberg, E.O., Sabatine, M.S., Murphy, S.A., Antman, E.M., *et al.* (2004) Serum levels of the interleukin-1 receptor family member ST2 predict mortality and clinical outcome in acute myocardial infarction. *Circulation*, **109** (18), 2186–2190.

62 Sabatine, M.S., Morrow, D.A., Higgins, L.J., MacGillivray, C., Guo, W., Bode, C., *et al.* (2008) Complementary roles for biomarkers of biomechanical strain ST2 and N-terminal prohormone B-type natriuretic peptide in patients with ST-elevation myocardial infarction. *Circulation*, **117** (15), 1936–1944.

63 Sanada, S., Hakuno, D., Higgins, L.J., Schreiter, E.R., McKenzie, A.N., and Lee, R.T. (2007) IL-33 and ST2 comprise a critical biomechanically induced and cardioprotective signaling system. *J. Clin. Invest.*, **117** (6), 1538–1549.

64 Arend, W.P., Palmer, G., and Gabay, C. (2008) IL-1, IL-18, and IL-33 families of cytokines. *Immunol. Rev.*, **223**, 20–38.

65 Li, P., Allen, H., Banerjee, S., Franklin, S., Herzog, L., Johnston, C., *et al.* (1995) Mice deficient in IL1β-converting enzyme are defective in production of mature IL-1β and resistant to endotoxic shock. *Cell*, **80**, 401–411.

66 Kuida, K., Lippke, J.A., Ku, G., Harding, M.W., Livingston, D.J., Su, M.S.S., *et al.* (1995) Altered cytokine export and apoptosis in mice deficient in interleukin-1b converting enzyme. *Science*, **267**, 2000–2003.

67 Tschopp, J., Martinon, F., and Burns, K. (2003) NALPs: a novel protein family involved in inflammation. *Nat. Rev. Mol. Cell. Biol.*, **4** (2), 95–104.

68 Singer, I.I., Scott, S., Chin, J., Bayne, E.K., Limjuco, G., Weidner, J., *et al.* (1995) The interleukin-1 beta-converting enzyme (ice) is localized on the external cell surface membranes and in the cytoplasmic ground substance of human monocytes by immune-electron microscopy. *J. Exp. Med*, **182**, 1447–1459.

69 Dinarello, C.A., Novick, D., Rubinstein, M., and Lonnemann, G. (2003) Interleukin 18 and interleukin 18 binding protein: possible role in immunosuppression of chronic renal failure. *Blood Purif.*, **21** (3), 258–270.

70 Agostini, L., Martinon, F., Burns, K., McDermott, M.F., Hawkins, P.N., and Tschopp, J. (2004) NALP3 forms an IL-1beta-processing inflammasome with increased activity in Muckle-Wells autoinflammatory disorder. *Immunity*, **20** (3), 319–325.

71 Huber, S.A., Sakkinen, P., Conze, D., Hardin, N., and Tracy, R. (1999) Interleukin-6 exacerbates early atherosclerosis in mice. *Arterioscler. Thromb. Vasc. Biol.*, **19** (10), 2364–2367.

72 Elhage, R., Clamens, S., Besnard, S., et al. (2001) Involvement of interleukin-6 in atherosclerosis but not in the prevention of fatty streak formation by 17beta-estradiol in apolipoprotein E-deficient mice. *Atherosclerosis*, **156** (2), 315–320.

73 Schieffer, B., Selle, T., Hilfiker, A., Hilfiker-Kleiner, D., Grote, K., Tietge, U.J., et al. (2004) Impact of interleukin-6 on plaque development and morphology in experimental atherosclerosis. *Circulation*, **110** (22), 3493–3500.

74 Tilg, H., Trehu, E., Atkins, M.B., Dinarello, C.A., and Mier, J.W. (1994) Interleukin-6 (IL-6) as an antiinflammatory cytokine: induction of circulating IL-1 receptor antagonist and soluble tumor necrosis factor receptor p55. *Blood*, **83**, 113–118.

75 Barton, B.E. (1996) The biological effects of interleukin 6. *Med. Res. Rev.*, **16** (1), 87–109.

76 Xing, Z., Gauldie, J., Cox, G., Baumann, H., Jordana, M., Lei, X.F., et al. (1998) IL-6 is an antiinflammatory cytokine required for controlling local or systemic acute inflammatory responses. *J. Clin. Invest.*, **101** (2), 311–320.

77 Liao, H.S., Matsumoto, A., Itakura, H., et al. (1999) Transcriptional inhibition by interleukin-6 of the class A macrophage scavenger receptor in macrophages derived from human peripheral monocytes and the THP-1 monocytic cell line. *Arterioscler. Thromb. Vasc. Biol.*, **19** (8), 1872–1880.

78 Luchtefeld, M., Schunkert, H., Stoll, M., Selle, T., Lorier, R., Grote, K., et al. (2007) Signal transducer of inflammation gp130 modulates atherosclerosis in mice and man. *J. Exp. Med.*, **204** (8), 1935–1944.

79 Kastelein, R.A., Hunter, C.A., and Cua, D.J. (2007) Discovery and biology of IL-23 and IL-27: related but functionally distinct regulators of inflammation. *Annu. Rev. Immunol.*, **25**, 221–242.

80 Buono, C., Binder, C.J., Stavrakis, G., Witztum, J.L., Glimcher, L.H., and Lichtman, A.H. (2005) T-bet deficiency reduces atherosclerosis and alters plaque antigen-specific immune responses. *Proc. Natl Acad. Sci. U. S. A.*, **102** (5), 1596–1601.

81 Davenport, P. and Tipping, P.G. (2003) The role of interleukin-4 and interleukin-12 in the progression of atherosclerosis in apolipoprotein E-deficient mice. *Am. J. Pathol.*, **163** (3), 1117–1125.

82 Zhao, L., Cuff, C.A., Moss, E., Wille, U., Cyrus, T., Klein, E.A., et al. (2002) Selective interleukin-12 synthesis defect in 12/15-lipoxygenase-deficient macrophages associated with reduced atherosclerosis in a mouse model of familial hypercholesterolemia. *J. Biol. Chem.*, **277** (38), 35350–35356.

83 Lee, T.S., Yen, H.C., Pan, C.C., and Chau, L.Y. (1999) The role of interleukin 12 in the development of atherosclerosis in ApoE-deficient mice. *Arterioscler. Thromb. Vasc. Biol.*, **19** (3), 734–742.

84 Ivanov, I.I., Zhou, L., and Littman, D.R. (2007) Transcriptional regulation of Th17 cell differentiation. *Semin. Immunol.*, **19** (6), 409–417.

85 Tableb, S., Romain, M., Ramkhelawon, B., Uyttenhove, C., Pasterkamp, G., Herbin, O., et al. (2009) Loss of SOCS3 expression in T cells reveals a regulatory role for interleukin-17 in atherosclerosis. *J. Exp. Med.*, **206**, 2067–2077.

86 Langrish, C.L., Chen, Y., Blumenschein, W.M., Mattson, J., Basham, B., Sedgwick, J.D., et al. (2005) IL23 drives a pathogenic T cell population that induces autoimmune inflammation. *J. Exp. Med.*, **201** (2), 233–240.

87 Duerr, R.H., Taylor, K.D., Brant, S.R., Rioux, J.D., Silverberg, M.S., Daly, M.J., et al. (2006) A genome-wide association study identifies IL23R as an inflammatory bowel disease gene. *Science*, **314** (5804), 1461–1463.

88 Mangino, M., Braund, P., Singh, R., et al. (2008) Association analysis of IL-12B and IL-23R polymorphisms in myocardial infarction. *J. Mol. Med.*, **86** (1), 99–103.

89 Chen, Q., Ghilardi, N., Wang, H., Baker, T., Xie, M.H., Gurney, A., et al. (2000) Development of Th1-type immune responses requires the type I cytokine receptor TCCR. *Nature*, **407** (6806), 916–920.

90 Pflanz, S., Timans, J.C., Cheung, J., Rosales R., *et al.* (2002) IL-27, a heterodimeric cytokine composed of EBI3 and p28 protein, induces proliferation of naive CD4+ T cells. *Immunity*, **16** (6), 779–790.

91 Villarino, A., Hibbert, L., Lieberman, L., Wilson, E., Mak, T., Yoshida, H., *et al.* (2003) The IL-27R (WSX-1) is required to suppress T cell hyperactivity during infection. *Immunity*, **19** (5), 645–655.

92 Awasthi, A., Carrier, Y., Peron, J.P., Bettelli, E., Kamanaka, M., Flavell, R.A., *et al.* (2007) A dominant function for interleukin 27 in generating interleukin 10-producing anti-inflammatory T cells. *Nat. Immunol.*, **8** (12), 1380–1389.

93 Stumhofer, J.S., Silver, J.S., Laurence, A., Porrett, P.M., Harris, T.H., Turka, L.A., *et al.* (2007) Interleukins 27 and 6 induce STAT3-mediated T cell production of interleukin 10. *Nat. Immunol.*, **8** (12), 1363–1371.

94 Fitzgerald, D.C., Zhang, G.X., El-Behi, M., Fonseca-Kelly, Z., Li, H., Yu, S., *et al.* (2007) Suppression of autoimmune inflammation of the central nervous system by interleukin 10 secreted by interleukin 27-stimulated T cells. *Nat. Immunol.*, **8** (12), 1372–1379.

95 Pflanz, S., Hibbert, L., Mattson, J., *et al.* (2004) WSX-1 and glycoprotein 130 constitute a signal-transducing receptor for IL-27. *J. Immunol.*, **172** (4), 2225–2231.

96 Artis, D., Villarino, A., Silverman, M., He, W., Thornton, E.M., Mu, S., *et al.* (2004) The IL-27 receptor (WSX-1) is an inhibitor of innate and adaptive elements of type 2 immunity. *J. Immunol.*, **173** (9), 5626–5634.

97 Holscher, C., Holscher, A., Ruckerl, D., Yoshimoto, T., Yoshida, H., Mak, T., *et al.* (2005) The IL-27 receptor chain WSX-1 differentially regulates antibacterial immunity and survival during experimental tuberculosis. *J. Immunol.*, **174** (6), 3534–3544.

98 Wang, S., Miyazaki, Y., Shinozaki, Y., and Yoshida, H. (2007) Augmentation of antigen-presenting and Th1-promoting functions of dendritic cells by

WSX-1(IL-27R) deficiency. *J. Immunol.*, **179** (10), 6421–6428.

99 Kalliolias, G.D. and Ivashkiv, L.B. (2008) IL-27 activates human monocytes via STAT1 and suppresses IL-10 production but the inflammatory functions of IL-27 are abrogated by TLRs and p38. *J. Immunol.*, **180** (9), 6325–6333.

100 Chen, W.Y., Cheng, B.C., Jiang, M.J., Hsieh, M.Y., and Chang, M.S. (2006) IL-20 is expressed in atherosclerosis plaques and promotes atherosclerosis in apolipoprotein E-deficient mice. *Arterioscler. Thromb. Vasc. Biol.*, **26** (9), 2090–2095.

101 Weber, C., Zernecke, A., and Libby, P. (2008) The multifaceted contributions of leukocyte subsets to atherosclerosis: lessons from mouse models. *Nat. Rev. Immunol.*, **8** (10), 802–815.

102 Gupta, S., Pablo, A.M., Jiang, X.C., Wang, N., Tall, A.R., and Schindler, C. (1997) IFN-gamma potentiates atherosclerosis in apoE knock-out mice. *J. Clin. Invest.*, **99** (11), 2752–2761.

103 Whitman, S.C., Ravisankar, P., Elam, H., and Daugherty, A. (2000) Exogenous interferon-gamma enhances atherosclerosis in apolipoprotein E-/-mice. *Am. J. Pathol.*, **157** (6), 1819–1824.

104 Niwa, T., Wada, H., Ohashi, H., Iwamoto, N., Ohta, H., Kirii, H., *et al.* (2004) Interferon-gamma produced by bone marrow-derived cells attenuates atherosclerotic lesion formation in LDLR-deficient mice. *J. Atheroscler. Thromb.*, **11** (2), 79–87.

105 Huber, S.A., Sakkinen, P., David, C., *et al.* (2001) T helper-cell phenotype regulates atherosclerosis in mice under conditions of mild hypercholesterolemia. *Circulation*, **103** (21), 2610–2616.

106 King, V.L., Cassis, L.A., and Daugherty, A. (2007) Interleukin-4 does not influence development of hypercholesterolemia or angiotensin II-induced atherosclerotic lesions in mice. *Am. J. Pathol.*, **171** (6), 2040–2047.

107 King, V.L., Szilvassy, S.J., and Daugherty, A. (2002) Interleukin-4 deficiency decreases atherosclerotic lesion formation in a site-specific

manner in female LDL receptor$^{-/-}$ mice. *Arterioscler. Thromb. Vasc. Biol.*, **22** (3), 456–461.

108 Stephens, G.L. and Shevach, E.M. (2007) Foxp3+ regulatory T cells: selfishness under scrutiny. *Immunity*, **27** (3), 417–419.

109 Mallat, Z., Gojova, A., Marchiol-Fourni-gault, C., Esposito, B., Kamate, C., Merval, R., *et al.* (2001) Inhibition of transforming growth factor-beta signaling accelerates atherosclerosis and induces an unstable plaque phenotype in mice. *Circ. Res.*, **89** (10), 930–934.

110 Lutgens, E. and Daemen, M.J. (2001) Transforming growth factor-beta: a local or systemic mediator of plaque stability?. *Circ. Res.*, **89** (10), 853–855.

111 Grainger, D.J., Mosedale, D.E., Metcalfe, J.C., and Bottinger, E.P. (2000) Dietary fat and reduced levels of TGFb1 act synergistically to promote activation of the vascular endothelium and formation of lipid lesions. *J. Cell Sci.*, **113** (Pt 13), 2355–2361.

112 Gojova, A., Brun, V., Esposito, B., Cottrez, F., Gourdy, P., Ardouin, P., *et al.* (2003) Specific abrogation of transforming growth factor-β signaling in T cells alters atherosclerotic lesion size and composition in mice. *Blood*, **102** (12), 4052–4058.

113 Robertson, A.K., Rudling, M., Zhou, X., Gorelik, L., Flavell, R.A., Hansson, G.K. (2003) Disruption of TGF-beta signaling in T cells accelerates atherosclerosis. *J. Clin. Invest.*, **112** (9), 1342–1350.

114 Ait-Oufella, H., Salomon, B.L., Potteaux, S., Robertson, A.K., Gourdy, P., Zoll, J., *et al.* (2006) Natural regulatory T cells control the development of atherosclerosis in mice. *Nat. Med.*, **12** (2), 178–180.

115 Gotsman, I., Grabie, N., Gupta, R., Dacosta, R., MacConmara, M., Lederer, J., *et al.* (2006) Impaired regulatory T-cell response and enhanced athero-sclerosis in the absence of inducible costimulatory molecule. *Circulation*, **114** (19), 2047–2055.

116 Mallat, Z., Besnard, S., Duriez, M., Deleuze, V., Emmanuel, F., Bureau, M.F., *et al.* (1999) Protective role of interleukin-10 in atherosclerosis. *Circ. Res.*, **85**, e17–e24.

117 Pinderski Oslund, L.J., Hedrick, C.C., Olvera, T., Hagenbaugh, A., Territo, M., Berliner, J.A., *et al.* (1999) Inter-leukin-10 blocks atherosclerotic events *in vitro* and *in vivo. Arterioscler. Thromb. Vasc. Biol.*, **19** (12), 2847–2853.

118 Caligiuri, G., Rudling, M., Ollivier, V., Jacob, M.P., Michel, J.B., Hansson, G.K., *et al.* (2003) Interleukin 10 deficiency increases atherosclerosis, thrombosis, and low-density lipopro-teins in apolipoprotein E knockout mice. *Mol. Med.*, **9** (1–2), 10–17.

119 Potteaux, S., Esposito, B., van Oostrom, O., *et al.* (2004) Leukocyte-derived interleukin 10 is required for protection against atherosclerosis in low-density lipoprotein receptor knockout mice. *Arterioscler. Thromb. Vasc. Biol.*, **24** (8), 1474–1478.

120 Von Der Thusen, J.H., Kuiper, J., Fekkes, M.L., De Vos, P., Van Berkel, T.J., and Biessen, E.A. (2001) Attenua-tion of atherogenesis by systemic and local adenovirus-mediated gene transfer of interleukin-10 in LDLr$^{-/-}$ mice. *FASEB J.*, **15**, 2730–2732.

121 Pinderski, L.J., Fischbein, M.P., Subbanagounder, G., *et al.* (2002) Overexpression of interleukin-10 by activated T lymphocytes inhibits atherosclerosis in LDL receptor-defi-cient mice by altering lymphocyte and macrophage phenotypes. *Circ. Res.*, **90** (10), 1064–1071.

122 Mallat, Z., Gojova, A., Brun, V., Esposito, B., Fournier, N., Cottrez, F., *et al.* (2003) Induction of a regulatory T cell type 1 response reduces the development of atherosclerosis in apolipoprotein E-knockout mice. *Circulation*, **108** (10), 1232–1237.

123 Ling, J.C., Freeman, A.F., Gharib, A.M., *et al.* (2007) Coronary artery aneurysms in patients with hyper IgE recurrent infection syndrome. *Clin. Immunol.*, **122**, 255–258.

5
Toll-Like Receptors in Atherosclerosis

Zhong-qun Yan, Kristina Edfeldt, and Anna M. Lundberg

5.1
Introduction

Inflammation is a major pathologic hallmark and precipitant of atherosclerosis [1], and current paradigms assert that it contributes to the progression of atherosclerosis. At later stages, chronic inflammation causes deterioration of the arterial atherosclerotic plaque, leading to plaque rupture and occlusive thrombosis, which constitute the underlying pathology of myocardial infarction and stroke.

Elucidating the molecular mechanisms that govern the activation and regulation of inflammation in atherosclerosis therefore is of paramount importance [2]. In recent decades, studies have identified modified low-density lipoprotein (LDL), microbial infections, and danger signals that are derived from damaged tissue as elicitors of inflammatory responses [3]. The molecular signatures that mediate the specific inflammatory signals that are induced by these stimuli, however, remain unknown.

A key advance in immunologic research has been the discovery of pattern recognition receptors (PRRs) in the human innate immune system. PRRs recognize structurally conserved moieties from foreign pathogens and endogenously derived inflammatory mediators and induce inflammatory responses [4]. Recently, the interest in the role of innate immunity in atherosclerosis has intensified. As a result, one family of PRRs in particular, the Toll-like receptors (TLRs), has been implicated in pro-atherogenic inflammatory reactions. In this chapter, we discuss the most recent evidence on the role of TLRs and TLR-signaling pathways in the pathogenesis of atherosclerosis.

5.2
Toll-Like Receptors, Molecular Pattern Recognition, and Signal Transduction Pathways

TLRs are the mammalian homologs of *Drosophila melanogaster* receptor Toll, which was shown to be essential for the immunity of *Drosophila* against fungal

Atherosclerosis: Molecular and Cellular Mechanisms. Edited by Sarah Jane George and Jason Johnson
Copyright © 2010 WILEY-VCH Verlag GmbH & Co. KGaA, Weinheim
ISBN: 978-3-527-32448-4

Table 5.1 Human Toll-like receptor and known ligands.

Receptor	Location	Adapter	Ligands	Origin
TLR2/TLR1 TLR2/TLR6	Cell surface	MyD88/MAL	Lipopeptides lipoteichoic acid	Bacteria
			Zymosan	Fungi
			Viral proteins	Virus
TLR3	Intracellular compartment	TRIF	Double-stranded RNA	Virus
			poly I:C	
			mRNA	Host
TLR4	Cell surface	MyD88/MAL/ TRIF/TRAM	Lipopolysaccharide	Bacteria
			Heat shock proteins	Bacteria and host
			Viral proteins	Virus
			Hyaluronic acid fragments	Host
			Fibrinogen	
			Heparan sulfate fragments	
TLR5	Cell surface	MyD88	Flagellin	Bacteria
TLR7	Intracellular compartment	MyD88	Single-stranded RNA	Virus and host
			Loxoribine	Synthetic
			Imidazoquinolines	
TLR8	Intracellular compartment	MyD88	Single-stranded RNA	Virus
TLR9	Intracellular compartment	MyD88	Unmethylated CpG DNA	Bacteria
TLR10	Unknown	MyD88	Unknown	

PolyI:C, polyinosine-polycytidylic acid.

infections [5]. To date, 13 mammalian TLRs have been described, based on their sequence homology to Toll, and 10 TLRs have been identified in humans (see Table 5.1).

TLRs are type I transmembrane receptors that belong to the TLR/IL-1 receptor superfamily, which also includes receptors for the proinflammatory cytokines interleukin-1 (IL-1) and IL-18 [4]. All members of this family have a TLR/IL-1 receptor homology (TIR) domain, a conserved cytoplasmic domain that is essential for signaling [6]. Although the cytoplasmic regions of TLR/IL-1 superfamily members share similarities, their extracellular domains are unrelated. IL-1R family members are characterized by the presence of Ig-like domains, whereas TLRs have extracellular leucine-rich repeats (LRRs), analogous to Toll.

TLRs are expressed on most cell types, including innate immune cells (APCs, mast cells, phagocytes, and natural killer cells), adaptive immune cells (T and B lymphocytes), and non-immune cells (endothelial cells, epithelial cells, fibroblasts, and smooth muscle cells) that contribute to the inflammatory response. All of these cells have different TLR expression profiles, which enables the receptors to induce cell type–specific innate immune responses. TLRs are also differentially distributed within a cell. TLR1, TLR2, TLR4, TLR5, and TLR6 are primarily restricted to the cell surface. In contrast, TLR3, TLR7, TLR8, and TLR9 are expressed in intracellular endosomes, placing them in a functionally distinct subfamily that detects ligands in endosomal and lysosomal compartments following acidification [7–9]. In addition, surface-bound TLR2 and TLR4 can be recruited to the endosomal compartment after ligand recognition [10–12].

TLRs recognize diverse molecular structures called pathogen-associated molecular patterns (PAMPs) [13], ranging from lipids, lipopeptides, to nucleic acids, that are common to many microorganisms [4]. TLR4 was the first characterized receptor of the mammalian TLR receptor family [14] and was shown to transduce signals on recognition of the Gram-negative cell wall component lipopolysaccharide (LPS) [15–17]. Although TLR4 is an essential receptor for LPS, several additional molecules participate in its recognition. LPS is captured by the serum protein LPS-binding protein (LBP), which delivers LPS to CD14 [18, 19]. CD14 is a glycosylphosphatidylinositol (GPI)-anchored cell surface glycoprotein that lacks a transmembrane domain and cytoplasmic regions, rendering it incapable of transducing signals [20]. CD14 also is present in serum as a soluble GPI-tail-less receptor that facilitates LPS-induced signal transduction in cells that lack membrane-bound CD14 [21].

In addition, the secreted molecule MD-2 is another coreceptor that associates with the extracellular portion of TLR4. It interacts specifically with the acylated lipid A core of LPS to enhance LPS responsiveness [22–25]. TLR4 has also been implicated in the recognition of other microbial components, such as bacterial toxins and structural envelope proteins of viruses [26–28].

The most well-characterized receptor in the TLR family is TLR2, which recognizes ligands, such as lipoproteins and lipopeptides. Unlike most of the TLRs, which bind only a few ligands, TLR2 interacts with many pathogenic structures, ranging from bacterial and fungal cell wall components to viral products. Furthermore, most TLRs homodimerize to bind ligands, whereas TLR2 recognizes microbial components in cooperation with other TLRs, namely TLR1 and TLR6. These two receptors are closely related to TLR2 [29]. Through heterodimerization with either TLR1 or TLR6, TLR2 increases its specificity for ligands, allowing it to discriminate between subtle differences in the ligands [30–33]. TLR2 also associates with other PRRs, such as the scavenger receptor CD36, a coreceptor that is involved in binding ligands to the TLR2/TLR6 complex [34, 35].

In contrast to TLR2 and TLR4, the remaining receptors that have been characterized bind only one type of microbial structure. TLR3 recognizes viral double-stranded RNA (dsRNA) [8, 36], TLR5 recognizes bacterial flagellin [37, 38], TLR7 and TLR8 bind single-stranded RNA (ssRNA) [39–41], and TLR9 recognizes hypomethylated CpG motifs in microbial DNA [42–44].

TLRs also mediate the recognition of certain endogenous tissue and cellular components that accumulate at sites of inflammation, are released during necrotic cell death, or are derived from the degradation of extracellular matrix (ECM). These endogenous TLR ligands have been termed damage-associated molecular patterns (DAMPs) and serve as danger signals to the innate immune system, indicating that TLRs are crucial for sensing both infection- and injury-induced tissue damages [45, 46]. The specific molecular patterns of endogenous ligands that are detected by TLRs, however, remain to be defined.

TLR stimulation induces synthesis of proinflammatory mediators by activating complex cytoplasmic signal transduction pathways (Figure 5.1). Ligand-induced dimerization of the extracellular domains is believed to orient the intracellular TIR domains properly, thereby recruiting TIR-containing adapters to the receptor complex [47], such as myeloid differentiation factor 88 (MyD88) [48, 49], MyD88 adapter–like (MAL) [50, 51], TIR-domain-containing adapter–inducing interferon β (TRIF) [52, 53], TRIF-related adapter molecule (TRAM) [54, 55], and sterile alpha motifs and beta-catenin/armadillo repeats (SARM) [56]. Multiple TIR–TIR interactions between TLRs and these adapters are important for activating intracellular signaling events. In addition, different combinations of these adapters can trigger unique or redundant pathways, leading to the activation of transcription factors, such as nuclear factor (NF)-κB, activator protein (AP)-1 and interferon regulatory factors (IRFs).

Despite recognizing diverse ligands, most TLRs, as well as the IL-1/IL-18 receptors, signal a common pathway via the adapter protein MyD88, which ultimately activates mitogen-activated protein kinases (MAPKs) and NF-κB, resulting in transcription of proinflammatory genes [57]. Through its N-terminal death domain (DD), MyD88 recruits serine-threonine kinases called IL-1 receptor–associated kinases (IRAKs) and TNF-associated factor 6 (TRAF6) to the receptor complex in homotypic DD interactions [58–63]. TRAF6, in turn, activates transforming growth factor β–associated kinase-1 (TAK1), which phosphorylates the IκB kinase (IKK) complex (IKKα/IKKβ/IKKγ) culminating in IκBα degradation, an essential step in NF-κB activation [64–70]. Certain TLRs, such as TLR2 and TLR4, also recruit MAL, whose principal function is to stabilize the interaction between MyD88 and the receptor complex [71, 72].

In contrast, TRIF is used by TLR3 and TLR4, for which TRAM acts as a bridging adapter between TRIF and TLR4 [55, 73]. Activation of TRIF-dependent pathways leads to TRAF3 recruitment, which is required for the association of TANK-binding kinase 1 (TBK1) and the subsequent activation of the IRF transcription factors [74, 75]. IRF3 and IRF7 regulate the production of type I interferons and are involved in viral recognition.

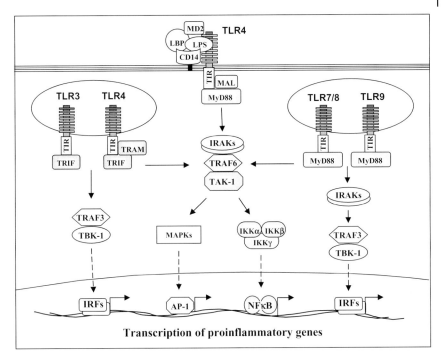

Figure 5.1 Simplified model of TLR signaling pathways. TLR4 activates the intracellular signaling cascade by recruiting MAL, MyD88, and IRAKs to the membrane. Once released, IRAKs can associate with and activate TRAF6 that causes activation of the IKK complex and MAPKs. The MAPKs activate the transcription factor AP-1, whereas the IKK complex cause activation of the transcription factor NF-κB, resulting in the transcription of proinflamma- tory genes. In the endosome, TLR4 also triggers a TRIF-dependent signaling pathway via TRAM to induce NF-κB and IRF transcrip- tion factors. The TRIF-dependent pathways are also initiated by TLR3. In contrast, TLR7, TLR8, and TLR9 activate IRFs through recruitment of MyD88 to the endosome. See text for further details, abbreviations and references.

Furthermore, TRIF also induces an alternate pathway toward NF-κB activation that is mediated by TRAF6 and the IKK complex. Recent data suggest that the cellular localization of the receptor determines whether the MyD88- or TRIF-dependent pathway is initiated on ligand binding [11, 12]. Membrane-bound recep- tors activate MAL and MyD88 to induce NF-κB activation, whereas endosomal TLR3 and TLR4 engage the TRIF-dependent pathway, leading to IRF activation and induction of type I IFNs. In contrast, endosomal TLR7, TLR8, and TLR9 induce type I IFN production in a MyD88-dependent manner. When MyD88 local- izes to the endosome instead of the plasma membrane, it instead associates with TRAF3, facilitating the formation of a complex with TBK1 and the activation of IRFs [74, 75]. Future research will clarify the mechanisms by which individual TLRs induce specific innate immune responses.

5.3
TLR Expression Patterns in Atherosclerosis

TLR expression has been evaluated in both human and mouse atherosclerotic lesions. TLR1, TLR2, and TLR4 are the chief TLRs in human lesions, and TLR5 has shown variable staining by immunohistochemistry [76, 77]. Further analysis has demonstrated that macrophages are the predominant TLR-positive cells within lesions. Endothelial cells that line the lesions also express TLRs, and basal expression of TLR1–TLR6 have been observed in endothelial cells in normal human vessels [76]. The latter finding was corroborated by a study that examined TLR expression patterns in medium and large human arteries, which observed that TLR2 and TLR4 were ubiquitously expressed in the vascular beds examined. In this study, TLR7 and TLR9 appeared infrequently, and other TLRs showed variable expression [78].

TLR4 is present in atherosclerotic lesions from apolipoprotein E–deficient (ApoE$^{-/-}$) mice, and oxidized LDL has been shown to upregulate TLR4 expression in monocyte-derived macrophages [77]. Furthermore, TLR4 expression is evident in smooth muscle cells from atherosclerotic plaques taken from patients with coronary artery disease [79], and TLR4 is expressed by cardiac myocytes in failing myocardium in mice and humans [80]. A recent study has documented TLR2 expression in murine vessels, preferentially at sites of disturbed flow [81]. Similarly, patients with coronary arteriosclerosis disease have higher numbers of TLR4-positive monocytes compared with healthy controls, as measured by flow cytometry [82–84].

In conclusion, several TLRs are highly expressed in atherosclerotic lesions, and patients with active cardiovascular disease harbor more circulating TLR4-positive monocytes than stable patients and healthy controls. Together, these data implicate a role for TLRs in atherogenesis.

5.4
TLRs in the Pathogenesis of Atherosclerosis

5.4.1
TLRs on Vascular Cells – *In Vitro* Studies

Endothelial cells play an important role in the development of atherosclerosis and have therefore been examined with regard to TLR expression and activation. An extensive analysis in cultured human endothelial cells revealed that TLR expression patterns differ between endothelial cells from various origins. Arterial endothelial cells, unlike venous cells, respond to a wide range of TLR ligands, including those for TLR2, TLR4, TLR5, and TLR9 [85]. Several other studies have observed TLR2 and TLR4 expression in human endothelial cells and induction of proinflammatory responses on receptor ligation [86, 87].

Atherosclerosis-associated factors, such as proinflammatory stimuli and low shear stress, induce TLR expression on endothelial cells [88]. Furthermore, TLRs may regulate monocyte adhesion to endothelial cells, a mechanism by which inflammatory cells can be recruited to the vessel wall [89, 90].

In a study of TLR expression on human aortic smooth muscle cells, Erridge and colleagues detected mRNA expression for all TLRs except TLR1 and TLR2 [85]. Furthermore, smooth muscle cells responded to TLR ligands, indicating that the receptors were active on these vascular cells. Other studies have supported these findings, reporting that TLR4 expression in vascular smooth muscle cells is induced by LPS [91]. Similarly, TLR2 is expressed in *Chlamydia pneumoniae*–infected smooth muscle cells, and the ensuing chemokine release is in part TLR2-dependent [92]. TLR3 activation by dsRNA evokes a proinflammatory phenotype in smooth muscle cells, and these cells proliferate in a TLR2- and TLR4-dependent manner in response to human heat shock protein 60 (HSP60) [93]. Taken together, there is increasing evidence for both constitutive and inducible expression of several TLRs in vascular cells, such as endothelial cells and smooth muscle cells. The precise contribution of vascular TLRs to atherosclerosis, however, remains unclear.

5.4.2
Genetic Variants of TLRs and Their Association with CAD

In genetic analyses of TLR polymorphisms, two missense mutations in the human TLR4 gene have received particular interest: the cosegregating Asp299Gly and Thr399Ile mutations. Initially, these polymorphisms were reported to encode a nonfunctional TLR4 molecule. A blunted response to inhaled LPS and hyporesponsiveness of primary human airway epithelial cells that were stimulated with LPS were observed for heterozygous carriers of the two mutant alleles [94]. Subsequent studies, however, have failed to recapitulate LPS hyporesponsiveness in monocytes from carriers of these polymorphisms [95, 96].

Several studies have sought to determine if there is an association between TLR4 polymorphisms and cardiovascular disease. An initial study found that carriers of the mutation had reduced atherosclerosis, as measured by intima–media thickness, compared with wild-type [97]. Subsequent studies of various sizes, however, have shown conflicting results; some have observed a reduction in atherosclerosis, whereas others did not confirm this association. Similarly, studies that have investigated the effects of TLR4 polymorphisms on the risk for myocardial infarction have generated inconsistent results [96].

With regard to polymorphisms of other TLRs, the TLR2 polymorphism Arg-753Gln generates a nonfunctional receptor [98] and is associated with increased restenosis after stent implantation, in contrast to TLR4 Asp299Gly, which shows no such association [99]. In two separate studies, however, there was no link between TLR2 haplotypes and carotid intima–media thickness [100] nor was there an association between TLR2 polymorphisms and myocardial infarction [101]. Two polymorphisms in the promoter region of human TLR9 have shown no association with restenosis or atherosclerosis [102].

In summary, the results of published association studies are inconclusive; larger studies will be needed to clarify the role of different TLR polymorphisms in cardiovascular disease.

5.4.3
The Role of TLRs in Atherosclerosis – Experimental Models

Experimental models of atherosclerosis have been instrumental in clarifying the role of TLRs in atherosclerosis. MyD88, the adapter protein downstream of most TLRs including TLR2 and TLR4, as well as the IL-1/IL-18 receptors, has been targeted to this effect. Bjorkbacka and colleagues demonstrated that lesion development was diminished in ApoE$^{-/-}$/MyD88$^{-/-}$ double-knockout mice, indicating that TLR signaling regulates the development of atherosclerosis [103]. These findings were corroborated by Michelsen and colleagues, who showed a reduction in atherosclerosis in the aortas of MyD88$^{-/-}$/ApoE$^{-/-}$ mice [104]. This decrease in lesion area was accompanied by decreased macrophage recruitment and lowered proinflammatory cytokine production. Because MyD88 is shared by other receptors, however, it cannot be ruled out that these effects are independent of TLR activation, especially given that IL-1/IL-18 signaling has been shown to promote atherosclerosis [105, 106].

Further insight was gained through the analysis of TLR4$^{-/-}$/ApoE$^{-/-}$ mice, in which atherosclerosis was reduced [104]. Therefore, the effect that is observed in MyD88$^{-/-}$/ApoE$^{-/-}$ mice may be partially attributed to TLR4. Although the mechanisms and putative ligands of the TLR response in atherosclerosis are unresolved, the TLR4-mediated effects on atherosclerosis may not only involve LPS, because abrogation of CD14 does not elicit reduction of atherosclerosis in CD14$^{-/-}$/ApoE$^{-/-}$ mice [103].

Further evidence that implicates TLRs in atherosclerosis comes from studies that have used TLR2 knockout mice. On the atherosclerosis-prone LDLR$^{-/-}$ background, these mice showed involvement of TLR2 in the development of atherosclerosis, as assessed by lesion size [107]. The degree of atherosclerosis was significantly reduced in these doubly deficient mice; in bone marrow transplantation experiments, these effects were attributed to resident cells, such as endothelial cells and possibly smooth muscle cells and fibroblasts, not to bone marrow–derived cells, such as monocytes and macrophages.

Interestingly, administration of a TLR2/TLR1 ligand, Pam3Cys, exacerbated disease severity, which was dependent on bone marrow–derived TLR2 expression. Liu and colleagues confirmed the involvement of TLR2 in atherosclerosis progression by comparing lesion areas in ApoE$^{-/-}$ and ApoE$^{-/-}$/TLR2$^{-/-}$ mice [108]; the TLR2-deficient mice showed reduced atherosclerosis progression and diminished inflammation compared with ApoE$^{-/-}$/TLR2$^{+/+}$ mice. Using rabbits that were fed a high-fat diet, others have demonstrated synergism between TLR2 and TLR4 in atherosclerosis development by local overexpression of the receptors [109]. In addition to their direct effects on lesion size, TLR2 and TLR4 have been implicated in neointima formation and outward arterial remodeling, respectively [110–112].

To date, several studies have established a role for TLR2 and TLR4 in athero-sclerotic lesion development; the underlying mechanisms of TLR signaling in disease progression, however, have not been determined. The involvement of other TLRs in atherosclerosis also awaits further exploration, as do the ligands that participate in TLR-mediated disease progression.

5.5
Exogenous TLR Ligands – Infections in Atherosclerosis

The role of infections in the pathogenesis of atherosclerosis is controversial. Nevertheless, there is evidence from both human and experimental atherosclerosis models that links certain infections or pathogenic burdens to disease [113]. In atherosclerosis-prone mouse models, several studies have associated increased atherosclerotic disease with infections, such as *C. pneumoniae* [114, 115], *Porphyromonas gingivalis* [116–119], and cytomegalovirus [120, 121]. Furthermore, several infectious organisms have been detected in human atherosclerotic lesions [122–124], and seroepidemiological studies have related pathogenic burden to increased risk of atherosclerosis [113, 125].

Multiple studies suggest that *C. pneumoniae* infection triggers proinflammatory cytokine production in a TLR2- and TLR4-dependent manner. *C. pneumoniae*–induced foam cell formation is TLR2-dependent, and potential *C. pneumoniae*–derived TLR ligands include LPS, HSP60, methylated DNA, and outer membrane protein 2. Similarly, *P. gingivalis* activates macrophages in a partially TLR2-dependent manner [126]. ApoE$^{-/-}$ mice that are infected with *P. gingivalis* experience increased atherosclerosis and upregulate both TLR2 and TLR4 in the aorta. In contrast, invasion-impaired *P. gingivalis* microorganisms do not induce atherosclerosis or TLR2 or TLR4 expression in atherosclerosis-prone ApoE$^{-/-}$ mice [127].

Cytomegalovirus infection increases lesion areas in ApoE$^{-/-}$ mice [115] and induces inflammatory cytokine production through TLR2-dependent activation of NF-κB [128]. The same group recently showed that this activation is mediated by TLR2/TLR1 recognition of two human cytomegalovirus envelope glycoproteins [129].

5.6
Endogenous TLR Ligands – Triggering Inflammation in the Absence of Exogenous Infectious Stimuli

The potential role for endogenous TLR ligands in the pathogenesis of atherosclerosis is compelling. Endogenous TLR agonists have been proposed to induce inflammation in the absence of exogenous infectious stimuli, but research on this topic has been hampered by the presence of highly potent bacterial contaminants in the preparations of host-derived TLR2- and TLR4-activating ligands. Despite this restriction, there is support for the involvement of host-derived TLR agonists in atherogenesis, such as heat shock proteins, fibrinogen, extra domain A of fibronectin, and soluble hyaluronan [130]. These host-derived ligands often are danger signals that develop from tissue damage and can concentrate at sites of chronic inflammation, such as the atherosclerotic vessel wall. Although they might not initiate disease, they represent pathways that sustain and prolong inflammation in atherosclerotic lesions. A role for heat shock proteins in atherosclerosis has been proposed in several studies, and TLR2-, TLR4-, and MyD88-dependent activation of various vascular cells and macrophages has been demonstrated [131–133]. Fibrinogen induces chemokine secretion by macrophages in a TLR4-dependent manner [134]. In an experimental model of atherosclerosis, extra domain A of fibronectin, a known TLR2 agonist [135], was involved in atherosclerotic lesion development [136]. Hyaluronan fragments are the breakdown products of extracellular matrix that are generated on tissue injury. Several studies have identified hyaluronan as a ligand for both TLR2 and TLR4 and shown that TLR-dependent activation by hyaluronan stimulates dendritic cell differentiation and endothelial cell activation [137–140].

High mobility group box 1 (HMGB1) has recently been found to induce proinflammatory macrophage and endothelial cell activity, and in addition to its DNA-binding capacities, it mediates leukocyte recruitment to sites of inflammation [141]. Its proinflammatory actions are mediated in part by TLR2 and TLR4 ligation, and HMGB1 expression increases with lesion progression in human atherosclerotic tissue [141–143].

Modified LDL is a particularly interesting component in the pathogenesis of atherosclerosis. Minimally modified LDL and its components have been shown to effect TLR4-mediated outcomes, such as cytokine secretion from and cytoskeletal rearrangement in macrophages [144, 145]. Furthermore, 1-palmitoyl-2-(5′-oxovaleroyl)-*sn*-glycero-3-phosphocholine (POVPC) and oxidized phosphocholine are ligands for CD36, a coreceptor of the TLR2/TLR6 heterodimer, and aortic endothelial cells are susceptible to POVPC-mediated induction of IL-8 and MCP-1 in a TLR4-dependent manner [146, 147]. Saturated fatty acids activate TLR4 and induce inflammatory changes in macrophages [148]. Surprisingly, phospholipid components of LDL have been shown to inhibit LPS-mediated effects [149, 150], and further studies are needed to ascertain the complex role of modified LDL and other potential endogenous ligands in TLR activation in atherosclerosis.

5.7
Summary

The identification of TLRs, their ligands, and their signaling pathways has led to significant progress in enhancing our understanding of the role of innate immunity in atherosclerosis. In the past several years, strong evidence has mounted that TLRs mediate innate immune responses in atherosclerosis.

Importantly, clinical and experimental investigations have demonstrated that TLR2-, TLR4-, and MyD88-mediated signaling regulate disease pathogenesis by controlling inflammatory responses. As increasing details of TLR signaling and interaction with other inflammatory pathways and new insights into TLR-dependent pathogen pattern recognition emerge, our historical view of innate immunity as a set of primitive and nonspecific responses is undergoing a dramatic evolution. We now understand innate immunity to be a sophisticated and highly evolved immune system that not only implements high specificity in pathogen pattern recognition but also determines the outcomes of immune responses through the regulation of a still poorly understood network of signaling pathways.

Furthermore, although it is not discussed in this chapter, accumulating evidence supports a vital role for TLRs in bridging innate and adaptive immunity. Given the established role for TLR signaling in the pathogenesis of atherosclerosis, the regulation of innate immunity represents new opportunities for clinical intervention. Strategies that selectively target TLRs or their coreceptors (such as MD2 for TLR4 or CD36 for TLR2/TLR6) can be more attractive than those that inhibit global signals, such as NF-κB.

It is possible, however, that a broad array of TLR ligands regulate different stages of atherosclerosis. Therefore, the identification of the features of these ligands and the elucidation of the corresponding TLR signaling pathways in atherosclerosis remain critical challenges in developing a specific therapy. Moreover, we are only beginning to investigate the functional role of TLRs other than TLR2 and TLR4. We are hopeful that the clarification of these molecular mechanisms will translate into improved patient care.

Acknowledgments

This work was supported by grants from the Swedish Research Council, Swedish Heart and Lung Foundation, and European Community project LSHM-CT-2006-037400. KE is a Wenner-Gren Fellow.

References

1 Hansson, G.K. (2005) Inflammation, atherosclerosis, and coronary artery disease. *N. Engl. J. Med.*, **352**, 1685–1695.

2 Hansson, G.K. and Libby, P. (2006) The immune response in atherosclerosis: a double-edged sword. *Nat. Rev. Immunol.*, **6**, 508–519.

3 Yan, Z.Q. and Hansson, G.K. (2007) Innate immunity, macrophage activation, and atherosclerosis. *Immunol. Rev.*, **219**, 187–203.

4 Akira, S., Uematsu, S., and Takeuchi, O. (2006) Pathogen recognition and innate immunity. *Cell*, **124**, 783–801.

5 Lemaitre, B., Nicolas, E., Michaut, L., Reichhart, J.M., and Hoffmann, J.A. (1996) The dorsoventral regulatory gene cassette spatzle/Toll/cactus controls the potent antifungal response in *Drosophila* adults. *Cell*, **86**, 973–983.

6 Bowie, A. and O'Neill, L.A. (2000) The interleukin-1 receptor/Toll-like receptor superfamily: signal generators for pro-inflammatory interleukins and microbial products. *J. Leukoc. Biol.*, **67**, 508–514.

7 Heil, F., Ahmad-Nejad, P., Hemmi, H., *et al.* (2003) The Toll-like receptor 7 (TLR7)-specific stimulus loxoribine

uncovers a strong relationship within the TLR7, 8 and 9 subfamily. *Eur. J. Immunol.*, **33**, 2987–2997.

8 Matsumoto, M., Funami, K., Tanabe, M., *et al.* (2003) Subcellular localization of toll-like receptor 3 in human dendritic cells. *J. Immunol.*, **171**, 3154–3162.

9 Latz, E., Schoenemeyer, A., Visintin, A., *et al.* (2004) TLR9 signals after translocating from the ER to CpG DNA in the lysosome. *Nat. Immunol.*, **5**, 190–198.

10 Underhill, D.M., Ozinsky, A., Hajjar, A.M., *et al.* (1999) The Toll-like receptor 2 is recruited to macrophage phagosomes and discriminates between pathogens. *Nature*, **401**, 811–815.

11 Tanimura, N., Saitoh, S., Matsumoto, F., Akashi-Takamura, S., and Miyake, K. (2008) Roles for LPS-dependent interaction and relocation of TLR4 and TRAM in TRIF-signaling. *Biochem. Biophys. Res. Commun.*, **368**, 94–99.

12 Kagan, J.C., Su, T., Horng, T., Chow, A., Akira, S., and Medzhitov, R. (2008) TRAM couples endocytosis of Toll-like receptor 4 to the induction of interferon-beta. *Nat. Immunol.*, **9**, 361–368.

13 Medzhitov, R. (2001) Toll-like receptors and innate immunity. *Nat. Rev. Immunol.*, **1**, 135–145.

14 Medzhitov, R., Preston-Hurlburt, P., and Janeway C.A., Jr. (1997) A human homologue of the *Drosophila* Toll protein signals activation of adaptive immunity. *Nature*, **388**, 394–397.

15 Poltorak, A., He, X., Smirnova, I., *et al.* (1998) Defective LPS signaling in C3H/HeJ and C57BL/10ScCr mice: mutations in Tlr4 gene. *Science*, **282**, 2085–2088.

16 Hoshino, K., Takeuchi, O., Kawai, T., *et al.* (1999) Cutting edge: toll-like receptor 4 (TLR4)-deficient mice are hyporesponsive to lipopolysaccharide: evidence for TLR4 as the Lps gene product. *J. Immunol.*, **162**, 3749–3752.

17 Chow, J.C., Young, D.W., Golenbock, D.T., Christ, W.J., and Gusovsky, F. (1999) Toll-like receptor 4 mediates lipopolysaccharide-induced signal transduction. *J. Biol. Chem.*, **274**, 10689–10692.

18 Wright, S.D., Ramos, R.A., Tobias, P.S., Ulevitch, R.J., and Mathison, J.C. (1990) CD14, a receptor for complexes of lipopolysaccharide (LPS) and LPS binding protein. *Science*, **249**, 1431–1433.

19 Schumann, R.R., Leong, S.R., Flaggs, G.W., *et al.* (1990) Structure and function of lipopolysaccharide binding protein. *Science*, **249**, 1429–1431.

20 Lee, J.D., Kravchenko, V., Kirkland, T.N., *et al.* (1993) Glycosyl-phosphatidylinositol-anchored or integral membrane forms of CD14 mediate identical cellular responses to endotoxin. *Proc. Natl Acad. Sci. U. S. A.*, **90**, 9930–9934.

21 Frey, E.A., Miller, D.S., Jahr, T.G., *et al.* (1992) Soluble CD14 participates in the response of cells to lipopolysaccharide. *J. Exp. Med.*, **176**, 1665–1671.

22 Shimazu, R., Akashi, S., Ogata, H., *et al.* (1999) MD-2, a molecule that confers lipopolysaccharide responsiveness on Toll-like receptor 4. *J. Exp. Med.*, **189**, 1777–1782.

23 Akashi, S., Nagai, Y., Ogata, H., *et al.* (2001) Human MD-2 confers on mouse Toll-like receptor 4 species-specific lipopolysaccharide recognition. *Int. Immunol.*, **13**, 1595–1599.

24 Viriyakosol, S., Tobias, P.S., Kitchens, R.L., and Kirkland, T.N. (2001) MD-2 binds to bacterial lipopolysaccharide. *J. Biol. Chem.*, **276**, 38044–38051.

25 Ohto, U., Fukase, K., Miyake, K., and Satow, Y. (2007) Crystal structures of human MD-2 and its complex with antiendotoxic lipid IVa. *Science*, **316**, 1632–1634.

26 Malley, R., Henneke, P., Morse, S.C., *et al.* (2003) Recognition of pneumolysin by Toll-like receptor 4 confers resistance to pneumococcal infection. *Proc. Natl Acad. Sci. U. S. A.*, **100**, 1966–1971.

27 Kurt-Jones, E.A., Popova, L., Kwinn, L., *et al.* (2000) Pattern recognition receptors TLR4 and CD14 mediate response to respiratory syncytial virus. *Nat. Immunol.*, **1**, 398–401.

28 Rassa, J.C., Meyers, J.L., Zhang, Y., Kudaravalli, R., and Ross, S.R. (2002) Murine retroviruses activate B cells via

interaction with toll-like receptor 4. *Proc. Natl Acad. Sci. U. S. A.*, **99**, 2281–2286.

29 Takeuchi, O., Kawai, T., Sanjo, H., *et al.* (1999) TLR6: a novel member of an expanding toll-like receptor family. *Gene*, **231**, 59–65.

30 Takeuchi, O., Kawai, T., Muhlradt, P.F., *et al.* (2001) Discrimination of bacterial lipoproteins by Toll-like receptor 6. *Int. Immunol.*, **13**, 933–940.

31 Takeuchi, O., Sato, S., Horiuchi, T., *et al.* (2002) Cutting edge: role of Toll-like receptor 1 in mediating immune response to microbial lipoproteins. *J. Immunol.*, **169**, 10–14.

32 Buwitt-Beckmann, U., Heine, H., Wiesmuller, K.H., *et al.* (2005) Toll-like receptor 6- independent signaling by diacylated lipopeptides. *Eur. J. Immunol.*, **35**, 282–289.

33 Shimizu, T., Kida, Y., and Kuwano, K. (2007) Triacylated lipoproteins derived from *Mycoplasma pneumoniae* activate nuclear factor-kappaB through toll-like receptors 1 and 2. *Immunology*, **121**, 473–483.

34 Hoebe, K., Georgel, P., Rutschmann, S., *et al.* (2005) CD36 is a sensor of diacylglycerides. *Nature*, **433**, 523–527.

35 Triantafilou, M., Gamper, F.G., Haston, R.M., *et al.* (2006) Membrane sorting of toll-like receptor (TLR)-2/6 and TLR2/1 heterodimers at the cell surface determines heterotypic associations with CD36 and intracellular targeting. *J. Biol. Chem.*, **281**, 31002–31011.

36 Alexopoulou, L., Holt, A.C., Medzhitov, R., and Flavell, R.A. (2001) Recognition of double- stranded RNA and activation of NF-kappaB by Toll-like receptor 3. *Nature*, **413**, 732–738.

37 Hayashi, F., Smith, K.D., Ozinsky, A., *et al.* (2001) The innate immune response to bacterial flagellin is mediated by Toll-like receptor 5. *Nature*, **410**, 1099–1103.

38 Smith, K.D., Andersen-Nissen, E., Hayashi, F., *et al.* (2003) Toll-like receptor 5 recognizes a conserved site on flagellin required for protofilament formation and bacterial motility. *Nat. Immunol.*, **4**, 1247–1253.

39 Diebold, S.S., Kaisho, T., Hemmi, H., Akira, S., and Reis e Sousa, C. (2004) Innate antiviral responses by means of TLR7-mediated recognition of single-stranded RNA. *Science*, **303**, 1529–1531.

40 Heil, F., Hemmi, H., Hochrein, H., *et al.* (2004) Species-specific recognition of single-stranded RNA via toll-like receptor 7 and 8. *Science*, **303**, 1526–1529.

41 Lund, J.M., Alexopoulou, L., Sato, A., *et al.* (2004) Recognition of single-stranded RNA viruses by Toll-like receptor 7. *Proc. Natl Acad. Sci. U. S. A.*, **101**, 5598–5603.

42 Hemmi, H., Takeuchi, O., Kawai, T., *et al.* (2000) A Toll-like receptor recognizes bacterial DNA. *Nature*, **408**, 740–745.

43 Krug, A., Luker, G.D., Barchet, W., Leib, D.A., Akira, S., and Colonna, M. (2004) Herpes simplex virus type 1 activates murine natural interferon-producing cells through toll-like receptor 9. *Blood*, **103**, 1433–1437.

44 Lund, J., Sato, A., Akira, S., Medzhitov, R., and Iwasaki, A. (2003) Toll-like receptor 9-mediated recognition of Herpes simplex virus-2 by plasmacytoid dendritic cells. *J. Exp. Med.*, **198**, 513–520.

45 Seong, S.Y. and Matzinger, P. (2004) Hydrophobicity: an ancient damage-associated molecular pattern that initiates innate immune responses. *Nat. Rev. Immunol.*, **4**, 469–478.

46 Matzinger, P. (1994) Tolerance, danger, and the extended family. *Annu. Rev. Immunol.*, **12**, 991–1045.

47 O'Neill, L.A. and Bowie, A.G. (2007) The family of five: TIR-domain-containing adaptors in Toll-like receptor signalling. *Nat. Rev. Immunol.*, **7**, 353–364.

48 Wesche, H., Henzel, W.J., Shillinglaw, W., Li, S., and Cao, Z. (1997) MyD88: an adapter that recruits IRAK to the IL-1 receptor complex. *Immunity*, **7**, 837–847.

49 Janssens, S. and Beyaert, R. (2002) A universal role for MyD88 in TLR/IL-1R-mediated signaling. *Trends Biochem. Sci.*, **27**, 474–482.

50 Fitzgerald, K.A., Palsson-McDermott, E.M., Bowie, A.G., *et al.* (2001) Mal (MyD88-adapter-like) is required for Toll-like receptor-4 signal transduction. *Nature*, **413**, 78–83.

51 Horng, T., Barton, G.M., and Medzhitov, R. (2001) TIRAP: an adapter molecule in the Toll signaling pathway. *Nat. Immunol.*, **2**, 835–841.

52 Yamamoto, M., Sato, S., Mori, K., *et al.* (2002) Cutting edge: a novel Toll/IL-1 receptor domain-containing adapter that preferentially activates the IFN-beta promoter in the Toll-like receptor signaling. *J. Immunol.*, **169**, 6668–6672.

53 Oshiumi, H., Matsumoto, M., Funami, K., Akazawa, T., and Seya, T. (2003) TICAM-1, an adaptor molecule that participates in Toll-like receptor 3-mediated interferon-beta induction. *Nat. Immunol.*, **4**, 161–167.

54 Fitzgerald, K.A., Rowe, D.C., Barnes, B.J., *et al.* (2003) LPS-TLR4 signaling to IRF-3/7 and NF-kappaB involves the toll adapters TRAM and TRIF. *J. Exp. Med.*, **198**, 1043–1055.

55 Oshiumi, H., Sasai, M., Shida, K., Fujita, T., Matsumoto, M., and Seya, T. (2003) TIR-containing adapter molecule (TICAM)-2, a bridging adapter recruiting to toll-like receptor 4 TICAM-1 that induces interferon-beta. *J. Biol. Chem.*, **278**, 49751–49762.

56 Mink, M., Fogelgren, B., Olszewski, K., Maroy, P., and Csiszar, K. (2001) A novel human gene (SARM) at chromosome 17q11 encodes a protein with a SAM motif and structural similarity to Armadillo/beta-catenin that is conserved in mouse, *Drosophila*, and *Caenorhabditis elegans*. *Genomics*, **74**, 234–244.

57 Adachi, O., Kawai, T., Takeda, K., *et al.* (1998) Targeted disruption of the MyD88 gene results in loss of IL-1- and IL-18-mediated function. *Immunity*, **9**, 143–150.

58 Kawagoe, T., Sato, S., Jung, A., *et al.* (2007) Essential role of IRAK-4 protein and its kinase activity in Toll-like receptor-mediated immune responses but not in TCR signaling. *J. Exp. Med.*, **204**, 1013–1024.

59 Kawagoe, T., Sato, S., Matsushita, K., *et al.* (2008) Sequential control of Toll-like receptor- dependent responses by IRAK1 and IRAK2. *Nat. Immunol.*, **9**, 684–691.

60 Lye, E., Mirtsos, C., Suzuki, N., Suzuki, S., and Yeh, W.C. (2004) The role of interleukin 1 receptor-associated kinase-4 (IRAK-4) kinase activity in IRAK-4-mediated signaling. *J. Biol. Chem.*, **279**, 40653–40658.

61 Qin, J., Jiang, Z., Qian, Y., Casanova, J.L., and Li, X. (2004) IRAK4 kinase activity is redundant for interleukin-1 (IL-1) receptor-associated kinase phosphorylation and IL-1 responsiveness. *J. Biol. Chem.*, **279**, 26748–26753.

62 Picard, C., Puel, A., Bonnet, M., *et al.* (2003) Pyogenic bacterial infections in humans with IRAK-4 deficiency. *Science*, **299**, 2076–2079.

63 Jiang, Z., Johnson, H.J., Nie, H., Qin, J., Bird, T.A., and Li, X. (2003) Pellino 1 is required for interleukin-1 (IL-1)-mediated signaling through its interaction with the IL-1 receptor-associated kinase 4 (IRAK4)-IRAK-tumor necrosis factor receptor- associated factor 6 (TRAF6) complex. *J. Biol. Chem.*, **278**, 10952–10956.

64 Kishida, S., Sanjo, H., Akira, S., Matsumoto, K., and Ninomiya-Tsuji, J. (2005) TAK1-binding protein 2 facilitates ubiquitination of TRAF6 and assembly of TRAF6 with IKK in the IL-1 signaling pathway. *Genes Cells*, **10**, 447–454.

65 Wang, C., Deng, L., Hong, M., Akkaraju, G.R., Inoue, J., and Chen, Z.J. (2001) TAK1 is a ubiquitin-dependent kinase of MKK and IKK. *Nature*, **412**, 346–351.

66 Kanayama, A., Seth, R.B., Sun, L., *et al.* (2004) TAB2 and TAB3 activate the NF-kappaB pathway through binding to polyubiquitin chains. *Mol. Cell*, **15**, 535–548.

67 Mercurio, F., Zhu, H., Murray, B.W., *et al.* (1997) IKK-1 and IKK-2: cytokine-activated IkappaB kinases essential for NF-kappaB activation. *Science*, **278**, 860–866.

68 Zandi, E., Chen, Y., and Karin, M. (1998) Direct phosphorylation of

IkappaB by IKKalpha and IKKbeta: discrimination between free and NF-kappaB-bound substrate. *Science*, **281**, 1360–1363.

69 Baldi, L., Brown, K., Franzoso, G., and Siebenlist, U. (1996) Critical role for lysines 21 and 22 in signal-induced, ubiquitin-mediated proteolysis of I kappa B-alpha. *J. Biol. Chem.*, **271**, 376–379.

70 Scherer, D.C., Brockman, J.A., Chen, Z., Maniatis, T., and Ballard, D.W. (1995) Signal-induced degradation of I kappa B alpha requires site-specific ubiquitination. *Proc. Natl Acad. Sci. U. S. A.*, **92**, 11259–11263.

71 Yamamoto, M., Sato, S., Hemmi, H., *et al.* (2002) Essential role for TIRAP in activation of the signalling cascade shared by TLR2 and TLR4. *Nature*, **420**, 324–329.

72 Kagan, J.C. and Medzhitov, R. (2006) Phosphoinositide-mediated adaptor recruitment controls Toll-like receptor signaling. *Cell*, **125**, 943–955.

73 Yamamoto, M., Sato, S., Hemmi, H., *et al.* (2003) Role of adaptor TRIF in the MyD88- independent toll-like receptor signaling pathway. *Science*, **301**, 640–643.

74 Hacker, H., Redecke, V., Blagoev, B., *et al.* (2006) Specificity in Toll-like receptor signalling through distinct effector functions of TRAF3 and TRAF6. *Nature*, **439**, 204–207.

75 Oganesyan, G., Saha, S.K., Guo, B., *et al.* (2006) Critical role of TRAF3 in the Toll-like receptor-dependent and -independent antiviral response. *Nature*, **439**, 208–211.

76 Edfeldt, K., Swedenborg, J., Hansson, G.K., and Yan, Z.Q. (2002) Expression of toll-like receptors in human atherosclerotic lesions: a possible pathway for plaque activation. *Circulation*, **105**, 1158–1161.

77 Xu, X.H., Shah, P.K., Faure, E., *et al.* (2001) Toll-like receptor-4 is expressed by macrophages in murine and human lipid-rich atherosclerotic plaques and upregulated by oxidized LDL. *Circulation*, **104**, 3103–3108.

78 Pryshchep, O., Ma-Krupa, W., Younge, B.R., Goronzy, J.J., and Weyand, C.M. (2008) Vessel-specific Toll-like receptor profiles in human medium and large arteries. *Circulation*, **118**, 1276–1284.

79 Otsui, K., Inoue, N., Kobayashi, S., *et al.* (2007) Enhanced expression of TLR4 in smooth muscle cells in human atherosclerotic coronary arteries. *Heart Vessels*, **22**, 416–422.

80 Frantz, S., Kobzik, L., Kim, Y.D., *et al.* (1999) Toll4 (TLR4) expression in cardiac myocytes in normal and failing myocardium. *J. Clin. Invest.*, **104**, 271–280.

81 Mullick, A.E., Soldau, K., Kiosses, W.B., Bell, T.A., III, Tobias, P.S., and Curtiss, L.K. (2008) Increased endothelial expression of Toll-like receptor 2 at sites of disturbed blood flow exacerbates early atherogenic events. *J. Exp. Med.*, **205**, 373–383.

82 Methe, H., Kim, J.O., Kofler, S., Weis, M., Nabauer, M., and Koglin, J. (2005) Expansion of circulating Toll-like receptor 4-positive monocytes in patients with acute coronary syndrome. *Circulation*, **111**, 2654–2661.

83 Shiraki, R., Inoue, N., Kobayashi, S., *et al.* (2006) Toll-like receptor 4 expressions on peripheral blood monocytes were enhanced in coronary artery disease even in patients with low C-reactive protein. *Life Sci.*, **80**, 59–66.

84 Geng, H.L., Lu, H.Q., Zhang, L.Z., *et al.* (2006) Increased expression of Toll like receptor 4 on peripheral-blood mononuclear cells in patients with coronary arteriosclerosis disease. *Clin. Exp. Immunol.*, **143**, 269–273.

85 Erridge, C., Burdess, A., Jackson, A.J., *et al.* (2008) Vascular cell responsiveness to Toll-like receptor ligands in carotid atheroma. *Eur. J. Clin. Invest.*, **38**, 713–720.

86 Erridge, C., Spickett, C.M., and Webb, D.J. (2007) Non-enterobacterial endotoxins stimulate human coronary artery but not venous endothelial cell activation via Toll-like receptor 2. *Cardiovasc. Res.*, **73**, 181–189.

87 Dunzendorfer, S., Lee, H.K., and Tobias, P.S. (2004) Flow-dependent regulation of endothelial Toll-like receptor 2 expression through inhibition of SP1 activity. *Circ. Res.*, **95**, 684–691.

88 Yang, Q.W., Mou, L., Lv, F.L., *et al.* (2005) Role of Toll-like receptor 4/ NF-kappaB pathway in monocyte-endothelial adhesion induced by low shear stress and ox-LDL. *Biorheology*, **42**, 225–236.

89 Kawakami, A., Osaka, M., Aikawa, M., *et al.* (2008) Toll-like receptor 2 mediates apolipoprotein CIII-induced monocyte activation. *Circ. Res.*, **103**, 1402–1409.

90 Harokopakis, E., Albzreh, M.H., Martin, M.H., and Hajishengallis, G. (2006) TLR2 transmodulates monocyte adhesion and transmigration via Rac1- and PI3K-mediated inside-out signaling in response to *Porphyromonas gingivalis* fimbriae. *J. Immunol.*, **176**, 7645–7656.

91 Li, H., He, Y., Zhang, J., Sun, S., and Sun, B. (2007) Lipopolysaccharide regulates toll-like receptor 4 expression in human aortic smooth muscle cells. *Cell Biol. Int.*, **31**, 831–835.

92 Yang, X., Coriolan, D., Schultz, K., Golenbock, D.T., and Beasley, D. (2005) Toll-like receptor 2 mediates persistent chemokine release by Chlamydia pneumoniae-infected vascular smooth muscle cells. *Arterioscler. Thromb. Vasc. Biol.*, **25**, 2308–2314.

93 de Graaf, G.R., Kloppenburg, G., Kitslaar, P.J., Bruggeman, C.A., and Stassen, F. (2006) Human heat shock protein 60 stimulates vascular smooth muscle cell proliferation through Toll-like receptors 2 and 4. *Microbes Infect.*, **8**, 1859–1865.

94 Arbour, N.C., Lorenz, E., Schutte, B.C., *et al.* (2000) TLR4 mutations are associated with endotoxin hyporesponsiveness in humans. *Nat. Genet.*, **25**, 187–191.

95 Erridge, C., Stewart, J., and Poxton, I.R. (2003) Monocytes heterozygous for the Asp299Gly and Thr399Ile mutations in the Toll-like receptor 4 gene show no deficit in lipopolysaccharide signalling. *J. Exp. Med.*, **197**, 1787–1791.

96 Frantz, S., Ertl, G., and Bauersachs, J. (2007) Mechanisms of disease: toll-like receptors in cardiovascular disease. *Nat.*

Clin. Pract. Cardiovasc. Med., **4**, 444–454.

97 Kiechl, S., Lorenz, E., Reindl, M., *et al.* (2002) Toll-like receptor 4 polymorphisms and atherogenesis. *N. Engl. J. Med.*, **347**, 185–192.

98 Lorenz, E., Mira, J.P., Cornish, K.L., Arbour, N.C., and Schwartz, D.A. (2000) A novel polymorphism in the toll-like receptor 2 gene and its potential association with staphylococcal infection. *Infect. Immun.*, **68**, 6398–6401.

99 Hamann, L., Gomma, A., Schroder, N.W., *et al.* (2005) A frequent toll-like receptor (TLR)-2 polymorphism is a risk factor for coronary restenosis. *J. Mol. Med.*, **83**, 478–485.

100 Labrum, R., Bevan, S., Sitzer, M., Lorenz, M., and Markus, H.S. (2007) Toll receptor polymorphisms and carotid artery intima-media thickness. *Stroke*, **38**, 1179–1184.

101 Balistreri, C.R., Candore, G., Mirabile, M., *et al.* (2008) TLR2 and age-related diseases: potential effects of Arg753Gln and Arg677Trp polymorphisms in acute myocardial infarction. *Rejuvenation Res.*, **11**, 293–296.

102 Hamann, L., Glaeser, C., Hamprecht, A., Gross, M., Gomma, A., and Schumann, R.R. (2006) Toll-like receptor (TLR)-9 promotor polymorphisms and atherosclerosis. *Clin. Chim. Acta*, **364**, 303–307.

103 Bjorkbacka, H., Kunjathoor, V.V., Moore, K.J., *et al.* (2004) Reduced atherosclerosis in MyD88-null mice links elevated serum cholesterol levels to activation of innate immunity signaling pathways. *Nat. Med.*, **10**, 416–421.

104 Michelsen, K.S., Wong, M.H., Shah, P.K., *et al.* (2004) Lack of Toll-like receptor 4 or myeloid differentiation factor 88 reduces atherosclerosis and alters plaque phenotype in mice deficient in apolipoprotein E. *Proc. Natl Acad. Sci. U. S. A.*, **101**, 10679–10684.

105 Kirii, H., Niwa, T., Yamada, Y., *et al.* (2003) Lack of interleukin-1beta decreases the severity of atherosclerosis in ApoE-deficient mice. *Arterioscler. Thromb. Vasc. Biol.*, **23**, 656–660.

106 Elhage, R., Jawien, J., Rudling, M., *et al.* (2003) Reduced atherosclerosis in interleukin-18 deficient apolipoprotein E-knockout mice. *Cardiovasc. Res.*, **59**, 234–240.

107 Mullick, A.E., Tobias, P.S., and Curtiss, L.K. (2005) Modulation of atherosclerosis in mice by Toll-like receptor 2. *J. Clin. Invest.*, **115**, 3149–3156.

108 Liu, X., Ukai, T., Yumoto, H., *et al.* (2008) Toll-like receptor 2 plays a critical role in the progression of atherosclerosis that is independent of dietary lipids. *Atherosclerosis*, **196**, 146–154.

109 Shinohara, M., Hirata, K., Yamashita, T., *et al.* (2007) Local overexpression of toll-like receptors at the vessel wall induces atherosclerotic lesion formation: synergism of TLR2 and TLR4. *Arterioscler. Thromb. Vasc. Biol.*, **27**, 2384–2391.

110 Shishido, T., Nozaki, N., Takahashi, H., *et al.* (2006) Central role of endogenous Toll-like receptor-2 activation in regulating inflammation, reactive oxygen species production, and subsequent neointimal formation after vascular injury. *Biochem. Biophys. Res. Commun.*, **345**, 1446–1453.

111 Schoneveld, A.H., Oude Nijhuis, M.M., van Middelaar, M.B., Laman, J.D., de Kleijn, D.P., and Pasterkamp, G. (2005) Toll-like receptor 2 stimulation induces intimal hyperplasia and atherosclerotic lesion development. *Cardiovasc. Res.*, **66**, 162–169.

112 Hollestelle, S.C., De Vries, M.R., Van Keulen, J.K., *et al.* (2004) Toll-like receptor 4 is involved in outward arterial remodeling. *Circulation*, **109**, 393–398.

113 Epstein, S.E., Zhu, J., Burnett, M.S., Zhou, Y.F., Vercellotti, G., and Hajjar, D. (2000) Infection and atherosclerosis: potential roles of pathogen burden and molecular mimicry. *Arterioscler. Thromb. Vasc. Biol.*, **20**, 1417–1420.

114 Liu, L., Hu, H., Ji, H., Murdin, A.D., Pierce, G.N., and Zhong, G. (2000) *Chlamydia pneumoniae* infection significantly exacerbates aortic atherosclerosis in an LDLR$^{-/-}$ mouse

model within six months. *Mol. Cell. Biochem.*, **215**, 123–128.

115 Moazed, T.C., Campbell, L.A., Rosenfeld, M.E., Grayston, J.T., and Kuo, C.C. (1999) *Chlamydia pneumoniae* infection accelerates the progression of atherosclerosis in apolipoprotein E-deficient mice. *J. Infect. Dis.*, **180**, 238–241.

116 Desvarieux, M., Demmer, R.T., Rundek, T., *et al.* (2005) Periodontal microbiota and carotid intima-media thickness: the Oral Infections and Vascular Disease Epidemiology Study (INVEST). *Circulation*, **111**, 576–582.

117 Lalla, E., Lamster, I.B., Hofmann, M.A., *et al.* (2003) Oral infection with a periodontal pathogen accelerates early atherosclerosis in apolipoprotein E-null mice. *Arterioscler. Thromb. Vasc. Biol.*, **23**, 1405–1411.

118 Li, L., Messas, E., Batista, E.L., Jr., Levine R.A., and Amar, S. (2002) Porphyromonas gingivalis infection accelerates the progression of atherosclerosis in a heterozygous apolipoprotein E-deficient murine model. *Circulation*, **105**, 861–867.

119 Chi, H., Messas, E., Levine, R.A., Graves, D.T., and Amar, S. (2004) Interleukin-1 receptor signaling mediates atherosclerosis associated with bacterial exposure and/or a high-fat diet in a murine apolipoprotein E heterozygote model: pharmacotherapeutic implications. *Circulation*, **110**, 1678–1685.

120 Hsich, E., Zhou, Y.F., Paigen, B., Johnson, T.M., Burnett, M.S., and Epstein, S.E. (2001) Cytomegalovirus infection increases development of atherosclerosis in apolipoprotein-E knockout mice. *Atherosclerosis*, **156**, 23–28.

121 Vliegen, I., Duijvestijn, A., Grauls, G., Herngreen, S., Bruggeman, C., and Stassen, F. (2004) Cytomegalovirus infection aggravates atherogenesis in apoE knockout mice by both local and systemic immune activation. *Microbes Infect.*, **6**, 17–24.

122 Kuo, C.C., Gown, A.M., Benditt, E.P., and Grayston, J.T. (1993) Detection of

Chlamydia pneumoniae in aortic lesions of atherosclerosis by immunocytochemical stain. *Arterioscler. Thromb.*, **13**, 1501–1504.

123 Hendrix, M.G., Salimans, M.M., van Boven, C.P., and Bruggeman, C.A. (1990) High prevalence of latently present cytomegalovirus in arterial walls of patients suffering from grade III atherosclerosis. *Am. J. Pathol.*, **136**, 23–28.

124 Kozarov, E.V., Dorn, B.R., Shelburne, C.E., Dunn, W.A., Jr., and Progulske-Fox A. (2005) Human atherosclerotic plaque contains viable invasive *Actinobacillus actinomycetemcomitans* and *Porphyromonas gingivalis*. *Arterioscler. Thromb. Vasc. Biol.*, **25**, e17–e18.

125 Saikku, P., Leinonen, M., Mattila, K., *et al.* (1988) Serological evidence of an association of a novel *Chlamydia*, TWAR, with chronic coronary heart disease and acute myocardial infarction. *Lancet*, **2**, 983–986.

126 Hajishengallis, G., Wang, M., Harokopakis, E., Triantafilou, M., and Triantafilou, K. (2006) *Porphyromonas gingivalis* fimbriae proactively modulate beta2 integrin adhesive activity and promote binding to and internalization by macrophages. *Infect. Immun.*, **74**, 5658–5666.

127 Gibson, F.C., III, Hong, C., Chou, H.H., *et al.* (2004) Innate immune recognition of invasive bacteria accelerates atherosclerosis in apolipoprotein E-deficient mice. *Circulation*, **109**, 2801–2806.

128 Compton, T., Kurt-Jones, E.A., Boehme, K.W., *et al.* (2003) Human cytomegalovirus activates inflammatory cytokine responses via CD14 and Toll-like receptor 2. *J. Virol.*, **77**, 4588–4596.

129 Boehme, K.W., Guerrero, M., and Compton, T. (2006) Human cytomegalovirus envelope glycoproteins B and H are necessary for TLR2 activation in permissive cells. *J. Immunol.*, **177**, 7094–7102.

130 Tsan, M.F. and Gao, B. (2004) Endogenous ligands of Toll-like receptors. *J. Leukoc. Biol.*, **76**, 514–519.

131 Ohashi, K., Burkart, V., Flohe, S., and Kolb, H. (2000) Cutting edge: heat shock protein 60 is a putative endogenous ligand of the toll-like receptor-4 complex. *J. Immunol.*, **164**, 558–561.

132 Vabulas, R.M., Braedel, S., Hilf, N., *et al.* (2002) The endoplasmic reticulum-resident heat shock protein Gp96 activates dendritic cells via the Toll-like receptor 2/4 pathway. *J. Biol. Chem.*, **277**, 20847–20853.

133 Liu, B., Dai, J., Zheng, H., Stoilova, D., Sun, S., and Li, Z. (2003) Cell surface expression of an endoplasmic reticulum resident heat shock protein gp96 triggers My D88- dependent systemic autoimmune diseases. *Proc. Natl Acad. Sci. U. S. A.*, **100**, 15824–15829.

134 Smiley, S.T., King, J.A., and Hancock, W.W. (2001) Fibrinogen stimulates macrophage chemokine secretion through toll-like receptor 4. *J. Immunol.*, **167**, 2887–2894.

135 Okamura, Y., Watari, M., Jerud, E.S., *et al.* (2001) The extra domain A of fibronectin activates Toll-like receptor 4. *J. Biol. Chem.*, **276**, 10229–10233.

136 Tan, M.H., Sun, Z., Opitz, S.L., Schmidt, T.E., Peters, J.H., and George, E.L. (2004) Deletion of the alternatively spliced fibronectin EIIIA domain in mice reduces atherosclerosis. *Blood*, **104**, 11–18.

137 Termeer, C., Benedix, F., Sleeman, J., *et al.* (2002) Oligosaccharides of hyaluronan activate dendritic cells via toll-like receptor 4. *J. Exp. Med.*, **195**, 99–111.

138 Scheibner, K.A., Lutz, M.A., Boodoo, S., Fenton, M.J., Powell, J.D., and Horton, M.R. (2006) Hyaluronan fragments act as an endogenous danger signal by engaging TLR2. *J. Immunol.*, **177**, 1272–1281.

139 Johnson, G.B., Brunn, G.J., Kodaira, Y., and Platt, J.L. (2002) Receptor-mediated monitoring of tissue well-being via detection of soluble heparan sulfate by Toll-like receptor 4. *J. Immunol.*, **168**, 5233–5239.

140 Taylor, K.R., Trowbridge, J.M., Rudisill, J.A., Termeer, C.C., Simon, J.C., and Gallo, R.L. (2004) Hyaluronan fragments stimulate endothelial recognition of injury through TLR4. *J. Biol. Chem.*, **279**, 17079–17084.

141 Lotze, M.T. and Tracey, K.J. (2005) High-mobility group box 1 protein (HMGB1): nuclear weapon in the immune arsenal. *Nat. Rev. Immunol.*, **5**, 331–342.

142 Kalinina, N., Agrotis, A., Antropova, Y., *et al.* (2004) Increased expression of the DNA- binding cytokine HMGB1 in human atherosclerotic lesions: role of activated macrophages and cytokines. *Arterioscler. Thromb. Vasc. Biol.*, **24**, 2320–2325.

143 Park, J.S., Svetkauskaite, D., He, Q., *et al.* (2004) Involvement of toll-like receptors 2 and 4 in cellular activation by high mobility group box 1 protein. *J. Biol. Chem.*, **279**, 7370–7377.

144 Miller, Y.I., Viriyakosol, S., Binder, C.J., Feramisco, J.R., Kirkland, T.N., and Witztum, J.L. (2003) Minimally modified LDL binds to CD14, induces macrophage spreading via TLR4/MD-2, and inhibits phagocytosis of apoptotic cells. *J. Biol. Chem.*, **278**, 1561–1568.

145 Miller, Y.I., Viriyakosol, S., Worrall, D.S., Boullier, A., Butler, S., and Witztum, J.L. (2005) Toll-like receptor 4-dependent and -independent cytokine secretion induced by minimally oxidized low-density lipoprotein in macrophages. *Arterioscler. Thromb. Vasc. Biol.*, **25**, 1213–1219.

146 Boullier, A., Friedman, P., Harkewicz, R., *et al.* (2005) Phosphocholine as a pattern recognition ligand for CD36. *J. Lipid Res.*, **46**, 969–976.

147 Walton, K.A., Hsieh, X., Gharavi, N., *et al.* (2003) Receptors involved in the oxidized 1- palmitoyl-2-arachidonoyl-sn-glycero-3-phosphorylcholine-mediated synthesis of interleukin-8. A role for Toll-like receptor 4 and a glycosylphos-phatidylinositol- anchored protein. *J. Biol. Chem.*, **278**, 29661–29666.

148 Suganami, T., Tanimoto-Koyama, K., Nishida, J., *et al.* (2007) Role of the Toll-like receptor 4/NF-kappaB pathway in saturated fatty acid-induced inflammatory changes in the interaction between adipocytes and macrophages. *Arterioscler. Thromb. Vasc. Biol.*, **27**, 84–91.

149 Bochkov, V.N., Kadl, A., Huber, J., Gruber, F., Binder, B.R., and Leitinger, N. (2002) Protective role of phospholipid oxidation products in endotoxin-induced tissue damage. *Nature*, **419**, 77–81.

150 Walton, K.A., Cole, A.L., Yeh, M., *et al.* (2003) Specific phospholipid oxidation products inhibit ligand activation of toll-like receptors 4 and 2. *Arterioscler. Thromb. Vasc. Biol.*, **23**, 1197–1203.

6

PPAR-Based Therapies for the Management of Atherosclerosis

Florence Gizard and Inés Pineda-Torra

Members of the nuclear receptor superfamily are transcription factors that play crucial roles in the regulation of developmental, reproductive, homeostatic, inflammatory, immune, and metabolic processes [1]. There are 48 genes that encode nuclear receptors in the human genome, and these receptors represent one of the most important targets for therapeutic drug development. Most nuclear receptors have a highly conserved structure consisting of an N-terminal domain that harbors a ligand-independent activation function (AF-1), a zinc finger DNA-binding domain, and a C-terminal ligand-binding domain containing a ligand-dependent transcriptional activation function (AF-2) [2].

Peroxisome proliferator–activated receptors (PPARs) are lipid-activated transcription factors that belong to a subclass of nuclear receptors that form heterodimers with the retinoid X receptor (RXR). In the absence of ligand, as for most RXR heterodimers, they are bound to chromatin in association with corepressors, histone deacetylases, and chromatin-modifying factors to maintain the active repression of target genes [3]. Upon ligand binding, a conformational change in the receptor occurs, prompting the exchange of corepressors for co-activators, resulting in the initiation of target gene expression. PPARs can activate or repress gene transcription in a ligand-specific and gene-specific fashion. Although these nuclear receptors were first described for their roles in lipid and glucose metabolism, they have subsequently emerged as important regulators of inflammation, immunity and vascular biology. This ability to link metabolism to inflammatory signaling has expanded their potential as molecular drug targets for the treatment of metabolic as well as inflammatory diseases such as atherosclerosis.

6.1

PPARs: Nuclear Receptors as Pleiotropic Transcriptional Regulators

The PPAR subfamily comprises PPARα, PPARδ (also known as PPARβ), and PPARγ (NR1C1, NR1C2, and NR1C3, respectively). Originally cloned from a *Xenopus* cDNA expression library [4], they are also present in mammals and

Atherosclerosis: Molecular and Cellular Mechanisms. Edited by Sarah Jane George and Jason Johnson
Copyright © 2010 WILEY-VCH Verlag GmbH & Co. KGaA, Weinheim
ISBN: 978-3-527-32448-4

humans. PPARα is the founding member and was identified as the target of fibrates and compounds that induce peroxisome proliferation in rodents [5]. PPARδ and PPARγ were subsequently cloned on the basis of sequence homology and their ability to control adipocyte tissue gene expression [6, 7].

The three PPARs have different tissue distributions and control different but overlapping biological processes. PPARα controls fatty acid oxidation and is a crucial regulator of hepatic energy homeostasis, while PPARδ plays an important role in energy homeostasis, thermogenesis, and fatty acid β-oxidation and PPARγ controls adipocyte differentiation and determines insulin sensitivity and glucose control.

The PPAR/RXR heterodimer binds to specific DNA sequences (termed PPAR response elements or PPREs) in the regulatory regions of their target genes. PPREs are degenerate versions of a direct repeat of the motif AGGTCA separated by one or two nucleotides [8]. Most PPREs identified to date are located in proximal promoter regions. However the advent of chromatin immunoprecipitation (ChIP)-on-chip technologies will likely uncover also PPREs present in intronic, downstream, and/or intergenic enhancers as has been shown for other nuclear receptors [9]. Endogenous ligands for PPARs include polyunsaturated fatty acids (FAs), eicosanoids and lipoprotein components (Table 6.1). Many of these bind with relatively low affinities ($\geq 1\,\mu M$) and thus the biological relevance of these interactions

Table 6.1 PPAR ligands.

Receptor	Endogenous ligands	Synthetic ligands
PPARα	Unsaturated fatty acids Saturated fatty acids Leukotriene B4 8-HETE	Clofibrate Fenofibrate Gemfibrozil GW7647 WY14643
PPARδ	Unsaturated fatty acids Saturated fatty acids Carbaprostacyclin ox-VLDL components	GW501516 L-165041 GW0742 GW7842
PPARγ	Unsaturated fatty acids 15dPG72, 15-HETE, 9-HODE, 13-HODE ox-LDL components	Ciglitazone Pioglitazone Rosiglitazone Troglitazone Tyrosine derivatives Farglitazar GW7845
PPARα + PPARγ	None	Muraglitazar Ragaglitazar Tesaglitazar

HETE, hydroxyeicosatetraenoic acid; HODE, hydroxyoctadecadienoic acid.

still remains to be demonstrated in most cases. It is thought that the high basal activity these receptors exhibit in mammalian cells compared with other members of the family could be at least partly due to endogenous ligand binding.

Another complication in extrapolating studies performed with PPAR natural ligands is that some act via both PPAR-dependent and PPAR-independent mechanisms [10, 11]. In addition, a number of synthetic ligands have been identified for PPARs. This molecular promiscuity is thought to be due to a large ligand-binding pocket that is able to accommodate a diverse range of natural and synthetic ligands [12].

6.2
PPARα

Mono- or polyunsaturated FAs and eicosanoids [13, 14], as well as long-chain fatty acyl-CoAs and saturated FAs, constitute PPARα endogenous ligands, the latter with EC_{50} values in the high-micromolar range [15, 16]. Consistent with this, liver-specific inactivation of the FA synthase (FAS) gene in mice results in a phenotype identical to that of fasting PPARα-deficient mice [17]. PPARα is highly expressed in tissues displaying a high catabolic rate of FAs, such as liver, heart, kidney, skeletal muscle (mostly in slow twitch, oxidative type I fibers), brown fat, and cells present in atherosclerotic lesions. In these tissues and cell types, activated PPARα regulates the expression of genes encoding either enzymes stimulating FA oxidation or proteins involved in lipid transport, thus improving lipoprotein metabolism. Fibrates are synthetic PPARα ligands with triglyceride (TG)-lowering and high-density lipoprotein (HDL)–raising properties that are currently administrated to patients with metabolic syndrome or type 2 diabetics in order to reduce cardiovascular diseases associated with dyslipidemia. Several studies have, however, highlighted their direct pleiotropic effects on the vascular wall as well, which may also account for their anti-atherogenic actions.

6.2.1
PPARα in Metabolic Control

The therapeutic benefit of fibrates is due in part to reduced very-low-density lipoprotein (VLDL) production and enhanced catabolism of TG-rich particles, which indirectly decreases small dense low-density lipoprotein (LDL) particles, enhancing HDL formation and hepatic elimination of excess cholesterol. Targeted disruption of the PPARα gene in mice has underscored the importance of this receptor in energy balance and provided many insights into the role of PPARα in lipid homeostasis. While the phenotype of PPARα$^{-/-}$ mice fed *ad libitum* is mild, severe impairment in FA uptake and FA β-oxidation were reported upon exposure to metabolic stress such as fasting [18], or inhibition of mitochondrial FA [19]. Further molecular studies have characterized multiple PPARα key target genes that are essential for lipid homeostasis. These include:

- **FA uptake, intracellular transport, and β-oxidation.** Notably, PPARα induces expression of genes encoding the FA transport protein, FA translocase, long-chain FA acetyl-coenzyme A synthase, and carnitine palmytoyl transferase I [20].

- **TG and LDL metabolism.** PPARα enhances the activity of the lipoprotein lipase (LPL), which hydrolyzes TG-rich lipoproteins, the major source of circulating FAs. Two mechanisms are involved: the direct upregulation of the LPL gene [21] and the repression of apolipoprotein (Apo) CIII expression, which is an endogenous inhibitor of LPL activity [22, 23]. Moreover, the gene encoding ApoA5, a major determinant of plasma TG levels, is also highly responsive to PPARα ligands [24].

- **HDL metabolism.** PPARα activation increases the transcription of apolipoproteins ApoAII [25] and ApoAI [26], the major source of HDL.

6.2.2
PPARα in the Vasculature

6.2.2.1 Vascular Endothelium
In addition to tissues with high FA catabolic rates, PPARα is also expressed in the vascular endothelium, where it can directly modulate endothelial responses [27, 28]. Activated PPARα represses gene expression of vascular cell adhesion molecule-1 (VCAM-1) [28], thereby limiting monocyte recruitment to nascent atherosclerotic lesions [29, 30]. In other cell types (see below), PPARα-dependent repression of inflammatory genes involves the inhibition of NF-κB activation in gene promoters. However, conflicting data still exist concerning the mechanistic basis of PPARα inhibition of chemokine expression in endothelial cells (ECs) [31, 32], suggesting that this effect may be gene-selective. In addition, in ECs, fibrates increase the Cu^{2+},Zn^{2+}-superoxide dismutase, which scavenges reactive oxygen species (ROS) and decreases NADPH oxidase, thus potentially limiting oxidative stress and LDL oxidation [33]. Whether the potential actions of PPARα on endothelial inflammation and oxidation have an effect on atherosclerosis development remain to be assessed in experimental models of the disease.

6.2.2.2 Monocytes/Macrophages/T Lymphocytes
PPARα is upregulated during monocyte differentiation into macrophages [34], where it increases ApoAI-mediated cholesterol efflux via inducing the expression of the cholesterol transporter ATP-binding cassette transporter-A1 (ABCA1) [35]. Conversely, PPARα ligands inhibit the expression of the ApoB48-remnant receptor [36] and reduce the uptake of glycated LDL and TG-rich remnant lipoproteins [37] in these cells. PPARα also inhibits the expression of various macrophage and T cell proinflammatory molecules, which contribute to the anti-atherothrombotic actions of this receptor. PPARα-deficiency provokes a chronic inflammatory response in mice providing genetic evidence for the anti-inflammatory activity of

receptor [38]. Moreover, PPARα induces macrophage ROS production, leading to a modification of LDL, which in turn acts as a PPARα ligand to inhibit the induction of inflammatory genes, such as inducible nitric oxide synthase (iNOS) [39]. PPARα activation also represses the expression osteopontin, a proinflammatory cytokine implicated in the chemoattraction of monocytes [40], and inhibits the expression of the procoagulant tissue factor [41, 42]. Additionally, PPARα activation reduces the expression and secretion of the matrix metalloproteinase (MMP)-9 [43], although the consequences of this repression remain to be investigated, as MMP-9 may increase vascular smooth muscle cell (SMC) migration, thereby altering the stability of atherosclerotic lesions. Finally, PPARα also limits the expression of inflammatory genes, such as interferon-γ (IFN-γ) and tumor necrosis factor-α (TNF-α) [44] in T lymphocytes, which contribute to the vascular inflammatory response during initial atherogenesis,

The mechanisms underlying PPARα negative regulation of inflammatory gene expression are not completely understood. Several processes are known to be involved, including direct interactions with the transcription factors NF-κB and AP-1, and regulation of cytokine and growth factor signaling [45]. Interestingly, this transrepression function of PPARα has been shown to be affected by the phosphorylation of the receptor by protein kinase C [46], although the mechanisms involved need to be defined. In contrast to PPARγ (see below), the implication of PPARα sumoylation in the transrepression activities of the receptor has not been explored to date.

6.2.2.3 SMCs

In SMCs, PPARα activation inhibits interleukin-6 (IL-6) and prostaglandin secretion and reduces the expression of cyclooxygenase-2 (COX-2) [47]. Consistent with this, fenofibrate decreases circulating levels of inflammatory markers and mediators, including IL-6 and C-reactive protein, in hyperlipidemic patients [47]. Likewise, IL-6 expression is repressed by fibrate treatment in mice aorta in a PPARα-dependent manner [48]. More recently, PPARα was shown to regulate SMC proliferation and migration, processes that are activated in response to vascular inflammation and are key in atherosclerosis development and its complications. Inhibition of SMC proliferation by PPARα activation occurs by interfering with $G_1 \rightarrow S$ cell cycle progression via the upregulation of p16[Ink4a] (p16) [49, 50], a cyclin-dependent kinase inhibitor that inhibits cell replication [51]. Furthermore, these effects on SMC growth involve the transcriptional repression of telomerase reverse transcriptase (TERT) activity in a retinoblastoma product (pRB)/E2F-dependent mechanism [50]. Interestingly, telomerase activity has been associated with atherosclerosis and neointima formation in animal models [52], suggesting that pharmacological inhibition of telomerase by PPARα agonists could represent an interesting strategy for the prevention and/or treatment of occlusive complications of atherosclerosis. Besides, PPARα activation also inhibits secretion of MMP-9, which contributes to the degradation of the extracellular matrix and thus to SMC migration [43]. However, the impact of SMC growth and migration, and thus of their regulation by fibrates, on cardiovascular morbidity may also depend on the stage of the atherosclerotic plaque development.

6.2.3
PPARα in Animal Models of Atherosclerosis and Associated Vascular Complications

A number of PPARα agonists have been tested in several mice models for atherosclerosis in studies that have yielded complex results (Table 6.2). It should be noted

Table 6.2 Effects of PPAR ligands on atherosclerosis.

Ligands	Dose (mg/kg bw/d)	Time (weeks)	Diet	Model/sex	Effect on lipids	Effect on lesions	Ref.
PPARα ligands							
Bezafibrate	0.05%w/w	13	chow	ApoE$^{-/-}$ (M/F)	↑	nd	[53]
	0.1 % w/w	8	1.25 chol	LDL-R$^{-/-}$ (M)	↑ HDL	↓	[54]
Ciprofibrate	0.05%w/w	13	chow	ApoE$^{-/-}$ (M/F)	↑	↑↑	[53]
	100	21	0.4% chol	LDLR$^{-/-}$ (F)	↑	↑ (nq)	[55]
Gemfibrozil	100	20	chow	ApoE$^{-/-}$	↓(LDL), nc (HDL)	↓↓(nq)	[56]
	100	20	chow	ApoE$^{-/-}$ (M)	↓ (LDL)	↓↓	[57]
Fenofibrate	100	8	0.2% chol	ApoE$^{-/-}$ (M)	nc	nc	[58]
	100	8	0.2% chol	hApoA1-ApoE$^{-/-}$ (M)	nc	↓↓	[58]
	200	14	0.2% chol	ApoE$^{-/-}$ (M)	↑	nc	[59]
	0.1% w/w	14	0.2% chol	ApoE$^{-/-}$ (M)	↑	nc	[60]
	10	11	chow	ApoE$^{-/-}$ (F)	+(V/LDL), nc HDL	nc	[61]
	0.05% w/w	13	chow	ApoE$^{-/-}$ (M/F)	↑	nd	[53]
	100	10	0.2% chol	hApoE2-KI (M)	↓ (V/LDL), ↑HDL	↓↓↓	[62]
	100	24	1.25% chol	LDLR$^{-/-}$ (M)	↓	↓↓↓	[63]
	30	18	0.5% chol	E*3L (F)	↓	↓↓↓	[64]
WY14643	0.2% w/w	13	chow	ApoE$^{-/-}$ (M/F)	↑	nd	[53]
GW7647	2.5	14	1.25% chol	LDLR$^{-/-}$ (M)	nc	↓↓	[65]
PPARγ ligands							
Pioglitazone	40	10	0.2% chol	hApoE2-KI (M)	nc	nc	[62]
	20	28	1.25% chol	LDLR$^{-/-}$ (M)	nc	↓	[66, 67]

Table 6.2 *Continued*

Ligands	Dose (mg/ kg bw/d)	Time (weeks)	Diet	Model/sex	Effect on lipids	Effect on lesions	Ref.
Rosiglitazone	20	12	0.02% chol	ApoE$^{-/-}$ (M)	↑	↓↓	[68]
	20	20	chow	ApoE$^{-/-}$ (M)	nc	↓↓↓↓	[69]
	10	11	chow	ApoE$^{-/-}$ (F)	↑ (V/ LDL), nc HDL	↓	[61]
	20	20	chow	ApoE$^{-/-}$ (M)	nc	↓↓↓	[57]
	10	10	0.2% chol	hApoE2-KI (M)	nc	nc	[62]
	20	10	1.25% chol	LDLR$^{-/-}$ (M/F)	nc	↓↓↓ (M), nc (F)	[70]
	100	24	1.25% chol	LDLR$^{-/-}$ (M)	↓	↓↓↓	[63]
Troglitazone	0.1% w/w	8	0.15% chol	ApoE$^{-/-}$ (M)	nc, ↑HDL	↓↓	[71]
	400	12	0.15% chol	LDLR$^{-/-}$ (M)	↓, ↓HDL	↓↓	[72]
GW7845	20	10	1.25% chol	LDLR$^{-/-}$ (M)	nc	↓↓↓	[70]
PPARδ ligands							
GW7842	5	14	1.25% chol	LDLR$^{-/-}$ (M)	nc	nc	[65]
GW0742X	6	10/16	0.25% chol	LDLR$^{-/-}$ (F)	↓VLDL	↓ (10), ↓↓(16)	[73]
	60	10/16	0.25% chol	LDLR$^{-/-}$ (F)	nc	↓↓ (10), ↓(16)	[73]
GW0742	1/10	4	1.25% chol	LDLR$^{-/-}$	↓TG	↓↓	[74]
GW501516	2	8	0.15% chol	ApoE$^{-/-}$	↑HDL	↓	[74]
Selective PPARγ modulator							
Telmisartan	1	12	1.5% chol	ApoE$^{-/-}$ (M)	nc	↓	[75]
	0.3	12	chow	ApoE$^{-/-}$ (M/F)	nc	↓	[76]
	3	12	chow	ApoE$^{-/-}$ (M/F)	nc	↓	[76]
	40	16	chow	ApoE$^{-/-}$ (M)	nc	↓	[77]
PPARα/γ dual ligands							
GW2331	5	11	chow	ApoE$^{-/-}$ (F)	↑	↓↓	[61]
LY465608	10	18	chow	ApoE$^{-/-}$ (M)	nc	↓↓	[78]
Tesaglitazar	20	28	1% chol	E*3L (F)	↓	↓↓↓	[79]
	20	12	0.15% chol	LDLR$^{-/-}$ (M/F)	nc	nc (M), ↓(F)	[80]
3q compound	3	20	chow	ApoE$^{-/-}$ (M)	↓LDL, ↓HDL	↑↑	[57]

nc, no change; nd, not determined; nq, not quantified. Adapted from Zadelaar *et al.* [263].

that low levels of PPARα expression in murine macrophages and the mouse-specific induction of hepatic peroxisome proliferation in response to fibrates, resulting in hepatomegaly and hepatocellular proliferation, may limit the extrapolation of those studies to humans [81, 82]. PPARα-deficient mice crossed with the ApoE-deficient (ApoE$^{-/-}$) atherosclerosis mouse on a high-fat diet are less susceptible to atherosclerotic lesion development and insulin resistance than their PPARα$^{+/+}$/ApoE$^{-/-}$ littermates [83]. In line with this, fenofibrate administration to other Western diet–fed murine models, such as the human ApoAI transgenic ApoE$^{-/-}$ (hApoAI Tg × ApoE$^{-/-}$) or ApoE2 knockin (ApoE2KI) mice, reduces atherosclerosis [58, 62]. Fenofibrate was also shown to reduce atherosclerosis development in LDL receptor–deficient (LDL-R$^{-/-}$) and ApoE3*Leiden mice fed a high-cholesterol diet [63, 64]. In these two studies, reduction of atherosclerosis was associated with a cholesterol-lowering effect of fenofibrate [63, 64]. On the other hand, several studies have indicated that fenofibrate administration to ApoE$^{-/-}$ mice does not reduce atherosclerotic lesions at the aortic origin [58–61]. Fenofibrate treatment in ApoE$^{-/-}$ mice did, however, reduce cholesterol content along the descending aorta in older mice [58].

Other PPARα agonists, such as gemfibrozil and bezafibrate, decrease atherosclerosis in ApoE$^{-/-}$ mice while reducing LDL-cholesterol or HDL-cholesterol levels, respectively [54, 56, 57]. Activation of PPARα by the highly specific and potent GW7647 agonist also inhibited atherosclerosis in hyperlipidemic LDL-R$^{-/-}$ mice, albeit without significantly altering hyperlipidemia [65]. This was in line with data from studies performed in rabbits [84, 85] and in the hApoAI Tg × ApoE$^{-/-}$ mouse model [58], indicating that fibrate treatment can inhibit atherosclerosis without affecting plasma lipid profiles. By contrast, ciprofibrate administration markedly increased plasma levels of atherogenic lipoproteins and aggravated atherosclerosis development both in ApoE$^{-/-}$ [53] and LDL-R$^{-/-}$ mice [55]. These discrepancies may be attributable to differences between agonists, duration of the experimental treatment, diet and animal models used in these studies. Notably, as highlighted in [65], we cannot exclude a role for ApoE in partly mediating the anti-atherogenic effects of PPARα. Moreover, fibrate effects on restenosis have been analyzed [86] and it was recently shown that fibrate inhibition of SMC hyperproliferation occurs via the p16/telomerase-pathway [49, 50]. These results further highlight that direct inhibitory effects of PPARα on the vasculature can affect both atherosclerosis as well as its occlusive complications.

6.2.4
PPARα in Cardiovascular Clinical Trials

PPARα signaling is critical to the progression of atherosclerotic lesion formation in humans. A reduction of coronary atherosclerosis progression was initially observed with bezafibrate in the Bezafibrate Coronary Atherosclerosis Intervention Trial (BECAIT) [87], with the PPAR pan-agonist gemfibrozil in the Lipid Coronary Angiography Trial (LOCAT) [88] and with fenofibrate in the Diabetes Atherosclerosis Intervention Study (DAIS) [81]. In addition, the influence of fibrate treatment on cardiovascular morbidity and mortality was studied in primary inter-vention trials, such as the Helsinki Heart Study (HHS) using gemfibrozil [89, 90] and more recently the Fenofibrate Intervention and Event Lowering in Diabetes (FIELD) trial [91], as well as in the secondary prevention studies Bezafibrate Infarc-tion Prevention (BIP) [92] and the Veterans Affairs High-Density Lipoprotein Cholesterol Intervention Trial (VA-HIT) with gemfibrozil [93, 94].

Fibrate treatment reduced the incidence of coronary heart disease in both the HHS [89] and VA-HIT trials [93]. Of note, BIP and VA-HIT subjects were not on statins, LDL-cholesterol-lowering drugs. Altogether, results from these three studies (BIP, HSS, and VA-HIT) indicate that fibrates are particularly useful in treating cardiovascular risk in insulin-resistant pre-diabetic individuals, and in diabetic patients with dyslipidemia. In addition, concurring with cellular and animal studies, the cardiovascular benefits of fibrates in the BECAIT, LOCAT, DAIS, and VA-HIT trials occurred through effects that were possibly independent or additional to their systemic action. Importantly, genetic studies on PPARα poly-morphisms have also indicated an association between PPARα variants and the risk of coronary atherosclerosis and ischemic heart diseases [95].

FIELD, the largest and most recent study, tested the effects of fenofibrate on first or recurrent cardiovascular events in type 2 diabetics with no indication of initial lipid-lowering therapy [91]. Primary end-points (death from coronary heart disease (CHD) or nonfatal myocardial infarction (MI)) in this study were decreased with treatment although this did not achieve a statistical significance. Various factors have been suggested as possible contributors to these results, including the relatively higher baseline levels of HDL in the fenofibrate-treated group and a higher drop-in rate of statin use in the placebo group during the course of the study. Since statins can decrease cardiovascular risk in type 2 diabetics [96], these differences may have led to an underestimation of fenofibrate effects. Several secondary end-points were significantly reduced however, including nonfatal MI and total cardiovascular events. It is worth noting that the beneficial effects of fenofibrate on cardiovascular events were concentrated on primary prevention [97].

Interestingly, fenofibrate also reduced microvascular complications (namely retinopathy and nephropathy), which are a major source of diabetic morbidity. It was suggested that this effect was due to a loss of LPL function, which generates PPAR ligands [98]. A statistically insignificant increase in cardiovascular mortality was also noted with fenofibrate, with a hazard ratio of 1.19. Moreover, fenofibrate administration was associated with a correction of type 2 diabetes-associated dysli-pidemia, including a reduction of LDL-cholesterol, suggesting that fenofibrate represents an interesting second-line alternative therapy to statins. Importantly, the FIELD trial does not indicate safety issues associated with fenofibrate–statin co-therapy. Whether fenofibrate treatment confers additional benefits when given on top of a statin will be addressed in the ongoing Action to Control Cardiovascular Risk in Diabetes (ACCORD) study, the results of which are expected in 2010.

PPARα activation has also been implicated on the hypertensive response in humans [98]. Additionally, association between PPARα gene variants and left ventricle hypertrophy in response to training and hypertension has been reported, demonstrating a link between PPARα, hypertension and cardiac function [99]. Thus, although until now hypertension has not been recorded in the fibrate clinical trials cited above, these putative side-effects should be considered during the current development of molecules with more potent PPARα activity [100, 101].

6.2.5
Perspectives for PPARα-Based Therapies

Over the last 15 years studies on the fibrate/PPARα pathway have demonstrated that expression and/or activation of this receptor reduces cholesterol accumulation, inflammation, oxidation, and proliferation. However, these PPARα-dependent effects may depend on the inflammatory status and stage of atherosclerotic plaque development. For instance monocyte chemotactic protein-1 (MCP-1) expression upon PPARα activation is enhanced in ECs and early atherosclerotic lesions [32], but inhibited in macrophages and late lesions [65]. The cardiovascular effects of PPARα activators may also be dependent on the PPARα agonist used, which may display selective PPARα regulation or PPARα-independent actions. These parameters will need to be taken into consideration for therapy improvement. Finally, ongoing studies analyzing the potential interaction of fibrates with statins suggest a mechanistic link between both classes of drugs to regulate vascular (dys)functions. Indeed, while PPARα can mediate the anti-inflammatory effects of statins [102], similar to fibrates, statins have been reported to limit SMC proliferation [103]. Results from clinical studies analyzing the cardiovascular effects of fibrate/statin combination therapies are thus eagerly anticipated. Finally, the benefits of fenofibrate administration on microvascular disease in combination therapy require further confirmation before their use as therapeutic agents.

6.3
PPARδ

PPARδ is the only PPAR subtype that is not a target of current drugs. Native and modified polyunsaturated fatty acids are endogenous ligands for this receptor [13, 16] as well as dietary fatty acids carried by VLDLs in the presence of enzymatically active lipoprotein lipase [104]. Although these findings initially suggested that PPARδ acts as a lipid sensor, the physiological role of this receptor remained somewhat obscure until recently, as a result of the identification of highly selective small molecule PPARδ agonists.

6.3.1
PPARδ in Metabolic Control

Although PPARδ deletion in mice results in frequent embryonic lethality, surviving pups showed reduced adipose tissue mass [105, 106]. PPARδ-deficient mice, however, do not exhibit differences in white and brown fat, and thus, unlike PPARγ, PPARδ does not appear to play a role as a master regulator of adipose tissue differentiation [105]. Overexpression of a constitutively active VP16-PPARδ transgene in mice adipose tissue improves lipid profiles and reduces adiposity while being resistant to obesity [107]. PPARδ also activates genes involved in FA oxidation and thermogenesis, suggesting the effect on obesity is due to an increase in energy expenditure and FA catabolism. Further supporting a role for PPARγ in obesity, administration of a PPARδ agonist to obese db/db mice abrogates lipid accumulation, while PPARδ-deficient mice on a high-fat diet are prone to obesity [107].

In addition to its functions in adipose tissue, studies have pointed to a role for PPARδ in muscle physiology and metabolism. In mice overexpressing PPARδ specifically in muscle, regardless of whether it is wild-type or VP-16-PPARδ, a switching in fiber type occurs with an increase in type I fibers (mitochondria-rich, which use oxidative metabolism for energy production) accompanied by an induction of mitochondria numbers and oxidative capacity [108, 109]. As a consequence these mice are lean and display improved metabolic status and increased running endurance on treadmills, which further induces PPARδ levels. Remarkably, muscle-specific PPARδ-deficient mice exhibited a reciprocal phenotype further highlighting the role of PPARδ in skeletal muscle function [110]. These studies also point to a role for PPARγ coactivator-1 (PGC-1) as a coactivator for PPARδ to control mitochondria biogenesis and muscle fiber type [107, 109, 110]. More recently PPARδ and AMP-activated protein kinase (AMPK) were shown to interact to control myofiber metabolism and exercise endurance [111]. In addition to skeletal muscle, PPARδ also controls FA metabolism in cardiomyocytes, which rely on FAs as their main energy source, and protects against cardiac lipotoxicity. Cardiomyocyte-specific PPARδ-deficient mice show reduced FA β-oxidation rates, which lead to lipid accumulation in the heart, with mice developing cardiac hypertrophy and myopathy, eventually leading to death [112].

6.3.1.1 PPARδ and Metabolic Syndrome

PPARδ activation with a high-affinity ligand like GW501516 has proven useful in ameliorating several features of the metabolic syndrome. This compound was shown to reduce weight gain and decrease circulating TGs in high-fat fed or leptin-deficient ob/ob mice by increasing FA catabolism [109, 113]. In addition, a number of studies using PPARδ agonists have demonstrated a role for this receptor in the regulation of HDL levels. Administration of GW501516 to obese primates resulted in a dramatic increase in HDL-cholesterol, while decreasing LDL-cholesterol and fasting TGs [114]. These effects were associated with the ability of PPARδ to induce ABCA1 expression and reverse cholesterol transport. An increase in circulating HDL-cholesterol levels was also achieved with the agonist L-165041 in insulin-resistant db/db mice. Surprisingly, however, PPARδ-deficient mice do not exhibit altered HDL or TG levels [105, 106]. In addition, neither ABCA1 expression nor cholesterol efflux is induced by GW501516 in mouse macrophages. These discrepancies point to a species-specific regulation or to a different mechanism altogether [115].

PPARδ is also a demonstrated key regulator of insulin sensitivity and hepatic steatosis [109, 113, 116]. Moreover, PPARδ-deficient mice also show a range of metabolic defects, including lower metabolic rates and glucose intolerance. The glucose-lowering effect of GW501516 is mediated in part by metabolizing glucose for lipogenesis in liver, leading to a moderate increase in hepatic TG levels. This hepatic lipogenic activity could not be confirmed in a mouse model of non-alcoholic steatohepatitis presenting massive lipid accumulation and inflammation in liver [117]. In this model GW501516 not only reduced hepatic TG levels but also fat accumulation, inflammatory cell infiltration, and expression of inflammatory markers, consistent with the ability of PPARδ ligands to modulate macrophage inflammation (see below). Interestingly, it has been suggested recently that certain PPARδ ligands may also confer their metabolic effects through mechanisms involving the posttranslational modification (sumoylation) of a second transcription factor: the Kruppel-like transcription factor 5 (KLF5) [118]. Sumoylated KLF5 appears to form a repressive complex with unliganded PPARδ in the promoters of genes involved in lipid oxidation and energy coupling [118]. Ligand activation of PPARδ promotes KLF5 desumoylation, which in turn triggers a dynamic exchange of cofactors, resulting in the transcriptional activation of these genes.

These findings highlight the crosstalk between two prominent families of transcription factors through a novel ligand-dependent mechanism. Altogether, results obtained to date suggest small molecules targeting PPARδ may have beneficial effects on lipid profiles, obesity, and type 2 diabetes.

6.3.2
PPARδ in the Vasculature

6.3.2.1 Macrophage Metabolism and Inflammation
Similar to PPARα, PPARδ expression is induced during differentiation of human macrophages [119]. The robust expression of adipocyte differentiation–related protein in a PPARδ-dependent fashion upon loading macrophages with VLDLs suggests a role for PPARδ also in macrophage TG accumulation [104]. Similar to PPARα and PPARγ agonists, the PPARδ ligand GW501516 induces ABCA1 expression in human monocytic cell lines [114]. However, other researchers have reported through gain- and loss-of function approaches that PPARδ does not directly regulate ABCA1 expression or cholesterol efflux in mouse macrophages [65, 115]. By contrast, other PPARδ agonists (compound F) were found to promote lipid accumulation via upregulation of the scavenger receptors CD36 and scavenger receptor A (SRA) and downregulation of cholesterol efflux genes, such as ApoE, in human macrophages [119]. Thus, the impact of PPARδ activation in macrophage cholesterol homeostasis remains controversial and appears to occur in a species-specific and compound-selective manner.

Results from PPARδ-deficient and PPARδ-overexpressing macrophages showed an unexpected proinflammatory role for the receptor, with PPARδ-deficient macrophages showing reduced expression of inflammatory mediators such as MCP-1 and IL-1β [115]. However, PPARδ activation suppresses the basal levels of these molecules. This apparent discrepancy appears to be partly mediated by the transcriptional repressor BCL-6. The interaction between unliganded PPARδ and BCL-6 was proposed to sequester BCL-6 away from promoters of inflammatory genes. Upon ligand activation of PPARδ corepressors are released from the complex and are able to bind to proinflammatory gene promoters, resulting in their repression. This model is in striking contrast to that proposed by Glass and colleagues for PPARγ and LXRα [120, 121], in which other transcriptional corepressors such as the silencing mediator of retinoic acid and thyroid hormone receptor (SMRT) and nuclear receptor corepressor (NCoR) are involved and post-translational modification of the receptors play a modulatory role (see below). Further studies are needed to elucidate whether this working model is restricted to BCL-6 or is also valid for other corepressors and whether the anti-inflammatory actions of PPARδ are modulated by modifications of the receptor, a process still poorly explored.

6.3.3
PPARδ in the Alternative Activation of Macrophages

Macrophages emerge from precursor cells in the bone marrow, circulate in blood as monocytes and migrate within tissues to remain in them (such as Kupffer cells in the liver) or infiltrating tissues as a defense mechanism in an inflammatory context. Macrophages classically activated (M1) by secreted molecules from T helper cell 1 (Th1) cells, such as IFN-γ, are proinflammatory and have microbicidal properties, whereas alternatively activated macrophages (M2) by exposure to Th2-secreted cytokines, such as IL-4 and IL-13, reduce proinflammatory cytokine secretion and promote endocytosis and tissue repair [122]. Recently, PPARδ, as was previously reported for PPARγ, has been implicated in the molecular switch towards the M2 phenotype and how this affects metabolic tissues like adipose tissue and liver. PPARδ activation regulates basal macrophage expression of alternative activation markers such as arginase 1 (Arg1) [123–125]. This suggests these genes may contain PPREs in their regulatory regions and at least for Arg1, a functional PPRE has been characterized [126]. Consistent with M2 activation PPARδ-induced macrophages diminish parasite clearance from these cells [123]. There are some discrepancies, however, concerning the regulation of IL-4-mediated induction of Arg1 by PPARδ. Odegaard and colleagues described that, similar to PPARγ, IL-4-induced Arg1 expression is diminished in PPARδ-deficient bone marrow-derived macrophages in a complementary fashion since deletion of either PPARγ or PPARδ exert this effect [126]. By contrast, a recent study reported no differences in the IL-4-induced levels of Arg1 and other M2 markers in PPARδ-deleted macrophages [127]. These conflicting results may stem from the different mouse strains or approaches used to generate PPARδ deficiency (see below).

It has become increasingly clear that inflammation mediated by tissue-resident macrophages is an essential component of cardiovascular as well metabolic disorders [128, 129]. Obesity is often accompanied by an increased infiltration of adipose tissue macrophages (ATMs) and lean mice present ATMs with a M2 profile, whereas obese animals exhibit infiltration of M1 macrophages [130]. As demonstrated for PPARγ, macrophage/myeloid cell-specific deletion of PPARδ affects obesity and insulin resistance metabolic profiles. Adipocytes, as well as hepatocytes, produce IL-4 and Il-13 cytokines, establishing a crosstalk between parenchymal cells and resident immune cells such as the Kupffer cells and ATMs [124]. Remarkably, IL-4/IL-13 signaling leads to PPARδ activation in macrophages, which is mediated through a STAT6-binding site located within the PPARδ gene promoter [124], but may also be partly mediated via ligand binding activation by an endogenous ligand produced upon macrophage exposure to IL-4/IL-13, as shown for PPARγ [131].

These studies show that macrophage PPARδ-deficiency *in vivo* leads to impaired glucose tolerance and insulin resistance. The effects of macrophage PPARδ on adiposity and adipocyte size are, however, inconsistent among studies. Moreover, a more recent report challenged previous observations and found no significant differences on glucose tolerance and insulin sensitivity in mice deficient in macrophage PPARδ [127]. A number of factors may have contributed to these different outcomes. The distinct mouse strains employed, C57BL/6 [124, 127] and 129 SvJ mice [125], exhibit important differences in cholesterol concentrations (see http://phenome.jax.org). In addition, the approaches used to delete PPARδ specifically in myeloid cells differ: either using the constitutively active LysM promoter-driven Cre recombinase system [124] or the interferon α/β-inducible Mx Cre promoter [127], which may account for differences in Cre recombination deficiencies resulting in incomplete deletion of PPAR expression [127]. In addition, by employing bone marrow transplantation from PPARδ$^{-/-}$ mice into wild-type mice [125, 127] all hematopoietic cells are affected by PPARδ deletion, which may also have a confounding effect on systemic inflammation and metabolic responses. Finally, there are differences in the diet used. Thus, these reports highlight the influence of experimental context and genetic background studies addressing parenchyma–macrophage crosstalk in metabolism. Clearly, further work is needed to more precisely define the role of PPARs in macrophage inflammatory signaling in the context of metabolic disease.

6.3.4
PPARδ in Atherosclerosis

A number of studies have now evaluated the consequences of PPARδ activation or deletion in atherogenesis in mice (see Table 6.2). Here again, the impact of PPARδ on the disease shows some discrepancies. Transplantation of PPARδ$^{-/-}$ bone marrow cells into LDLR$^{-/-}$ mice resulted in less atherosclerosis compared with LDLR$^{-/-}$ mice receiving wild-type cells, suggesting that PPARδ is pro-atherogenic [115]. It was postulated these results were confounded by the repressor sequestering activity of unliganded PPARδ (see above). In a different study the PPARδ ligand GW7842 did not affect atherosclerosis progression in hyperlipidemic LDLR$^{-/-}$ mice compared with untreated mice, regardless of a decrease in inflammatory cytokine expression [65]. However, using the same mouse model under moderate hypercholesterolemic conditions the PPARδ agonist GW0742 reduced atherosclerosis [73]. This compound also has anti-inflammatory effects albeit without affecting insulin and HDL-cholesterol levels. These discrepancies have been partly explained by differences in the efficacies of these compounds [73, 114].

Based on recent work using GW501516, which possesses potent lipid-modifying effects, it has been postulated that the anti-inflammatory effects of PPARδ agonists *per se* may be insufficient to attenuate atherosclerosis progression without ameliorating the metabolic abnormalities associated. Consistently, GW501516 reduces atherosclerosis significantly in the ApoE-deficient mice model while inhibiting chemoattractant expression and transendothelial migration of monocytes [132]. Furthermore, the atheroprotective effect of GW0742 has been recently confirmed in a model of angiotensin II (Ang-II)-accelerated atherosclerosis in the LDLR$^{-/-}$ mice [74]. These beneficial effects were accompanied by an increase in the expression of BCL-6 and regulators of G protein-coupled signaling and the suppression of Ang-II-induced kinase activation [74]. In this report, the metabolic effect shown by GW0742 may have also contributed to the overall atheroprotective outcome by reducing plasma levels of TGs and total cholesterol, as well as insulin and glucose at the highest dose. In concert, these studies support the concept that PPARδ activation not only controls lipid and glucose metabolism but also exerts important actions in cells of the arterial wall that determine the progression of atherosclerosis and thus highlight the potential therapeutic value for PPARδ agonists for the management of this disease.

6.3.5
PPARδ in Human Studies

PPARδ activation exerts many favorable activities on energy utilization and metabolic control, cardiovascular function and atherogenic inflammation. Recently the relevance of these findings in human pathophysiology have begun to be explored. PPARδ polymorphisms have shown significant associations with various features of the metabolic syndrome, including dyslipidemia, risk of CHD, body mass index [133–136], as well as with the response to statin treatment [137], glucose levels, and insulin sensitivity [138, 139]. It has also been reported that PPARδ variation could be associated with metabolic syndrome risk [140], and metabolic changes induced by lifestyle intervention [141].

In addition to the aforementioned genetic studies, clinical trials are underway to investigate the utility of PPARδ ligands as therapies against metabolic diseases. In a first short-term clinical safety trial, a small cohort was given placebo or GW501516 while sedentary [142]. No toxicity was observed during this period and all doses tested improved TG clearance post feeding and prevented the decline of HDL-cholesterol levels due to the lack of physical activity. A second report shows that administration of this PPARδ ligand for two weeks to moderately overweight subjects significantly reduces fasting plasma TGs, apolipoprotein B, LDL-cholesterol, and insulin [143]. This is accompanied by a reduction in liver fat content and increased expression of FA oxidation enzymes in skeletal muscle.

6.3.6
Perspectives for PPARδ-Based Therapies

A growing body of evidence indicates that development of PPARδ agonists may be useful for the treatment of several abnormalities observed in metabolic disease. Several studies have revealed that PPARδ controls an array of metabolic genes involved in glucose homeostasis, fatty acid synthesis, storage, mobilization, and catabolism in a tissue-specific manner [144]. PPARδ synthetic ligands also ameliorate the symptoms of metabolic disorders, such as dyslipidemia, insulin resistance, hyperglycemia, and obesity in animal models [145]. Thus, some of these agonists have progressed into (early phase) clinical trials in an attempt to demonstrate that pharmacological activation of PPARδ could be a good therapeutic approach to improve some aspects of the metabolic syndrome. Despite the improved metabolic profile observed in subjects treated with certain PPARδ ligands like GW501516, the effects of these compounds over longer periods of time need to be determined, as well as its actions on weight gain, adiposity, and insulin resistance in humans. There is also concern about the reported effects of PPARδ agonists on tumor angiogenesis and colorectal carcinogenesis [146, 147], which may represent a major challenge for the development of these agonists as therapeutic drugs.

6.4
PPARγ

Two isoforms of PPARγ exist (PPARγ1 and PPARγ2) which are derived from the same gene through alternative promoter usage. PPARγ is modulated *in vitro* by the prostaglandin D2 derivative 15-deoxy-$\Delta^{12,14}$-prostaglandin J2 (15d-PGJ2) [148], although at supraphysiological concentrations it can have receptor-independent effects that should be considered. Forms of oxidized linoleic acid, 9- and 13(S)-HODE [149], and FAs also activate PPARγ [149, 150]. Moreover, thiazolidinediones (TZDs), such as troglitazone (TRO), rosiglitazone (RSG), and pioglitazone (PIO) are high-affinity synthetic ligands for PPARγ [151, 152]. Based on their efficiency in improving insulin sensitivity, the TZD PPARγ ligands RSG and PIO are currently used in clinical practice to treat insulin resistance in type 2 diabetics [151, 152], although the mechanism of action underlying their insulin-sensitizing effects still needs to be definitely determined.

In addition to its expression in adipose tissue, PPARγ is present in all major vascular cells [47, 153–156], inflammatory cells [44], as well as in atherosclerotic lesions [157, 158]. Thus, besides studying its major role in adipose tissue, during the past decade research has been increasingly directed towards elucidating the actions of PPARγ in inflammation, vascular biology, and atherosclerosis.

6.4.1
PPARγ Actions in the Vasculature

6.4.1.1 Vascular Endothelium

PPARγ is expressed in ECs [153, 159] and is activated in response to laminar flow [160]. Experiments in human ECs indicate that activation of PPARγ, but not PPARα, inhibits the expression of IFN-inducible protein of 10 kDa (IP-10), leading to a decrease in lymphocyte chemotaxis [31]. In addition, ligand-induced activation of PPARγ in ECs suppresses the expression of genes encoding vascular adhesion molecules including VCAM-1 [29] and ICAM-1 [161]. As a result, PPARγ activators may reduce monocyte recruitment into atherosclerotic plaques, as shown with TRO in the ApoE$^{-/-}$ model [162]. PPARγ ligands also prevent endothelial INF-γ-induced expression of major histocompatibility complex class II (MHC-II), which is directly involved in the activation of T lymphocytes and controls immune responses [163].

In ECs, PPARγ may also act as a vasorelaxant as it promotes the release of nitric oxide (NO) [164]. For certain PPARγ ligands, these effects are mediated by an increased interaction between heat shock protein 90 and endothelial NOS (eNOS), and eNOS phosphorylation through a PPARγ-dependent mechanism [165]. Alternatively, PPARγ agonists may increase NO availability by repressing the NADPH oxidase enzyme complex [33, 166]. As observed with PPARα, PPARγ activation also inhibits the expression of endothelin-1 (ET-1), a vasoactive peptide involved in the regulation of vascular tone [159]. Interestingly, laminar flow, but not PPARγ activators, induces eNOS expression [167], suggesting that PPARγ activation could participate in and amplify the anti-inflammatory endothelial response to shear stress.

The role of PPARγ on ECs growth has been analyzed but still requires further investigation. Two groups have reported opposite pro- and anti-apoptotic functions for PPARγ [168, 169]. It is worth noting that they used different PPARγ activators (15d-PGJ2 and ciglitazone versus TRO) and ECs from different species [168, 169]. PPARγ-independent effects have been reported for both 15d-PGJ2 and TRO [170, 171] and the expression level of PPARγ may vary in ECs from different vascular sites, which may modify the susceptibility towards PPARγ activators. Finally, PPARγ activation decreases the expression of certain growth factors in ECs thus inhibiting angiogenesis. Indeed, treatment of ECs with PPARγ activators inhibit vascular endothelial growth factor (VEGF) receptor 1 and 2 expression, VEGF- and leptin-induced EC migration and the *in vitro* endothelial tube formation [153, 172]. Administration of 15d-PGJ2 also inhibits VEGF-induced angiogenesis in rat cornea [153].

Considering both the potential benefits and drawbacks of inhibiting angiogenesis in the treatment of cardiovascular diseases [173], the definitive role of PPARγ activation in this context remains to be determined.

6.4.1.2 **Monocytes/Macrophages/T Lymphocytes**

Anti-Inflammatory Actions and Transrepression Mechanisms   PPARγ is expressed in macrophages and foam cells in the lipid core of human atherosclerotic lesions [158, 174]. As shown for PPARα, PPARγ expression increases upon monocyte differentiation into macrophages [34, 158]. Several studies on implicate broad anti-inflammatory and potential anti-atherogenic effects of PPARγ activation in primary human monocytes [154], in the undifferentiated monocytic THP-1 cell line [163, 175], in RAW 264.7 macrophage-like cells, macrophages differentiated from bone marrow progenitor cells and/or peritoneal macrophages [174, 176]. In these cells, PPARγ ligands inhibit genes that promote native and acquired immunity, such as IFN-inducible protein (IP-10), monokine induced by IFN-γ (MIG), and IL-12 [176]. Moreover, PPARγ activation inhibits numerous genes underlying macrophage activation, including MMP-9 [174], as well as T lymphocyte activation, which may reduce their proinflammatory activity towards monocytes and ECs [177].

IL-4, an anti-inflammatory cytokine and inducer of macrophage alternative activation, stimulates the generation of natural PPARγ ligands by the activation of the 12/15-lipoxygenase pathway in these cells, thus enhancing the inhibition of iNOS expression [131]. As for PPARδ, PPARγ polarizes macrophages to the alternative activation state, which display a more pronounced anti-inflammatory phenotype [126, 178]. Whether inducing PPARγ macrophage actions affects insulin sensitivity and glucose tolerance is currently a matter of debate [126, 127, 179] (see above).

To date, it has been difficult to identify a single unifying mechanism to explain the inhibition of inflammatory gene expression by PPARs. It appears there are possibly multiple mechanisms depending on the signal, cell type and even the repressed gene [45, 180]. Recent studies, however, have substantially advanced our understanding of the processes underlying PPARγ transrepression. In their basal state, proinflammatory genes are repressed by protein complexes bound to their promoters which contain corepressors such as NCoR or SMRT. Upon cell activation, the corepressor complex is degraded and transcription factors, such as NF-κB, in association with coactivator complexes bind those promoters and drive gene transcription. Pascual and colleagues proposed that PPARγ activation inhibits the expression of proinflammatory genes by preventing the signal-induced removal of corepressor complexes through a mechanism involving the sumoylation of PPARγ [121]. Whether this process is biologically relevant still needs to be explored.

Macrophage Cholesterol Homeostasis   Similar to PPARα, PPARγ activators play a major role in the regulation of macrophage cholesterol homeostasis. PPARγ directs macrophage expression of a number of genes involved in cholesterol efflux, such as the HDL receptor CLA-I/SRB-I [181], ApoE [182] as well as the ABCA1 and ABCG1 transporters [35, 183]. Interestingly, the induction of the PPARγ/ABCA1 pathway is mediated by activation of a different nuclear receptor: LXRα [183]. PPARγ is also able to stimulate mobilization of cellular membrane cholesterol by activating caveolin-1 expression [184], which induces cholesterol association with lipid rafts thus promoting its efflux [185]. PPARγ also positively regulated the expression and activity of macrophage CYP27A1 (sterol 27-hydroxylase), which converts cholesterol into downstream water-soluble products [186] that are considered weak endogenous ligands for LXRα [187].

Moreover, in addition to cholesterol elimination, overexpression of CYP27A1 promoted cholesterol efflux to ApoAI, independent of ABCA1 expression [188]. Similar to PPARα, PPARγ activation decreased glycated LDL uptake [37], as well as ApoB48 receptor expression and TG accumulation [36] in these cells.

Although initially thought to display deleterious effects by increasing the expression of the scavenger receptor CD36 [150], subsequent work indicated that PPARγ is not required for foam cell formation [189], and either has no effect [35] or inhibits [65, 190] macrophage transformation into foam cells in a species-specific manner. Furthermore, administration of TZD ligands to obese ob/ob and ob/ob × LDL-R$^{-/-}$ mice reverses the increased macrophage expression of CD36 associated with insulin resistance [191]. *Ex vivo* treatment of foam cells from hypercholesterolemic LDL-R$^{-/-}$ mice with PPARγ ligands reduced cholesterol esterification, increased HDL-dependent cholesterol efflux and induced expression of ABCG1.

Taken together, these data indicate an overall beneficial effect of synthetic PPARγ activators on fatty streak formation. Further studies using models with a conditional disruption of macrophage PPARγ will be useful to decipher the role of PPARγ on lipid accumulation in these cells during atherosclerotic development.

6.4.1.3 SMCs

Evidence that patients treated with TZDs showed lowered systolic blood pressure stimulated studies on the function of PPARγ in SMCs. Notably, PPARγ activation has been shown to blunt SMC responses to Ang-II, which plays a major role in the vascular remodeling underlying hypertension, partly by downregulating the Ang-II type I receptor [192, 193]. Interestingly, in these studies the possibility that some angiotensin receptor blockers may activate PPARγ was also raised [194, 195]. In SMCs PPARγ ligands also directly inhibit growth factor- and insulin-stimulated mitogenic signaling and migration [156, 196].

As observed for PPARα, TZD and non-TZD PPARγ agonists inhibit SMC proliferation by preventing the degradation of the kinase inhibitor p27kip1 (p27), which in turn inhibits pRB phosphorylation, the critical checkpoint of $G_1 \rightarrow S$ cell cycle transition [197–199]. PPARγ activation also downregulates gene expression of the mini-chromosome maintenance (MCM) 6 and 7 [200], which play a central role in the initiation of DNA replication [201].

Finally, in SMCs PPARγ-dependent inhibition of telomerase activity [202] and TERT gene expression, which is required for the anti-proliferative effects of PPARγ activators, is mediated through a negative crosstalk between PPARγ and the transcription factor Ets-1 on the TERT promoter. This finding, also observed with PPARα activators, places the inhibition of telomerase in a central position for the growth-inhibitory effects of fibrates and TZDs. In addition to promoting SMC cell cycle arrest, PPARγ agonists trigger the upregulation of smooth muscle myosin heavy chain and smooth muscle α actin, two specific markers of differentiated SMCs [203].

PPARγ activators have also been shown to inhibit SMC migration [156], which has been associated with a decrease in the expression and activity of MMP-9, a matrix-remodeling enzyme implicated in plaque rupture [155]. Finally, different studies underscore the ability of PPARγ to promote SMC apoptosis [204–208], via upregulation of p53 [204], the growth-arrest and DNA damage-inducible gene (GADD) GADD45 [204, 206], as well as IFN regulatory factor-1 (IRF-1) [209] expression. The transforming growth factor-b1 (TGF-β1)/Smad2 pathway is also involved [207, 208]. In addition, inhibition of telomerase activity could also contribute to the induction of SMC apoptosis [52, 202].

Altogether, these data indicate that PPARγ activation promotes maintenance of SMCs in a quiescent and differentiated state, which is associated with decreased migration and increased susceptibility to apoptosis. Interestingly, as occurs with PPARγ anti-inflammatory properties, PPARγ actions on SMC growth involve interactions between PPARγ and transcription factors at response elements other than canonical PPREs. The impact of the SMC growth regulatory effects of TZDs on clinical risk associated with atherosclerosis development need to be clarified. Notably, SMC apoptosis can accelerate atherosclerosis, decrease the stability of the atherosclerotic lesion and predispose them to rupture [210, 211]. On the other hand, PPARγ activation in SMCs may attenuate the arterial remodeling following coronary interventions.

6.4.2
PPARγ in Animal Models of Atherosclerosis and Associated Vascular Complications

PPARγ activators were first hypothesized to be pro-atherogenic because of their positive effects on CD36 expression and ox-LDL uptake [149, 150]. However, it is currently postulated that PPARγ activation could protect against atherosclerosis through the regulation of macrophage homeostasis and endothelial function. In agreement with this, TZD PPARγ ligands (PIO, RSG, TRO, and the highly specific GW7845) have been shown to prevent the development of atherosclerosis in murine models of atherosclerosis: the LDLR$^{-/-}$ [63, 66, 70, 72] and the ApoE$^{-/-}$ [57, 61, 68, 69, 71] (see Table 6.2). This anti-atherogenic effect may be sex-specific, as the PPARγ agonists RSG and GW7845 are only effective in male LDL-R$^{-/-}$ mice [70]. Of note, RSG and PIO did not affect plasma cholesterol or atherosclerosis in ApoE2 knockin mice, a mouse model of dyslipidemia responsive to fenofibrate [62]. Inhibition of atherosclerosis by TZD ligands was also observed in both insulin-sensitive [61, 71, 72] and insulin-resistant [68, 69, 72] mice, indicating that these actions are independent of their efficacy as insulin sensitizers. More generally, the preventive effect of TZDs on hyperlipidemia-induced atherosclerosis occurs independently of their beneficial effects on circulating lipids, blood pressure, glucose metabolism, and insulin [61, 68–72], implicating direct pleiotropic effects on the vascular wall.

Bone marrow transplantation from PPARγ$^{-/-}$ or macrophage-specific PPARγ$^{-/-}$ mice to LDL-R$^{-/-}$ mice resulted in significantly larger atherosclerotic lesions than when wild-type bone marrow was used [183, 212]. PPARγ$^{-/-}$ macrophages exhibit increased levels and chemokine receptor 2 expression, induced cell migration [212], and impaired alternative macrophage activation [126]. These data demonstrate the involvement of macrophage-driven processes in the prevention of murine atherosclerosis by TZDs. Moreover, in a rat model of hypertension, TZDs were shown to reduce hypertension progression, an effect associated with the prevention of endothelial dysfunction and vascular remodeling [213]. In line with this, specific deletion of PPARγ in ECs has been shown to increase blood pressure in mice [214]. Thus, inhibition of hypertension through activation of endothelial PPARγ could also contribute to the beneficial effects of TZDs on atherosclerosis.

Consistent with the observation that TZDs ligands limit SMC proliferation *in vitro*, troglitazone reduces intimal hyperplasia in a rat artery balloon injury model [215]. Furthermore, RSG and PIO inhibit intimal hyperplasia as well in models of restenosis regardless of insulin sensitivity [216–219]. TZDs may exert anti-proliferative effects independent of PPARγ [171]. However, the observation that transfer of the PPARγ gene into rat arteries *in vivo* also inhibits SMC proliferation and neointima formation while sustaining apoptosis clearly demonstrates the protective role of PPARγ expression against restenosis [220]. In a hypercholesterolemic rabbit model of atherosclerosis PIO was shown to suppress in-stent neointimal growth disease, which was associated with neointimal reduction of macrophage accumulation and expression of proinflammatory cytokines MCP-1 and TGF-β [221]. Therefore, local anti-inflammatory effects of TZDs at the site of vascular injury could also be protective against the hyperproliferative intimal response. Additional beneficial effects of TZDs on neointima formation include accelerated re-endothelialization [169, 222], extension of allograft survival, and attenuation of neointimal hyperplasia through suppression of SMC proliferation [223], which further extend the protective effects of PPARγ agonists on complications of atherosclerosis.

6.4.3
PPARγ in Human Cardiovascular Clinical Diseases

TZDs are demonstrated efficacious drugs in the management of insulin resistance and type 2 diabetes [224, 225]. Although TRO was withdrawn, because of a rare but severe hepatoxicity, TZDs PIO and RSG have been increasingly prescribed to type 2 diabetics and may be potentially hepatoprotective against fatty liver disease and non-alcoholic steatohepatitis [226, 227]. Given the expression of PPARγ in human atherosclerotic lesions [157, 158], and their favorable vascular effects *in vitro* and in mice, ongoing studies are analyzing their clinical potential in the prevention and/or treatment of cardiovascular diseases.

A decade ago the first study assessing the anti-atherogenic potential of TZDs analyzed the effects of TRO treatment on carotid artery intima–media thickness (CIMT) in type 2 diabetic patients [228]. In this study, TRO administration resulted in a significant decrease in CIMT, which was observed as early as three months after treatment [228]. A similar outcome was observed in two subsequent studies performed with PIO [229, 230]. Other studies compared the effects of PIO versus glimeripide on CIMT progression [231, 232]. In a randomized control study with type 2 diabetic patients, a reduction of CIMT was observed in the PIO group independently of glycemic control [231]. The double-blind randomized Carotid Intimal-medial Thickness in Atherosclerosis using Pioglitazone (CHICAGO) trial extended these findings over a longer follow-up period, showing that the PIO group had significantly less progression of CIMT compared with glimepiride [232].

Interestingly, comparable results have been reported with RSG in patients without type 2 diabetes, but with angiographically documented coronary artery disease [233]. In line with these results were also those of the Study of Atherosclerosis with Ramipril and Rosiglitazone (STARR), a substudy of the Diabetes Reduction Assessment with Ramipril and Rosiglitazone Medication (DREAM) [234], which enrolled patients with impaired fasting glucose without known cardiovascular disease. The effects of RSG versus metformin in type 2 diabetics were also compared [235]. Interestingly, a regression of maximal CIMT was found in the RSG group, while CIMT progressed in the metformin group. These effects were independent of the glycemic control improvement.

The larger Rosiglitazone Atherosclerosis Study analyzed the effect of RSG treatment in a mixed patient cohort of subjects with insulin resistance and type 2 diabetes [236]. In this study RSG treatment had no effect on CMIT in the mixed population, however in separate analyses a significant effect was found for type 2 diabetic patients. These reports indicate that TZDs, PIO, and TSG are beneficial against the development of atherosclerosis in type 2 diabetics, however their benefit for non-diabetic patients may require further investigation.

CIMT is a well-described surrogate marker for cardiovascular risk and correlates not only with the presence of cardiovascular risk factors but also with the risk of future macrovascular events [237, 238]. The beneficial effects observed with TZDs provided the rationale for larger cardiovascular trials. The Prospective Pioglitazone Clinical Trial in Macrovascular Events (PROactive) evaluated the influence of PIO on CVD in 5238 patients with type 2 diabetes and prior macrovascular disease [243]. PIO or placebo was administered on top of current type 2 diabetes and cardiovascular medications. Despite the broad primary end-point (composite of all-cause mortality) not being significantly different groups, a 16% reduction in secondary end-points MI and stroke was reported in patients treated with PIO. Further post hoc analysis showed that PIO significantly lowered the risk of recurrent fatal and nonfatal stroke [239] as well as recurrent MI and acute coronary syndrome in patients with prior MI [240]. Cardiovascular outcome studies are currently ongoing with RSG: the Action to Control Cardiovascular Risk in Diabetes (ACCORD) [241], the Bypass Angioplasty Revascularization Investigation 2 Diabetes (BARI 2D) [90], and the Rosiglitazone Evaluation for Cardiac Outcomes and Regulation of Glycemia in Diabetes (RECORD) [242] trials.

TZD administration has been associated, however, with a number of adverse effects, including peripheral edema which may precipitate congestive heart failure, as observed in the PROactive trial [243], and in the interim analysis of the RECORD trial [244]. However, TZDs may be used safely in individuals with stable heart failure [245]. Moreover, a recent meta-analysis indicated that treatment with TZD (PIO or RSG) does increase the risk of congestive heart failure but does not increase cardiovascular death [246]. Another independent meta-analysis suggested that RSG use may be associated with a risk of MI and heart failure [247, 248] which triggered a significant discussion in the field. However, in the published individual large trials on low-risk patients included in the study (specifically the DREAM [224] and the A Diabetes Outcome Progression (ADOPT) [225]), no increase in the rates of MI or cardiovascular death was observed. Moreover, the intermediary safety analysis of the RECORD reported no significant changes in morbidity and mortality after RSG treatment [244]. Completion of clinical outcome studies will shed more light on the long-term cardiovascular effects of RSG compared with PIO.

Investigating the clinical potential of TZDs in the prevention of occlusive complications of atherosclerosis, Takagi and colleagues first analyzed the effects of TZDs on post-angiography restenosis. They demonstrated that treatment with TRO and/or PIO resulted in a significant reduction in neointimal tissue proliferation after coronary stent implantation in non-diabetic patients [249], as well as in patients with non-insulin-dependent diabetes [250] or type 2 diabetes mellitus [251, 252]. More recently, in type 2 diabetic patients, RSG was also shown to prevent restenosis after coronary stent implantation [253]. Therefore, TZDs could be useful as an adjunctive therapy in diabetic patients undergoing percutaneous interventions.

Finally, the relevance of PPARγ actions in human atherosclerosis has also been addressed in a number of genetic studies. Mutations in the human PPARγ gene, Pro12Ala and C161T, have been associated with a lower incidence of CHD independently of metabolic changes [254–256], Surprisingly, these two mutations have been characterized *in vitro* to lead to a loss-of-function although their functional status in humans remains to be established [255].

6.5
PPARs: Conclusion and Perspectives

As indicated above, results from clinical studies demonstrate that treatment with fibrate PPARα ligands decreases the risk of cardiovascular disease in patients with type 2 diabetes without pre-existing cardiovascular disease or metabolic syndrome, and PIO improves cardiovascular outcomes in secondary prevention. Unarguably these results demonstrate that PPARα and PPARγ activators are interesting therapeutic targets for cardiovascular diseases, despite safety concerns that have recently led to late stage development failures of various PPAR agonists [101]. Currently, trials analyzing the cardiovascular outcomes of therapies with fibrate/statin combination, RSG, or PPARβ/δ activators are in development. Current strategies also include the development of PPAR activators that act on multiple receptors [100, 101]. The initial interest in PPARα/γ co-agonists such as tesaglitazar [79, 80, 100] (see Table 6.2) was tempered by the emergence of deleterious effects of certain PPARα/γ co-agonists and by their modest effect on weight gain. Given the recent discovery of the beneficial effects of PPARδ on several features of the metabolic syndrome, research interest has shifted towards the development of dual PPARα/δ or PPARγ/δ activators, pan-activators or, even more promising, based on the selective modulator concept [257], compounds that would be devoid of adverse effects while maintaining the desired biological efficacy of PPARs [258, 259].

To date, promising results have been reported for the dual PPARα/δ agonist T0913659 in a primate model of atherosclerosis [260], showing more efficient increases in HDL-cholesterol than with fenofibrate and with similar potency to that of PPARδ agonist GW501516. Moreover, in preclinical models, the dual PPARγ/δ propionic acid derivative was shown to improve insulin sensitivity and reverse diabetic hyperglycemia with less weight gain relative to RSG [261]. Selective PPAR modulators (SPPARMs) currently in development include metaglidasen, nTZDpa, SPPARM12, and the T131 molecule (Amgen (Tularik)) [101]. Finally, clinical studies are also warranted to investigate the therapeutic interest of PPAR activation in the prevention of SMC proliferation, which underlies the failure of many therapies [86, 262].

Acknowledgements

Florence Gizard was supported by a Postdoctoral Fellowship from the American Heart Association, Great Rivers Affiliate (0725313B). Inés Pineda Jorra is supported by a Medical Research Council New Investigator Grant (G0801278). We thank Dr. Bruemmer (MD, PhD), B. Staels (PhD) and A. Daugherty (PhD) for their support. Florence Gizard also thanks T. Claudel (PhD) for his constant help.

References

1 Mangelsdorf, D.J., Thummel, C., Beato, M., Herrlich, P., *et al.* (1995) *Cell*, **83**, 835–839.

2 Castrillo, A., Tontonoz, P., and Rev, A. (2004) *Cell Dev. Biol.*, **20**, 455–480.

3 Glass, C.K. and Rosenfeld, M.G. (2000) *Genes Dev.*, **14**, 121–141.

4 Dreyer, C., Krey, G., Keller, H., Givel, F., *et al.* (1992) *Cell*, **68**, 897–887.

5 Issemann, I. and Green, S. (1990) *Nature*, **347**, 645–650.

6 Kliewer, S.A., Forman, B.M., Blumberg, B., Ong, E.S., *et al.* (1994) *Proc. Natl. Acad. Sci. U. S. A.*, **91**, 7355–7359.

7 Tontonoz, P., Hu, E., Graves, R.A., Budavari, A.I., and Spiegelman, B.M. (1994) *Genes Dev.*, **8**, 1224–1234.

8 Kliewer, S.A., Umesono, K., Noonan, D.J., Heyman, R.A., and Evans, R.M. (1992) *Nature*, **358**, 771–774.

9 Carroll, J.S. and Brown, M. (2006) *Mol. Endocrinol.*, **20**, 1707–1714.

10 Rossi, A., Kapahi, P., Natoli, G., Takahashi, T., *et al.* (2000) *Nature*, **403**, 103–108.

11 Straus, D.S., Pascual, G., Li, M., Welch, J.S., *et al.* (2000) *Proc. Natl Acad. Sci. U. S. A.*, **97**, 4844–4849.

12 Gampe, R.T., Montana, V.G., Lambert, M.H., Miller, A.B., *et al.* (2000) *Mol. Cell*, **5**, 545–555.

13 Forman, B.M., Chen, J., and Evans, R.M. (1997) *Proc. Natl Acad. Sci. U. S. A.*, **94**, 4312–4317.

14 Kliewer, S.A., Sundseth, S.S., Jones, S.A., Brown, P.J., *et al.* (1997) *Proc. Natl Acad. Sci. U. S. A.*, **94**, 4318–4323.

15 Hostetler, H.A., Petrescu, A.D., Kier, A.B., and Schroeder, F. (2005) *J. Biol. Chem.*, **280**, 18667–18682.

16 Xu, H.E., Lambert, M.H., Montana, V.G., Parks, D.J., *et al.* (1999) *Mol. Cell*, **3**, 397–403.

17 Chakravarthy, M.V., Pan, Z., Zhu, Y., Tordjman, K., *et al.* (2005) *Cell Metab.*, **1**, 309–322.

18 Kersten, S., Seydoux, J., Peters, J.M., Gonzalez, F.J., *et al.* (1999) *J. Clin. Invest.*, **103**, 1489–1498.

19 Djouadi, F., Weinheimer, C.J., Saffitz, J.E., Pitchford, C., *et al.* (1998) *J. Clin. Invest.*, **102**, 1083–1091.

20 Lefebvre, P., Chinetti, G., Fruchart, J.C., and Staels, B. (2006) *J. Clin. Invest.*, **116**, 571–580.

21 Schoonjans, K., Peinado-Onsurbe, J., Lefebvre, A.M., Heyman, R.A., *et al.* (1996) *EMBO J.*, **15**, 5336–5348.

22 Haubenwallner, S., Essenburg, A.D., Barnett, B.C., Pape, M.E., *et al.* (1995) *J. Lipid Res.*, **36**, 2541–2551.

23 Frederiksen, K.S., Wulff, E.M., Sauerberg, P., Mogensen, J.P., *et al.* (2004) *J. Lipid Res.*, **45**, 592–601.

24 Vu-Dac, N., Gervois, P., Jakel, H., Nowak, M., *et al.* (2003) *J. Biol. Chem.*, **278**, 17982–17985.

25 Schultze, A.E., Alborn, W.E., Newton, R.K., and Konrad, R.J. (2005) *J. Lipid Res.*, **46**, 1591–1595.

26 Duez, H., Lefebvre, B., Poulain, P., Torra, I.P., *et al.* (2005) *Arterioscler. Thromb. Vasc. Biol.*, **25**, 585–591.

27 Inoue, I., Shino, K., Noji, S., Awata, T., Katayama, S., and Biophys, B. (1998) *Res. Commun.*, **246**, 370–374.

28 Marx, N., Sukhova, G.K., Collins, T., Libby, P., and Plutzky, J. (1999) *Circulation*, **99**, 3125–3131.

29 Jackson, S.M., Parhami, F., Xi, X.P., Berliner, J.A., *et al.* (1999) *Arterioscler. Thromb. Vasc. Biol.*, **19**, 2094–2104.

30 Sethi, S., Ziouzenkova, O., Ni, H., Wagner, D.D., *et al.* (2002) *Blood*, **100**, 1340–1346.

31 Marx, N., Mach, F., Sauty, A., Leung, J.H., *et al.* (2000) *J. Immunol.*, **164**, 6503–6508.

32 Lee, H., Shi, W., Tontonoz, P., Wang, S., *et al.* (2000) *Circ. Res.*, **87**, 516–521.

33 Inoue, I., Goto, S., Matsunaga, T., Nakajima, T., *et al.* (2001) *Metabolism*, **50**, 3–11.

34 Chinetti, G., Griglio, S., Antonucci, M., Torra, I.P., *et al.* (1998) *J. Biol. Chem.*, **273**, 25573–25580.

35 Chinetti, G., Lestavel, S., Bocher, V., Remaley, A.T., *et al.* (2001) *Nat. Med.*, **7**, 53–58.

36 Haraguchi, G., Kobayashi, Y., Brown, M.L., Tanaka, A., *et al.* (2003) *J. Lipid Res.*, **44**, 1224–1231.

37 Gbaguidi, F.G., Chinetti, G., Milosav-ljevic, D., Teissier, E., *et al.* (2002) *FEBS Lett.*, **512**, 85–90.

38 Devchand, P.R., Keller, H., Peters, J.M., Vazquez, M., *et al.* (1996) *Nature*, **384**, 39–43.

39 Teissier, E., Nohara, A., Chinetti, G., Paumelle, R., *et al.* (2004) *Circ. Res.*, **95**, 1174–1182.

40 Nakamachi, T., Nomiyama, T., Gizard, F., Heywood, E.B., *et al.* (2007) *Diabetes*, **56**, 1662–1670.

41 Marx, N., Mackman, N., Schonbeck, U., Yilmaz, N., *et al.* (2001) *Circulation*, **103**, 213–219.

42 Neve, B.P., Corseaux, D., Chinetti, G., Zawadzki, C., *et al.* (2001) *Circulation*, **103**, 207–212.

43 Shu, H., Wong, B., Zhou, G., Li, Y., *et al.* (2000) *Biochem. Biophys. Res. Commun.*, **267**, 345–349.

44 Marx, N., Kehrle, B., Kohlhammer, K., Grub, M., *et al.* (2002) *Circ. Res.*, **90**, 703–710.

45 Pascual, G. and Glass, C.K. (2006) *Trends Endocrinol. Metab.*, **17**, 321–327.

46 Blanquart, C., Mansouri, R., Paumelle, R., Fruchart, J.C., *et al.* (2004) *Mol. Endocrinol.*, **18**, 1906–1918.

47 Staels, B., Koenig, W., Habib, A., Merval, R., *et al.* (1998) *Nature*, **393**, 790–793.

48 Delerive, P., De Bosscher, K., Besnard, S., Vanden Berghe, W., *et al.* (1999) *J. Biol. Chem.*, **274**, 32048–32054.

49 Gizard, F., Amant, C., Barbier, O., Bellosta, S., *et al.* (2005) *J. Clin. Invest.*, **115**, 3228–3238.

50 Gizard, F., Nomiyama, T., Zhao, Y., Findeisen, H.M., *et al.* (2008) *Circ. Res.*, **103**, 1155–1163.

51 Kim, W.Y. and Sharpless, N.E. (2006) *Cell*, **127**, 265–275.

52 Fuster, J.J. and Andres, V. (2006) *Circ. Res.*, **99**, 1167–1180.

53 Fu, T., Kashireddy, P., and Borensztajn, J. (2003) *Biochem. J.*, **373**, 941–947.

54 Inaba, T., Yagyu, H., Itabashi, N., Tazoe, F., *et al.* (2008) *Hypertens. Res.*, **31**, 999–1005.

55 Fu, T., Mukhopadhyay, D., Davidson, N.O., and Borensztajn, J. (2004) *J. Biol. Chem.*, **279**, 28662–28669.

56 Calkin, A.C., Cooper, M.E., Jandeleit-Dahm, K.A., and Allen, T.J. (2006) *Diabetologia*, **49**, 766–774.

57 Calkin, A.C., Allen, T.J., Lassila, M., Tikellis, C., *et al.* (2007) *Atherosclerosis*, **195**, 17–22.

58 Duez, H., Chao, Y.S., Hernandez, M., Torpier, G., *et al.* (2002) *J. Biol. Chem.*, **277**, 48051–48057.

59 Declercq, V., Yeganeh, B., Moshtaghi-Kashanian, G.R., Khademi, H., *et al.* (2005) *J. Cardiovasc. Pharmacol.*, **46**, 18–24.

60 Yeganeh, B., Moshtaghi-Kashanian, G.R., Declercq, V., and Moghadasian, M.H. (2005) *J. Nutr. Biochem.*, **16**, 222–228.

61 Claudel, T., Leibowitz, M.D., Fievet, C., Tailleux, A., *et al.* (2001) *Proc. Natl Acad. Sci. U. S. A.*, **98**, 2610–2615.

62 Hennuyer, N., Tailleux, A., Torpier, G., Mezdour, H., *et al.* (2005) *Arterioscler. Thromb. Vasc. Biol.*, **25**, 1897–1902.

63 Srivastava, R.A., Jahagirdar, R., Azhar, S., Sharma, S., and Bisgaier, C.L. (2006) *Mol. Cell. Biochem.*, **285**, 35–50.

64 Kooistra, T., Verschuren, L., de Vries-van der Weij, J., Koenig, W., *et al.* (2006) *Arterioscler. Thromb. Vasc. Biol.*, **26**, 2322–2330.

65 Li, A.C., Binder, C.J., Gutierrez, A., Brown, K.K., *et al.* (2004) *J. Clin. Invest.*, **114**, 1564–1576.

66 He, L., Game, B.A., Nareika, A., Garvey, W.T., and Huang, Y. (2006) *J. Cardiovasc. Pharmacol.*, **48**, 212–222.

67 Game, B.A., He, L., Jarido, V., Nareika, A., *et al.* (2007) *Atherosclerosis*, **192**, 85–91.

68 Levi, Z., Shaish, A., Yacov, N., Levkovitz, H., *et al.* (2003) *Diabetes Obes. Metab.*, **5**, 45–50.

69 Calkin, A.C., Forbes, J.M., Smith, C.M., Lassila, M., *et al.* (2005) *Arterioscler. Thromb. Vasc. Biol.*, **25**, 1903–1909.

70 Li, A.C., Brown, K.K., Silvestre, M.J., Willson, T.M., *et al.* (2000) *J. Clin. Invest.*, **106**, 523–531.

71 Chen, Z., Ishibashi, S., Perrey, S., Osuga, J., *et al.* (2001) *Arterioscler. Thromb. Vasc. Biol.*, **21**, 372–377.

72 Collins, A.R., Meehan, W.P., Kintscher, U., Jackson, S., *et al.* (2001) *Arterioscler. Thromb. Vasc. Biol.*, **21**, 365–371.

73 Graham, T.L., Mookherjee, C., Suckling, K.E., Palmer, C.N., and Patel, L. (2005) *Atherosclerosis*, **181**, 29–37.

74 Takata, Y., Liu, J., Yin, F., Collins, A.R., *et al.* (2008) *Proc. Natl Acad. Sci. U. S. A.*, **105**, 4277–4282.

75 Grothusen, C., Bley, S., Selle, T., Luchtefeld, M., *et al.* (2005) *Atherosclerosis*, **182**, 57–69.

76 Takaya, T., Kawashima, S., Shinohara, M., Yamashita, T., *et al.* (2006) *Atherosclerosis*, **186**, 402–410.

77 Blessing, E., Preusch, M., Kranzhofer, R., Kinscherf, R., *et al.* (2008) *Atherosclerosis*, **199**, 295–303.

78 Zuckerman, S.H., Kauffman, R.F., and Evans, G.F. (2002) *Lipids*, **37**, 487–494.

79 Zadelaar, A.S., Boesten, L.S., Jukema, J.W., van Vlijmen, B.J., *et al.* (2006) *Arterioscler. Thromb. Vasc. Biol.*, **26**, 2560–2566.

80 Chira, E.C., McMillen, T.S., Wang, S., Haw, A., 3rd, *et al.* (2007) *Atherosclerosis*, **195**, 100–109,

81 Steiner, G., Hamsten, A., Hosking, A., Stewart, J., *et al.* (2001) *Lancet*, **357**, 905–910.

82 Willson, T.M., Brown, P.J., Sternbach, D.D., and Henke, B.R. (2000) *J. Med. Chem.*, **43**, 527–550.

83 Tordjman, K., Bernal-Mizrachi, C., Zemany, L., Weng, S., *et al.* (2001) *J. Clin. Invest.*, **107**, 1025–1034.

84 Saitoh, K., Mori, T., Kasai, H., Nagayama, T., *et al.* (1995) *Nippon Yakurigaku Zasshi*, **106**, 41–50.

85 Saitoh, K., Mori, T., Kasai, H., Nagayama, T., and Ohbayashi, S. (1995) *Nippon Yakurigaku Zasshi*, **106**, 51–60.

86 Dzau, V.J., Braun-Dullaeus, R.C., and Sedding, D.G. (2002) *Nat. Med.*, **8**, 1249–1256.

87 Ericsson, C.G., Nilsson, J., Grip, L., Svane, B., and Hamsten, A. (1997) *Am. J. Cardiol.*, **80**, 1125–1129.

88 Frick, M.H., Syvanne, M., Nieminen, M.S., Kauma, H., *et al.* (1997) *Circulation*, **96**, 2137–2143.

89 Frick, M.H., Elo, O., Haapa, K., Heinonen, O.P., *et al.* (1987) *N. Engl. J. Med.*, **317**, 1237–1245.

90 Brooks, M.M., Frye, R.L., Genuth, S., Detre, K.M., *et al.* (2006) *Am. J. Cardiol.*, **97**, 9–19.

91 Keech, A., Simes, R.J., Barter, P., Best, J., *et al.* (2005) *Lancet*, **366**, 1849–1861.

92 Anon (2000) *Circulation*, **102**, 21–27.

93 Rubins, H.B., Robins, S.J., Collins, D., Fye, C.L., *et al.* (1999) *N. Engl. J. Med.*, **341**, 410–418.

94 Robins, S.J., Collins, D., Wittes, J.T., Papademetriou, V., *et al.* (2001) *JAMA*, **285**, 1585–1591.

95 Flavell, D.M., Jamshidi, Y., Hawe, E., Pineda Torra, I., *et al.* (2002) *Circulation*, **105**, 1440–1445.

96 Colhoun, H.M., Betteridge, D.J., Durrington, P.N., Hitman, G.A., *et al.* (2004) *Lancet*, **364**, 685–696.

97 Wierzbicki, A.S. (2006) *Int. J. Clin. Pract.*, **60**, 442–449.

98 Brown, J.D. and Plutzky, J. (2007) *Circulation*, **115**, 518–533.

99 Jamshidi, Y., Montgomery, H.E., Hense, H.W., Myerson, S.G., *et al.* (2002) *Circulation*, **105**, 950–955.

100 Gross, B., Staels, B. and Pract, B. (2007) *Res. Clin. Endocrinol. Metab.*, **21**, 687–710.

101 Rubenstrunk, A., Hanf, R., Hum, D.W., Fruchart, J.C., and Staels, B. (2007) *Biochim. Biophys. Acta*, **1771**, 1065–1081.

102 Paumelle, R. and Staels, B. (2008) *Trends Cardiovasc. Med.*, **18**, 73–78.

103 Laufs, U., Marra, D., Node, K., and Liao, J.K. (1999) *J. Biol. Chem.*, **274**, 21926–21931.

104 Chawla, A., Lee, C.H., Barak, Y., He, W., *et al.* (2003) *Proc. Natl Acad. Sci. U. S. A.*, **100**, 1268–1273.

105 Barak, Y., Liao, D., He, W., Ong, E.S., *et al.* (2002) *Proc. Natl Acad. Sci. U. S. A.*, **99**, 303–308.

106 Peters, J.M., Lee, S.S., Li, W., Ward, J.M., *et al.* (2000) *Mol. Cell Biol.*, **20**, 5119–5128.

107 Wang, Y.X., Lee, C.H., Tiep, S., Yu, R.T., *et al.* (2003) *Cell*, **113**, 159–170.

108 Luquet, S., Lopez-Soriano, J., Holst, D., Fredenrich, A., *et al.* (2003) *FASEB J.*, **17**, 2299–2301.

109 Wang, Y.X., Zhang, C.L., Yu, R.T., Cho, H.K., *et al.* (2004) *PLoS Biol.*, **2**, e294.

110 Schuler, M., Ali, F., Chambon, C., Duteil, D., *et al.* (2006) *Cell Metab.*, **4**, 407–414.

111 Narkar, V.A., Downes, M., Yu, R.T., Embler, E., *et al.* (2008) *Cell*, **134**, 405–415.

112 Cheng, L., Ding, G., Qin, Q., Huang, Y., *et al.* (2004) *Nat. Med.*, **10**, 1245–1250.

113 Tanaka, T., Yamamoto, J., Iwasaki, S., Asaba, H., *et al.* (2003) *Proc. Natl Acad. Sci. U. S. A.*, **100**, 15924–15929.

114 Oliver, W.R., Shenk, J.L., Snaith, M.R., Russell, C.S., *et al.* (2001) *Proc. Natl Acad. Sci. U. S. A.*, **98**, 5306–5311.

115 Lee, C.H., Chawla, A., Urbiztondo, N., Liao, D., *et al.* (2003) *Science*, **302**, 453–457.

116 Lee, C.H., Kang, K., Mehl, I.R., Nofsinger, R., *et al.* (2006) *Proc. Natl Acad. Sci. U. S. A.*, **103**, 2434–2439.

117 Nagasawa, T., Inada, Y., Nakano, S., Tamura, T., *et al.* (2006) *Eur. J. Pharmacol.*, **536**, 182–191.

118 Oishi, Y., Manabe, I., Tobe, K., Ohsugi, M., *et al.* (2008) *Nat. Med.*, **14**, 656–666.

119 Vosper, H., Patel, L., Graham, T.L., Khoudoli, G.A., *et al.* (2001) *J. Biol. Chem.*, **276**, 44258–44265.

120 Ghisletti, S., Huang, W., Ogawa, S., Pascual, G., *et al.* (2007) *Mol. Cell*, **25**, 57–70.

121 Pascual, G., Fong, A.L., Ogawa, S., Gamliel, A., *et al.* (2005) *Nature*, **437**, 759–763.

122 Gordon, S. (2003) *Nat. Rev. Immunol.*, **3**, 23–35.

123 Gallardo-Soler, A., Gomez-Nieto, C., Luisa Campo, M., Marathe, C., *et al.*

(2008) *Mol. Endocrinol.*, **22** (6), 1394–1402.

124 Kang, K., Reilly, S.M., Karabacak, V., Gangl, M.R., *et al.* (2008) *Cell Metab.*, **7**, 485–495.

125 Odegaard, J.I., Ricardo-Gonzalez, R.R., Red Eagle, A., Vats, D., *et al.* (2008) *Cell Metab.*, **7**, 496–507.

126 Odegaard, J.I., Ricardo-Gonzalez, R.R., Goforth, M.H., Morel, C.R., *et al.* (2007) *Nature*, **447**, 1116–1120.

127 Marathe, C., Bradley, M.N., Hong, C., Chao, L., *et al.* (2009) *J. Lipid Res.*, **50**, 214–224.

128 Ferrante, A.W., Jr. (2007) *J. Intern. Med.*, **262**, 408–414.

129 Hotamisligil, G.S. (2006) *Nature*, **444**, 860–867.

130 Lumeng, C.N., Bodzin, J.L., and Saltiel, A.R. (2007) *J. Clin. Invest.*, **117**, 175–184.

131 Huang, J.T., Welch, J.S., Ricote, M., Binder, C.J., *et al.* (1999) *Nature*, **400**, 378–382.

132 Barish, G.D., Atkins, A.R., Downes, M., Olson, P., *et al.* (2008) *Proc. Natl Acad. Sci. U. S. A.*, **105**, 4271–4276.

133 Aberle, J., Hopfer, I., Beil, F.U., and Seedorf, U. (2006) *Int. J. Obes. (Lond.)*, **30**, 1709–1713.

134 Aberle, J., Hopfer, I., Beil, F.U., and Seedorf, U. (2006) *Int. J. Med. Sci.*, **3**, 108–111.

135 Skogsberg, J., Kannisto, K., Cassel, T.N., Hamsten, A., *et al.* (2003) *Arterioscler. Thromb. Vasc. Biol.*, **23**, 637–643.

136 Skogsberg, J., McMahon, A.D., Karpe, F., Hamsten, A., *et al.* (2003) *J. Intern. Med.*, **254**, 597–604.

137 Chen, S., Tsybouleva, N., Ballantyne, C.M., Gotto, A.M., Jr., and Marian, A.J. (2004) *Pharmacogenetics*, **14**, 61–71.

138 Vanttinen, M., Nuutila, P., Kuulasmaa, T., Pihlajamaki, J., *et al.* (2005) *Diabetes*, **54**, 3587–3591.

139 Shin, H.D., Park, B.L., Kim, L.H., Jung, H.S., *et al.* (2004) *Diabetes*, **53**, 847–851.

140 Robitaille, J., Gaudet, D., Perusse, L., and Vohl, M.C. (2007) *Int. J. Obes. (Lond.)*, **31**, 411–417.

141 Thamer, C., Machann, J., Stefan, N., Schafer, S.A., *et al.* (2008) *J. Clin. Endocrinol. Metab.*, **93**, 1497–1500.

142 Sprecher, D.L., Massien, C., Pearce, G., Billin, A.N., *et al.* (2007) *Arterioscler. Thromb. Vasc. Biol.*, **27**, 359–365.

143 Riserus, U., Sprecher, D., Johnson, T., Olson, E., *et al.* (2008) *Diabetes*, **57**, 332–339.

144 Reilly, S.M. and Lee, C.H. (2008) *FEBS Lett.*, **582**, 26–31.

145 Barish, G.D. (2006) *J. Nutr.*, **136**, 690–694.

146 Gupta, R.A., Wang, D., Katkuri, S., Wang, H., *et al.* (2004) *Nat. Med.*, **10**, 245–247.

147 Wang, D., Wang, H., Guo, Y., Ning, W., *et al.* (2006) *Proc. Natl Acad. Sci. U. S. A.*, **103**, 19069–19074.

148 Kliewer, S.A., Lenhard, J.M., Willson, T.M., Patel, I., *et al.* (1995) *Cell*, **83**, 813–819.

149 Nagy, L., Tontonoz, P., Alvarez, J.G., Chen, H., and Evans, R.M. (1998) *Cell*, **93**, 229–240.

150 Tontonoz, P., Nagy, L., Alvarez, J.G., Thomazy, V.A., and Evans, R.M. (1998) *Cell*, **93**, 241–252.

151 Lehmann, J.M., Moore, L.B., Smith-Oliver, T.A., Wilkison, W.O., *et al.* (1995) *J. Biol. Chem.*, **270**, 12953–12956.

152 Yki-Jarvinen, H. (2004) *N. Engl. J. Med.*, **351**, 1106–1118.

153 Xin, X., Yang, S., Kowalski, J., and Gerritsen, M.E. (1999) *J. Biol. Chem.*, **274**, 9116–9121.

154 Jiang, C., Ting, A.T., and Seed, B. (1998) *Nature*, **391**, 82–86.

155 Marx, N., Schonbeck, U., Lazar, M.A., Libby, P., and Plutzky, J. (1998) *Circ. Res.*, **83**, 1097–1103.

156 Goetze, S., Xi, X.P., Kawano, H., Gotlibowski, T., *et al.* (1999) *J. Cardiovasc. Pharmacol.*, **33**, 798–806.

157 Ricote, M., Huang, J., Fajas, L., Li, A., *et al.* (1998) *Proc. Natl Acad. Sci. U. S. A.*, **95**, 7614–7619.

158 Marx, N., Sukhova, G., Murphy, C., Libby, P., and Plutzky, J. (1998) *Am. J. Pathol.*, **153**, 17–23.

159 Delerive, P., Martin-Nizard, F., Chinetti, G., Trottein, F., *et al.* (1999) *Circ. Res.*, **85**, 394–402.

160 Liu, Y., Zhu, Y., Rameou, F., Lee, T.S., *et al.* (2004) *Circulation*, **110**, 1128–1133.

161 Wang, N., Verna, L., Chen, N.-G., Chen, J., *et al.* (2002) *J. Biol. Chem.*, **277**, 34176–34181.

162 Pasceri, V., Wu, H.D., Willerson, J.T., and Yeh, E.T. (2000) *Circulation*, **101**, 235–238.

163 Kwak, B.R., Myit, S., Mulhaupt, F., Veillard, N., *et al.* (2002) *Circ. Res.*, **90**, 356–362.

164 Calnek, D.S., Mazzella, L., Roser, S., Roman, J., and Hart, C.M. (2003) *Arterioscler. Thromb. Vasc. Biol.*, **23**, 52–57.

165 Polikandriotis, J.A., Mazzella, L.J., Rupnow, H.L., and Hart, C.M. (2005) *Arterioscler. Thromb. Vasc. Biol.*, **25**, 1810–1816.

166 Hwang, J., Kleinhenz, D.J., Lassegue, B., Griendling, K.K., *et al.* (2005) *Am. J. Physiol. Cell Physiol.*, **288**, C899–C905.

167 Topper, J.N., Cai, J., Falb, D., and Gimbrone, M.A., Jr. (1996) *Proc. Natl Acad. Sci. U. S. A.*, **93**, 10417–10422.

168 Bishop-Bailey, D. and Hla, T. (1999) *J. Biol. Chem.*, **274**, 17042–17048.

169 Hannan, K.M., Dilley, R.J., de Dios, S.T., and Little, P.J. (2003) *Arterioscler. Thromb. Vasc. Biol.*, **23**, 762–768.

170 Vaidya, S., Somers, E.P., Wright, S.D., Detmers, P.A., and Bansal, V.S. (1999) *J. Immunol.*, **163**, 6187–6192.

171 Turturro, F., Friday, E., Fowler, R., Surie, D., and Welbourne, T. (2004) *Clin. Cancer Res.*, **10**, 7022–7030.

172 Goetze, S., Eilers, F., Bungenstock, A., Kintscher, U., *et al.* (2002) *Biochem. Biophys. Res. Commun.*, **293**, 1431–1437.

173 Hoefer, I.E., Timmers, L., and Piek, J.J. (2007) *Curr. Pharm. Des.*, **13**, 1803–1810.

174 Ricote, M., Li, A.C., Willson, T.M., Kelly, C.J., and Glass, C.K. (1998) *Nature*, **391**, 79–82.

175 Meier, C.A., Chicheportiche, R., Juge-Aubry, C.E., Dreyer, M.G., and Dayer, J.M. (2002) *Cytokine*, **18**, 320–328.

176 Welch, J.S., Ricote, M., Akiyama, T.E., Gonzalez, F.J., and Glass, C.K. (2003) *Proc. Natl Acad. Sci. U. S. A.*, **100**, 6712–6717.

177 Matarese, G. and La Cava, A. (2004) *Trends Immunol.*, **25**, 193–200.

178 Bouhlel, M.A., Derudas, B., Rigamonti, E., Dievart, R., *et al.* (2007) *Cell Metab.*, **6**, 137–143.

179 Hevener, A.L., Olefsky, J.M., Reichart, D., Nguyen, M.T., *et al.* (2007) *J. Clin. Invest.*, **117**, 1658–1669.

180 Glass, C.K. and Ogawa, S. (2006) *Nat. Rev. Immunol.*, **6**, 44–55.

181 Chinetti, G., Gbaguidi, F.G., Griglio, S., Mallat, Z., *et al.* (2000) *Circulation*, **101**, 2411–2417.

182 Akiyama, T.E., Sakai, S., Lambert, G., Nicol, C.J., *et al.* (2002) *Mol. Cell Biol.*, **22**, 2607–2619.

183 Chawla, A., Boisvert, W.A., Lee, C.H., Laffitte, B.A., *et al.* (2001) *Mol. Cell*, **7**, 161–171.

184 Llaverias, G., Vazquez-Carrera, M., Sanchez, R.M., Noe, V., *et al.* (2004) *J. Lipid Res.*, **45**, 2015–2024.

185 Frank, P.G., Hassan, G.S., Rodriguez-Feo, J.A., and Lisanti, M.P. (2007) *Curr. Pharm. Des.*, **13**, 1761–1769.

186 Quinn, C.M., Jessup, W., Wong, J., Kritharides, L., and Brown, A.J. (2005) *Biochem. J.*, **385**, 823–830.

187 Javitt, N.B. (2002) *J. Lipid Res.*, **43**, 665–670.

188 Escher, G., Krozowski, Z., Croft, K.D., and Sviridov, D. (2003) *J. Biol. Chem.*, **278**, 11015–11019.

189 Moore, K.J., Rosen, E.D., Fitzgerald, M.L., Randow, F., *et al.* (2001) *Nat. Med.*, **7**, 41–47.

190 Argmann, C.A., Sawyez, C.G., McNeil, C.J., Hegele, R.A., and Huff, M.W. (2003) *Arterioscler. Thromb. Vasc. Biol.*, **23**, 475–482.

191 Liang, C.P., Han, S., Okamoto, H., Carnemolla, R., *et al.* (2004) *J. Clin. Invest.*, **113**, 764–773.

192 Takeda, K., Ichiki, T., Tokunou, T., Funakoshi, Y., *et al.* (2000) *Circulation*, **102**, 1834–1839.

193 Diep, Q.N., El Mabrouk, M., Cohn, J.S., Endemann, D., *et al.* (2002) *Circulation*, **105**, 2296–2302.

194 Kurtz, T.W. and Pravenec, M. (2004) *J. Hypertens.*, **22**, 2253–2261.

195 Schupp, M., Clemenz, M., Gineste, R., Witt, H., *et al.* (2005) *Diabetes*, **54**, 3442–3452.

196 Goetze, S., Kim, S., Xi, X.P., Graf, K., *et al.* (2000) *J. Cardiovasc. Pharmacol.*, **35**, 749–757.

197 Wakino, S., Kintscher, U., Kim, S., Yin, F., *et al.* (2000) *J. Biol. Chem.*, **275**, 22435–22441.

198 Bruemmer, D., Berger, J.P., Liu, J., Kintscher, U., *et al.* (2003) *Eur. J. Pharmacol.*, **466**, 225–234.

199 de Dios, S.T., Bruemmer, D., Dilley, R.J., Ivey, M.E., *et al.* (2003) *Circulation*, **107**, 2548–2550.

200 Bruemmer, D., Yin, F., Liu, J., Berger, J.P., *et al.* (2003) *Mol. Endocrinol.*, **17**, 1005–1018.

201 Maiorano, D., Lutzmann, M., Mechali, M., and Opin, C. (2006) *Cell Biol.*, **18**, 130–136.

202 Ogawa, D., Nomiyama, T., Nakamachi, T., Heywood, E.B., *et al.* (2006) *Circ. Res.*, **98**, e50–e59.

203 Abe, M., Hasegawa, K., Wada, H., Morimoto, T., *et al.* (2003) *Arterioscler. Thromb. Vasc. Biol.*, **23**, 404–410.

204 Okura, T., Nakamura, M., Takata, Y., Watanabe, S., *et al.* (2000) *Eur. J. Pharmacol.*, **407**, 227–235.

205 Bishop-Bailey, D., Hla, T., and Warner, T.D. (2002) *Circ. Res.*, **91**, 210–217.

206 Bruemmer, D., Yin, F., Liu, J., Berger, J.P., *et al.* (2003) *Circ. Res.*, **93**, e38–e47.

207 Redondo, S., Ruiz, E., Santos-Gallego, C.G., Padilla, E., and Tejerina, T. (2005) *Diabetes*, **54**, 811–817.

208 Ruiz, E., Redondo, S., Gordillo-Moscoso, A., and Tejerina, T. (2007) *J. Pharmacol. Exp. Ther.*, **321**, 431–438.

209 Lin, Y., Zhu, X., McLntee, F.L., Xiao, H., *et al.* (2004) *Arterioscler. Thromb. Vasc. Biol.*, **24**, 257–263.

210 Clarke, M.C., Figg, N., Maguire, J.J., Davenport, A.P., *et al.* (2006) *Nat. Med.*, **12**, 1075–1080.

211 Clarke, M.C., Littlewood, T.D., Figg, N., Maguire, J.J., *et al.* (2008) *Circ. Res.*, **102**, 1529–1538.

212 Babaev, V.R., Yancey, P.G., Ryzhov, S.V., Kon, V., *et al.* (2005) *Arterioscler. Thromb. Vasc. Biol.*, **25**, 1647–1653.

213 Iglarz, M., Touyz, R.M., Amiri, F., Lavoie, M.F., *et al.* (2003) *Arterioscler. Thromb. Vasc. Biol.*, **23**, 45–51.

214 Nicol, C.J., Adachi, M., Akiyama, T.E., and Gonzalez, F.J. (2005) *Am. J. Hypertens.*, **18**, 549–556.

215 Law, R.E., Meehan, W.P., Xi, X.P., Graf, K., *et al.* (1996) *J. Clin. Invest.*, **98**, 1897–1905.

216 Phillips, J.W., Barringhaus, K.G., Sanders, J.M., Yang, Z., *et al.* (2003) *Circulation*, **108**, 1994–1999.

217 Aizawa, Y., Kawabe, J., Hasebe, N., Takehara, N., and Kikuchi, K. (2001) *Circulation*, **104**, 455–460.

218 Igarashi, M., Takeda, Y., Ishibashi, N., Takahashi, K., *et al.* (1997) *Horm. Metab. Res.*, **29**, 444–449.

219 Yoshimoto, T., Naruse, M., Shizume, H., Naruse, K., *et al.* (1999) *Atherosclerosis*, **145**, 333–340.

220 Lim, S., Jin, C.J., Kim, M., Chung, S.S., *et al.* (2006) *Arterioscler. Thromb. Vasc. Biol.*, **26**, 808–813.

221 Joner, M., Farb, A., Cheng, Q., Finn, A.V., *et al.* (2007) *Arterioscler. Thromb. Vasc. Biol.*, **27**, 182–189.

222 Wang, C.H., Ciliberti, N., Li, S.H., Szmitko, P.E., *et al.* (2004) *Circulation*, **109**, 1392–1400.

223 Kosuge, H., Haraguchi, G., Koga, N., Maejima, Y., *et al.* (2006) *Circulation*, **113**, 2613–2622.

224 Gerstein, H.C., Yusuf, S., Bosch, J., Pogue, J., *et al.* (2006) *Lancet*, **368**, 1096–1105.

225 Kahn, S.E., Haffner, S.M., Heise, M.A., Herman, W.H., *et al.* (2006) *N. Engl. J. Med.*, **355**, 2427–2443.

226 Belfort, R., Harrison, S.A., Brown, K., Darland, C., *et al.* (2006) *N. Engl. J. Med.*, **355**, 2297–2307.

227 Ratziu, V., Giral, P., Jacqueminet, S., Charlotte, F., *et al.* (2008) *Gastroenterology*, **135**, 100–110.

228 Minamikawa, J., Tanaka, S., Yamauchi, M., Inoue, D., and Koshiyama, H. (1998) *J. Clin. Endocrinol. Metab.*, **83**, 1818–1820.

229 Koshiyama, H., Shimono, D., Kuwamura, N., Minamikawa, J., and Nakamura, Y. (2001) *J. Clin. Endocrinol. Metab.*, **86**, 3452–3456.

230 Nakamura, T., Matsuda, T., Kawagoe, Y., Ogawa, H., *et al.* (2004) *Metabolism*, **53**, 1382–1386.

231 Langenfeld, M.R., Forst, T., Hohberg, C., Kann, P., *et al.* (2005) *Circulation*, **111**, 2525–2531.

232 Mazzone, T., Meyer, P.M., Feinstein, S.B., Davidson, M.H., *et al.* (2006) *JAMA*, **296**, 2572–2581.

233 Sidhu, J.S., Kaposzta, Z., Markus, H.S., and Kaski, J.C. (2004) *Arterioscler. Thromb. Vasc. Biol.*, **24**, 930–934.

234 Lonn, E.P.M., Gerstein, H.C, Sheridan, P., Smith, S., *et al.* (2009) *J. Am. Coll. Cardiol.*, **53**, 2028–2035.

235 Stocker, D.J., Taylor, A.J., Langley, R.W., Jezior, M.R., and Vigersky, R.A. (2007) *Am. Heart J.*, **153**, 445, e441–e446.

236 Hedblad, B., Zambanini, A., Nilsson, P., Janzon, L., and Berglund, G. (2007) *J. Intern. Med.*, **261**, 293–305.

237 Bots, M.L., Hoes, A.W., Koudstaal, P.J., Hofman, A., and Grobbee, D.E. (1997) *Circulation*, **96**, 1432–1437.

238 Touboul, P.J., Elbaz, A., Koller, C., Lucas, C., *et al.* (2000) *Circulation*, **102**, 313–318.

239 Wilcox, R., Bousser, M.G., Betteridge, D.J., Schernthaner, G., *et al.* (2007) *Stroke*, **38**, 865–873.

240 Erdmann, E., Dormandy, J.A., Charbonnel, B., Massi-Benedetti, M., *et al.* (2007) *J. Am. Coll. Cardiol.*, **49**, 1772–1780.

241 Buse, J.B. (2007) *Am. J. Cardiol.*, **99**, S21–S33.

242 Home, P.D., Pocock, S.J., Beck-Nielsen, H., Gomis, R., *et al.* (2005) *Diabetologia*, **48**, 1726–1735.

243 Dormandy, J.A., Charbonnel, B., Eckland, D.J., Erdmann, E., *et al.* (2005) *Lancet*, **366**, 1279–1289.

244 Home, P.D., Pocock, S.J., Beck-Nielsen, H., Gomis, R., *et al.* (2007) *N. Engl. J. Med.*, **357**, 28–38.

245 Aguilar, D., Bozkurt, B., Pritchett, A., Petersen, N.J., and Deswal, A. (2007) *J. Am. Coll. Cardiol.*, **50**, 32–36.

246 Lago, R.M., Singh, P.P., and Nesto, R.W. (2007) *Lancet*, **370**, 1129–1136.

247 Nissen, S.E. and Wolski, K. (2007) *N. Engl. J. Med.*, **356** (24), 2457–2471.

248 Singh, S., Loke, Y.K., and Furberg, C.D. (2007) *JAMA*, **298**, 1189–1195.

249 Takagi, T., Akasaka, T., Yamamuro, A., Honda, Y., *et al.* (2002) *J. Diabetes Complications*, **16**, 50–55.

250 Takagi, T., Akasaka, T., Yamamuro, A., Honda, Y., *et al.* (2000) *J. Am. Coll. Cardiol.*, **36**, 1529–1535.

251 Takagi, T., Yamamuro, A., Tamita, K., Yamabe, K., *et al.* (2002) *Am. J. Cardiol.*, **89**, 318–322.

252 Takagi, T., Yamamuro, A., Tamita, K., Yamabe, K., *et al.* (2003) *Am. Heart J.*, **146**, E5.

253 Choi, D., Kim, S.K., Choi, S.H., Ko, Y.G., *et al.* (2004) *Diabetes Care*, **27**, 2654–2660.

254 Wang, X.L., Oosterhof, J., and Duarte, N. (1999) *Cardiovasc. Res.*, **44**, 588–594.

255 Ridker, P.M., Cook, N.R., Cheng, S., Erlich, H.A., *et al.* (2003) *Arterioscler. Thromb. Vasc. Biol.*, **23**, 859–863.

256 Liu, Y., Yuan, Z., Zhang, J., Yin, P., *et al.* (2007) *Am. Heart J.*, **154**, 718–724.

257 Olefsky, J.M. (2000) *J. Clin. Invest.*, **106**, 467–472.

258 Kasuga, J., Yamasaki, D., Araya, Y., Nakagawa, A., *et al.* (2006) *Bioorg. Med. Chem.*, **14**, 8405–8414.

259 Feldman, P.L., Lambert, M.H., Henke, B.R., and Top, C. (2008) *Med. Chem.*, **8**, 728–749.

260 Wallace, J.M., Schwarz, M., Coward, P., Houze, J., *et al.* (2005) *J. Lipid Res.*, **46**, 1009–1016.

261 Xu, Y., Etgen, G.J., Broderick, C.L., Canada, E., *et al.* (2006) *J. Med. Chem.*, **49**, 5649–5652.

262 Marsboom, G. and Archer, S.L. (2008) *Circ. Res.*, **103**, 1047–1049.

263 Zadelaar, S., Kleemann, R., Verschuren, L., de Vries-Van der Weij, J., van der Hoorn, J., Princen, H.M., and Kooistra, J. (2007) *Arterioscler. Thromb. Vasc. Biol.*, **27**, 1706–1721.

7

Pentraxins in Vascular Pathology: The Role of PTX3

Alberto Mantovani, Cecilia Garlanda, Barbara Bottazzi, Fabiola Molla,
and Roberto Latini

7.1
Introduction

The innate immune system consists of a cellular and a humoral arm. The humoral innate immune response includes members of the complement cascade and soluble pattern recognition receptors (PRRs), such as collectins (surfactant protein-A (SP-A) and SP-D), ficolins, and pentraxins [1, 2]. Therefore fluid phase PRRs are a heterogeneous group of molecular families, which represent functional ancestors of antibodies and play a key role as effectors and modulators of innate resistance in animals and humans.

   Here we will review the key properties of the pentraxin superfamily in relation to vascular pathology with emphasis on C-reactive protein (CRP) and PTX3, as prototypic members of the short and long pentraxin family, respectively. We will focus in particular on PTX3 because conservation has allowed stringent testing of its function using genetic approaches, and on its role as a mediator and marker of cardiovascular pathology.

7.2
The Pentraxin Superfamily

Proteins belonging to the pentraxin family are characterized by a high degree of amino acid sequence conservation, as well as by a multimeric organization. A common structural feature of all the family members is the presence of a conserved sequence 8 amino acids long, the so-called "pentraxin signature" (HxCxS/TWxS, where x is any amino acid). CRP is the prototype of the family: it was identified in the 1930s as an acute phase response protein in human serum [3]. Human serum amyloid P–component (SAP) was subsequently identified as a relative of CRP on the basis of amino acid sequence identity (51%) and for the similar appearance in electron microscopy (annular disk–like structure with pentameric symmetry) [4–6]. CRP and SAP orthologs in different mammal species share

Atherosclerosis: Molecular and Cellular Mechanisms. Edited by Sarah Jane George and Jason Johnson
Copyright © 2010 WILEY-VCH Verlag GmbH & Co. KGaA, Weinheim
ISBN: 978-3-527-32448-4

substantial sequence similarity, with notable differences including serum basal levels and changes during the acute phase response, CRP and SAP being the main acute phase reactants in human and mouse, respectively. In the arthropod *Limulus polyphemus*, different forms of CRP and SAP were identified as abundant constituents of the hemolymph [7–9] involved in recognizing and destroying pathogens.

During the early 1990s, a new secreted protein containing a pentraxin domain was identified as an IL-1-inducible gene (PTX3) in endothelial cells (ECs) or as a tumor necrosis factor (TNF)–stimulated gene (TSG-14) in fibroblasts [10, 11]. Because of the presence of a long N-terminal portion associated to a C-terminal pentraxin-like domain, PTX3 was considered the prototype of the long pentraxin subfamily. Other proteins sharing the same general organization were identified after PTX3, including guinea-pig apexin, neuronal pentraxin-1 (NP1) or NPTX1, NP2 (also called Narp or NPTX2), and neuronal pentraxin receptor (NPR), a transmembrane molecule (reviewed in [1]). The amino acid sequence identity among members of this subfamily is relatively high in the C-terminal pentraxin domain and ranges from 28% between human PTX3 (hPTX3) and hNP1 to 68% between hNP1 and hNP2. By contrast, in the N-terminal domain a low level of similarity is found (about 10% for PTX3 versus NP1). However, identity in the N-terminal domain among the neuronal pentraxins is higher and ranges between 28% and 38%, suggesting the existence of subclasses of molecules among the long pentraxins.

Ortholog molecules have been found so far for PTX3, NP1, NP2, and NPR not only in human, mouse, and rat, but also in lower vertebrates such as zebrafish and puffer-fish (Y. Martinez, unpublished results). Long pentraxins have also been identified in *Xenopus laevis* (XL-PXN1) [12].

Finally, the pentraxin domain has also been found in multidomain proteins, such as in the extracellular protein polydom [13], and in few adhesion G protein–coupled receptors, in particular GPR144 [14]. Pentraxin domain–containing multidomain proteins have been also found in insects (for instance Y17570 in *D. melanogaster* [15]). The function of these proteins has not been defined yet, nor has the role of the pentraxin domain in multidomain proteins been determined.

7.3
Gene and Protein Production and Regulation

The human PTX3 gene is localized on chromosome 3q25 and is organized in three exons separated by two introns. The first two exons code for the leader peptide and the N-terminal domain of the protein, respectively, and the third exon encodes for the pentraxin domain [10]. Thus, the PTX3 protein consists of a C-terminal 203-amino-acid pentraxin-like domain coupled with an N-terminal portion of 178 amino acids (including the 17-amino-acid leader peptide) unrelated to other known proteins.

Analysis of human PTX3 sequence indicates the presence of an *N*-linked glycosylation site in the C-terminal domain at Asn220. Endo- and exoglycosidases digestion and mass spectrometry revealed that fucosylated and sialylated complex-type sugars are *N*-linked to PTX3. Heterogeneity in the relative amount of bi-, tri-, and tetraantennary structures in the glycosidic moiety depends on the cell type and inflammatory stimulus inducing PTX3 production [16].

PTX3 C-terminal domain contains a canonical pentraxin signature and two cysteines at amino acid positions 210 and 271, conserved in all members of the family. Recent data indicate that human recombinant PTX3 is mainly composed of covalently linked octamers. The network of disulfide bonds supporting this octameric assembly was resolved by mass spectrometry and Cys-to-Ser site-directed mutagenesis. Cysteine residues at positions 47, 49, and 103 in the N-terminal domain form three symmetric interchain disulfide bonds stabilizing four protein subunits in a tetrameric arrangement. PTX3 tetramers are linked into octamers thanks to additional interchain disulfide bonds involving cysteines Cys317 and Cys318, located in the C-terminal domain of the protein [17].

In contrast to CRP and SAP, which are produced primarily in the liver in response to IL-6, PTX3 is induced by several stimuli but not IL-6, in various cell types but not hepatocytes. Among human peripheral blood leukocytes, only monocytes release PTX3 in response to proinflammatory cytokines (IL-1 and TNF-α) or following stimulation with microbial components, including lipopolysaccharide (LPS), lipoarabinomannan, and outer membrane proteins (OMP) [18–20]. Engagement of Toll-like receptors (TLRs) by outer membrane protein A or peptidoglycan (TLR2), poly(I):(C) (TLR3), LPS, or *Candida* (TLR4) and flagellin (TLR5), can induce PTX3 production not only in monocytes, but also in monocyte-derived dendritic cells (DCs), which appear to produce the highest amounts of PTX3 *in vitro* [21]. Production of PTX3 is restricted to myeloid DCs, while plasmacytoid DCs are unable to produce the protein irrespective of the stimulus used [21]. Polymorphonuclear cells (PMNs) do not express PTX3 mRNA in the resting condition nor after stimulation [18], however intracellular storage of functional protein is present in specific lactoferrin-positive granules of human neutrophils [22].

Following microbial recognition, PTX3 is promptly released from PMN granules and localized to extracellular traps extruded from activated PMN called neutrophil extracellular traps (NETs), where it can contribute to the generation of a antimicrobial microenvironment essential to trap and kill microbes [23, 24].

Other cell types can produce PTX3 in response to appropriate stimulation, such as fibroblasts, adipocytes, chondrocytes, stromal, mesangial, endothelial, and epithelial cells (reviewed in [1]). Granulosa cells are a peculiar cell type which express PTX3 mRNA under ovulatory stimuli [25]. The protein localizes in the extracellular matrix and plays a non-redundant role as a structural constituent of the cumulus oophorus extracellular matrix essential for female fertility. Expression of PTX3 mRNA and protein by human cumulus cells [25, 26] suggests that this molecule might have the same role in murine and human female fertility.

This wide distribution of cells able to produce PTX3 likely reflects the local role exerted by the protein, compared with the systemic role exerted by the liver-derived classical short pentraxins CRP and SAP.

As mentioned above, vascular ECs are a major source of PTX3 in response to inflammatory signals. By contrast, lymphatic ECs constitutively express the protein, as revealed by gene profiling analysis performed on both murine and human lymphatic ECs [27–29]. Analysis of advanced atherosclerotic plaques also revealed a strong expression of PTX3 in macrophages and ECs within the lesions, as well as in subendothelial smooth muscle cells (SMCs) and in foam cells within the lipid-rich area of the plaques [30]. SMCs, which actively participate in the atherogenic process, can be induced *in vitro* to produce large amounts of PTX3 in response to inflammatory signals, notably the pro-atherogenic oxidized low-density lipoproteins (ox-LDLs) or enzymatically degraded LDLs (E-LDLs) [31]. Under the same conditions, native LDLs had no effect on PTX3 expression.

The lipoprotein-driven expression of PTX3 by SMCs may suggest a potential pathogenic role for PTX3 in mediating atherosclerosis progression, however more recent data show that high-density lipoproteins (HDLs), which are well known for their anti-inflammatory and anti-atherogenic role, induce PTX3 production by vascular ECs [32]. This finding has been also confirmed *in vivo*, where HDL injection increases PTX3 expression in the aorta in wild-type or ApoAI-overexpressing mice [32]. This is not the only example of PTX3 production in response to an atheroprotective signal. Indeed IL-10, an anti-inflammatory cytokine with atheroprotective properties [33, 34], is also a mild inducer of PTX3 in monocytes and DCs, as observed by both transcriptional profiling and protein levels in culture supernatants of IL-10-stimulated cells [21, 35]. In addition, IL-10 can amplify PTX3 production induced by LPS. Given the role of IL-10 in the chronic and resolution phase of inflammation [36], the induction of PTX3 likely reflects the role in the orchestration of matrix deposition, tissue repair, and remodeling, suggested by several observations in *ptx3*-deficient animals. The induction of PTX3 by anti-inflammatory and atheroprotective signals, such as HDL and IL-10, may therefore reflect a potential regulatory role of this protein in the innate and adaptive immune responses, as well as a possible anti-atherogenic function of PTX3.

To understand the regulation of protein production, the proximal promoter of both human and murine PTX3 gene has been characterized [37, 38], revealing the presence of numerous potential enhancer binding elements, including Pu1, AP-1, NF-κB, SP1 and NF-IL6 sites. The NF-κB site is functionally active in response to IL-1β and TNF-α, while Sp1 and AP-1 sites do not seem to play a major role in terms of responsiveness to cytokines. The involvement of NF-κB pathway is also demonstrated by recent data obtained in a model of acute myocardial ischemia and reperfusion in the mice [39]. In this model, induction of ischemia results in upregulation of PTX3 production, an effect almost completely abolished in IL-1RI-(type I receptor for IL-1) or MyD88-deficient mice. Different pathways can be also involved, depending on the cell type and/or the stimuli. Induction of PTX3 by TNF-α in alveolar epithelial cells from human lungs is mediated by activation of the JNK pathway [40], while induction of PTX3 by HDL requires activation of the PI3K/Akt pathway through a G-coupled lysosphingolipid receptor [32].

An unexpected mechanism for regulation of PTX3 expression has been recently identified in the FUS/CHOP translocation involved in the pathogenesis of a subset of soft tissue sarcomas [41, 42]. Two independent studies have shown that PTX3 is a major gene downstream of the FUS/CHOP translocation, even if PTX3 expression in the liposarcoma tissues is not related only to the presence of this type of translocation [41, 42].

Glucocorticoid hormones (GCs), which play a complex regulatory activity on components of innate and adaptive immunity, have divergent regulatory effects on PTX3 expression and production in hematopoietic versus non-hematopoietic cells [43]. In fact GCs inhibit PTX3 production in myeloid DCs while inducing, and enhancing under inflammatory conditions, the production of the protein in fibroblasts and ECs. The divergent effect of GCs on PTX3 regulation is likely due to differences in the action of the GC receptor (GR) in different cell populations. GR could act as a ligand-dependent transcription factor through direct DNA binding (dimerization-dependent), or as a gene transcription repressor through the protein–protein interference with the action of other signaling pathways (dimerization-independent). The results obtained so far show that in non-hematopoietic cells the stimulation of PTX3 gene expression and production is dimerization-dependent, whereas suppression of PTX3 production in cells of hematopoietic origin is dimerization-independent and likely mediated by interference by the NF-κB pathway [43]. Modulation of PTX3 production by GCs is also evident *in vivo* in humans, where administration of GCs results in increased serum levels of PTX3. Endogenous GCs have a similar effect, since patients with Cushing's syndrome had increased levels of circulating PTX3, whereas PTX3 levels are reduced in people affected by iatrogenic hypocortisolism [43].

The data summarized above indicate the complex regulation of PTX3 production from various cell types in response to different stimuli, and likely reflect the differing roles of this multifunctional protein in the innate immune response and as a constituent of the extracellular matrix. In addition, the induction of PTX3 by anti-inflammatory and atheroprotective signals, such as HDL and IL-10, may reflect a potential regulatory role of this protein on the innate and adaptive immune responses, as well as a possible protective function of PTX3 on ECs and vascular integrity.

7.4
Role of PTX3 in Innate Immunity and Inflammation

The multifunctional properties of PTX3 can be at least in part explained by its capacity to interact with a number of different ligands (Figure 7.1), a characteristic shared with CRP and SAP. In particular, PTX3 binds to the complement component C1q, interacting with C1q globular head [44, 45]. The role of PTX3 on the modulation of complement activation is a complex subject and is not completely defined yet. The interaction of PTX3 with C1q results in activation of the classical complement cascade only when C1q is plastic-immobilized, a situation which mimics C1q bound to a microbial surface. In contrast, when interaction occurs in the fluid phase, a dose-dependent inhibition of C1q hemolytic activity is observed, suggesting a possible inhibitory effect by competitive blocking of relevant sites

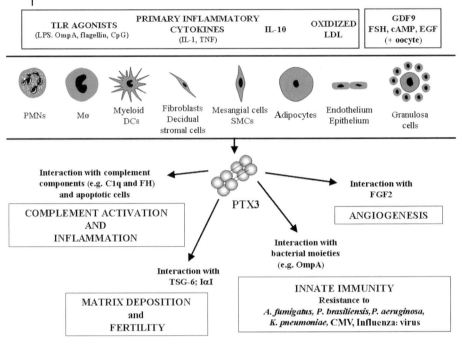

Figure 7.1 Cellular sources and main functions of the long pentraxin PTX3. PTX3 is produced by different cells in response to Toll-like receptor (TLR) agonists, primary inflammatory cytokines, modified low-density lipoprotein (LDL), and hormones. The released protein can interact with different ligands, playing a crucial role at the interface between innate immunity, inflammation, angiogenesis, and matrix deposition.

[45]. Modulation of activation of the classical complement cascade is also obtained through glycosylation: in fact, enzymatically deglycosylated PTX3 binds more efficiently to C1q and induces a higher level of complement activation [16]. As PTX3 glycosidic moieties may vary depending on the cell type and/or inflammatory context in which the protein is produced, glycosylation of PTX3 could act as a tuning mechanism of its biological functions. In addition, PTX3 can interact with factor H (FH), the main soluble regulator of the alternative pathway of complement activation [46]. Like CRP, interaction of PTX3 with FH modulates complement activation, promoting FH deposition on PTX3-coated surfaces and preventing exaggerated complement activation [46].

PTX3 binds specific pathogens, such as fungi, bacteria, and viruses, promoting phagocytosis and consequent clearance of the pathogen. Specific binding has been observed to the conidia of *Aspergillus fumigatus*, *Paracoccidioides brasiliensis*, and zymosan; to selected Gram-positive and Gram-negative bacteria (*Staphylococcus aureus*, *Pseudomonas aeruginosa*, *Salmonella thyphimurium*, *Neisseria meningitides*) ([20, 47] and B. Bottazzi, unpublished); and finally to some viruses such as cytomegalovirus and influenza virus type A [48, 49]. Accordingly, *in vivo* studies with gene-modified mice indicated that PTX3 plays non-redundant roles in recognition and resistance against these microorganisms [47–49] (Figure 7.1).

PTX3 recognition of microbial components can trigger a proinflammatory response, as demonstrated by the results obtained on the interaction with a member of the OMP family from *Klebsiella pneumoniae* (OMP-A). OMP-A is recognized by monocytes and DCs through scavenger receptors and induces a TLR2-dependent proinflammatory response. This inflammatory response includes PTX3 production, which in turn binds OMP-A. The interaction of PTX3 with OMP-A activates a program of amplification of the inflammatory response *in vivo*, which is dependent on complement activation [20, 50].

The first genetic evidence of the functional role of PTX3 in innate resistance in humans was observed by Olesen *et al.*, who analyzed the role of polymorphisms within PTX3 in pulmonary tuberculosis and found that two of them showed a significant association with tuberculosis disease [51].

Altogether, these studies suggest that PTX3, released by PMNs and produced by DCs, macrophages, and other cell types upon TLR engagement or pathogen recognition, recognizes microbial moieties, opsonizes fungi, selected bacteria, and viruses, and activates complement. Opsonization results in facilitated pathogen recognition (increased phagocytosis and killing) and in innate immune cell activation (increased cytokine and nitric oxide production); moreover, opsonization by PTX3 is likely to be involved in the activation of an appropriate adaptive immune response (DC maturation and polarization; see, for example [52]). All these properties suggest that this long pentraxin behaves as a functional ancestor of antibodies.

Amplification of the inflammatory response by PTX3 is not a general rule, at least in sterile inflammation. In a model of experimental myocardial infarction (MI), *ptx3* deficiency caused greater myocardial lesions associated with increased leukocyte infiltration, cell death, and C3 deposition in ischemic myocardium. Thus, PTX3 plays a non-redundant cardioprotective function in mice, modulating the reperfusion-associated inflammation and tissue damage, possibly by modulating the classical and the alternative pathways of complement activation [39, 44–46].

Like CRP and SAP, PTX3 binds to apoptotic cells during late phases of apoptosis [53]. Phosphoethanolamine, phosphocholine, small nuclear ribonucleoproteins, and chromatin/nucleolar components, recognized by CRP and/or SAP, redistribute to the plasma membrane during late apoptosis. PTX3 can bind histones, raising the possibility that interaction with nuclear components could actually occur, whereas it does not bind to phosphoethanolamine and phosphocholine [44].

All together, these results outline the delicate role exerted by PTX3 in the inflammatory context, which possibly differs depending on the model examined (i.e., sterile versus non-sterile conditions). In addition, while in some circumstances PTX3 promotes removal of selected pathogens by professional phagocytes, it inhibits removal of apoptotic cells by immature DCs and macrophages, thus preventing inflammatory uptake of late apoptotic cells and antigen presentation by antigen-presenting cells [53].

7.5
PTX3 in Models of Vascular Pathology

Among the ligands recognized by PTX3 is fibroblast growth factor 2 (FGF2) [54]. FGF2 plays important roles *in vivo* by promoting angiogenesis and neovascularization during wound healing, inflammation, atherosclerosis, and tumor growth. All these pathologic conditions are also characterized by accumulation of macrophages that together with ECs may represent a major local source of both FGF2 and PTX3. The specific interaction between PTX3 and FGF2 results in the inhibition of FGF-dependent proliferation *in vitro*. Moreover, PTX3 overexpression in FGF2-transformed ECs inhibits both their proliferation *in vitro* and their capacity to generate vascular lesions *in vivo*. FGF2 plays a key role in the induction of proliferation, migration, and survival of SMCs and excessive growth of SMCs is an important component in atherosclerosis and restenosis. Interaction between PTX3 and FGF2 inhibits SMC proliferation *in vitro*; in addition, PTX3 overexpression in transduced SMCs reduces intimal hyperplasia after arterial injury as a result of direct binding to FGF2 [55]. Thus, PTX3 could act as an "FGF2 decoy" able to sequester the growth factor in an inactive form. The interaction between PTX3 and FGF2 may modulate angiogenesis in various physiopathological conditions, affecting the crosstalk between inflammatory cells and endothelium. In addition, the potent inhibitory effect on FGF2-mediated activation of SMCs suggests that PTX3 may modulate SMC activation after arterial injury [56].

Recent studies have provided further evidences on the roles played by PTX3 in atherosclerosis. Bosutti *et al.* [57] have shown that PTX3 mRNA levels were elevated in white blood cells and in adipose tissue samples of patients with high levels of LDL compared to those with low levels. This finding suggests that PTX3 may be involved in the mechanisms through which LDL triggers inflammation and oxidative stress associated with atherosclerosis. Another study has extended these findings by showing that both LDL/HDL ratio and fibrinogen were independent determinants of PTX3 expression in visceral fat of obese subjects, thus supporting the involvement of PTX3 in the mechanisms underlying the development of atherosclerosis [58]. A secretome of ECs treated with lysophosphatidic acid (LPA) revealed a marked upregulation of PTX3 both in terms of mRNA and protein level [59]. In addition, an immunohistochemical study of human atherosclerotic lesions obtained at autopsy using a novel monoclonal antibody against PTX3 showed that macrophages and polymorphonuclear cells infiltrating the atherosclerotic plaques were positive for PTX3 [60]. All together these data support the involvement of PTX3 in atherogenesis.

The recent observation that HDL induced PTX3 mRNA expression in human ECs, whereas no effect was observed on CRP and SAP expression [32], further elucidates the possible roles of PTX3 in immune and inflammatory response of endothelium. These data suggest that part of the atheroprotective effects of HDL could result from the modulation of molecules such as PTX3 that act as sensors of the immuno-inflammatory balance in the vascular wall.

The presence of PTX3 in infarcted hearts from deceased patients and the higher degree of ischemic injury in hearts of mice deficient for PTX3 support a pathophysiologic role of the protein in myocardial damage and repair [39]. Unlike the classic short pentraxins CRP and SAP, whose sequence and regulation have diverged from mouse to humans, PTX3 is highly conserved in evolution. Thus, results obtained using genetic approaches in the mouse are likely to be informative for the function of PTX3 in humans. Based on these facts, the role of PTX3 in experimental MI was assessed using *ptx3*-deficient mice [39]. Ptx3 peaked in the serum of wild-type mice after ischemia followed by reperfusion, similar to humans, and infarcts were significantly larger in *ptx3*-deficient mice than in wild-type mice. A possible protective cardiovascular role for PTX3 has been hypothesized to take place through modulation of complement activation by PTX3: in this context a higher deposition of C3 was observed in *ptx3*-deficient mice in the infarct area. In addition, the interaction between PTX3 and FH, resulting in promotion of FH deposition on PTX3-coated surfaces, could represent a mechanism for preventing tissue damage associated with an exaggerated complement activation [46]. It remains to be assessed whether PTX3 plays a role in the progression of post-infarction left ventricular dysfunction and failure.

7.6
PTX3 as Marker of Vascular Pathology

The data obtained with gene-modified animals and the relationship with a classical marker of inflammation such as CRP, prompted us to evaluate PTX3 as a possible marker of human pathology. The observation that PTX3, in contrast to the liver-produced CRP, is produced by different cell types, suggests that it may represent a rapid marker reflecting the local activation of innate immunity and inflammation. Indeed, the general characteristics emerging from studies on PTX3 blood levels in human pathologies are (i) the rapidity of its increase compared with CRP, consistent with the identification of PTX3 as an early induced gene [10, 61], and (ii) the lack of correlation between levels of CRP and PTX3.

In light of the data obtained in preclinical studies in the mouse and on the basis of the pattern of cells producing PTX3 and stimuli able to induce the protein *in vitro*, attention was focused in particular on pathologic conditions with an inflammatory and/or infectious origin as well as on pathologies involving the vascular compartment.

In humans and mice PTX3 behaves as an acute phase protein and its blood levels increase rapidly (peak at 6–8 h) from less than 2 ng/ml in normal conditions up to even 200–800 ng/ml during endotoxin shock, sepsis, and other inflammatory or infectious conditions. A dramatic increase in PTX3 plasma levels was observed in patients with sepsis and septic shock, tuberculosis, dengue virus, and meningococcal disease [62–65]. In all the cases a significant correlation was found between PTX3 plasma levels and severity of disease; in addition, the lack of correlation between PTX3 and CRP levels could be potentially useful to discriminate between shock and absence of shock, as it seems to occur in a small cohort of patients with meningococcal disease [65].

Inflammation is a critical component of ischemic heart disorders and CRP has emerged as a valuable prognostic and diagnostic tool (for a critical appraisal, see [66, 67]). Lately the results of the large multicenter trial JUPITER have shown a remarkable reduction of cardiac events with rosuvastatin when compared with placebo in 17 802 apparently healthy men and women, with normal LDL (<130 mg/dl) but CRP >2 mg/l [68]. Though this is not a definite proof of the etiologic role of inflammatory activation in cardiac disease, it extends the indication for preventive treatment aimed at decreasing CRP levels to a large class of subjects for whom there was no indication in the past for a statin.

Because of the homology between PTX3 and CRP, the relevance of PTX3 measurement as diagnostic tool has been particularly investigated in cardiovascular pathology. Prompted by the increasing relevance of circulating PTX3 as marker of severity and outcomes of cardiovascular diseases, a high-sensitivity specific ELISA assay has been developed with a lower detection limit of 0.1 ng/ml, about tenfold lower than levels found in healthy volunteers [69]. Circulating PTX3 was significantly elevated in patients who underwent percutaneous coronary intervention (PCI) and in those with unstable angina, suggesting that PTX3 may be a candidate marker to predict occurrence of unstable angina. Patients with arterial inflammation, eligible for coronary intervention, exhibited high concentrations of plasma PTX3; in particular, patients with unstable angina pectoris exhibited PTX3 levels three times higher than the normal range [69].

A role for PTX3 has been recently proposed in patients with coronary artery disease after PCI [70]: in a group of 20 patients undergoing coronary stenting, PTX3 levels increase both in peripheral blood and in the coronary sinus, and correlate with CD11b/CD18 expression on neutrophil surface. Furthermore, the relative PTX3 increase observed at 24h is the most powerful predictor of late lumen loss, suggesting that PTX3 may be a useful marker for the evaluation of an inflammatory response and neointimal thickening after vascular injury. Of interest is the positive transcardiac gradient of 0.40 ± 0.64 ng/ml in patients with post-PCI restenosis, but not in those without, suggesting a production of PTX3 by the coronary vasculature [70].

The high levels of expression in the heart during inflammatory reactions, the production by vascular cells in response to inflammatory signals [10] and ox-LDL [31], and the occurrence in atherosclerotic lesions [30] prompted studies on PTX3 levels in acute myocardial infarction [61, 71]. A first study showed that PTX3 peaked in plasma early, within 6–8h from symptom onset, and was not correlated with CRP [61]. In a cohort of 748 patients with MI and ST elevation, PTX3, measured within the first day from the onset of symptoms along with established markers including CRP, NT-proBNP, and troponin-T, emerged as the only independent predictor of three-month mortality [71]. It remains to be elucidated whether this impressive relation with fatal outcome actually reflects a role in the pathogenesis of damage, for instance by amplifying the complement and coagulation cascades [72, 73] or a marked protective response to severe cardiac injury.

Inflammatory activation is currently found in patients with chronic heart failure, as evidenced by higher than normal circulating concentrations of inflammatory cytokines and CRP. Two studies reported on PTX3 in a total of 233 patients with heart failure overall. PTX3 was higher than in healthy controls, its levels increasing with severity of heart failure, but more interesting was the observation that PTX3 independently predicted outcome in these patients, in whom BNP had also been assayed and included in the multivariable analysis [70, 74]. These findings have been extended by two studies respectively on 37 and 164 patients with heart failure, where it was found that best risk prediction was attained by the combined use of three biomarkers: BNP, H-FABP, and PTX3 [75, 76]. PTX3 was measured at baseline in 1233 stable patients with chronic HF enrolled in the GISSI-Heart Failure trial. PTX3 was found to be elevated in patients with more severe disease, in particular in those with advanced age, ventricular dysfunction, worse symptoms, and comorbidities such as atrial fibrillation or diabetes [77].

Increased levels of PTX3 have been observed in a restricted set of autoimmune disorders (e.g., in the blood in small vessel vasculitis, in the synovial fluid in rheumatoid arthritis), but not in others (e.g., systemic lupus erythematosus) [78, 79]. In small vessel vasculitis, PTX3 levels correlate with clinical activity of the disease and represent a candidate marker for monitoring the disease [78]. Immunohistochemistry performed on skin sections at sites of vasculitis shows that ECs are responsible for PTX3 production [80]. Moreover, in these patients, PTX3 is abundantly present at sites of leukocytoclastic infiltration; the finding that PTX3, in contrast to the short pentraxin SAP, inhibits the uptake of apoptotic PMNs by macrophages [81], suggests that PTX3 is a key factor in the incomplete clearance of apoptotic and secondary necrotic PMNs observed in small vessel vasculitis [80].

Patients with chronic kidney disease (CKD) also show increase in PTX3 plasma levels [82], with the highest concentrations observed in the group of patients with a more severe disease. In a parallel study, it has been observed that patients undergoing chronic hemodialysis have higher plasma PTX3 levels compared with those undergoing peritoneal hemodialysis [83]; in addition, the presence of peripheral or coronary artery disease results in significantly higher levels of PTX3. Finally, patients with high PTX3 levels had higher all-cause mortality and cardiovascular mortality, suggesting that PTX3 could have a predictive value of mortality in CKD patients [82]. The latter finding is reminiscent of MI data [71]. In a cohort of 207 patients with CKD stage 5 and in another of patients with type 2 diabetes, PTX3 was independently associated with proteinuria; PTX3 and proteinuria were associated with endothelial dysfunction in type 2 diabetes [84], thus suggesting that PTX3 is more than an additional marker of inflammation in chronic kidney disease [85].

Recent results show that pregnancy itself, a condition associated with relevant involvement of inflammatory molecules at the implantation site [86], is associated with slight increase in maternal circulating PTX3 levels compared with the non-pregnant condition. Higher maternal PTX3 levels were observed in pregnancies complicated by preeclampsia [87, 88], which represents the clinical manifestation of an endothelial dysfunction as part of an excessive maternal inflammatory response to pregnancy [87, 89, 90].

7.7
Concluding Remarks

Pentraxins are multifunctional fluid phase (with one exception) pattern recognition molecules. CRP and PTX3 are prototypic molecules representative of the short and the long pentraxin family respectively. The lack of strict evolutionary conservation of CRP between mouse and human, in terms of sequence, ligand recognized, and regulation has prevented assessment of its function using genetic approaches. In contrast, PTX3 is strictly conserved in evolution. Available genetic evidence suggests that PTX3 is a key component of the humoral innate immune system and has a regulatory role in the pathogenesis of atherosclerosis and ischemic heart disorders. In addition, data obtained so far suggest that PTX3 may represent a useful marker of cardiovascular pathology, preceding CRP and reflecting the vascular involvement by inflammatory process. CRP and PTX3 are components of the humoral arm of innate immune system, with complementary functions in terms of cellular sources, ligands and time of production. Available information suggests that CRP and PTX3 may also be complementary as markers of human vascular pathology.

Acknowledgment

The contribution of the European Commission ("MUGEN" LSHG-CT-2005-005203, "MUVAPRED" LSHP-CT-2003-503240, "TOLERAGE" 2008-02156, FLUINNATE 2007-044161, "EVGN" LSHM-CT-2003-503254), Ministero dell'Istruzione, Università e della Ricerca (MIUR) (project FIRB), Telethon (Telethon grant n. GGP05095), fondazione CARIPLO (project Nobel), Ministero della Salute (Ricerca finalizzata), and the Italian Association for Cancer Research (AIRC) is gratefully acknowledged.

References

1 Garlanda, C., Bottazzi, B., Bastone, A., and Mantovani, A. (2005) Pentraxins at the crossroads between innate immunity, inflammation, matrix deposition, and female fertility. *Annu. Rev. Immunol.*, **23**, 337–366.

2 Bottazzi, B., Garlanda, C., Cotena, A., Moalli, F., Jaillon, S., Deban, L., and Mantovani, A. (2009) The long pentraxin PTX3 as a prototypic humoral pattern recognition receptor: interplay with cellular innate immunity. *Immunol. Rev.*, **227**, 9–18.

3 Tillet, W.S. and Francis, T., Jr. (1930) Serological reactions in pneumonia with a non protein somatic fraction of pneumococcus. *J. Exp. Med.*, **52**, 561–571.

4 Srinivasan, N., White, H.E., Emsley, J., Wood, S.P., Pepys, M.B., and Blundell, T.L. (1994) Comparative analyses of pentraxins: implications for protomer assembly and ligand binding. *Structure*, **2**, 1017–1027.

5 Emsley, J., White, H.E., O'Hara, B.P., Oliva, G., Srinivasan, N., Tickle, I.J., Blundell, T.L., Pepys, M.B., and Wood, S.P. (1994) Structure of pentameric human serum amyloid P component. *Nature*, **367**, 338–345.

6 Gewurz, H., Zhang, X.H., and Lint, T.F. (1995) Structure and function of the pentraxins. *Curr. Opin. Immunol.*, 7, 54–64.

7 Shrive, A.K., Metcalfe, A.M., Cartwright, J.R., and Greenhough, T.J. (1999) C-reactive protein and SAP-like pentraxin are both present in *Limulus polyphemus* haemolymph: crystal structure of *Limulus* SAP. *J. Mol. Biol.*, 290, 997–1008.

8 Armstrong, P.B., Swarnakar, S., Srimal, S., Misquith, S., Hahn, E.A., Aimes, R.T., and Quigley, J.P. (1996) A cytolytic function for a sialic acid-binding lectin that is a member of the pentraxin family of proteins. *J. Biol. Chem.*, 271, 14717–14721.

9 Robey, F.A. and Liu, T.Y. (1981) Limulin: a C-reactive protein from *Limulus polyphemus*. *J. Biol. Chem.*, 256, 969–975.

10 Breviario, F., d'Aniello, E.M., Golay, J., Peri, G., Bottazzi, B., Bairoch, A., Saccone, S., Marzella, R., Predazzi, V., Rocchi, M., Della Valle, G., Dejana, E., Mantovani, A., and Introna, M. (1992) Interleukin-1-inducible genes in endothelial cells. Cloning of a new gene related to C-reactive protein and serum amyloid P component. *J. Biol. Chem.*, 267, 22190–22197.

11 Lee, G.W., Lee, T.H., and Vilcek, J. (1993) TSG-14, a tumor necrosis factor- and IL-1-inducible protein, is a novel member of the pentaxin family of acute phase proteins. *J. Immunol.*, 150, 1804–1812.

12 Seery, L.T., Schoenberg, D.R., Barbaux, S., Sharp, P.M., and Whitehead, A.S. (1993) Identification of a novel member of the pentraxin family in Xenopus laevis. *Proc. R. Soc. Lond. B Biol. Sci.*, 253, 263–270.

13 Gilges, D., Vinit, M.A., Callebaut, I., Coulombel, L., Cacheux, V., Romeo, P.H., and Vigon, I. (2000) Polydom: a secreted protein with pentraxin, complement control protein, epidermal growth factor and von Willebrand factor A domains. *Biochem. J.*, 352 (Pt 1), 49–59.

14 Bjarnadottir, T.K., Fredriksson, R., Hoglund, P.J., Gloriam, D.E., Lagerstrom, M.C., and Schioth, H.B. (2004) The human and mouse repertoire of the adhesion family of G-protein-coupled receptors. *Genomics*, 84, 23–33.

15 Schwartz, Y.B., Boykova, T., Belyaeva, E.S., Ashburner, M., and Zhimulev, I.F. (2004) Molecular characterization of the singed wings locus of Drosophila melanogaster. *BMC Genet.*, 5, 15.

16 Inforzato, A., Peri, G., Doni, A., Garlanda, C., Mantovani, A., Bastone, A., Carpentieri, A., Amoresano, A., Pucci, P., Roos, A., Daha, M.R., Vincenti, S., Gallo, G., Carminati, P., De Santis, R., and Salvatori, G. (2006) Structure and function of the long pentraxin PTX3 glycosidic moiety: fine-tuning of the interaction with C1q and complement activation. *Biochemistry*, 45, 11540–11551.

17 Inforzato, A., Rivieccio, V., Morreale, A.P., Bastone, A., Salustri, A., Scarchilli, L., Verdoliva, A., Vincenti, S., Gallo, G., Chiapparino, C., Pacello, L., Nucera, E., Serlupi-Crescenzi, O., Day, A.J., Bottazzi, B., Mantovani, A., De Santis, R., and Salvatori, G. (2008) Structural characterization of PTX3 disulfide bond network and its multimeric status in cumulus matrix organization. *J. Biol. Chem.*, 283, 10147–10161.

18 Vidal Alles, V., Bottazzi, B., Peri, G., Golay, J., Introna, M., and Mantovani, A. (1994) Inducible expression of PTX3, a new member of the pentraxin family, in human mononuclear phagocytes. *Blood*, 84, 3483–3493.

19 Vouret-Craviari, V., Matteucci, C., Peri, G., Poli, G., Introna, M., and Mantovani, A. (1997) Expression of a long pentraxin, PTX3, by monocytes exposed to the mycobacterial cell wall component lipoarabinomannan. *Infect. Immun.*, 65, 1345–1350.

20 Jeannin, P., Bottazzi, B., Sironi, M., Doni, A., Rusnati, M., Presta, M., Maina, V., Magistrelli, G., Haeuw, J.F., Hoeffel, G., Thieblemont, N., Corvaia, N., Garlanda, C., Delneste, Y., and Mantovani, A. (2005) Complexity and complementarity of outer membrane protein A recognition by cellular and humoral innate immunity receptors. *Immunity*, 22, 551–560.

21 Doni, A., Peri, G., Chieppa, M., Allavena, P., Pasqualini, F., Vago, L., Romani, L., Garlanda, C., and Mantovani, A. (2003)

Production of the soluble pattern recognition receptor PTX3 by myeloid, but not plasmacytoid, dendritic cells. *Eur. J. Immunol.*, **33**, 2886–2893.

22 Jaillon, S., Peri, G., Delneste, Y., Fremaux, I., Doni, A., Moalli, F., Garlanda, C., Romani, L., Gascan, H., Bellocchio, S., Bozza, S., Cassatella, M.A., Jeannin, P., and Mantovani, A. (2007) The humoral pattern recognition receptor PTX3 is stored in neutrophil granules and localizes in extracellular traps. *J. Exp. Med.*, **204**, 793–804.

23 Brinkmann, V., Reichard, U., Goos-mann, C., Fauler, B., Uhlemann, Y., Weiss, D.S., Weinrauch, Y., and Zychlinsky, A. (2004) Neutrophil extracellular traps kill bacteria. *Science*, **303**, 1532–1535.

24 Fuchs, T.A., Abed, U., Goosmann, C., Hurwitz, R., Schulze, I., Wahn, V., Weinrauch, Y., Brinkmann, V., and Zychlinsky, A. (2007) Novel cell death program leads to neutrophil extracellular traps. *J. Cell Biol.*, **176**, 231–241.

25 Salustri, A., Garlanda, C., Hirsch, E., De Acetis, M., Maccagno, A., Bottazzi, B., Doni, A., Bastone, A., Mantovani, G., Beck Peccoz, P., Salvatori, G., Mahoney, D.J., Day, A.J., Siracusa, G., Romani, L., and Mantovani, A. (2004) PTX3 plays a key role in the organization of the cumulus oophorus extracellular matrix and in *in vivo* fertilization. *Development*, **131**, 1577–1586.

26 Paffoni, A., Ragni, G., Doni, A., Somigliana, E., Pasqualini, F., Restelli, L., Pardi, G., Mantovani, A., and Garlanda, C. (2006) Follicular fluid levels of the long pentraxin PTX3. *J. Soc. Gynecol. Investig.*, **13**, 226–231.

27 Sironi, M., Conti, A., Bernasconi, S., Fra, A.M., Pasqualini, F., Nebuloni, M., Lauri, E., De Bortoli, M., Mantovani, A., Dejana, E., and Vecchi, A. (2006) Generation and characterization of a mouse lymphatic endothelial cell line. *Cell Tissue Res.*, **325**, 91–100.

28 Wick, N., Saharinen, P., Saharinen, J., Gurnhofer, E., Steiner, C.W., Raab, I., Stokić, D., Giovanoli, P., Buchsbaum, S., Burchard, A., Thurner, S., Alitalo, K., and Kerjaschki, D. (2007) Transcriptomal comparison of human dermal lymphatic

endothelial cells ex vivo and in vitro. *Physiol. Genomics*, **28**, 179–192.

29 Amatschek, S., Kriehuber, E., Bauer, W., Reininger, B., Meraner, P., Wolpl, A., Schweifer, N., Haslinger, C., Stingl, G., and Maurer, D. (2007) Blood and lymphatic endothelial cell-specific differentiation programs are stringently controlled by the tissue environment. *Blood*, **109**, 4777–4785.

30 Rolph, M.S., Zimmer, S., Bottazzi, B., Garlanda, C., Mantovani, A., and Hansson, G.K. (2002) Production of the long pentraxin PTX3 in advanced atherosclerotic plaques. *Arterioscler. Thromb. Vasc. Biol.*, **22**, e10–14.

31 Klouche, M., Peri, G., Knabbe, C., Eckstein, H.H., Schmid, F.X., Schmitz, G., and Mantovani, A. (2004) Modified atherogenic lipoproteins induce expression of pentraxin-3 by human vascular smooth muscle cells. *Atheroscle-rosis*, **175**, 221–228.

32 Norata, G.D., Marchesi, P., Pirillo, A., Uboldi, P., Chiesa, G., Maina, V., Garlanda, C., Mantovani, A., and Catapano, A.L. (2008) Long pentraxin 3, a key component of innate immunity, is modulated by high-density lipoproteins in endothelial cells. *Arterioscler. Thromb. Vasc. Biol.*, **28**, 925–931.

33 Mallat, Z., Besnard, S., Duriez, M., Deleuze, V., Emmanuel, F., Bureau, M.F., Soubrier, F., Esposito, B., Duez, H., Fievet, C., Staels, B., Duverger, N., Scherman, D., and Tedgui, A. (1999) Protective role of interleukin-10 in atherosclerosis. *Circ. Res.*, **85**, e17–e24.

34 Tedgui, A., and Mallat, Z. (2006) Cytokines in atherosclerosis: pathogenic and regulatory pathways. *Physiol. Rev.*, **86**, 515–581.

35 Perrier, P., Martinez, F.O., Locati, M., Bianchi, G., Nebuloni, M., Vago, G., Bazzoni, F., Sozzani, S., Allavena, P., and Mantovani, A. (2004) Distinct transcriptional programs activated by interleukin-10 with or without lipopoly-saccharide in dendritic cells: induction of the B cell-activating chemokine, CXC chemokine ligand 13. *J. Immunol.*, **172**, 7031–7042.

36 Moore, K.W., de Waal Malefyt, R., Coffman, R.L., and O'Garra, A. (2001)

Interleukin-10 and the interleukin-10 receptor. *Annu. Rev. Immunol.*, **19**, 683–765.

37 Altmeyer, A., Klampfer, L., Goodman, A.R., and Vilcek, J. (1995) Promoter structure and transcriptional activation of the murine TSG-14 gene encoding a tumor necrosis factor/interleukin-1-inducible pentraxin protein. *J. Biol. Chem.*, **270**, 25584–25590.

38 Basile, A., Sica, A., d'Aniello, E., Breviario, F., Garrido, G., Castellano, M., Mantovani, A., and Introna, M. (1997) Characterization of the promoter for the human long pentraxin PTX3. Role of NF-kappaB in tumor necrosis factor-alpha and interleukin-1beta regulation. *J. Biol. Chem.*, **272**, 8172–8178.

39 Salio, M., Chimenti, S., De Angelis, N., Molla, F., Maina, V., Nebuloni, M., Pasqualini, F., Latini, R., Garlanda, C., and Mantovani, A. (2008) Cardioprotective function of the long pentraxin PTX3 in acute myocardial infarction. *Circulation*, **117**, 1055–1064.

40 Han, B., Mura, M., Andrade, C.F., Okutani, D., Lodyga, M., dos Santos, C.C., Keshavjee, S., Matthay, M., and Liu, M. (2005) TNFalpha-induced long pentraxin PTX3 expression in human lung epithelial cells via JNK. *J. Immunol.*, **175**, 8303–8311.

41 Willeke, F., Assad, A., Findeisen, P., Schromm, E., Grobholz, R., von Gerstenbergk, B., Mantovani, A., Peri, S., Friess, H.H., Post, S., von Knebel Doeberitz, M., and Schwarzbach, M.H. (2006) Overexpression of a member of the pentraxin family (PTX3) in human soft tissue liposarcoma. *Eur. J. Cancer*, **42**, 2639–2646.

42 Kuroda, M., Ishida, T., Horiuchi, H., Kida, N., Uozaki, H., Takeuchi, H., Tsuji, K., Imamura, T., Mori, S., Machinami, R. *et al.* (1995) Chimeric TLS/FUS-CHOP gene expression and the heterogeneity of its junction in human myxoid and round cell liposarcoma. *Am. J. Pathol.*, **147**, 1221–1227.

43 Doni, A., Mantovani, G., Porta, C., Tuckermann, J., Reichardt, H.M., Kleiman, A., Sironi, M., Rubino, L., Pasqualini, F., Nebuloni, M., Signorini, S., Peri, G., Sica, A., Beck-Peccoz, P., Bottazzi, B., and Mantovani, A. (2008) Cell-specific regulation of PTX3 by glucocorticoid hormones in hematopoietic and non-hematopoietic cells. *J. Biol. Chem.*, **283**, 29983–29992.

44 Bottazzi, B., Vouret-Craviari, V., Bastone, A., De Gioia, L., Matteucci, C., Peri, G., Spreafico, F., Pausa, M., D'Ettorre, C., Gianazza, E., Tagliabue, A., Salmona, M., Tedesco, F., Introna, M., and Mantovani, A. (1997) Multimer formation and ligand recognition by the long pentraxin PTX3. Similarities and differences with the short pentraxins C-reactive protein and serum amyloid P component. *J. Biol. Chem.*, **272**, 32817–32823.

45 Nauta, A.J., Bottazzi, B., Mantovani, A., Salvatori, G., Kishore, U., Schwaeble, W.J., Gingras, A.R., Tzima, S., Vivanco, F., Egido, J., Tijsma, O., Hack, E.C., Daha, M.R., and Roos, A. (2003) Biochemical and functional characterization of the interaction between pentraxin 3 and C1q. *Eur. J. Immunol.*, **33**, 465–473.

46 Deban, L., Jarva, H., Lehtinen, M.J., Bottazzi, B., Bastone, A., Doni, A., Jokiranta, T.S., Mantovani, A., and Meri, S. (2008) Binding of the long pentraxin PTX3 to Factor H: interacting domains and function in the regulation of complement activation. *J. Immunol.*, **181**, 8433–8440.

47 Garlanda, C., Hirsch, E., Bozza, S., Salustri, A., De Acetis, M., Nota, R., Maccagno, A., Riva, F., Bottazzi, B., Peri, G., Doni, A., Vago, L., Botto, M., De Santis, R., Carminati, P., Siracusa, G., Altruda, F., Vecchi, A., Romani, L., and Mantovani, A. (2002) Non-redundant role of the long pentraxin PTX3 in anti-fungal innate immune response. *Nature*, **420**, 182–186.

48 Bozza, S., Bistoni, F., Gaziano, R., Pitzurra, L., Zelante, T., Bonifazi, P., Perruccio, K., Bellocchio, S., Neri, M., Iorio, A.M., Salvatori, G., De Santis, R., Calvitti, M., Doni, A., Garlanda, C., Mantovani, A., and Romani, L. (2006) Pentraxin 3 protects from MCMV infection and reactivation through TLR sensing pathways leading to IRF3 activation. *Blood*, **108**, 3387–3396.

49 Reading, P.C., Bozza, S., Gilbertson, B., Tate, M., Moretti, S., Job, E.R., Crouch, E.C., Brooks, A.G., Brown, L.E., Bottazzi, B., Romani, L., and Mantovani, A. (2008) Antiviral activity of the long chain pentraxin PTX3 against influenza viruses. *J. Immunol.*, **180**, 3391–3398.

50 Cotena, A., Maina, V., Sironi, M., Bottazzi, B., Jeannin, P., Vecchi, A., Corvaia, N., Daha, M.R., Mantovani, A., and Garlanda, C. (2007) Complement dependent amplification of the innate response to a cognate microbial ligand by the long pentraxin PTX3. *J. Immunol.*, **179**, 6311–6317.

51 Olesen, R., Wejse, C., Velez, D.R., Bisseye, C., Sodemann, M., Aaby, P., Rabna, P., Worwui, A., Chapman, H., Diatta, M., Adegbola, R.A., Hill, P.C., Ostergaard, L., Williams, S.M., and Sirugo, G. (2007) DC-SIGN (CD209), Pentraxin 3 and vitamin D receptor gene variants associate with pulmonary tuberculosis risk in West Africans. *Genes Immun.*, **8** (6), 456–467.

52 Baruah, P., Propato, A., Dumitriu, I.E., Rovere-Querini, P., Russo, V., Fontana, R., Accapezzato, D., Peri, G., Mantovani, A., Barnaba, V., and Manfredi, A.A. (2006) The pattern recognition receptor PTX3 is recruited at the synapse between dying and dendritic cells, and edits the cross-presentation of self, viral, and tumor antigens. *Blood*, **107**, 151–158.

53 Manfredi, A.A., Rovere-Querini, P., Bottazzi, B., Garlanda, C., and Mantovani, A. (2008) Pentraxins, humoral innate immunity and tissue injury. *Curr. Opin. Immunol.*, **20** (5), 538–544.

54 Rusnati, M., Camozzi, M., Moroni, E., Bottazzi, B., Peri, G., Indraccolo, S., Amadori, A., Mantovani, A., and Presta, M. (2004) Selective recognition of fibroblast growth factor-2 by the long pentraxin PTX3 inhibits angiogenesis. *Blood*, **104** (1), 92–99.

55 Camozzi, M., Zacchigna, S., Rusnati, M., Coltrini, D., Ramirez-Correa, G., Bottazzi, B., Mantovani, A., Giacca, M., and Presta, M. (2005) Pentraxin 3 inhibits fibroblast growth factor 2-dependent activation of smooth muscle cells *in vitro* and neointima formation *in vivo*. *Arterioscler. Thromb. Vasc. Biol.*, **25**, 1837–1842.

56 Presta, M., Camozzi, M., Salvatori, G., and Rusnati, M. (2007) Role of the soluble pattern recognition receptor PTX3 in vascular biology. *J. Cell. Mol. Med.*, **11**, 723–738.

57 Bosutti, A., Grassi, G., Zanetti, M., Aleksova, A., Zecchin, M., Sinagra, G., Biolo, G., and Guarnieri, G. (2007) Relation between the plasma levels of LDL-cholesterol and the expression of the early marker of inflammation long pentraxin PTX3 and the stress response gene p66ShcA in pacemaker-implanted patients. *Clin. Exp. Med.*, **7**, 16–23.

58 Alberti, L., Gilardini, L., Zulian, A., Micheletto, G., Peri, G., Doni, A., Mantovani, A., and Invitti, C. (2008) Expression of long pentraxin PTX3 in human adipose tissue and its relation with cardiovascular risk factors. *Atherosclerosis*, **202** (2), 455–460.

59 Gustin, C., Delaive, E., Dieu, M., Calay, D., and Raes, M. (2008) Upregulation of pentraxin-3 in human endothelial cells after lysophosphatidic acid exposure. *Arterioscler. Thromb. Vasc. Biol.*, **28**, 491–497.

60 Savchenko, A., Imamura, M., Ohashi, R., Jiang, S., Kawasaki, T., Hasegawa, G., Emura, I., Iwanari, H., Sagara, M., Tanaka, T., Hamakubo, T., Kodama, T., and Naito, M. (2008) Expression of pentraxin 3 (PTX3) in human atherosclerotic lesions. *J. Pathol.*, **215**, 48–55.

61 Peri, G., Introna, M., Corradi, D., Iacuitti, G., Signorini, S., Avanzini, F., Pizzetti, F., Maggioni, A.P., Moccetti, T., Metra, M., Cas, L.D., Ghezzi, P., Sipe, J.D., Re, G., Olivetti, G., Mantovani, A., and Latini, R. (2000) PTX3, A prototypical long pentraxin, is an early indicator of acute myocardial infarction in humans. *Circulation*, **102**, 636–641.

62 Muller, B., Peri, G., Doni, A., Torri, V., Landmann, R., Bottazzi, B., and Mantovani, A. (2001) Circulating levels of the long pentraxin PTX3 correlate with severity of infection in critically ill patients. *Crit. Care Med.*, **29**, 1404–1407.

63 Mairuhu, A.T., Peri, G., Setiati, T.E., Hack, C.E., Koraka, P., Soemantri, A., Osterhaus, A.D., Brandjes, D.P., van der Meer, J.W., Mantovani, A., and van Gorp, E.C. (2005) Elevated plasma levels of the long pentraxin, pentraxin 3, in severe dengue virus infections. *J. Med. Virol.*, **76**, 547–552.

64 Azzurri, A., Sow, O.Y., Amedei, A., Bah, B., Diallo, S., Peri, G., Benagiano, M., D'Elios, M.M., Mantovani, A., and Del Prete, G. (2005) IFN-gamma-inducible protein 10 and pentraxin 3 plasma levels are tools for monitoring inflammation and disease activity in *Mycobacterium tuberculosis* infection. *Microbes Infect.*, **7**, 1–8.

65 Sprong, T., Peri, G., Neeleman, C., Mantovani, A., Signorini, S., van der Meer, J.W., and van Deuren, M. (2009) Ptx3 and C-reactive protein in severe meningococcal disease. *Shock*, **31** (1), 28–32.

66 Pepys, M.B. and Hirschfield, G.M. (2003) C-reactive protein: a critical update. *J. Clin. Invest.*, **111**, 1805–1812.

67 Casas, J.P., Shah, T., Hingorani, A.D., Danesh, J., and Pepys, M.B. (2008) C-reactive protein and coronary heart disease: a critical review. *J. Intern. Med.*, **264**, 295–314.

68 Ridker, P.M., Danielson, E., Fonseca, F.A., Genest, J., Gotto, A.M., Jr., Kastelein, J.J., Koenig, W., Libby, P., Lorenzatti, A.J., MacFadyen, J.G., Nordestgaard, B.G., Shepherd, J., Willerson, J.T., and Glynn, R.J. (2008) Rosuvastatin to prevent vascular events in men and women with elevated C-reactive protein. *N. Engl. J. Med.*, **359**, 2195–2207.

69 Inoue, K., Sugiyama, A., Reid, P.C., Ito, Y., Miyauchi, K., Mukai, S., Sagara, M., Miyamoto, K., Satoh, H., Kohno, I., Kurata, T., Ota, H., Mantovani, A., Hamakubo, T., Daida, H., and Kodama, T. (2007) Establishment of a high sensitivity plasma assay for human pentraxin3 as a marker for unstable angina pectoris. *Arterioscler. Thromb. Vasc. Biol.*, **27**, 161–167.

70 Kotooka, N., Inoue, T., Fujimatsu, D., Morooka, T., Hashimoto, S., Hikichi, Y., Uchida, T., Sugiyama, A., and Node, K. (2008) Pentraxin3 is a novel marker for stent-induced inflammation and neointimal thickening. *Atherosclerosis*, **197**, 368–374.

71 Latini, R., Maggioni, A.P., Peri, G., Gonzini, L., Lucci, D., Mocarelli, P., Vago, L., Pasqualini, F., Signorini, S.,

Soldateschi, D., Tarli, L., Schweiger, C., Fresco, C., Cecere, R., Tognoni, G., and Mantovani, A. (2004) Prognostic significance of the long pentraxin PTX3 in acute myocardial infarction. *Circulation*, **110**, 2349–2354.

72 Napoleone, E., Di Santo, A., Peri, G., Mantovani, A., de Gaetano, G., Donati, M.B., and Lorenzet, R. (2004) The long pentraxin PTX3 up-regulates tissue factor in activated monocytes: another link between inflammation and clotting activation. *J. Leukoc. Biol.*, **76**, 203–209.

73 Napoleone, E., Di Santo, A., Bastone, A., Peri, G., Mantovani, A., de Gaetano, G., Donati, M.B., and Lorenzet, R. (2002) Long pentraxin PTX3 upregulates tissue factor expression in human endothelial cells: a novel link between vascular inflammation and clotting activation. *Arterioscler. Thromb. Vasc. Biol.*, **22**, 782–787.

74 Suzuki, S., Takeishi, Y., Niizeki, T., Koyama, Y., Kitahara, T., Sasaki, T., Sagara, M., and Kubota, I. (2008) Pentraxin 3, a new marker for vascular inflammation, predicts adverse clinical outcomes in patients with heart failure. *Am. Heart J.*, **155**, 75–81.

75 Ishino, M., Takeishi, Y., Niizeki, T., Watanabe, T., Nitobe, J., Miyamoto, T., Miyashita, T., Kitahara, T., Suzuki, S., Sasaki, T., Bilim, O., and Kubota, I. (2008) Risk stratification of chronic heart failure patients by multiple biomarkers: implications of BNP, H-FABP, and PTX3. *Circ. J.*, **72**, 1800–1805.

76 Kotooka, N., Inoue, T., Aoki, S., Anan, M., Komoda, H., and Node, K. (2008) Prognostic value of pentraxin 3 in patients with chronic heart failure. *Int. J. Cardiol.*, **130**, 19–22.

77 Latini, R., Peri, G., Masson, D., Lucci, M., Sagara, M., Yamada, T., Mantovani, A., Maggioni, A., Tognoni, G., and Tavazzi, L. (2008) Clinical determinants of Pentraxin-3 (PTX3), a novel marker of vascular inflammation, in patients with chronic HF. Data from GISSI-HF study. *Eur. Heart J.*, **29** (Suppl. 1), 393 (abs).

78 Fazzini, F., Peri, G., Doni, A., Dell'Antonio, G., Dal Cin, E., Bozzolo, E., D'Auria, F., Praderio, L., Ciboddo, G., Sabbadini, M.G., Manfredi, A.A., Mantovani, A., and Querini, P.R. (2001)

PTX3 in small-vessel vasculitides: an independent indicator of disease activity produced at sites of inflammation. *Arthritis Rheum.*, **44**, 2841–2850.

79 Luchetti, M.M., Piccinini, G., Mantovani, A., Peri, G., Matteucci, C., Pomponio, G., Fratini, M., Fraticelli, P., Sambo, P., Di Loreto, C., Doni, A., Introna, M., and Gabrielli, A. (2000) Expression and production of the long pentraxin PTX3 in rheumatoid arthritis (RA). *Clin. Exp. Immunol.*, **119**, 196–202.

80 van Rossum, A.P., Pas, H.H., Fazzini, F., Huitema, M.G., Limburg, P.C., Jonkman, M.F., and Kallenberg, C.G. (2006) Abundance of the long pentraxin PTX3 at sites of leukocytoclastic lesions in patients with small-vessel vasculitis. *Arthritis Rheum.*, **54**, 986–991.

81 van Rossum, A.P., Fazzini, F., Limburg, P.C., Manfredi, A.A., Rovere-Querini, P., Mantovani, A., and Kallenberg, C.G. (2004) The prototypic tissue pentraxin PTX3, in contrast to the short pentraxin serum amyloid P, inhibits phagocytosis of late apoptotic neutrophils by macrophages. *Arthritis Rheum.*, **50**, 2667–2674.

82 Tong, M., Carrero, J.J., Qureshi, A.R., Anderstam, B., Heimburger, O., Barany, P., Axelsson, J., Alverstrand, A., Stenvinkel, P., Lindholm, B., and Suliman, M.E. (2007) Plasma pentraxin 3 in chronic kidney disease patients: association with renal function, protein-energy wasting, cardiovascular disease and mortality. *Clin. J. Am. Soc. Nephrol.*, **2**, 889–897.

83 Boehme, M., Kaehne, F., Kuehne, A., Bernhardt, W., Schroder, M., Pommer, W., Fischer, C., Becker, H., Muller, C., and Schindler, R. (2007) Pentraxin 3 is elevated in haemodialysis patients and is associated with cardiovascular disease. *Nephrol. Dial. Transplant.*, **22** (8), 2224–2229.

84 Suliman, M.E., Yilmaz, M.I., Carrero, J.J., Qureshi, A.R., Saglam, M., Ipcioglu, O.M., Yenicesu, M., Tong, M., Heimburger, O., Barany, P., Alvestrand, A.,

Lindholm, B., and Stenvinkel, P. (2008) Novel links between the long pentraxin 3, endothelial dysfunction, and albuminuria in early and advanced chronic kidney disease. *Clin. J. Am. Soc. Nephrol.*, **3**, 976–985.

85 Malaponte, G., Libra, M., Bevelacqua, Y., Merito, P., Fatuzzo, P., Rapisarda, F., Cristina, M., Naselli, G., Stivala, F., Mazzarino, M.C., and Castellino, P. (2007) Inflammatory status in patients with chronic renal failure: the role of PTX3 and pro-inflammatory cytokines. *Int. J. Mol. Med.*, **20**, 471–481.

86 Redman, C.W., Sacks, G.P., and Sargent, I.L. (1999) Preeclampsia: an excessive maternal inflammatory response to pregnancy. *Am. J. Obstet. Gynecol.*, **180**, 499–506.

87 Cetin, I., Cozzi, V., Pasqualini, F., Nebuloni, M., Garlanda, C., Vago, L., Pardi, G., and Mantovani, A. (2006) Elevated maternal levels of the long pentraxin 3 (PTX3) in preeclampsia and intrauterine growth restriction. *Am. J. Obstet. Gynecol.*, **194**, 1347–1353.

88 Rovere-Querini, P., Antonacci, S., Dell'antonio, G., Angeli, A., Almirante, G., Cin, E.D., Valsecchi, L., Lanzani, C., Sabbadini, M.G., Doglioni, C., Manfredi, A.A., and Castiglioni, M.T. (2006) Plasma and tissue expression of the long pentraxin 3 during normal pregnancy and preeclampsia. *Obstet. Gynecol.*, **108**, 148–155.

89 Benyo, D.F., Smarason, A., Redman, C.W., Sims, C., and Conrad, K.P. (2001) Expression of inflammatory cytokines in placentas from women with preeclampsia. *J. Clin. Endocrinol. Metab.*, **86**, 2505–2512.

90 Rinehart, B.K., Terrone, D.A., Lagoo-Deenadayalan, S., Barber, W.H., Hale, E.A., Martin, J.N., Jr., and Bennett, W.A. (1999) Expression of the placental cytokines tumor necrosis factor alpha, interleukin 1beta, and interleukin 10 is increased in preeclampsia. *Am. J. Obstet. Gynecol.*, **181**, 915–920.

PART III
Proteases

8
Metalloproteinases, the Endothelium, and Atherosclerosis

Stephen J. White and Andrew C. Newby

8.1
Introduction and Scope of this Review

Atherosclerosis develops silently in humans over several decades. It only becomes evident clinically (for example as angina pectoris and intermittent claudication) when conduit arteries are narrowed by more than 70%. Interestingly, as atherosclerotic plaque size increases there is an initial tendency for the artery to enlarge in a process called compensatory remodeling [1, 2]. Narrowing of the artery can therefore be seen as an eventual failure of the artery to remodel. Atherosclerosis can become evident suddenly when thrombosis occurs on the surface of a disrupted atherosclerotic plaque [3]. This gives rise to symptoms such as unstable angina pectoris or myocardial infarction (MI) if it occurs in the coronary arteries, and transient ischemic attacks or cerebral infarction (thrombotic stroke) in the carotid arteries. Atherosclerosis is also a primary cause of certain types of aneurysms (excessive outward dilatation), particularly those in the abdominal aorta [4].

MI is the most prevalent life-threatening consequence of atherosclerosis; there are more than 200 000 MIs in the UK alone each year. About 75% result from rupture of a thin fibrous cap overlying a highly inflamed atherosclerotic plaque that has a large lipid core [5]. The remaining 25% have a drastically different etiology. They result from loss of a large sheet of endothelial cells from the surface of highly narrowed, fibrous plaques, in a process known as "erosion" [5]. Plaque rupture and erosion may be the result of different risk factors; men, for example, are more likely to suffer ruptures and women erosions [6]. Intriguingly, plaque ruptures are associated with outward arterial remodeling and erosions with inward remodeling (Figure 8.1).

Expression of extracellular proteinases is prominent in human atherosclerotic plaques at all stages of development. The matrix metalloproteinases (MMPs) are a family of 23 genetically related proteins, which directly degrade extracellular matrix (ECM) components and are efficient at neutral pH. They share several common properties (Figure 8.2), including a Zn^{2+}-containing catalytic site, production as pro-forms that require proteolytic activation either inside or outside the cell, and inhibition by one or more of the four tissue inhibitors of metalloprotei-

Atherosclerosis: Molecular and Cellular Mechanisms. Edited by Sarah Jane George and Jason Johnson
Copyright © 2010 WILEY-VCH Verlag GmbH & Co. KGaA, Weinheim
ISBN: 978-3-527-32448-4

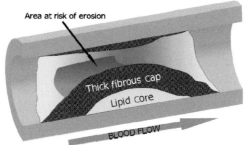

High shear stress => Remodeling phenotype

Oxidative stress
Peroxynitrite production
MMP production
Matrix turnover

Endothelial cell adhesion?

Figure 8.1 Factors potentially involved in endothelial erosion. Stenotic thick fibrous capped atherosclerotic plaques may chronically induce a remodeling phenotype in endothelial cells. An elevation of oxidative stress and peroxynitrite formation would be sustained, increasing matrix metalloprotein- ase (MMP) production and matrix remod- eling. It remains to be proven whether chronic exposure to elevated shear stress in conjunction with other factors that induce endothelial dysfunction compromise endothelial adhesion, predisposing these sites to endothelial erosion.

nases (TIMPs) [7]. Most MMPs are secreted as pro-proteins and then become activated proteolytically by other MMPs, by serine proteases such as plasmin, kalikrienes, chymase, and tryptase or by cysteine proteinases including cathepsins [7]. Chemical activation by reactive oxygen species and nitric oxide is also possible [8]. In addition, the membrane-type MMPs (MT-MMPs) and a few other MMPs are activated by furins in the endosome; the MT-MMPs remain attached to the outer plasma membrane surface. MMPs degrade collagens, elastins, the core proteins of proteoglycans, and ECM glycoproteins such as fibronectin and laminin. Extracellular proteases can also cleave cell surface molecules and signaling mol- ecules such as growth factors and cytokines [7], and this in turn influences the migration, proliferation, and death of vascular cells [9].

Studies in genetically modified mice and rabbits directly demonstrate the par- ticipation of MMPs in plaque growth and stability (Table 8.1). When MMPs favor plaque disruption (see Table 8.1), the most obvious mechanism is degradation of the ECM. Loss of collagen in the plaque cap reduces tensile strength, while absence of collagen from the lipid core promotes transfer of hydrodynamic stress to the cap during the cardiac cycle. Both favor plaque rupture. Only MMP-1, MMP-2, MMP-8, MMP-13, and MMP-14 actively degrade intact fibrillar collagens and may therefore have a special role in weakening plaques. Loss of elastin leads to outward remodeling and ultimately aneurysms. The elastolytic action of MMP-9 and MMP-12 appears to have a role in outward remodeling and aneurysm forma- tion [24]. Degradation of some ECM and non-matrix substrates might favor plaque

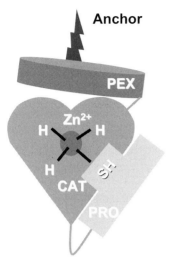

Figure 8.2 The common domain structure of matrix metalloproteinases (MMPs). All MMPs have an N-terminal pro-domain (PRO). Except in the case of MMP-23, this contains a cysteine group that blocks access to substrates to the Zn^{2+} ion, which is coordinated by three hydrogen bonds (H) in the ubiquitous catalytic (CAT) subunit. The CAT domain has extra fibronectin-like inserts in MMP-2 and MMP-9. Pro-form activation occurs by initial proteolytic cleavage of the linker between the PRO and CAT domains. This occurs extracellularly in most MMPs, except MMP-11, MMP-14 to MMP-17, MMP-21, MMP-23, and MMP-28, which have a furin site for endosomal activation. Most MMPs (except MMP-7, MMP-23, and MMP-26) have a PEX domain that modifies substrate recognition and pro-form activation. The membrane-type MMPs (MMP-14 to MMP-17, MMP-24, and MMP-25) have either a transmembrane domain or a glycophosphatidylcholine anchor, which localizes them to the external membrane surface.

stability and others rupture. For example, MMPs promote smooth muscle cell (VSMC) migration and proliferation, which underlies growth of the plaque fibrous cap and therefore plaque stability [9]. On the other hand MMPs promote immigration of monocytes [25, 26], heightening plaque inflammation. MMPs also promote angiogenesis, which is also associated with plaque rupture [5].

Several previous reviews have focused on the apparently opposing effects of MMPs on plaque stability mediated through actions on VSMCs and macrophages [27, 28]. However, the role of MMPs in endothelial cells (ECs) during atherosclerosis has been relatively neglected. The present chapter aims to rectify that situation by concentrating on the production and function of MMPs in ECs in the context of atherosclerosis formation and its adverse clinical consequences.

8.2
General Comments on the Role of MMP Production from ECs in Atherosclerosis

A range of MMPs can be produced by ECs under varying conditions, enabling digestion or remodeling of all the matrix components of the vessel wall. Previous

Table 8.1 Effects of matrix metalloproteinase interventions in atherosclerotic plaque development and stability.

Number	Site	Size	SMC	Macrophages	ECM integrity	Overall stability	Ref.
MMP-1 ++	Root/arch	↓	=	=	↓	?	[10]
MMP-2 null	Root/arch	↓	↓	=	=	↓	[11]
MMP-3 null	Ao, BCA	↑, ↑	?, ↓	↓, =	↑, ?	↑, ↓	[12, 13]
MMP-7 null	BCA	=	↑	=	?	=	[13]
MMP-9 null	Ao, BCA	↓, ↑	?, ↓	↓, ↑	↓↑, =	↑, ↓	[13, 14]
MMP-9 ++	Arch, collar	=, =	=, =	=, =	=, =	=, ↓	[15, 16]
MMP-9 ++a	Arch, BCA			=	↓	↓	[15]
MMP-12 null	Ao, BCA	=, ↓	=, ↑	=, ↓	=, ?	↑, ↑	[13, 14]
MMP-12 ++a	Ao (rabbit)	↑	↑	↑	?	?	[17]
MMP-13 null	Root	=	=	=	↑	↑	[18]
MMP-14 null	Root	=	=	=	↑	↑	[19]
TIMP-1 null	Root, Aorta	=, ↓	=, ?	=, ↑	↓, ↓	↓, ↓	[20, 21]
TIMP-1 ++	Root, BCA	↓, =	?, =	↓, =	↑, ↑	↑, =	[22, 23]
TIMP-2 ++	BCA	↓	↑	↓	↑	↑	[23]

Results of overexpression (++) overexpression of activated enzymes (++a) and knockout (null) on the specified parameters. Unless specified data were obtained in mice.
Increases (↑), decreases (↓) or no change (=). Missing or uncertain data is indicated as (?).
Ao, aorta; arch, aortic arch; BCA, brachiocephalic artery; root, aortic root.

publications have demonstrated MMP-1, MMP-2, MMP-3, MMP-7, MMP-8, MMP-9, MMP-10, MMP-11, MMP-13, MMP-14 (MT-1 MMP), MMP-15 (MT-2 MMP), and MMP-16 (MT-3 MMP) expression in ECs [29–39]. In addition, ECs can express a number of MMP modifiers, including TIMPs 1–4, urokinase plasminogen activator receptor (UPAR), tissue plasminogen activator (TPA), CD44, reversion-inducing cysteine-rich protein with Kazal motifs (RECK), and neutrophil gelatinase-associated lipocalin (N-GAL) [40–47]. The coordination of these genes, both in their expression and localization, should allow endothelial cells to accurately control proteolysis. However, dysfunction in these mechanisms could lead to pathology. Most obviously, excessive production of MMPs could weaken the interaction between ECs and their underlying basement membrane and cause plaque erosion. The evidence to support this attractive hypothesis is, however, limited (see below).

The analysis of MMP expression by endothelial cells is complicated by observations of ECs from different vascular beds and different species. Clearly (human) large arterial ECs are most relevant to atherosclerosis but most studies have been conducted with more abundant sources from other vessels. The possible pitfalls were clearly demonstrated in a study by Viemann *et al.*, which examined the effects of tumor necrosis factor-α (TNF-α) in human microvascular endothelial cells (HMVECs) and the commonly used human umbilical vein endothelial cells (HUVECs) [48]. There was a marked difference in response, which included differences in MMP expression, highlighting the importance of EC origin. In addition, exposure of ECs to either shear stress or cyclic strain induces significant changes in gene expression and behavior that is further complicated by an initial adaptive phase during which many proinflammatory genes are upregulated and subsequently repressed at normal physiological levels of shear and cyclic strain [49].

Vascular endothelial cells are mechanosensitive and respond to both shear stress (the frictional force caused by the flow of blood over the surface of the cell) and cyclic strain (the rhythmical stretching force caused by the pulsatile nature of blood flow) [50]. High average shear stress is associated with protection from atherosclerosis, while low average shear stress and disturbed flow are associated with high risk for developing atherosclerosis [51]. These flow-dependent effects are believed to be major determinants of the focal distribution of atherosclerotic plaques and a wealth of effort has therefore been devoted to their study. ECs *in vivo* under normal arterial laminar shear are in a "quiescent" phenotype, with low active protease production. The quiescent state can be simulated *in vitro* by culturing ECs under arterial laminar shear stress for longer than 16 h. Cells cultured under these conditions are refractory to signals (e.g., TNF-α) that would otherwise shift them from this quiescent state into an activated state [52, 53] when cultured *in vitro* under static conditions. ECs activated by inflammatory mediators or disturbed flow show widespread genomic changes compared to quiescent ECs, including upregulation of MMP expression (Figure 8.3). In addition, they show increased expression of adhesion molecules for leukocytes and chemokines, and decreased

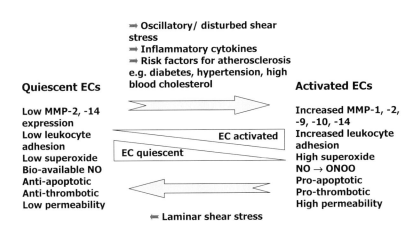

Figure 8.3 Endothelial cell (EC) phenotype. Laminar shear stress promotes maintenance of ECs in a quiescent state and reduces activation of ECs by inflammatory cytokines. ECs in oscillatory or disturbed flow are primed for activation by inflammatory cytokines and other risk factors for atherosclerosis, such as high blood glucose and cholesterol. Activated ECs increase the expression of MMPs and leukocyte adhesion molecules and elevate production of super oxide radicals which combine with nitric oxide (NO) to form peroxynitrite (ONOO). Activated ECs are also more prone to apoptosis and their anti-thrombotic properties are attenuated.

formation of nitric oxide. These changes provide a rationale for the enhanced recruitment of inflammatory cells that characterizes the progression and rupture of atherosclerotic plaques. Increased MMP production and activation by ECs could therefore contribute to both growth and instability of atherosclerosis plaques. Angiogenesis leading to expansion of adventitial vasa vasorum is associated with plaque growth in animal models. Invasion of ECs into the base of human plaques from the vasa vasorum is also associated with intra-plaque hemorrhage and lipid core expansion. These events are therefore relevant to the pathogenesis of plaque rupture.

At a cellular level, ECs can form podosomes or podosome-type adhesions (PTAs), which are a type of actin-rich adhesive structure. PTAs localize MMP-2, MMP-9, and MMP-14 on the surface and focus MMP degradation of the underlying matrix (reviewed in [54]), through association with the transmembrane receptor CD44 [55]. This may be important in the angiogenic response of ECs [56], allowing focused degradation of the underlying basement membrane and vessel wall to allow tube formation and the establishment of new collateral vessels. It has also been shown that ECs shed vesicles (300–600 nm) containing both the active and pro-enzymes of MMP-2 and MMP-9, with the MMP-14 pro-enzyme together with TIMP-1 and TIMP-2 [57]. Serum and the angiogenic factors basic fibroblast growth factor-2 (FGF-2) and vascular endothelial growth factor (VEGF) stimulate the shedding of MMP-containing vesicles, which are able to enhance EC invasion. The ability of ECs to form PTAs and secrete MMP-containing vesicles highlights that the regulation of MMP expression is only part of the story and that their localized secretion and activation are also orchestrated to direct proteolysis of the underlying matrix.

8.3
Endothelial Responses to Laminar and Oscillatory Shear Stress

Endothelial cells under normal laminar flow adopt a quiescent phenotype which confers resistance to the development of atherosclerosis [51]. Three studies have shown differential regulation of MMPs by laminar shear, comparing them to cells grown in static culture. Brooks *et al.* compared the changes in gene expression in human coronary artery endothelial cells (HCAECs) and showed laminar shear stress increased MMP-1 and MMP-8, while it decreased MMP-7 and MMP-14 expression [58]. The increase in MMP-1 was confirmed by Chen and colleagues [59]. Shear-induced phosphorylation of transcription factor Sp1 was responsible for the downregulation of MMP-14 [60]. Many of the effects of laminar shear are mediated by upregulation and action of the transcription factors KLF-2 [61–64] and Nrf2 [65], which upregulate antioxidant genes and suppress inflammatory cytokine signaling [52, 53]. To unpick the effects of shear Dekker *et al.* overexpressed KLF-2 in endothelial cells and performed a gene array identifying increased ADAM15 and TIMP-1 (fourfold and twofold respectively) [63]. The signaling mechanisms responsible for the changes of MMP-1, MMP-7, and MMP-8 have yet to be identified.

In contrast to laminar shear stress, abrupt changes in shear stress induce an inflammatory response in ECs, through a burst of MAP kinase signaling [66]. Similarly, oscillatory or disturbed patterns of shear stress precipitate a number of processes that activate endothelial cells, and can sensitize the cell to apoptosis [51, 64, 67–72]. Changes include NADPH oxidase upregulation, excessive reactive oxygen species (ROS) production, which scavenges nitric oxide (NO) and forms peroxynitrite, induces NF-κB translocation to the nucleus and upregulates adhesion molecule expression. This results in an activated/inflammatory cell phenotype. NF-κB and peroxynitrite can induce the production of a number of MMPs (e.g., MMP-1, MMP-3, MMP-9) [29, 73, 74], but only MMP-10 and MMP-14 have been shown to be upregulated in oscillatory shear compared with laminar shear stress *in vitro* (see [75] and S.J. White, E.M. Hayes and A.C. Newby, unpublished observations).

8.4
Response to Step Changes in Shear Stress

Chronic changes in blood flow induce compensatory vascular remodeling, either reducing or enlarging vessel diameter tending towards a point where the shear stress is returned to normal. The remodeling process is endothelial dependent. If the endothelium is removed from the vessel wall, the remodeling process is prevented [76, 77]. Endothelial dysfunction, as demonstrated by a reduction in flow-mediated dilatation and P-selectin expression, reduces expansive remodeling in humans [78], as does L-NAME inhibition of nitric oxide synthase in the rabbit [79], or genetic ablation of endothelial nitric oxide synthase (eNOS) in the mouse [29]. These studies show that expansive remodeling is dependent on EC-derived factors including NO, demonstrating that ECs are the master regulators of vessel remodeling. Arterial remodeling requires inflammation and macrophages [80], but the direction of remodeling (inwards due to a reduction of shear stress, or outward as a response to increased shear stress) is driven by vascular tone [81].

The role of MMPs in arterial remodeling has been studied in the model of carotid artery ligation, where the contralateral carotid artery experiences a ~44% increase in blood flow sufficient to induce outward remodeling in the unligated artery [82]. MMP-9 was upregulated in ECs in the contralateral artery, with maximal expression 3 days post ligation [83]. Remodeling has also been studied more acutely in the carotid-jugular fistula [29] model, which experiences a very large increase in shear stress that induces rapid arterial expansion. From this model Castier *et al.* demonstrated the signaling pathways that induce remodeling require both the production of NO and ROS from NADPH oxidase. Increased NO and ROS combine to form peroxynitrite and induce the expression of MMP-9 [29, 73]. Endothelial expression of MMP-9 was also upregulated during the adaptive remodeling of rabbit vein grafts but not in arterial grafts, which maintained their size [84]. Hence in the majority of animal models, endothelial expression of MMP-9 seems to be increased in the adaptive remodeling observed due to chronic changes in blood flow. In humans, MMP-10 may play a role in arterial remodeling, as it has been associated with abdominal aortic aneurysms through genetic single nucleotide polymorphism screens along with TIMP-1 and TIMP-3 [85]. MMP-10 is also expressed by plaque endothelium [86].

8.5
Response to Cyclic Strain

ECs adapt to continuous exposure to cyclic strain, analogous to the adaptation to shear stress, normalizing their behavior after approximately 24 h (reviewed by [87]). MMP-2 and MMP-14 were increased by cyclic strain in HUVECs, with a maximal increase in mRNA at 12 h, which reduced by 24 h [88]. ECs exposed to cyclic strain produced TNF-α and the upregulation of MMP-2 and MMP-14 could be blocked by a neutralizing antibody to TNF-α, suggesting that the observed increase in MMPs was due to an autocrine inflammatory reaction upon induction of cyclic strain. It would be interesting to determine what the effect of 48 h of cyclic stretch would be in this system, to determine if the increase in MMPs was simply due to the adaption to cyclic strain as opposed to the natural reaction of ECs to sustained cyclic strain. As previously mentioned, TNF-α signaling is inhibited under laminar shear stress, so the *in vivo* significance of this finding is yet to be ascertained. Bovine aortic endothelial cells were also shown to upregulate MMP-2 after 24 h of cyclic strain [89], although it was not demonstrated if TNF-α was the mediator in this case. A role for MMP-9 in strain-enhanced EC migration has also been demonstrated [90].

8.6
Response to Inflammatory Mediators

Endothelial cells respond to a number of inflammatory mediators and increase MMP expression. TNF-α affects EC behavior in a context and vascular bed-specific manner. MMP-1 and MMP-12 are upregulated in HMEC-1 after TNF-α stimulation, although TNF-α did not increase the expression of the same MMPs in HUVECs [48]. As previously stated, TNF-α upregulated MMP-2 and MMP-14 in HUVECs in the context of cyclic strain [88]. TNF-α, interleukin-1α (IL-1α), or IL-1β increase MMP-14 mRNA levels, as does oxidized LDL (ox-LDL), with the combination of TNF-α and ox-LDL eliciting an additive increase in MMP-14 expression [91]. Ox-LDL also upregulated MMP-1 in HUVECs and HCAECs, while repressing the expression of TIMP-1 [92]. C-reactive protein (CRP) increased MMP-1 and MMP-10 in ECs [86], a finding that was also observed *in vivo*. Ligation of CD40 on ECs induced *de novo* expression of MMP-1, MMP-3, and MMP-9 and increased the activation of MMP-2 [93]. IL-18 upregulated MMP-1 and MMP-13 expression and also increased IL-8 production [94], with the IL-18 receptor being induced by TNF-α and IL-1β. IL-8 itself increased MMP-2 and MMP-9 and directly enhanced EC survival and proliferation and promoted angiogenesis [95].

8.7
Response to Angiogenic Signals

Burbridge *et al.* examined the role of MMPs in angiogenesis in the rat aortic ring angiogenesis assay in both collagen and fibrin matrices [38]. The MMP inhibitor marimastat inhibited angiogenesis and demonstrated the requirement of MMPs for new vessel growth. The profile of MMP expression was seen to be modified by both matrix composition and exogenous growth factors, with MMP-2 and MMP-3 found at high levels in fibrin culture and MMP-11 and MMP-14 in collagen. In separate studies MMP-2 and MMP-14 were also shown to be upregulated in rat microvascular ECs in three-dimensional collagen matrices [96, 97]. Burbridge *et al.* demonstrated that the pro-angiogenic FGF-2 upregulated the expression of multiple MMPs (MMP-2, MMP-3, MMP-9, MMP-10, MMP-11, and MMP-13), whereas VEGF led to a marked increase in expression of MMP-2 only [38]. FGF-2 was additionally shown to increase the expression of MMP-7 through activation of AP-1 transcription factors [98]; MMP-7 is able to cleave VE-cadherin from the cell surface increasing the nuclear localization of β-catenin, promoting proliferation and further increasing MMP-7 expression [99]. FGF-1 has been shown to upregulate MMP-1 and promote endothelial migration [100].

The membrane-bound MMPs – MMP-14 (and MMP-16) – play an important role in angiogenesis [35, 101] (reviewed by [102]) and migration of ECs [103], due to both their ability to perform pericellular proteolysis and through the ability of MMP-14 to activate MMP-2. Thrombin also has pro-angiogenic effects upregulating MMP-1 and MMP-3 [30] and enhancing MMP-2 activation via MMP-14 [104]. Hepatocyte growth factor is another pro-angiogenic growth factor that increases MMP-14 expression and MMP-2 activation [105].

8.8
Signaling by MMPs

In addition to their extracellular matrix-degrading abilities, MMPs can participate in signaling pathways that effect EC function. MMP-1, MMP-3, and MMP-13 can cleave connective tissue growth factor (CTGF), which binds and sequesters VEGF. Cleavage of CTGF by MMPs releases the bound VEGF and makes it accessible to bind to its receptor, thus stimulating angiogenesis [106]. In a similar manner, MMP-14 proteolytically cleaves the latent TGF-β-binding protein 1, releasing TGF-β1 from the matrix [107], a possible mechanism for the observed activation of the extracellular signal-regulated protein kinase (ERK) cascade by MMP-14 [108].

Fragments of the extracellular matrix which are generated by the actions of MMPs can induce signaling in ECs. MMP-9 generates a fragment of collagen IV α3-chain termed tumstatin, which suppresses angiogenesis via $\alpha_v\beta_3$-integrin [109]. Similarly, the collagen XVIII fragment endostatin inhibits EC invasion by blocking the activation and catalytic activity of MMP-2 [110]. Conversely, elastin-derived peptides accelerate angiogenesis in the chick chorioallantoic membrane model and stimulated pseudotube formation in human vascular and microvascular ECs, as well as cell migration in an *in vitro* wound-healing assay [111].

8.9
Conclusions

ECs overlying atherosclerotic plaques are constantly exposed to a plethora of mechanical and chemical stimuli, which modify their behavior. Inflammatory activation of ECs is, for example, an essential prerequisite for leukocyte recruitment into plaques, which is the basis of plaque expansion and is strongly associated with plaque rupture. EC responses to disturbed and oscillatory blood flow patterns dictate the propensity of particular regions of the artery to suffer atherogenesis. As part of these processes, ECs express a selection of MMPs that are equipped to digest or remodel all the extracellular components of the vessel wall. A summary of the changes in MMP expression by ECs are listed in Table 8.2. ECs may prove to be an important source of MMPs in advanced atherosclerotic plaques, although the contribution of EC-derived MMPs in acute coronary syndromes has so far attracted less attention than those from VSMCs or macrophages.

Expanding atherosclerotic plaques impinge on the vessel lumen, elevate shear stress and can precipitate "beneficial" compensatory outward remodeling of the vessel [2, 112]. Of the human plaques assayed by Varnava *et al.*, 60% showed no loss of lumen, demonstrating that they had undergone outward remodeling to maintain lumen size (expansively remodeled) [2]. On the other hand, there may be adverse consequences of expansive remodeling because it is strongly associated with inflammation, calcification, and medial thinning, which are features of plaque instability [113]. As discussed above, expansive remodeling is endothelium-dependent and involves endothelium-derived MMPs. The other 40% of atherosclerotic plaques studied by Varnava *et al.* showed a reduction in luminal area (constrictive remodeled), which was associated with a more fibrotic plaque phenotype with a lower level of inflammatory cell infiltrate [2]. The existence of these constrictively remodeled plaques suggests a level of endothelial dysfunction leading to the failure of the expansive remodeling process to maintain the lumen.

Both plaque ruptures [114] and erosions [115] occur most frequently in regions of arteries that experience elevated shear stress. Hence the EC remodeling response and the proteases that are upregulated in this process could make a major contribution to both causes of myocardial infarction. Since the proteolytic degradation of the fibrous cap in plaque rupture is most likely mediated by macrophages and VSMCs, the contribution of ECs may be mainly indirect, for example by promoting macrophage infiltration. However, endothelial erosion is primarily an endothelial pathology. Human atherosclerotic lesions display a large degree of heterogeneity, however there are two commonly observed features of eroded plaques; these are a fibrotic plaque phenotype and a high degree of stenosis [6, 116–121]. Because of the stenosis, ECs would experience high shear stress,

Table 8.2 Changes in MMP expression elicited by different stimuli. ↑ increase, ↓ decrease.

MMP	Response to stimuli					Ref.
	Short term shear stress (<16 h), or changes in shear stress	Laminar shear vs. oscillatory or static control >16 h	Cyclic strain	Inflammatory stimuli	Angiogenic stimuli	
MMP-1	↑	↑		↑	↑	[30, 58, 59, 86, 93, 94]
MMP-2			↑	↑	↑	[38, 88, 89, 93, 95, 104, 105]
MMP-3				↑	↑	[30, 38, 93]
MMP-7		↓			↑	[58, 98, 99]
MMP-8		↑				[58]
MMP-9	↑	↓	↑	↑	↑	[29, 38, 84, 90, 93, 95]
MMP-10		↓		↑	↑	[38, 75, 86]
MMP-11					↑	[38]
MMP-13				↑	↑	[38, 94]
MMP-14	↓	↓	↑	↑	↑	[38, 58, 60, 88, 91, 97, 101, 103, 105]
MMP-15	↑					[39]
MMP-16					↑	[35]

which may chronically activate the remodeling response. In addition the fibrous nature of the plaque would act as an internal stent, thereby reducing the cyclic strain experienced by ECs. These features could produce a specific phenotype in ECs that promotes endothelial loss. Among other processes, the chronic "remodeling" state would be expected to sustain the production of peroxynitrite, which in combination with other factors that promote oxidative stress and EC dysfunction could upregulate MMPs and reduce EC adhesion to its basement membrane, precipitating EC loss and thrombus formation. This hypothesis is attractive but currently lacks firm evidence. Further work is required, including the development of novel animal models.

Acknowledgments

The authors' work is supported by The British Heart Foundation.

References

1 Glagov, S., Weisenberg, E., Zarins, C.K., Stankunavicius, R., and Kolettis, G. (1987) *N. Engl. J. Med.*, **316**, 1371–1375.

2 Varnava, A.M., Mills, P.G., and Davies, M.J. (2002) *Circulation*, 105.

3 Davies, M.J. (2000) *Heart*, **83**, 361–366.

4 Reed, D., Reed, C., Stemmermann, G., and Hayashi, T. (1992) *Circulation*, **85**, 205–211.

5 Virmani, R., Burke, A.P., Farb, A., and Kolodgie, F.D. (2006) *J. Am. Coll. Cardiol.*, **47**, C13–C18.

6 Virmani, R., Kolodgie, F.D., Burke, A.P., Farb, A., and Schwartz, S.M. (2000) *Arterioscler. Thromb. Vasc. Biol.*, **20**, 1262–1275.

7 Nagase, H., Visse, R., and Murphy, G. (2006) *Cardiovasc. Res.*, **69**, 562–573.

8 Dollery, C.M. and Libby, P. (2006) *Cardiovasc. Res.*, **69**, 625–635.

9 Newby, A.C. (2006) *Cardiovasc. Res.*, **69**, 614–624.

10 Lemaître, V., O'Byrne, T.K., Borczuk, A.C., Okada, Y., Tall, A.R., and D'Armiento, J. (2001) *J. Clin. Invest.*, **107**, 1227–1234.

11 Kuzuya, M., Nakamura, K., Sasaki, T., Cheng, X.W., Itohara, S., and Iguchi, A. (2006) *Arterioscler. Thromb. Vasc. Biol.*, **26**, 1120–1125.

12 Silence, J., Lupu, F., Collen, D., and Lijnen, H.R. (2001) *Arterioscler. Thromb. Vasc. Biol.*, **21**, 1440–1445.

13 Johnson, J.L., George, S.J., Newby, A.C., and Jackson, C.L. (2005) *Proc. Natl Acad. Sci. U. S. A.*, **102**, 15575–15580.

14 Luttun, A., Lutgens, E., Manderveld, A., Maris, K., Collen, D., Carmeliet, P., and Moons, L. (2004) *Circulation*, **109**, 1408–1414.

15 Gough, P.J., Gomez, I.G., Wille, P.T., and Raines, E.W. (2006) *J. Clin. Invest.*, **116**, 59–69.

16 de Nooijer, R., Verkleij, C.J., von der Thuesen, J.H., Jukema, W.J., van der Wall, E.E., van Berkel, T.J., Baker, A.H., and Biessen, E.A. (2006) *Arterioscler. Thromb. Vasc. Biol.*, **26**, 340–346.

17 Liang, J., Liu, E., Yu, Y., Kitajima, S., Koike, T., Jin, Y., Morimoto, M., Hatakeyama, K., Asada, Y., Watanabe, T., Sasaguri, Y., Watanabe, S., and Fan, J. (2006) *Circulation*, **113**, 1993–2001.

18 Deguchi, J.O., Aikawa, E., Libby, P., Vachon, J.R., Inada, M., Krane, S.M., Whittaker, P., and Aikawa, M. (2005) *Circulation*, **112**, 2708–2715.

19 Schneider, F., Sukhova, G.K., Aikawa, M., Canner, J., Gerdes, N., Tang, S.-M.T., Shi, G.-P., Apte, S.S., and Libby, P. (2008) *Circulation*, **117**, 931–939.

20 Lemaître, V., Soloway, P.D., and D'Armiento, J. (2003) *Circulation*, **107**, 333–338.

21 Silence, J., Collen, D., and Lijnen, H.R. (2002) *Circ. Res.*, **90**, 897–903.

22 Rouis, M., Adamy, C., Duverger, N., Lesnik, P., Horellou, P., Moreau, M., Emmanuel, F., Caillaud, J.M., Laplaud, P.M., Dachet, C., and Chapman, M.J. (1999) *Circulation*, **100**, 533–540.

23 Johnson, J.L., Baker, A.H., Oka, K., Chan, L., Newby, A.C., Jackson, C.L., and George, S.J. (2006) *Circulation*, **113**, 2435–2444.

24 Shimizu, K., Mitchell, R.N., and Libby, P. (2006) *Arterioscler. Thromb. Vasc. Biol.*, **26**, 987–994.

25 Shipley, J.M., Wesselschmidt, R.L., Kobayashi, D.K., Ley, T.J., and Shapiro, S.D. (1996) *Proc. Natl Acad. Sci. U. S. A.*, **93**, 3942–3946.

26 Sithu, S.D., English, W.R., Olson, P., Krubasik, D., Baker, A.H., Murphy, G., and D'Souza, S.E. (2007) *J. Biol. Chem.*, **282**, 25010–25019.

27 Galis, Z.S. and Khatri, J.J. (2002) *Circ. Res.*, **90**, 251–262.

28 Newby, A.C. (2005) *Physiol. Rev.*, **85**, 1–31.

29 Castier, Y., Brandes, R.P., Leseche, G., Tedgui, A., and Lehoux, S. (2005) *Circ. Res.*, **97**, 533–540.

30 Duhamel-Clérin, E., Orvain, C., Lanza, F., Cazenave, J.-P., and Klein-Soyer, C. (1997) *Arterioscler. Thromb. Vasc. Biol.*, **17**, 1931–1938.

31 Montero, I., Orbe, J., Varo, N., Beloqui, O., Monreal, J.I., Rodriguez, J.A., Diez, J., Libby, P., and Paramo, J.A. (2006) *J. Am. Coll. Cardiol.*, **47**, 1369–1378.

32 Nagashima, Y., Hasegawa, S., Koshikawa, N., Taki, A., Ichikawa, Y., Kitamura, H., Misugi, K., Kihira, Y., Matuo, Y., Yasumitsu, H., and Miyazaki, K. (1997) *Int. J. Cancer*, **72**, 441–445.

33 Peracchia, F., Tamburro, A., Prontera, C., Mariani, B., and Rotilio, D. (1997) *Arterioscler. Thromb. Vasc. Biol.*, **17**, 3185–3190.

34 Puyraimond, A., Fridman, R., Lemesle, M., Arbeille, B., and Menashi, S. (2001) *Exp. Cell Res.*, **262**, 28–36.

35 Plaisier, M., Kapiteijn, K., Koolwijk, P., Fijten, C., Hanemaaijer, R., Grimbergen, J.M., Mulder-Stapel, A., Quax, P.H., Helmerhorst, F.M., and van Hinsbergh, V.W. (2004) *Clin. Endocrinol. Metab.*, **89**, 5828–5836., J

36 Dollery, C.M., Owen, C.A., Sukhova, G.K., Krettek, A., Shapiro, S.D., and Libby, P. (2003) *Circ. Res.*, **107**, 2829–2836.

37 Hattori, Y., Nerusu, K.C., Bhagavathula, N., Brennan, M., Hattori, N., Murphy, H.S., Su, L.D., Wang, T.S., Johnson, T.M., and Varani, J. (2003) *Exp. Mol. Pathol.*, **74**, 230–237.

38 Burbridge, M.F., Cogé, F., Galizzi, J.P., Boutin, J.A., West, D.C., and Tucker, G.C. (2002) *Angiogenesis*, **5**, 215–226.

39 Wesselman, J.P.M., Kuijs, R., Hermans, J.J.R., Janssen, G.M.J., Fazzi, G.E., van Essen, H., Evelo, C.T.A., Struijker-Boudier, H.A.J., and De Mey, J.G.R. (2004) *J. Vasc. Res.*, **41**, 277–290.

40 Polette, M., Clavel, C., Birembaut, P., and Declerck, Y.A. (1993) *Pathol. Res. Pract.*, **189**, 1052–1057.

41 Musso, O., Theret, N., Campion, J.P., Turlin, B., Milani, S., Grappone, C., and Clement, B. (1997) *J. Hepatol.*, **26**, 593–605.

42 Qi, J.H., Ebrahem, Q., Moore, N., Murphy, G., Claesson-Welsh, L., Bond, M., Baker, A., and Anand-Apte, B. (2003) *Nat. Med.*, **9**, 407–415.

43 Oh, J., Seo, D.W., Diaz, T., Wei, B.Y., Ward, Y., Ray, J.M., Morioka, Y., Shi, S.L., Kitayama, H., Takahashi, C., Noda, M., and Stetler-Stevenson, W.G. (2004) *Cancer Res.*, **64**, 9062–9069.

44 Mandriota, S.J., Seghezzi, G., Vassalli, J.D., Ferrara, N., Wasi, S., Mazzieri, R., Mignatti, P., and Pepper, M.S. (1995) *J. Biol. Chem.*, **270**, 9709–9716.

45 Schleef, R.R., Bevilacqua, M.P., Sawdey, M., Gimbrone, M.A., and Loskutoff, D.J. (1988) *J. Biol. Chem.*, **263**, 5797–5803.

46 Aruffo, A., Stamenkovic, I., Melnick, M., Underhill, C.B., and Seed, B. (1990) *Cell*, **61**, 1303–1313.

47 Hemdahl, A.-L., Gabrielsen, A., Zhu, C., Eriksson, P., Hedin, U., Kastrup, J., Thoren, P., and Hansson, G.K. (2006) *Arterioscler. Thromb. Vasc. Biol.*, **26**, 136–142.

48 Viemann, D., Goebeler, M., Schmid, S., Nordhues, U., Klimmek, K., Sorg, C., and Roth, J. (2006) *J. Leukoc. Biol.*, **80**, 174–185.

49 Hahn, C. and Schwartz, M.A. (2009) *Nat. Rev. Mol. Cell Biol.*, **10**, 53–62.

50 Andersson, M., Karlsson, L., Svensson, P.A., Ulfhammer, E., Ekman, M., Jernas, M., Carlsson, L.M.S., and Jern, S. (2005) *J. Vasc. Res.*, **42**, 441–452.

51 Dai, G.H., Kaazempur-Mofrad, M.R., Natarajan, S., Zhang, Y.Z., Vaughn, S., Blackman, B.R., Kamm, R.D., Garcia-Cardena, G., and Gimbrone, M.A. (2004) *Proc. Natl Acad. Sci. U. S. A.*, **101**, 14871–14876.

52 Yamawaki, H., Lehoux, S., and Berk, B.C. (2003) *Circulation*, **108**, 1619–1625.

53 Chiu, J.-J., Lee, P.-L., Chen, C.-N., Lee, C.-I., Chang, S.-F., Chen, L.-J., Lien, S.-C., Usami, Y.-C. Ko, S., and Chien, S. (2004) *Arterioscler. Thromb. Vasc. Biol.*, **24**, 73–79.

54 Linder, S. (2007) *Trends Cell Biol.*, **17**, 107–117.

55 Chabadel, A., Banon-Rodriguez, I., Cluet, D., Rudkin, B.B., Wehrle-Haller, B., Genot, E., Jurdic, P., Anton, I.M., and Saltel, F. (2007) *Mol. Biol. Cell*, **18**, 4899–4910.

56 Wang, J., Taba, Y., Pang, J., Yin, G., Yan, C., and Berk, B.C. (2009) *Arterioscler. Thromb. Vasc. Biol.*, **29**, 202–208.

57 Taraboletti, G., D'Ascenzo, S., Borsotti, P., Giavazzi, R., Pavan, A., and Dolo, V. (2002) *Am. J. Pathol.*, **160**, 673–680.

58 Brooks, A.R., Lelkes, P.I., and Rubanyi, G.M. (2002) *Physiol. Genomics*, **9**, 27–41.

59 Chen, B.P.C., Li, Y.-S., Zhao, Y., Chen, K.-D., Li, S., Lao, J., Yuan, S., Shyy, J.Y.J., and Chien, S.H.U. (2001) *Physiol. Genomics*, 7, 55–63.

60 Yun, S., Dardik, A., Haga, M., Yamashita, A., Yamaguchi, S., Koh, Y., Madri, J.A., and Sumpio, B.E. (2002) *J. Biol. Chem.*, 277, 34808–34814.

61 Parmar, K.M., Larman, H.B., Dai, G.H., Zhang, Y.H., Wang, E.T., Moorthy, S.N., Kratz, J.R., Lin, Z.Y., Jain, M.K., Gimbrone, M.A., and Garcia-Cardena, G. (2006) *J. Clin. Invest.*, 116, 49–58.

62 Lin, Z.Y., Kumar, A., SenBanerjee, S., Staniszewski, K., Parmar, K., Vaughan, D.E., Gimbrone, M.A., Balasubramanian, V., Garcia-Cardena, G., and Jain, M.K. (2005) *Circ. Res.*, 96, E48–E57.

63 Dekker, R.J., Boon, R.A., Rondaij, M.G., Kragt, A., Volger, O.L., Elderkamp, Y.W., Meijers, J.C.M., Voorberg, J., Pannekoek, H., and Horrevoets, A.J.G. (2006) *Blood*, 107, 4354–4363.

64 Dekker, R.J., van Soest, S., Fontijn, R.D., Salamanca, S., de Groot, P.G., VanBavel, E., Pannekoek, H., and Horrevoets, A.J.G. (2002) *Blood*, 100, 1689–1698.

65 Dai, G., Vaughn, S., Zhang, Y., Wang, E.T., Garcia-Cardena, G., and Gimbrone, M.A., Jr. (2007) *Circ. Res.*, 101, 723–733.

66 Bao, X., Clark, C.B., and Frangos, J.A. (2000) *Am. J. Physiol. Heart Circ. Physiol.*, 278, H1598–H1605.

67 Mohan, S., Mohan, N., and Sprague, E.A. (1997) *Am. J. Physiol. Cell Physiol.*, 42, C572–C578.

68 Duerrschmidt, N., Stielow, C., Muller, G., Pagano, P.J., and Morawietz, H. (2006) *J. Physiol.*, 576, 557–567.

69 Cai, H. (2005) *Circ. Res.*, 96, 818–822.

70 Tricot, O., Mallat, Z., Heymes, C., Belmin, J., Leseche, G., and Tedgui, A. (2000) *Circulation*, 101, 2450–2453.

71 Suo, J., Ferrara, D.E., Sorescu, D., Guldberg, R.E., Taylor, W.R., and Giddens, D.P. (2007) *Arterioscler. Thromb. Vasc. Biol.*, 27, 346–351.

72 Aoki, M., Nata, T., Morishita, R., Matsushita, H., Nakagami, H., Yamamoto, K., Yamazaki, K., Nakabayashi, M., Ogihara, T., and Kaneda, Y. (2001) *Hypertension*, 38, 48–55.

73 Dumont, O., Loufrani, L., and Henrion, D. (2007) *Arterioscler. Thromb. Vasc. Biol.*, 27, 317–324.

74 Chase, A.J., Bond, M., Crook, M.F., and Newby, A.C. (2002) *Arterioscler. Thromb. Vasc. Biol.*, 22, 765–771.

75 Whalen, A.M., Chen, X.L., Qiu, F.H., Cook, C.K., Sinibaldi, D., Thomas, S., Varner, S.E., Rao, A.S., Wasserman, M.A., and Medford, R.M. (2003) *FASEB J.*, 17, A274–A274.

76 Langille, B.L. and O'Donnell, F. (1986) *Science*, 231, 405–407.

77 Tohda, K., Masuda, H., Kawamura, K., and Shozawa, T. (1992) *Arterioscler. Thromb.*, 12, 519–528.

78 Vita, J.A., Holbrook, M., Palmisano, J., Shenouda, S.M., Chung, W.B., Hamburg, N.M., Eskenazi, B.R., Joseph, L., and Shapira, O.M. (2008) *Circulation*, 117, 3126–3133.

79 Tronc, F., Wassef, M., Esposito, B., Henrion, D., Glagov, S., and Tedgui, A. (1996) *Arterioscler. Thromb. Vasc. Biol.*, 16, 1256–1262.

80 Ivan, E., Khatri, J.J., Johnson, C., Magid, R., Godin, D., Nandi, S., Lessner, S., and Galis, Z.S. (2002) *Circulation*, 105, 2686–2691.

81 Bakker, E.N.T.P., Matlung, H.L., Bonta, P., de Vries, C.J., van Rooijen, N., and VanBavel, E. (2008) *Cardiovasc. Res.*, 78, 341–348.

82 Jiang, Z.H., Berceli, S.A., Pfahnl, C.L., Wu, L.H., Killingsworth, C.D., Vieira, F.G., and Ozaki, C.K. (2004) *Surgery*, 136, 478–482.

83 Godin, D., Ivan, E., Johnson, C., Magid, R., and Galis, Z.S. (2000) *Circulation*, 102, 2861–2866.

84 Berceli, S.A., Jiang, Z.H., Klingman, N.V., Schultz, G.S., and Ozaki, C.K. (2006) *J. Surg. Res.*, 134, 327–334.

85 Ogata, T., Shibamura, H., Tromp, G., Sinha, M., Goddard, K.A.B., Sakalihasan, N., Limet, R., MacKean, G.L., Arthur, C., Sueda, T., Land, S., and Kuivaniemi, H. (2005) *J. Vasc. Surg.*, 41, 1036–1042.

86 Montero, I., Orbe, J., Varo, N., Beloqui, O., Monreal, J.I., Rodríguez, J.A., Díez, J., Libby, P., and Páramo, J.A. (2006) *J. Am. Coll. Cardiol.*, 47, 1369–1378.

87 Cummins, P.M., von Offenberg Sweeney, N., Killeen, M.T., Birney, Y.A., Redmond, E.M., and Cahill, P.A. (2007) *Am. J. Physiol. Heart Circ. Physiol.*, **292**, H28–H42.

88 Wang, B.-W., Chang, H., Lin, S., Kuan, P., and Shyu, K.-G. (2003) *Cardiovasc. Res.*, **59**, 460–469.

89 von Offenberg Sweeney, N., Cummins, P.M., Birney, Y.A., Cullen, J.P., Redmond, E.M., and Cahill, P.A. (2004) *Cardiovasc. Res.*, **63**, 625–634.

90 Sweeney, N.V., Cummins, P.M., Cotter, E.J., Fitzpatrick, P.A., Birney, Y.A., Redmond, E.M., and Cahill, P.A. (2005) *Biochem. Biophys. Res. Commun.*, **329**, 573–582.

91 Rajavashisth, T.B., Liao, J.K., Galis, Z.S., Tripathi, S., Laufs, U., Tripathi, J., Chai, N.N., Xu, X.P., Jovinge, S., Shah, P.K., and Libby, P. (1999) *J. Biol. Chem.*, **274**, 11924–11929.

92 Huang, Y., Song, L.X., Fan, S.W.u, F., and Lopes-Virella, M.F. (2001) *Atherosclerosis*, **156**, 119–125.

93 Mach, F., Schonbeck, U., Fabunmi, R.P., Murphy, C., Atkinson, E., Bonnefoy, J.Y., Graber, P., and Libby, P. (1999) *Am. J. Pathol.*, **154**, 229–238.

94 Gerdes, N., Sukhova, G.K., Libby, P., Reynolds, R.S., Young, J.L., and Schonbeck, U. (2002) *J. Exp. Med.*, **195**, 245–257.

95 Li, A.H., Dubey, S., Varney, M.L., Dave, B.J., and Singh, R.K. (2003) *J. Immunol.*, **170**, 3369–3376.

96 Han, X.Y., Boyd, P.J., Colgan, S., Madri, J.A., and Haas, T.L. (2003) *J. Biol. Chem.*, **278**, 47785–47791.

97 Haas, T.L., Stitelman, D., Davis, S.J., Apte, S.S., and Madri, J.A. (1999) *J. Biol. Chem.*, **274**, 22679–22685.

98 Holnthoner, W., Kerenyi, M., Groger, M., Kratochvill, F., and Petzelbauer, P. (2006) *Biochem. Biophys. Res. Commun.*, **342**, 725–733.

99 Ichikawa, Y., Ishikawa, T., Momiyama, N., Kamiyama, M., Sakurada, H., Matsuyama, R., Hasegawa, S., Chishima, T., Hamaguchi, Y., Fujii, S., Saito, S., Kubota, K., Hasegawa, S., Ike, H., Oki, S., and Shimada, H. (2006) *Oncol. Rep.*, **15**, 311–315.

100 Partridge, C.R., Hawker, J.R., and Forough, R. (2000) *J. Cell Biochem.*, **78**, 487–499.

101 Yana, I., Sagara, H., Takaki, S., Takatsu, K., Nakamura, K., Nakao, K., Katsuki, M., Aoki, S.-i. Taniguchi, T., Sato, H., Weiss, S.J., and Seiki, M. (2007) *J. Cell Sci.*, **120**, 1607–1614.

102 van Hinsbergh, V.W.M., and Koolwijk, P. (2008) *Cardiovasc. Res.*, **78**, 203–212.

103 Itoh, Y. (2006) *IUBMB Life*, **58**, 589–596.

104 Lafleur, M.A., Hollenberg, M.D., Atkinson, S.J., Knauper, V., Murphy, G., and Edwards, D.R. (2001) *Biochem. J.*, **357**, 107–115.

105 Wang, H., and Keiser, J.A. (2000) *Biochem. Biophys. Res. Commun.*, **272**, 900–905.

106 Hashimoto, G., Inoki, I., Fujii, Y., Aoki, T., Ikeda, E., and Okada, Y. (2002) *J. Biol. Chem.*, **277**, 36288–36295.

107 Tatti, O., Vehvilainen, P., Lehti, K., and Keski-Oja, J. (2008) *Exp. Cell Res.*, **314**, 2501–2514.

108 Gingras, D., Bousquet-Gagnon, N., Langlois, S., Lachambre, M.-P., Annabi, B., and Béliveau, R. (2001) *FEBS Lett.*, **507**, 231–236.

109 Hamano, Y., Zeisberg, M., Sugimoto, H., Lively, J.C., Maeshima, Y., Yang, C.Q., Hynes, R.O., Werb, Z., Sudhakar, A., and Kalluri, R. (2003) *Cancer Cell*, **3**, 589–601.

110 Kim, Y.M., Jang, J.W., Lee, O.H., Yeon, J., Choi, E.Y., Kim, K.W., Lee, S.T., and Kwon, Y.G. (2000) *Cancer Res.*, **60**, 5410–5413.

111 Robinet, A., Fahem, A., Cauchard, J.H., Huet, E., Vincent, L., Lorimier, S., Antonicelli, F., Soria, C., Crepin, M., Hornebeck, W., and Bellon, G. (2005) *J. Cell. Sci.*, **118**, 343–356.

112 Hasegawa, T., Ehara, S., Kobayashi, Y., Kataoka, T., Yamashita, H., Nishioka, H., Asawa, K., Yamagishi, H., Yoshiyama, M., Takeuchi, K., Yoshikawa, J., and Ueda, M. (2006) *Am. Heart J.*, **151**, 332–337.

113 Burke, A.P., Kolodgie, F.D., Farb, A., Weber, D., and Virmani, R. (2002) *Circulation*, **105**, 297–303.

114 Fukumoto, Y., Hiro, T., Fujii, T., Hashimoto, G., Fujimura, T., Yamada,

J., Okamura, T., and Matsuzaki, M. (2008) *J. Am. Coll. Cardiol.*, **51**, 645–650.

115 Lovett, J.K. and Rothwell, P.M. (2003) *Cerebrovasc. Dis.*, **16**, 369–375.

116 Arbustini, E., Dal Bello, B., Morbini, P., Burke, A.P., Bocciarelli, M., Specchia, G., and Virmani, R. (1999) *Heart*, **82**, 269–272.

117 Farb, A., Burke, A.P., Tang, A.L., Liang, Y.H., Mannan, P., Smialek, J., and Virmani, R. (1996) *Circulation*, **93**, 1354–1363.

118 Kolodgie, F.D., Burke, A.P., Farb, A., Weber, D.K., Kutys, R., Wight, T.N., and Virmani, R. (2002) *Arterioscler. Thromb. Vasc. Biol.*, **22**, 1642–1648.

119 Kolodgie, F.D., Burke, A.P., Wight, T.N., and Virmani, R. (2004) *Curr. Opin. Lipidol.*, **15**, 575–582.

120 Sato, Y., Hatakeyama, K., Yamashita, A., Marutsuka, K., Sumiyoshi, A., and Asada, Y. (2005) *Heart*, **91**, 526–530.

121 Shah, P.K. (2002) *Prog. Cardiovasc. Dis.*, **44**, 357–368.

9
Cathepsins in Atherosclerosis

Lili Bai, Esther Lutgens, and Sylvia Heeneman

9.1
Introduction

Cathepsins are a family of proteolytic enzymes that degrade the extracellular matrix (ECM). The ECM consists of elastins, collagens, and proteoglycans. It gives anchorage, support, and structure to tissue and functions as an adhesive substrate for vascular endothelial cells and smooth muscle cells. ECM remodeling is critically involved in many physiological and pathological processes such as wound healing, tumor growth [1], chronic inflammatory diseases such as rheumatoid arthritis [2], neurological disorders [3], and also in cardiovascular pathologies such as atherosclerosis. Degradation of ECM in the vessel wall enables smooth muscle cells (SMCs) to migrate from the media into the intima and inflammatory cells to infiltrate from the circulation into the arterial wall, processes critical in the pathogenesis of atherosclerosis. Moreover, within the plaque, ECM degradation causes thinning of the fibrous cap, often resulting in plaque rupture and thrombosis. When an occlusive thrombus forms, clinical complications of atherosclerosis with a high morbidity and mortality, such as myocardial infarction and stroke occur.

The term "cathepsin" represents lysosomal proteolytic enzymes irrespective of their enzyme category. Based on their catalytic actions, cathepsins are classified into cysteine (cathepsins B, C, F, H, K, L, O, S, V, X, and W), serine (cathepsins A and G) and aspartate (cathepsins D and E) proteases [3, 4].

The function of cathepsins and their most well-known inhibitor, cystatin, in ECM degradation, vascular remodeling, and atherosclerosis has been demonstrated in human and animal models. This chapter will emphasize and discuss the importance of cathepsins in atherosclerosis.

9.2
Synthesis and Activity

Under physiologic conditions, cathepsins are localized intralysosomally. Although cathepsins were originally shown to be active in lysosomes and endosomes and

Atherosclerosis: Molecular and Cellular Mechanisms. Edited by Sarah Jane George and Jason Johnson
Copyright © 2010 WILEY-VCH Verlag GmbH & Co. KGaA, Weinheim
ISBN: 978-3-527-32448-4

to execute unspecific proteolysis [5], there is growing evidence that cathepsins can function outside lysosomes or endosomes [6]. In response to certain signals such as inflammatory cytokines (interleukin-1β (IL-1β), tumor necrosis factor-α (TNF-α), interferon-γ (IFN-γ) [7]), angiotensin II [8], or oxidized low-density lipoprotein (ox-LDL) [9], they are released from the lysosomes into the cytoplasm where they activate various biological and pathological pathways including ECM degradation, inflammation, and apoptosis. For example, disruption of lysosomes results in translocation of cathepsins to the cytosol and induction of apoptosis through a caspase-dependent mechanism [10]. Cathepsins can also exert specific functions in the nucleus, and even in the mitochondrion [11, 12].

9.3
Regulation

Control and regulation of proteolytic activity are indispensable processes. Failure of these processes inevitably results in various fatal pathologies such as metastasis of cancer cells or inflammation [13]. Cathepsins are regulated by endogenous cathepsin inhibitors called cystatins. In general, cystatins function as a protection against the irregular release of peptidases such as cathepsins from the lysosome during apoptosis or phagocyte degranulation [14]. They also serve as defense against proteases secreted by proliferating cancer cells or by invading organisms, such as parasites [15].

The cystatin family is divided into four subfamilies [15, 16]: the stefins (type 1), the cystatins (type 2), the kininogens (type 3), and various structurally related but non-inhibitory proteins (type 4). Type 1 cystatins are cystatins A (stefin A) and B (stefin B), which are present mainly intracellularly, but can also appear in body fluids at significant concentrations. Type 2 cystatins, including cystatin C, are found in most body fluids, and mainly operate extracellularly. Type 3 cystatins include kininogens, which are present in blood proteins. Among these cathepsin inhibitors, cystatin C is the most potent inhibitor [15, 17], with the greatest inhibitory properties to cathepsins L and S, followed by cathepsins B and H [14].

9.4
Cathepsins in (Patho)physiological Tissue Remodeling

Cathepsins not only function in intralysosomal protein degradation, but also contribute to tissue remodeling by degrading the ECM. As cathepsins are expressed in several cell types, they participate in various tissue remodeling processes (Table 9.1). Cathepsin K is the most abundant cysteine protease expressed in osteoclasts and is instrumental in bone matrix degradation necessary for bone resorption [18]. Cathepsin S is expressed and secreted by the human adipose tissue and is upregulated in obesity [19]. The increase of cathepsin S in adipose tissue causes local degradation of fibronectin network, a key preadipocyte–ECM component, support-

Table 9.1 Cathepsins in tissue remodeling.

Tissue remodeling	Description
Bone resorption	Cathepsin K
Fat tissue turnover	Cathepsin S
Cardiac remodeling	Cathepsins B and S
Tumor	Cathepsins B, D, L, and S
Neurodegeneration	Cathepsin D
Renal pathology	Cathepsin B, H and L

ing the development of fat mass [20]. In addition, circulating levels of cathepsin S were also increased in obese subjects [19]. Cathepsin S mRNA or protein in the left ventricular tissues was more abundant in rats or humans with heart failure compared to control, suggesting that cathepsin S participates in pathological left ventricular remodeling [21]. Patients with heart failure were found to have increased expression of cathepsin B in the myocardium, suggesting that cathepsin B plays a role in the development of heart failure [22]. Furthermore, enhanced expression of several cathepsins (B, D, L, and S) was observed in several types of tumors, and is thought to contribute to tumor growth and metastasis [23, 24]. Increased cathepsin D expression was present in activated astrocytes in neurodegenerative diseases, such as scrapie in mice and Alzheimer's disease in humans. The increase in cathepsin D may be an ongoing response to the deposition of abnormal aggregated proteins that have neurotoxic effects [25].

Some *in vitro* and *in vivo* studies suggested that cysteine proteinases play an important role in renal pathophysiology. Cathepsins B, H, and L were present in glomeruli and other fractions prepared from normal rat kidney and were able to degrade the intact glomerular basement membrane *in vitro* [26].

9.5
Cathepsins in Atherosclerosis

9.5.1
Expression of Cathepsins in Human Atherosclerosis

Considering the widespread functions of cathepsins in the different pathologies, it is not surprising that cathepsins play a key role in cardiovascular disease. In atherosclerotic arteries, cathepsins are expressed by most of the plaque cell types, including macrophages, endothelial cells (ECs) and SMCs [27, 28]. Various studies have shown that cathepsin B [29], F [30], L [31], K [32], and S [7] mRNA or protein level were increased in either human or mouse atherosclerotic lesions, whereas they were only weakly expressed in normal arteries. Cathepsins F, L, K, and S were mainly found in macrophages, SMCs and ECs. Specifically, macrophages in the shoulder region of human atheroma contained abundant expression of cathepsin K and S [7]. Likewise, SMCs that appeared to traverse the internal elastic laminae and the fibrous cap also expressed cathepsins K and S [7], suggesting that SMCs and macrophages utilize these cathepsins to enter the atherosclerotic plaque. Endothelial cells lining the lumen of the vessel itself and the intraplaque microvessels in human atheroma lesions expressed cathepsin S [28], suggesting an involvement of this protein in neovascularization.

The enhanced cathepsin expression in human and murine atherosclerotic lesions suggests an involvement of cathepsins in the process of atherosclerosis. The next paragraphs outline the consequences of cathepsin deficiency on plaque progression and plaque phenotype and describe the potential mechanisms how the different cathepsins affect atherogenesis (Figure 9.1).

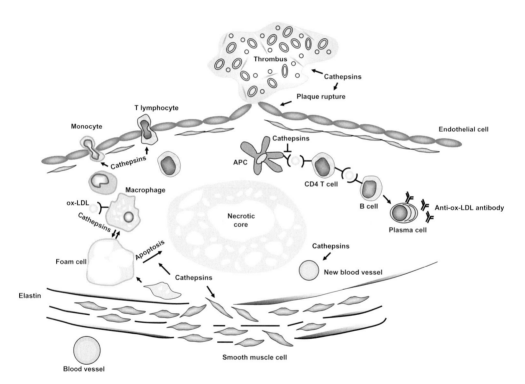

Figure 9.1 Overview of potential roles of cathepsins in atherosclerosis formation. Cathepsins are expressed in endothelial cells, smooth muscle cells, macrophages, and foam cells. Cathepsins are able to degrade extracellular matrix (ECM) containing elastin and collagen, which facilitate smooth muscle cell to migrate from media and monocyte and T lymphocytes from circulation to intima. ECM degradation by cathepsins might facilitate neovessel formation. Both macro-phages and smooth muscle cells (SMCs) take up oxidized low-density lipoprotein (ox-LDL) and become foam cells. These processes also involve cathepsins. The apoptosis of foam cells contributes to necrotic core formation and cathepsins might participate in this process. Cathepsins might compromise anti-ox-LDL antibody production by interfering with antigen-presentation process. Cathepsin may also induce plaque rupture and thrombosis formation.

9.5.2
The Role of Cathepsins in Atherogenesis: Lessons Learned from Animal Models

Cathepsin S deficiency in LDL receptor–deficient (LDLr$^{-/-}$) mice protected against atherosclerosis, by reducing atherosclerotic plaque area, plaque progression, the number of elastin breaks, and elastase activity. Furthermore, cathepsin S deficiency induced a reduction in SMCs and collagen content and decreased fibrous cap thickness [33]. Cathepsin S deficiency in apolipoprotein E–deficient (ApoE$^{-/-}$) mice was also found to reduce plaque ruptures, defined as visible defects in the cap of atheromatous lesions and accompanied by intrusion of erythrocytes into the region below [7]. In addition, cathepsin S expression in macrophages colocalized with areas of elastin fragmentation in mice [34].

Cathepsin K deficiency was found to protect against ECM remodeling in atherosclerosis as well. Cathepsin K deficiency in ApoE$^{-/-}$ mice resulted in a 42% reduction in atherosclerotic plaque area; although the total number of plaques remained unchanged, there was a relative increase in early lesions and a relative decrease of the number of advanced lesions when cathepsin K was absent [32]. Furthermore, cathepsin K deficiency led to an increase in collagen content and macrophage size, and a decrease in elastin breaks [32]. These results were confirmed by Samokhin *et al.* who demonstrated that cathepsin K deficiency inhibited plaque progression and increased fibrous cap thickness in the brachiocephalic artery after high fat feeding [35]. Interestingly, opposite results were found when only leukocyte cathepsin K was absent. In a bone marrow transplantation model in which cathepsin K–deficient bone marrow was transplanted into irradiated atherosclerotic LDLR$^{-/-}$ mice, atherosclerotic lesions had a vulnerable atherosclerotic plaque phenotype with reduced collagen levels, an increase in macrophage content and an accelerated necrotic core formation [36]. However, these data are supported by the paper of Lutgens *et al.*, in which cathepsin K–deficient macrophages showed an increase in lipid uptake and foam cell formation [32].

Atherosclerotic lesions in either human or ApoE$^{-/-}$ mice have comparatively low levels of cystatin C (cathepsin inhibitor), whereas normal arteries have abundant cystatin C expression in medial SMCs and in ECs [28, 37]. Deficiency of cystatin C in ApoE$^{-/-}$ mice significantly increased plaque size [38] and increased elastolytic activity [39], suggesting an important role played by cystatin C in atherosclerosis development.

All the above findings demonstrate the importance of cysteine proteases and the cathepsin inhibitor cystatin C in atherosclerosis.

9.5.3
Cathepsins in Atherosclerosis: Potential Mechanisms

9.5.3.1   Cathepsins are Regulated by Shear Stress
The mechanic force (shear stress) that is constantly present over vascular endothelial cells regulates the structure and function of endothelial cells [40]. This mechanic force is generated by blood flow over the vascular endothelium and tends to develop different patterns of hemodynamic force based on the region and curva-

ture of the artery. Oscillatory shear stress, which occurs in branched or curved regions, is associated with pro-atherogenic events. Laminar shear stress occurs in "straight" arteries, and is associated with low susceptibility to atherosclerosis [40]. Platt *et al.* showed that atheroprotective laminar shear stress decreased cathepsin L activity, while pro-atherogenic oscillatory shear stress significantly enhanced cathepsin L activity compared with laminar shear stress. These observations suggested that cathepsin L could participate in the development of vascular remodeling and atherosclerosis partly by a shear stress sensitive response [41]. Cathepsin K was also found to be mechanosensitive. Laminar shear stress decreased cathepsin K protein expression in endothelial cells *in vitro* compared with oscillatory shear stress, suggesting a potential role for cathepsin K at sites of disturbed flow associated with vascular pathology. This study also showed a positive correlation between cathepsin K levels in endothelium and human atherosclerotic lesion development, supporting the evidence for a role of cathepsin K in atherosclerosis formation. The above findings suggest that cathepsin L and K expression in endothelial cells are regulated by shear stress and partially contribute to the shear-dependent regulation of the ECM protease activity, leading to vascular remodeling and atherosclerosis.

9.5.3.2 Cathepsins Mediate Leukocyte and SMC Migration

One of the first signs of atherosclerosis is recruitment of leukocytes from the vascular lumen into the arterial wall, followed by adhesion and migration through the endothelial layer and arterial basement membrane [42]. Blood-borne monocytes use cathepsins to degrade ECM and migrate through endothelial layers. For example, cathepsin S–deficient monocytes displayed less migration through a SMC layer than wild-type monocytes *in vitro* [33].

The role of cathepsins in leukocyte migration has also been illustrated in Jurkat T cells [43]. The migration activity of T lymphocytes through ECM did not require ECM degradation but was mediated by adhesion molecules and cytoskeletal rearrangements. Cathepsin X overexpressing T lymphocytes displayed polarized migration-associated morphology and enhanced migration on two- and three-dimensional models using intercellular adhesion molecule-1 (ICAM-1) and Matrigel-coated surfaces. The proteolytic activity of cathepsin X was not involved in the increased invasiveness of cathepsin X–overexpressing cells. Instead, an active form of cathepsin X colocalized with lymphocyte function-associated antigen-1 (LFA-1) in migrating cells. LFA-1 is found on leukocytes and binds to ICAM-1 on antigen-presenting cells and functions as an adhesion molecule [44].

In addition to the migration of monocytes into the arterial wall, migration of SMCs from the media into the intima also plays a key role in atherosclerotic lesion formation, as well as in restenosis [45, 46]. Migration of SMCs from the medial to the subendothelial space requires degradation of ECM. The interaction of cathepsin S with ECM components during SMC migration was examined by Cheng *et al.* A selective cathepsin S inhibitor (morpholinurea-leucine-homophenylalanine-vinylsulfonephenyl (LHVS) and the endogenous inhibitor cystatin C significantly attenuated SMC invasion through ECM, suggesting that cathepsins released from SMCs contribute to SMC invasion through collagen and elastin substrates. Western blot analysis on subcellular fractions showed that the active form of cathepsin S was present on the SMC plasma membrane but not in the cytosol. In contrast, the active forms of cathepsins K, B, L, and D were not expressed on the SMC plasma membranes. These findings suggested that membrane-bound cathepsin S facilitated SMC proteolytic activity and thereby SMC invasion through ECM.

This notion was further confirmed by the finding of partial colocalization of cathepsin S and $\alpha_v\beta_3$-integrin at the cell surfaces. $\alpha_v\beta_3$-Integrin is a receptor for several proteases such as MMPs [47–49] and involved in SMC adhesion and migration. Thus cathepsin S might cooperate with $\alpha_v\beta_3$-integrin to facilitate SMC invasion. In the study by Cheng *et al.*, the inhibition of cathepsin S reduced SMC invasion through ECM but had no effect on adhesion or on migration through ECM. It suggests that cathepsin S likely contributes to the proteolytic process during SMC invasion through ECM but does not play a role in mediating the migration process itself [50].

9.5.3.3 Cathepsin-Induced ECM Degradation in Atherosclerosis

ECM degradation plays an important role in the development and destabilization of the atherosclerotic plaque [51]. Modifications in ECM homeostasis, as a consequence of alteration in the degradation and/or synthesis of the vessel wall ECM, have been associated with vascular diseases. Plaque formation occurs as a result of SMC migration and proliferation accompanied by degradation of ECM. In the latter stages of atherosclerosis, thrombotic complications frequently develop from rupture of the fibrous cap or superficial erosion of the endothelium, both of which are the result of ECM degradation.

Cathepsins were found to exert strong elastinolytic and collagenolytic activity in culture media conditioned by various cell types. Cultured SMCs stimulated with cytokines secreted active cathepsins B and L and degraded substantial amounts of insoluble elastin [7, 31]. Monocyte-derived macrophages synthesized not only metalloproteinases, but also cathepsins B, L, and S. However, only the cathepsins were detected in the extracellular milieu, and macrophage-mediated elastinolytic activity was completely abrogated by inhibiting cathepsins L and S but not by MMP inhibitors [44]. Cultured ECs stimulated with proinflammatory cytokines or growth factors induced the expression of cathepsin L, and enabled cathepsin L–dependent degradation of extracellular collagen and elastin [52]. These findings suggested that SMCs, macrophages, and ECs use cysteine proteinase–dependent processes for ECM remodeling and this contributes to atherosclerosis formation.

Besides the above *in vitro* data, one *in vivo* study also suggested that cathepsins play important roles in ECM degradation. In addition to abundant expression of cathepsins K and S in macrophages and intima SMCs, extracts from human atheroma showed a twofold increase in elastinolytic activity compared with normal vessels, which could be inhibited significantly by a cysteine protease inhibitor E64 [7].

9.5.3.4 Cathepsin-Mediated Neovascularization

The microvascular network of vasa vasorum is found in the adventitia and the outer media of normal vessels. However, during atherosclerotic plaque development, these networks grow and expand into the plaque [53]. Plaque neovascularization is thought to contribute to lesion progression in various ways. It offers a port of entry for leukocyte and plasma constituents such as lipoproteins into atherosclerotic lesions [54, 55]. Furthermore, the fragile nature of these neovessels can lead to focal intraplaque hemorrhage, which further promotes inflammation and thrombotic complications of atherosclerosis [56]. The effects of cathepsins on neovascularization have been studied intensively in tumors and wound healing, but less in cardiovascular diseases. Cathepsin activity has been shown to be critical during tumor invasion and angiogenesis [57], which requires lysis of the ECM to pave the way for neovasculature [58] and proteolysis of the endothelial basement membrane [59].

The role of cathepsins has been further investigated in the context of endothelial progenitor cell (EPC)–mediated neovascularization. EPCs have been shown to improve neovascularization in ischemic tissues [60, 61]. Gene expression profiling of EPCs and ECs showed that cathepsin L was highly expressed in EPCs compared with ECs. Mice receiving cathepsin L–deficient bone marrow cells showed impaired functional recovery following hind limb ischemia. However, inhibition of other proteases such as cathepsins S and D, and MMPs did not affect EPC metalloproteinase activity [62], suggesting that cathepsin L is specifically required for EPC-mediated neovascularization. Endothelium-derived cathepsins may also contribute to angiogenesis as well. Cathepsin S inhibition reduced microtubule formation in ECs by impairing cell invasion. Furthermore, cathepsin S–deficient mice displayed fewer microvessels in healing wounds [63].

Taken together, these data suggested that cathepsins are involved in the processes of neovascularization. However, until now, direct evidence for a significant role for cathepsins in plaque neovascularization is lacking.

9.5.3.5 Cathepsins Mediate Inflammation

Atherosclerosis is an inflammatory disease in which both innate and adaptive immune responses are involved [64]. Innate, but also adaptive immune responses such as antigen presentation occur within the atherosclerotic plaque, which is rich in T lymphocytes, but also in antigen-presenting macrophages and dendritic cells [65]. In addition to proteolytic activity, cathepsins also play roles in inflammatory processes involving both innate immunity, such as Toll-like receptor-9 (TLR9) signaling and processes of adaptive immunity, including antigen presentation.

Cathepsins in Innate Immunity: TLR9 Signaling   Matsumot *et al.* found that cathepsin B and L inhibitors suppress the interaction of CpG-B (TLR ligand) with TLR9 in 293T cells, suggesting a role for cathepsins in regulating CpG-B–TLR9 interaction, and thus innate immunity [66]. Moreover, cathepsins S and F, but not H, are able to complement TLR9 responses in Ba/F3 cells, which are defective in TLR9 responses (unable to activate NF-κB in response to CpG-B) [66].

The involvement of cathepsins in TLR9 signaling was also present in dendritic cells [67]. Administration of a potent orally active cathepsin K inhibitor named NC-2300 in rats with adjuvant-induced arthritis not only suppressed bone resorption by osteoclasts, but also showed anti-inflammatory effects, resulting in reduced paw swelling. Further studies showed that cathepsin K inactivation led to the blockade of essentially all the downstream pathways of TLR9 signaling in dendritic cells, showing a crucial role for cathepsin K in TLR9 signaling.

TLRs, especially TLR1, 2, and 4, have been recognized for their roles in atherosclerotic lesion development and progression [68]. Unfortunately, the involvement of TLR9 signaling in atherosclerosis is not yet clear. Further research on the role of cathepsin-mediated TLR9 response in the development of atherosclerosis is warranted.

Cathepsins and Antigen Presentation The function of cathepsins in antigen presentation has been intensively studied. MHC class II molecules present antigen peptides on the surface of antigen-presenting cells (APCs), which are recognized by CD4$^+$ T cells. The cooperation of invariant (Ii) chain with MHC class II dimers is required for proper antigen presentation to CD4$^+$ T cells. The Li cytoplasmic tail targets the MHC class II–Ii complex to the endosomal pathway and prevents early loading of antigenic protein on MHC class II with class II–associated invariant chain peptide (CLIP). Degradation of Ii is an important regulatory step in the maturation of MHC class II dimers. Maturation of the endosome leads to the activation of lysosomal enzymes, which degrade Li from class II–Ii complexes and induce subsequent maturation of class II molecules. Cathepsins F, L, S, K, and V have all degraded MHC II–associated Ii in professional APCs and are thus required for antigen processing [69].

Compared with other lysosomal cysteine proteases, cathepsin S has displayed some unique characteristics in that it can not only degrade MHC II–associated Ii in professional APCs, but also in nonprofessional APCs such as intestinal epithelial cells [70, 71]. Several studies describe the functional significance of cathepsin inhibition in antigen presentation, thus inhibition of cathepsin S reduced autantigen presentation and development of organ-specific autoimmunity in mice [72, 73].

In atherosclerosis, a well-known autoantigen in the cellular immune response of atherosclerosis is ox-LDL [74]. Anti-ox-LDL antibodies have been detected within rabbit and human atherosclerotic lesions [75]. In addition, anti-ox-LDL antibodies titers significantly correlated with the extent of atherosclerosis in mice [76]. Moreover, immunization with splenic B cells led to the production of anti-ox-LDL antibodies, which conferred protection against atherosclerosis in mice [77]. Interestingly, cathepsin S–deficient LDLr$^{-/-}$ mice showed significant reduction in atherosclerotic lesion size and lower titers of autoantibody against both malondialdehyde–ox-LDL and copper–ox-LDL epitopes [28].

Cathepsins Mediate TGF-β Signaling Cathepsins might be involved in anti-inflammatory responses via the TGF-β pathway. Lutgens *et al.* showed that besides reduced atherosclerotic lesion size, deficiency of cathepsin K–induced expression of genes involved TGF-β signaling in atherosclerotic lesions [78]. As inhibition of TGF-β in ApoE$^{-/-}$ mice initiated an inflammatory plaque phenotype [79], induction of TFG-β signaling might partly explain the protective effect of cathepsin K deficiency in atherosclerosis formation.

9.5.3.6 Apoptosis

Apoptosis of foam cells contributes to necrotic core formation, a hallmark of plaque severity. Several cysteine proteases have been described to participate in this process [80–82].

Z-FAFMK (a selective inhibitor of cathepsins B and L) prevents oxysterol-induced apoptosis of mononuclear cells [83]. The expression of cathepsin L in human atherosclerotic plaques was correlated with apoptosis, suggesting that cathepsin L is involved in macrophage apoptosis. Moreover, macrophage apoptosis in atherosclerotic coronary artery specimens was significantly correlated with expression of cathepsin L in cell membranes and nuclei [84]. Furthermore, selective cathepsin S inhibition by 05141 and a null mutation of cathepsin S, ameliorated IFN-γ-induced apoptosis as manifested by reduced activation of caspases-3, -8, and -9 and reduced expression of Bim, Bid, Fas, Fas ligand (FasL), TNF-α, TNF-related apoptosis-inducing ligand (TRAIL), and protein kinase C-δ (PKCδ) in lung. These observations highlight the suppressive effects of cathepsin S inhibition in IFN-γ-induced apoptosis activation via both the intrinsic/mitochondrial and extrinsic pathways [85].

The mechanism by which cysteine proteases induce apoptosis is thought to involve both caspase-dependent and caspase-independent cell death pathways [86]. Cathepsins are located in the lysosomes under physiological conditions. Nonetheless, oxidative stress causes lysosomal leakage and rapidly initiates cathepsin translocation from lysosomes to the cytosol and participation in both apoptotic pathways. The caspase-dependent pathway includes a direct cleavage of Bid and/ or Bak/Bim. Translocation of these pro-apoptotic proteins to the mitochondrial outer membrane leads to release of apoptotic factors such as cytochrome C. This results in an indirect activation of caspases and subsequent apoptosis. Cathepsins are also able to directly cleave caspases, which is followed by cleavage of Bid and/ or Bax, translocation of these proteins to mitochondria and subsequent downstream events that cause apoptosis. Cathepsins are also able to trigger the release of apoptosis-inducing factor (AIF) and cause caspase-independent cell death [83, 87, 88].

9.5.3.7 Lipid Metabolism

Lipid retention in macrophage-derived foam cells is one of the hallmarks of atherosclerotic plaque development and progression. Cathepsins have been shown to affect foam cell formation bivalently. Some cathepsin family members enhance foam cell formation, while others impair the formation of foam cells [27].

Cathepsins Reduce Foam Cell Formation Cathepsins are able to degrade (modified) LDL. Cathepsin B inhibition reduced modified LDL degradation in human aortic SMC lysates. Decreased lysosomal degradation may cause LDL accumulation in SMCs and subsequent foam cell formation [89]. Several studies showed that cathepsin K deficiency was able to increase lipid uptake and consequently foam cell formation. Bone marrow–derived macrophages from cathepsin K–deficient ApoE$^{-/-}$ mice showed an increased uptake of modified LDL [32]. Pathway analysis revealed that the increased lipid uptake was regulated by both CD36 and caveolins [78]. Furthermore, bone marrow–derived macrophages from cathepsin K–deficient ApoE$^{-/-}$ mice showed an increase in cholesterol ester accumulation compared with ApoE$^{-/-}$ bone marrow–derived macrophages, which was stored in lysosomal compartments [32]. These data indicate that cathepsin B and K are able to reduce foam cell formation by degrading (modified) LDL and inhibiting lipid uptake.

Cathepsins Stimulate Extracellular Lipid Accumulation and Foam Cell Formation *In vitro* studies showed that recombinant cathepsin F strongly degraded ApoB100 [30]. Degradation of ApoB100 by cathepsin F induced accumulation and fusion of LDL particles and enhanced the ability of LDL to bind proteoglycans, leading to the accumulation of extracellular lipid droplets [30]. This study indicates that cathepsins could contribute to extracellular lipid accumulation in the arterial wall, an important characteristic of atherosclerosis. Degradation of cholesterol acceptors by cathepsins not only induced extracellular lipid accumulation, but also inhibited lipid efflux. Cathepsin S completely (100%) degraded apolipoprotein AI (ApoAI), which caused total loss of the ability of ApoAI to induce cholesterol efflux. Cathepsins F and K incompletely degraded ApoAI, leading to a reduction of cholesterol efflux by 30% and 15%, respectively [90]. These data suggest that cathepsin-mediated degradation of cholesterol acceptors can inhibit lipid efflux and thereby aggravate foam cell formation.

9.5.3.8 Thrombus Formation

Cardiovascular disease is characterized by end-stage thrombotic complications, either due to rupture of the fibrous cap or superficial erosion. Cathepsin G is known to be involved in thrombosis. After release from neutrophils, cathepsin G leads to calcium mobilization via the cathepsin G platelet receptor (protease-activated receptor 4) [91], thereby inducing platelet aggregation [92]. Cathepsin G can also affect the morphological integrity of ECs through a mechanism involving vitronectin-bound plasminogen activator inhibitor-1 (PAI-1) detachment. The detachment of PAI-1 from the subendothelial matrix induces F-actin cytoskeleton rearrangement with consequent changes in morphological integrity of EC. These events expose a highly thrombogenic surface to which platelets can adhere and become activated [93].

Despite large amount of data implying a functional role of cathepsin G in thrombosis formation, cathepsin S seems to have antithrombotic property. Cathepsin S–deficient mice had accelerated thrombotic responses to artery injury and shorter plasma clotting time [28].The mechanism of cathepsin S on thrombus formation in atherosclerosis development, however, needs to be further investigated.

9.5.4
Cathepsins in Plaque Instability

Rupture of atherosclerotic plaques is the most common cause for thrombotic complications, resulting in high morbidity and mortality [94]. Rupture-prone lesions are characterized by a large lipid core and a thin fibrous cap. Morphological studies showed that cathepsins were particularly expressed in macrophages in the shoulder regions of plaques, an area prone to rupture [34]. In a human study, comparison of gene expression in a stable versus a ruptured (an ulcerated surface with or without thrombosis or hemorrhage) atherosclerotic plaque obtained from the same patient, identified significantly greater amounts of cathepsin B (mRNA and protein) and cathepsin S (mRNA) in ruptured segments compared with stable segments [95]. Furthermore, cathepsin L was significantly increased in atherosclerotic plaques containing a large necrotic core and a ruptured plaque, suggesting that cathepsin L was involved in the development of unstable plaques [94]. It is likely that the presence of cathepsins in atherosclerotic lesions can degrade ECMs and thereby contribute to plaque vulnerability. In contrast, upregulation of cathepsin K was identified in advanced stable plaques compared with plaques containing thrombus [96].

9.6
Therapeutic Potential

Given their role in bone resorption, cathepsins have been considered valuable therapeutic targets in osteoporosis by pharmaceutical companies. Several cathepsin K–inhibiting compounds used in clinical trials to study their effects on osteoporosis or osteoarthritis have inhibited bone resorption and improved bone formation [97, 98]. The recent discovery of the involvement of cathepsins in atherosclerosis development directed attention of pharmaceutical companies to investigate the therapeutic potential of cathepsin inhibitors on the initiation and/or progression of atherosclerosis. Until now, one cathepsin K inhibitor compound has been patented and is claimed to be useful for treating or preventing atherosclerosis and atherosclerotic cardiovascular disease (http://www.wipo.int/pctdb/en/wo.jsp?WO=2006076797). However, before cathepsin K inhibitors are suitable for treating patients with cardiovascular disease, the mechanisms involved in its detrimental effects on foam cell formation should be further investigated.

9.7
Conclusions

In recent years, several studies have indicated important roles for cathepsins in the process of atherosclerosis. Their capacity to degrade ECMs paves the way for monocyte recruitment from the circulation into the intima and for SMCs to migrate from the media into the intima. In addition, cathepsins cooperate with adhesion molecules such as LFA-1 and $\alpha_v\beta_3$-integrin during leukocyte invasion and modulate cytoskeletal rearragengement. Within the atherosclerotic lesion itself, ECM degradation induces rupture of the fibrous cap. Cathepsins are also implicated in several inflammatory responses. Cathepsins are needed for processes involved in both innate immune reactions such as TLR3, 7, and 9 responses and adaptive responses by participating in antigen presentation, processes crucial in atherosclerosis. Furthermore, cathepsins might affect sprouting of neovessels of the existing vasa vasorum into the plaque, apoptosis, foam cell formation, and thrombus formation and thereby contribute to atherosclerotic plaque vulnerability.

As shown in this review, the different members of the cathepsin family are very versatile molecules, modulating proteolysis, inflammation, lipid uptake by macrophages, apoptosis, and coagulation, all processes known to work in concert during the process of atherosclerosis. Therefore, cathepsin antagonists are considered an attractive treatment for atherosclerosis. Several of these antagonists have been patented and further patient-related studies need to be performed to reveal the value of cathepsin antagonists in preventing morbidity and mortality as a result of clinical complications of atherosclerosis.

References

1 Ohtani, H. (1998) Stromal reaction in cancer tissue: pathophysiologic significance of the expression of matrix-degrading enzymes in relation to matrix turnover and immune/inflammatory reactions. *Pathol. Int.*, **48**, 1–9.

2 Weber, A.J. and De Bandt, M. (2000) Angiogenesis: general mechanisms and implications for rheumatoid arthritis. *Joint Bone Spine*, **67**, 366–383.

3 Chwieralski, C.E., Welte, T., and Buhling, F. (2006) Cathepsin-regulated apoptosis. *Apoptosis*, **11**, 143–149.

4 Chapman, H.A., Riese, R.J., and Shi, G.P. (1997) Emerging roles for cysteine proteases in human biology. *Annu. Rev. Physiol.*, **59**, 63–88.

5 Barrett, A.J.(1992) Cellular proteolysis. An overview. *Ann. N. Y. Acad. Sci.*, **674**, 1–15.

6 Mohamed, M.M. and Sloane, B.F. (2006) Cysteine cathepsins: multifunctional enzymes in cancer. *Nat. Rev. Cancer*, **6**, 764–775.

7 Sukhova, G.K., Shi, G.P., Simon, D.I., Chapman, H.A., and Libby, P. (1998) Expression of the elastolytic cathepsins S and K in human atheroma and regulation of their production in smooth muscle cells. *J. Clin. Invest.*, **102**, 576–583.

8 Kaakinen, R., Lindstedt, K.A., Sneck, M., Kovanen, P.T., and Oorni, K. (2007) Angiotensin II increases expression and secretion of cathepsin F in cultured human monocyte-derived macrophages: an angiotensin II type 2 receptor-mediated effect. *Atherosclerosis*, **192**, 323–327.

9 Li, W., Yuan, X.M., Olsson, A.G., and Brunk, U.T. (1998) Uptake of oxidized LDL by macrophages results in partial lysosomal enzyme inactivation and relocation. *Arterioscler. Thromb. Vasc. Biol.*, **18**, 177–184.

10 Cirman, T., Oresic, K., Mazovec, G.D., Turk, V., Reed, J.C., Myers, R.M., Salvesen, G.S., and Turk, B. (2004) Selective disruption of lysosomes in HeLa cells triggers apoptosis mediated by cleavage of Bid by multiple papain-like lysosomal cathepsins. *J. Biol. Chem.*, **279**, 3578–3587.

11 Goulet, B., Baruch, A., Moon, N.S., Poirier, M., Sansregret, L.L., Erickson, A., Bogyo, M., and Nepveu, A. (2004) A cathepsin L isoform that is devoid of a signal peptide localizes to the nucleus in S phase and processes the CDP/Cux transcription factor. *Mol. Cell*, **14**, 207–219.

12 Muntener, K., Zwicky, R., Csucs, G., Rohrer, J., and Baici, A. (2004) Exon skipping of cathepsin B: mitochondrial targeting of a lysosomal peptidase provokes cell death. *J. Biol. Chem.*, **279**, 41012–41017.

13 Abrahamson, M., Alvarez-Fernandez, M., and Nathanson, C.M. (2003) Cystatins. *Biochem. Soc. Symp.*, 179–199.

14 Hall, A., Ekiel, I., Mason, R.W., Kasprzykowski, F., Grubb, A., and Abrahamson, M. (1998) Structural basis for different inhibitory specificities of human cystatins C and D. *Biochemistry*, **37**, 4071–4079.

15 Dubin, G. (2005) Proteinaceous cysteine protease inhibitors. *Cell Mol. Life Sci.*, **62**, 653–669.

16 Kos, J., Krasovec, M., Cimerman, N., Nielsen, H.J., Christensen, I.J., and Brunner, N. (2000) Cysteine proteinase inhibitors stefin A, stefin B, and cystatin C in sera from patients with colorectal cancer: relation to prognosis. *Clin. Cancer Res.*, **6**, 505–511.

17 Lindahl, P., Nycander, M., Ylinenjarvi, K., Pol, E., and Bjork, I. (1992) Characterization by rapid-kinetic and equilibrium methods of the interaction between N-terminally truncated forms of chicken cystatin and the cysteine proteinases papain and actinidin. *Biochem. J.*, **286** (Pt 1), 165–171.

18 Stoch, S.A., Wagner, J.A., and Cathepsin, K. (2008) inhibitors: a novel target for osteoporosis therapy. *Clin. Pharmacol. Ther.*, **83**, 172–176.

19 Taleb, S., Lacasa, D., Bastard, J.P., Poitou, C., Cancello, R., Pelloux, V., Viguerie, N., Benis, A., Zucker, J.D., Bouillot, J.L., Coussieu, C., Basdevant, A., Langin, D., and Clement, K. (2005) Cathepsin S, a novel biomarker of adiposity: relevance to atherogenesis. *FASEB J.*, **19**, 1540–1542.

20 Taleb, S., Cancello, R., Clement, K., and Lacasa, D. (2006) Cathepsin s promotes human preadipocyte differentiation: possible involvement of fibronectin degradation. *Endocrinology*, **147**, 4950–4959.

21 Cheng, X.W., Obata, K., Kuzuya, M., Izawa, H., Nakamura, K., Asai, E., Nagasaka, T., Saka, M., Kimata, T., Noda, A., Nagata, K., Jin, H., Shi, G.P., Iguchi, A., Murohara, T., and Yokota, M. (2006) Elastolytic cathepsin induction/activation system exists in myocardium and is upregulated in hypertensive heart failure. *Hypertension*, **48**, 979–987.

22 Ge, J., Zhao, G., Chen, R., Li, S., Wang, S., Zhang, X., Zhuang, Y., Du, J., Yu, X., Li, G., and Yang, Y. (2006) Enhanced myocardial cathepsin B expression in patients with dilated cardiomyopathy. *Eur. J. Heart Fail.*, **8**, 284–289.

23 Nomura, T. and Katunuma, N. (2005) Involvement of cathepsins in the invasion, metastasis and proliferation of cancer cells. *J. Med. Invest.*, **52**, 1–9.

24 Gocheva, V., Zeng, W., Ke, D., Klimstra, D., Reinheckel, T., Peters, C., Hanahan, D., and Joyce, J.A. (2006) Distinct roles for cysteine cathepsin genes in multistage tumorigenesis. *Genes Dev.*, **20**, 543–556.

25 Diedrich, J.F., Minnigan, H., Carp, R.I., Whitaker, J.N., Race, R., Frey, W., 2nd, and Haase, A.T. (1991) Neuropathological changes in scrapie and Alzheimer's disease are associated with increased expression of apolipoprotein E and cathepsin D in astrocytes. *J. Virol.*, **65**, 4759–4768.

26 Baricos, W.H., Cortez, S.L., Le, Q.C., Zhou, Y.W., Dicarlo, R.M., O'Connor, S.E., and Shah, S.V. (1990) Glomerular basement membrane degradation by endogenous cysteine proteinases in isolated rat glomeruli. *Kidney Int.*, **38**, 395–401.

27 Lutgens, S.P., Cleutjens, K.B., Daemen, M.J., and Heeneman, S. (2007) Cathepsin cysteine proteases in cardiovascular disease. *FASEB J.*, **21**, 3029–3041.

28 Liu, J., Sukhova, G.K., Sun, J.S., Xu, W.H., Libby, P., and Shi, G.P. (2004) Lysosomal cysteine proteases in atherosclerosis. *Arterioscler. Thromb. Vasc. Biol.*, **24**, 1359–1366.

29 Chen, J., Tung, C.H., Mahmood, U., Ntziachristos, V., Gyurko, R., Fishman, M.C., Huang, P.L., and Weissleder, R. (2002) *In vivo* imaging of proteolytic activity in atherosclerosis. *Circulation*, **105**, 2766–2771.

30 Oorni, K., Sneck, M., Bromme, D., Pentikainen, M.O., Lindstedt, K.A., Mayranpaa, M., Aitio, H., and Kovanen, P.T. (2004) Cysteine protease cathepsin F is expressed in human atherosclerotic lesions, is secreted by cultured macrophages, and modifies low density lipoprotein particles *in vitro*. *J. Biol. Chem.*, **279**, 34776–34784.

31 Liu, J., Sukhova, G.K., Yang, J.T., Sun, J., Ma, L., Ren, A., Xu, W.H., Fu, H., Dolganov, G.M., Hu, C., Libby, P., and Shi, G.P. (2006) Cathepsin L expression and regulation in human abdominal aortic aneurysm, atherosclerosis, and vascular cells. *Atherosclerosis*, **184**, 302–311.

32 Lutgens, E., Lutgens, S.P., Faber, B.C., Heeneman, S., Gijbels, M.M., de Winther, M.P., Frederik, P., van der Made, I., Daugherty, A., Sijbers, A.M., Fisher, A., Long, C.J., Saftig, P., Black, D., Daemen, M.J., and Cleutjens, K.B. (2006) Disruption of the cathepsin K gene reduces atherosclerosis progression and induces plaque fibrosis but accelerates macrophage foam cell formation. *Circulation*, **113**, 98–107.

33 Sukhova, G.K., Zhang, Y., Pan, J.H., Wada, Y., Yamamoto, T., Naito, M., Kodama, T., Tsimikas, S., Witztum, J.L., Lu, M.L., Sakara, Y., Chin, M.T., Libby, P., and Shi, G.P. (2003) Deficiency of cathepsin S reduces atherosclerosis in LDL receptor-deficient mice. *J. Clin. Invest.*, **111**, 897–906.

34 Rodgers, K.J., Watkins, D.J., Miller, A.L., Chan, P.Y., Karanam, S., Brissette, W.H., Long, C.J., and Jackson, C.L. (2006) Destabilizing role of cathepsin S in murine atherosclerotic plaques. *Arterioscler. Thromb. Vasc. Biol.*, **26**, 851–856.

35 Samokhin, A.O., Wong, A., Saftig, P., and Bromme, D. (2008) Role of cathepsin K in structural changes in brachiocephalic artery during progression of atherosclerosis in apoE-deficient mice. *Atherosclerosis*, **200**, 58–68.

36 Guo, J., Bot, I., de Nooijer, R., Hoffman, S.J., Stroup, G.B., Biessen, E.A., Benson, G.M., Groot, P.H., Van Eck, M., Van Berkel, T.J. (2009) Leucocyte cathepsin K affects atherosclerotic lesion composition and bone mineral density in low-density lipoprotein receptor deficient mice. *Cardiovasc. Res.*, **81**, 278–285.

37 Shi, G.P., Sukhova, G.K., Grubb, A., Ducharme, A., Rhode, L.H., Lee, R.T., Ridker, P.M., Libby, P., and Chapman, H.A. (1999) Cystatin C deficiency in human atherosclerosis and aortic aneurysms. *J. Clin. Invest.*, **104**, 1191–1197.

38 Bengtsson, E., To, F., Hakansson, K., Grubb, A., Branen, L., Nilsson, J., and Jovinge, S. (2005) Lack of the cysteine protease inhibitor cystatin C promotes atherosclerosis in apolipoprotein E-deficient mice. *Arterioscler. Thromb. Vasc. Biol.*, **25**, 2151–2156.

39 Sukhova, G.K., Wang, B., Libby, P., Pan, J.H., Zhang, Y., Grubb, A., Fang, K., Chapman, H.A., and Shi, G.P. (2005) Cystatin C deficiency increases elastic lamina degradation and aortic dilatation in apolipoprotein E-null mice. *Circ. Res.*, **96**, 368–375.

40 Traub, O. and Berk, B.C. (1998) Laminar shear stress: mechanisms by which endothelial cells transduce an atheroprotective force. *Arterioscler. Thromb. Vasc. Biol.*, **18**, 677–685.

41 Platt, M.O., Ankeny, R.F., Shi, G.P., Weiss, D., Vega, J.D., Taylor, W.R., and Jo, H. (2007) Expression of cathepsin K is regulated by shear stress in cultured endothelial cells and is increased in endothelium in human atherosclerosis. *Am. J. Physiol. Heart Circ. Physiol.*, **292**, H1479–86.

42 Libby, P. (2000) Changing concepts of atherogenesis. *J. Intern. Med.*, **247**, 349–358.

43 Jevnikar, Z., Obermajer, N., Bogyo, M., and Kos, J. (2008) The role of cathepsin X in the migration and invasiveness of T lymphocytes. *J. Cell Sci.*, **121**, 2652–2661.

44 Davignon, D., Martz, E., Reynolds, T., Kurzinger, K., and Springer, T.A. (1981) Lymphocyte function-associated antigen 1 (LFA-1): a surface antigen distinct from Lyt-2,3 that participates in T lymphocyte-

mediated killing. *Proc. Natl Acad. Sci. U. S. A.*, **78**, 4535–4539.

45 Ross, R. (1993) The pathogenesis of atherosclerosis: a perspective for the 1990s. *Nature*, **362**, 801–809.

46 Yoshida, Y., Mitsumata, M., Ling, G., Jiang, J., and Shu, Q. (1997) Migration of medial smooth muscle cells to the intima after balloon injury. *Ann. N. Y. Acad. Sci.*, **811**, 459–470.

47 Kanda, S., Kuzuya, M., Ramos, M.A., Koike, T., Yoshino, K., Ikeda, S., and Iguchi, A. (2000) Matrix metalloproteinase and alphavbeta3 integrin-dependent vascular smooth muscle cell invasion through a type I collagen lattice. *Arterioscler. Thromb. Vasc. Biol.*, **20**, 998–1005.

48 Deryugina, E.I., Ratnikov, B., Monosov, E., Postnova, T.I., DiScipio, R., Smith, J.W., and Strongin, A.Y. (2001) MT1-MMP initiates activation of pro-MMP-2 and integrin alphavbeta3 promotes maturation of MMP-2 in breast carcinoma cells. *Exp. Cell Res.*, **263**, 209–223.

49 Ellerbroek, S.M., Fishman, D.A., Kearns, A.S., Bafetti, L.M., and Stack, M.S. (1999) Ovarian carcinoma regulation of matrix metalloproteinase-2 and membrane type 1 matrix metalloproteinase through beta1 integrin. *Cancer Res.*, **59**, 1635–1641.

50 Cheng, X.W., Kuzuya, M., Nakamura, K., Di, Q., Liu, Z., Sasaki, T., Kanda, S., Jin, H., Shi, G.P., Murohara, T., Yokota, M., and Iguchi, A. (2006) Localization of cysteine protease, cathepsin S, to the surface of vascular smooth muscle cells by association with integrin alphanu-beta3. *Am. J. Pathol.*, **168**, 685–694.

51 Katsuda, S. and Kaji, T. (2003) Atherosclerosis and extracellular matrix. *J. Atheroscler. Thromb.*, **10**, 267–274.

52 Reddy, V.Y., Zhang, Q.Y., and Weiss, S.J. (1995) Pericellular mobilization of the tissue-destructive cysteine proteinases, cathepsins B, L, and S, by human monocyte-derived macrophages. *Proc. Natl Acad. Sci. U. S. A.*, **92**, 3849–3853.

53 Barger, A.C., Beeuwkes, R., 3rd, Lainey, L.L., and Silverman, K.J. (1984) Hypothesis: vasa vasorum and neovascularization of human coronary arteries. A possible role in the pathophysiology of atherosclerosis. *N. Engl. J. Med.*, **310**, 175–177.

54 Moulton, K.S., Vakili, K., Zurakowski, D., Soliman, M., Butterfield, C., Sylvin, E., Lo, K.M., Gillies, S., Javaherian, K., and Folkman, J. (2003) Inhibition of plaque neovascularization reduces macrophage accumulation and progression of advanced atherosclerosis. *Proc. Natl Acad. Sci. U. S. A.*, **100**, 4736–4741.

55 Zhang, Y., Cliff, W.J., Schoefl, G.I., and Higgins, G. (1993) Immunohistochemical study of intimal microvessels in coronary atherosclerosis. *Am. J. Pathol.*, **143**, 164–172.

56 Packard, R.R. and Libby, P. (2008) Inflammation in atherosclerosis: from vascular biology to biomarker discovery and risk prediction. *Clin. Chem.*, **54**, 24–38.

57 Wang, B., Sun, J., Kitamoto, S., Yang, M., Grubb, A., Chapman, H.A., Kalluri, R., and Shi, G.P. (2006) Cathepsin S controls angiogenesis and tumor growth via matrix-derived angiogenic factors. *J. Biol. Chem.*, **281**, 6020–6029.

58 Sukhova, G.K., Schonbeck, U., Rabkin, E., Schoen, F.J., Poole, A.R., Billinghurst, R.C., and Libby, P. (1999) Evidence for increased collagenolysis by interstitial collagenases-1 and -3 in vulnerable human atheromatous plaques. *Circulation*, **99**, 2503–2509.

59 Joyce, J.A., Baruch, A., Chehade, K., Meyer-Morse, N., Giraudo, E., Tsai, F.Y., Greenbaum, D.C., Hager, J.H., Bogyo, M., and Hanahan, D. (2004) Cathepsin cysteine proteases are effectors of invasive growth and angiogenesis during multistage tumorigenesis. *Cancer Cell*, **5**, 443–453.

60 Shi, Q., Rafii, S., Wu, M.H., Wijelath, E.S., Yu, C., Ishida, A., Fujita, Y., Kothari, S., Mohle, R., Sauvage, L.R., Moore, M.A., Storb, R.F., and Hammond, W.P. (1998) Evidence for circulating bone marrow-derived endothelial cells. *Blood*, **92**, 362–367.

61 Kalka, C., Masuda, H., Takahashi, T., Kalka-Moll, W.M., Silver, M., Kearney, M., Li, T., Isner, J.M., and Asahara, T. (2000) Transplantation of ex vivo

expanded endothelial progenitor cells for therapeutic neovascularization. *Proc. Natl. Acad. Sci. U. S. A.*, **97**, 3422–3427.

62 Urbich, C., Heeschen, C., Aicher, A., Sasaki, K., Bruhl, T., Farhadi, M.R., Vajkoczy, P., Hofmann, W.K., Peters, C., Pennacchio, L.A., Abolmaali, N.D., Chavakis, E., Reinheckel, T., Zeiher, A.M., and Dimmeler, S. (2005) Cathepsin L is required for endothelial progenitor cell-induced neovascularization. *Nat. Med.*, **11**, 206–213.

63 Shi, G.P., Sukhova, G.K., Kuzuya, M., Ye, Q., Du, J., Zhang, Y., Pan, J.H., Lu, M.L., Cheng, X.W., Iguchi, A., Perrey, S., Lee, A.M., Chapman, H.A., and Libby, P. (2003) Deficiency of the cysteine protease cathepsin S impairs microvessel growth. *Circ. Res.*, **92**, 493–500.

64 Nilsson, J., Hansson, G.K., and Shah, P.K. (2005) Immunomodulation of atherosclerosis: implications for vaccine development. *Arterioscler. Thromb. Vasc. Biol.*, **25**, 18–28.

65 Tedgui, A. and Mallat, Z. (2006) Cytokines in atherosclerosis: pathogenic and regulatory pathways. *Physiol. Rev.*, **86**, 515–581.

66 Matsumoto, F., Saitoh, S., Fukui, R., Kobayashi, T., Tanimura, N., Konno, K., Kusumoto, Y., Akashi-Takamura, S., and Miyake, K. (2008) Cathepsins are required for Toll-like receptor 9 responses. *Biochem. Biophys. Res. Commun.*, **367**, 693–699.

67 Asagiri, M., Hirai, T., Kunigami, T., Kamano, S., Gober, H.J., Okamoto, K., Nishikawa, K., Latz, E., Golenbock, D.T., Aoki, K., Ohya, K., Imai, Y., Morishita, Y., Miyazono, K., Kato, S., Saftig, P., and Takayanagi, H. (2008) Cathepsin K-dependent toll-like receptor 9 signaling revealed in experimental arthritis. *Science*, **319**, 624–627.

68 Edfeldt, K., Swedenborg, J., Hansson, G.K., and Yan, Z.Q. (2002) Expression of toll-like receptors in human atherosclerotic lesions: a possible pathway for plaque activation. *Circulation*, **105**, 1158–1161.

69 Riese, R.J., and Chapman, H.A. (2000) Cathepsins and compartmentalization in antigen presentation. *Curr. Opin. Immunol.*, **12**, 107–113.

70 Bania, J., Gatti, E., Lelouard, H., David, A., Cappello, F., Weber, E., Camosseto, V., and Pierre, P. (2003) Human cathepsin S, but not cathepsin L, degrades efficiently MHC class II-associated invariant chain in non-professional APCs. *Proc. Natl Acad. Sci. U. S. A.*, **100**, 6664–6669.

71 Beers, C., Burich, A., Kleijmeer, M.J., Griffith, J.M., Wong, P., and Rudensky, A.Y. (2005) Cathepsin S controls MHC class II-mediated antigen presentation by epithelial cells *in vivo*. *J. Immunol.*, **174**, 1205–1212.

72 Riese, R.J., Mitchell, R.N., Villadangos, J.A., Shi, G.P., Palmer, J.T., Karp, E.R., De Sanctis, G.T., Ploegh, H.L., and Chapman, H.A. (1998) Cathepsin S activity regulates antigen presentation and immunity. *J. Clin. Invest.*, **101**, 2351–2363.

73 Saegusa, K., Ishimaru, N., Yanagi, K., Arakaki, R., Ogawa, K., Saito, I., Katunuma, N., and Hayashi, Y. (2002) Cathepsin S inhibitor prevents autoantigen presentation and autoimmunity. *J. Clin. Invest.*, **110**, 361–369.

74 Hansson, G.K., Libby, P., Schonbeck, U., and Yan, Z.Q. (2002) Innate and adaptive immunity in the pathogenesis of atherosclerosis. *Circ. Res.*, **91**, 281–291.

75 Yla-Herttuala, S., Palinski, W., Butler, S.W., Picard, S., Steinberg, D., and Witztum, J.L. (1994) Rabbit and human atherosclerotic lesions contain IgG that recognizes epitopes of oxidized LDL. *Arterioscler. Thromb.*, **14**, 32–40.

76 Palinski, W., Tangirala, R.K., Miller, E., Young, S.G., and Witztum, J.L. (1995) Increased autoantibody titers against epitopes of oxidized LDL in LDL receptor-deficient mice with increased atherosclerosis. *Arterioscler. Thromb. Vasc. Biol.*, **15**, 1569–1576.

77 Caligiuri, G., Nicoletti, A., Poirier, B., and Hansson, G.K. (2002) Protective immunity against atherosclerosis carried by B cells of hypercholesterolemic mice. *J. Clin. Invest.*, **109**, 745–753.

78 Lutgens, S.P., Kisters, N., Lutgens, E., van Haaften, R.I., Evelo, C.T., de Winther, M.P., Saftig, P., Daemen, M.J., Heeneman, S., and Cleutjens, K.B. (2006) Gene profiling of cathepsin K deficiency in atherogenesis: profibrotic but lipogenic. *J. Pathol.*, **210**, 334–343.

79 Lutgens, E., Gijbels, M., Smook, M., Heeringa, P., Gotwals, P., Koteliansky, V.E., and Daemen, M.J. (2002) Transforming growth factor-beta mediates balance between inflammation and fibrosis during plaque progression. *Arterioscler. Thromb. Vasc. Biol.*, **22**, 975–982.

80 Zdolsek, J.M., Olsson, G.M., and Brunk, U.T. (1990) Photooxidative damage to lysosomes of cultured macrophages by acridine orange. *Photochem. Photobiol.*, **51**, 67–76.

81 Vancompernolle, K., Van Herreweghe, F., Pynaert, G., Van de Craen, M., De Vos, K., Totty, N., Sterling, A., Fiers, W., Vandenabeele, P., and Grooten, J. (1998) Atractyloside-induced release of cathepsin B, a protease with caspase-processing activity. *FEBS Lett.*, **438**, 150–158.

82 Ishisaka, R., Utsumi, T., Kanno, T., Arita, K., Katunuma, N., Akiyama, J., and Utsumi, K. (1999) Participation of a cathepsin L-type protease in the activation of caspase-3. *Cell Struct. Funct.*, **24**, 465–470.

83 Li, W., Dalen, H., Eaton, J.W., and Yuan, X.M. (2001) Apoptotic death of inflammatory cells in human atheroma. *Arterioscler. Thromb. Vasc. Biol.*, **21**, 1124–1130.

84 Li, W., Kornmark, L., Jonasson, L., Forssell, C., and Yuan, X.M. (2009) Cathepsin L is significantly associated with apoptosis and plaque destabilization in human atherosclerosis. *Atherosclerosis*, **202**, 92–102.

85 Zheng, T., Kang, M.J., Crothers, K., Zhu, Z., Liu, W., Lee, C.G., Rabach, L.A., Chapman, H.A., Homer, R.J., Aldous, D., De Sanctis, G.T., Underwood, S., Graupe, M., Flavell, R.A., Schmidt, J.A., and Elias, J.A. (2005) Role of cathepsin S-dependent epithelial cell apoptosis in IFN-gamma-induced alveolar remodeling and pulmonary emphysema. *J. Immunol.*, **174**, 8106–8115.

86 Conus, S. and Simon, H.U. (2008) Cathepsins: key modulators of cell death and inflammatory responses. *Biochem. Pharmacol.*, **76**, 1374–1382.

87 Kagedal, K., Johansson, U., and Ollinger, K. (2001) The lysosomal protease cathepsin D mediates apoptosis induced by oxidative stress. *FASEB J.*, **15**, 1592–1594.

88 Li, W. and Yuan, X.M. (2004) Increased expression and translocation of lysosomal cathepsins contribute to macrophage apoptosis in atherogenesis. *Ann. N. Y. Acad. Sci.*, **1030**, 427–433.

89 Tertov, V.V. and Orekhov, A.N. (1997) Metabolism of native and naturally occurring multiple modified low density lipoprotein in smooth muscle cells of human aortic intima. *Exp. Mol. Pathol.*, **64**, 127–145.

90 Lindstedt, L., Lee, M., Oorni, K., Bromme, D., and Kovanen, P.T. (2003) Cathepsins F and S block HDL3-induced cholesterol efflux from macrophage foam cells. *Biochem. Biophys. Res. Commun.*, **312**, 1019–1024.

91 Sambrano, G.R., Huang, W., Faruqi, T., Mahrus, S., Craik, C., and Coughlin, S.R. (2000) Cathepsin G activates protease-activated receptor-4 in human platelets. *J. Biol. Chem.*, **275**, 6819–6823.

92 Herrmann, S.M., Funke-Kaiser, H., Schmidt-Petersen, K., Nicaud, V., Gautier-Bertrand, M., Evans, A., Kee, F., Arveiler, D., Morrison, C., Orzechowski, H.D., Elbaz, A., Amarenco, P., Cambien, F., and Paul, M. (2001) Characterization of polymorphic structure of cathepsin G gene: role in cardiovascular and cerebrovascular diseases. *Arterioscler. Thromb. Vasc. Biol.*, **21**, 1538–1543.

93 Iacoviello, L., Kolpakov, V., Salvatore, L., Amore, C., Pintucci, G., de Gaetano, G., and Donati, M.B. (1995) Human endothelial cell damage by neutrophil-derived cathepsin G. Role of cytoskeleton rearrangement and matrix-bound plasminogen activator inhibitor-1. *Arterioscler. Thromb. Vasc. Biol.*, **15**, 2037–2046.

94 Fuster, V., Moreno, P.R., Fayad, Z.A., Corti, R., and Badimon, J.J. (2005) Atherothrombosis and high-risk plaque: part I: evolving concepts. *J. Am. Coll. Cardiol.*, **46**, 937–954.

95 Papaspyridonos, M., Smith, A., Burnand, K.G., Taylor, P., Padayachee, S., Suckling, K.E., James, C.H., Greaves,

D.R., and Patel, L. (2006) Novel candidate genes in unstable areas of human atherosclerotic plaques. *Arterioscler. Thromb. Vasc. Biol.*, **26**, 1837–1844.

96 Faber, B.C., Cleutjens, K.B., Niessen, R.L., Aarts, P.L., Boon, W., Greenberg, A.S., Kitslaar, P.J., Tordoir, J.H., and Daemen, M.J. (2001) Identification of genes potentially involved in rupture of human atherosclerotic plaques. *Circ. Res.*, **89**, 547–554.

97 Abbenante, G. and Fairlie, D.P. (2005) Protease inhibitors in the clinic. *Med. Chem.*, **1**, 71–104.

98 Yasuda, Y., Kaleta, J., and Bromme, D. (2005) The role of cathepsins in osteoporosis and arthritis: rationale for the design of new therapeutics. *Adv. Drug. Deliv. Rev.*, **57**, 973–993.

10
The Plasmin System and Atherosclerosis

Christopher L. Jackson and Kevin G.S. Carson

10.1
Introduction

Activation of the fibrinolytic system is dependent on the conversion of the abundant plasma protein plasminogen to the serine proteinase plasmin. This conversion is achieved through the physiological activators urokinase-type plasminogen activator (uPA) or tissue-type plasminogen activator (tPA). The primary *in vivo* function of plasmin is to protect vascular patency by degrading fibrin within thrombi. However, the identification of receptors for plasminogen and uPA, and the ability of plasmin to degrade proteins in the extracellular matrix, have implicated plasmin in other physiological and pathological processes. This chapter deals with the involvement of the plasmin system in atherosclerosis.

10.2
Components of the Plasmin System

10.2.1
Tissue-Type Plasminogen Activator

tPA is produced by endothelial cells and smooth muscle cells. In its native form it is a single-chain glycoprotein of molecular mass ~70 kDa [1]. As is true of uPA, tPA can be cleaved by plasmin to a two-chain form linked by a single disulfide bond (Figure 10.1). The N-terminal region has a finger domain followed by a growth factor domain and two kringles. The C-terminal region contains the serine protease domain. Unlike uPA, single-chain tPA has significant enzymatic activity [2]. The finger and kringle domains are important for the binding of tPA to fibrin, and this binding increases the plasminogen-activating activity of tPA up to 1000-fold [3].

Atherosclerosis: Molecular and Cellular Mechanisms. Edited by Sarah Jane George and Jason Johnson
Copyright © 2010 WILEY-VCH Verlag GmbH & Co. KGaA, Weinheim
ISBN: 978-3-527-32448-4

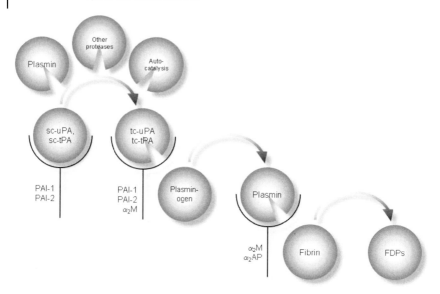

Figure 10.1 Schematic diagram of the plasmin system. Single-chain urokinase-type plasminogen activator (uPA) and tissue-type plasminogen activator (tPA) can be converted to the active two-chain forms by plasmin, many other proteinases, and also through autocatalysis. The two-chain forms can cleave plasminogen to form plasmin, which degrades fibrin. sc, single-chain; tc, two-chain; α_2AP, α_2-antiplasmin; α_2M, α_2-macroglobulin; FDPs, fibrin-degradation products.

10.2.2
Urokinase-Type Plasminogen Activator

In its native form, uPA is a single-chain glycoprotein of ~54 kDa, also known as pro-uPA. It is activated to the two-chain form by cleavage of a single peptide bond, which can be catalyzed by plasmin, kallikrein, or other proteinases [4, 5]. The tertiary structure of uPA comprises three different domains. The N-terminal region comprises a growth factor domain followed by a kringle domain, which together form the 135-amino-acid N-terminal fragment (ATF). The binding site for uPA to its receptor is located within the growth factor domain of the ATF fragment. In the C-terminal region uPA has a classical serine proteinase catalytic domain. Two-chain uPA exists in a high-molecular-mass form (~54 kDa) and a low-molecular-mass form (~33 kDa), the latter lacking the ATF and thus not able to bind to the uPA receptor.

10.2.3
uPA Receptor

Both single-chain and high-molecular-mass uPA bind to the uPA receptor, uPAR, with the same subnanomolar affinity [6]. uPAR also binds vitronectin with similar affinity.

10.2.4
Plasminogen

Human plasminogen is synthesized as an 810-amino-acid polypeptide. Cleavage of a 19-amino-acid leader peptide during secretion yields the 791-amino-acid mature form of the protein [7]. The liver is the primary source of synthesis of plasminogen. The conversion of plasminogen to plasmin usually involves cleavage at Arg561–Val562, resulting in the generation of an N-terminal heavy chain of 561 amino acids and a disulfide-linked C-terminal light chain of 230 amino acids.

10.2.5
Plasmin

The serine proteinase plasmin, of molecular mass ~90 kDa, comprises two disulfide bond-linked polypeptide chains. The N-terminal A chain contains five kringle domains while the C-terminal B chain contains a typical serine proteinase domain, which is responsible for the catalytic activity of plasmin [8]. Plasmin cleaves fibrin to generate soluble degradation products, exposing C-terminal lysine residues which bind to kringles in tPA and plasminogen. This leads to further binding to fibrin, promoting increased plasmin generation and fibrinolysis.

10.2.6
Plasminogen Activator Inhibitor-1

Plasminogen activator inhibitor-1 (PAI-1) has a molecular mass of ~54 kDa. It can exist in a number of forms: its native inhibitory form; an inactive latent form, in complexes with proteinases; and in a cleaved substrate form [3]. PAI-1 forms complexes with both single-chain and two-chain tPA, and with two-chain uPA [9]. Receptor-bound active uPA is inhibited by both PAI-1 and PAI-2 [10], which leads to the inhibition of plasminogen activation. PAI-1 also directly inhibits plasmin [11]. PAI-1 is stabilized by binding to vitronectin [12].

10.2.7
Plasminogen Activator Inhibitor-2

Plasminogen activator inhibitor-2 (PAI-2) protein exists in two forms: a nonglycosylated intracellular form and a glycosylated, secreted extracellular form [3]. The molecular mass of the intracellular form is ~42 kDa and that of the secreted form ~60 kDa [80]. PAI-2 inhibits uPA rapidly but tPA very slowly [13].

10.2.8
Endogenous Plasmin Inhibitors

The plasmin inhibitor α_2-antiplasmin circulates in plasma at high concentration. It consists of 452 amino acids with two disulfide bridges [14], and is also a constituent of platelet α-granules [15]. Plasmin released in the vicinity of a platelet-rich thrombus is therefore rapidly neutralized by $α_2$-antiplasmin.

$α_2$-Macroglobulin is a large dimeric protein synthesized by endothelial cells and macrophages, and like $α_2$-antiplasmin is found in platelet α-granules. It forms noncovalent complexes with plasmin and is about 10% as efficient at inhibiting plasmin as $α_2$-antiplasmin [16].

10.2.9
Thrombin-Activatable Fibrinolysis Inhibitor

Thrombin-activatable fibrinolysis inhibitor (TAFI) is synthesized by the liver and is also present in platelets: its activation by thrombin is accelerated in the presence of thrombomodulin more than 1000-fold [5]. Activated TAFI has high specificity for C-terminal lysine residues. Because these are binding sites for plasminogen and tPA on fibrin, activated TAFI is a potent attenuator of fibrinolysis [17].

10.3
Involvement of the Plasmin System in Atherosclerosis

Studies of the involvement of plasmin system components in atherosclerosis have been hampered by the lack of specific inhibitory drugs. Therefore, most progress in animal studies has been made by the use of genetic alteration, and in human studies by investigation of the plasma concentrations of plasmin system components in patients suffering from conditions caused by atherosclerosis, such as coronary artery disease. However, early interest in the field was directed mainly at examination of the phenotypic effects in humans of spontaneous gene polymorphisms.

10.3.1
Human Gene Polymorphisms

There are no functional mutations of the plasmin system that are clearly associated with a change in the severity or rate of progression of atherosclerosis. Though several of these polymorphisms are characterized by an increased incidence of either bleeding or inappropriate thrombosis [18–20], confirming their functional importance, these do not seem to translate into a measurable change in the incidence of atherosclerosis-related clinical events. To some degree this mirrors what has been seen in otherwise normal mice with null mutations to plasmin system components. In general, these animals develop normally and their phenotypes relate to thrombosis and fibrinolysis much more clearly than they do to atherosclerosis.

10.3.2
Clinical Association Studies

The FINRISK'92 study reported an independent association between elevated plasma plasminogen levels and future cardiovascular events [21], but a subsequent case–control study failed to confirm this surprising and difficult to explain finding [22].

Patients with coronary artery disease have, also surprisingly, elevated plasma levels of antigenically detectable tPA [23]. This may be due to an increase in tPA/PAI-1 complexes secondary to increased PAI-1 plasma levels [24], since the local PAI-1 concentration can increase tenfold because of release from platelets at sites of thrombus formation [25]. The first suggestion of a link between increased PAI-1 levels and cardiovascular disease came from the Northwick Park Heart Study, which showed a strong correlation between low fibrinolytic activity at enrollment and the subsequent occurrence of coronary artery disease [26]. PAI-1 expression is also increased in human atherosclerotic lesions [27, 28], and it has been suggested that this promotes the progression of atherosclerosis by inhibiting the clearance of fibrin that has become incorporated into plaques [29].

In contrast to tPA, very little is known about associations between plasma levels of uPA and the incidence of coronary artery disease [30], most likely because uPA is produced in tissues and little is present in the circulation. Consistent with this, the level of expression of uPA in human coronary arteries is correlated with atherosclerosis severity [31].

Can we gain information about the possible roles of plasmin system components in atherosclerosis by examining outcomes after therapeutic thrombolysis? For instance, could therapeutic administration of tPA, uPA, or synthetic plasminogen activators threaten the stability of other vulnerable plaques in the same patient? There is some evidence that thrombolytic therapy does indeed cause significantly higher rates of early recurrent myocardial ischemia and infarction than angioplasty and stenting [32–34], suggesting that plasminogen activators are potential mediators of plaque destabilization.

10.3.3
Studies in Mice

Data from studies in mice are summarized in Table 10.1.

10.3.3.1 Plasminogen
Complete plasminogen deficiency in mice, achieved by targeted null mutation, is not lethal and in fact the majority of these animals survive into adulthood [45, 46]. Phenotypes associated with a lack of plasminogen include reduced growth, spontaneous fibrin deposition in some organs (though not arteries), gastric ulceration, and rectal prolapse. (It must be noted that descriptions of fibrin deposition in mice generally should be interpreted as depositions of fibrinogen and/or fibrin, as most antibodies do not distinguish between the two murine proteins and electron

Table 10.1 Summary of studies of the plasmin system in mouse models of atherosclerosis.

Gene	Modification	Model	Diet	Time (weeks)	Site	Plaque size	Plaque stability	Reference
Plasminogen	Knockout	ApoE−/−	Chow	18–25	Aortic arch	↑	ND	[35]
					Aortic sinus	↑	ND	
tPA	Knockout	ApoE3-Leiden	Atherogenic	12	Aortic sinus	→	↑	[36]
	Knockout	ApoE−/−	Atherogenic	8	Brachiocephalic artery	↔	↔	a
				40		↔	↑	
uPA	Knockout	ApoE−/−	Atherogenic	30	Aortic sinus	↔	↔	[37]
	Knockout	ApoE3-Leiden	Atherogenic	12	Aortic sinus	↔	ND	[36]
	Transgenic	ApoE−/−	Atherogenic	10	Aortic sinus	↑	↔	[38]
					Aorta	↑	↔	
	Knockout	ApoE−/−	Atherogenic	8	Brachiocephalic artery	→	↑	b
				40		↑	↑	
	Knockout	ApoE−/−	Chow	NA	Brachiocephalic artery	↑	→	[39]
PAI-1	Knockout	ApoE−/−	Atherogenic	25	Aorta	↑	→	[40]
	Knockout	ApoE−/−	Atherogenic	25	Brachiocephalic artery	↔	→	c
	Knockout	ApoE3-Leiden	Atherogenic	12	Aortic sinus	↔	ND	[36]
	Knockout	ApoE−/−	Atherogenic	6, 15, 30	Aortic sinus	↔	↔	[41]
	Transgenic					↔	↔	
	Knockout	ApoE−/−	Chow	52	Carotid artery	→	ND	[42]
					Aortic arch	↔	ND	
	Knockout	ApoE−/−	Atherogenic	25	Aorta	ND	→	[43]
	Transgenic	ApoE−/−	Atherogenic	20	Aortic sinus	↔	→	[44]

a) Unpublished observations (K.G.S. Carson and C.L. Jackson).
b) Unpublished observations (K.G.S. Carson, P. Carmeliet, and C.L. Jackson).
c) Unpublished observations (K.G.S. Carson, A. Luttun, P. Carmeliet, and C.L. Jackson).

microscopic confirmation of the typical fibrillar structure of fibrin is rarely performed.)

In the apolipoprotein E (ApoE) knockout mouse model of atherosclerosis, accelerated lesion formation was reported in the aortic arch and aortic sinus in mice that were additionally deficient in plasminogen [35]. However, analysis of the aortic arch was restricted to subjective scoring of lesion severity, and in the aortic sinus to quantitative morphometry of Oil Red O-stained sections. It is clear in the published images that major lesions were present in the brachiocephalic arteries of the double knockouts, so it is disappointing that careful histological examination of this region was not carried out. All we can conclude with certainty is that lipid deposition in the aortic arch was increased about 15-fold in the double knockouts. It is worth mentioning as an aside that this paper represents the first description of a complex atherosclerotic plaque in the brachiocephalic artery of an ApoE knockout mouse.

10.3.3.2 tPA
In fat-fed ApoE3-Leiden mice, a transgenic model of atherosclerosis, deletion of tPA resulted in a marked decrease in lesion area and lesion severity at the aortic root [36].

We have carried out our own studies in this area (K.G.S. Carson, P. Carmeliet, and C.L. Jackson, unpublished observations: Figure 10.2). These show that the effects of deletion of tPA in fat-fed ApoE knockout mice were manifested most strongly over a longer time-course. In animals terminated after 40 weeks of fat-feeding, plaque size was unchanged but there were significant reductions in the number of acute plaque ruptures and healed plaque ruptures. These changes were accompanied by a doubling of fibrous cap thickness.

These data strongly suggest that tPA is involved in plaque destablization, and support the notion that exogenous tPA might contribute to the destabilization of "bystander" plaques during therapeutic thrombolysis.

10.3.3.3 uPA
Mice deficient in uPA develop rectal prolapses and have a tendency towards poor healing of minor skin wounds, but are otherwise grossly normal [47]. ApoE knockout mice, when fed a high-fat, high-cholesterol diet, develop small defects at the outer border of the tunica media that have been described as micro-aneurysms: gene deletion of uPA significantly reduces the occurrence of these structures but has no effect on the size or complexity of lesions in the aortic root [37].

Similar conclusions were reached in a study of aortic root lesions in fat-fed ApoE3-Leiden mice [36]: these data suggest that uPA is unimportant in the processes of atherogenesis and lesion development, insofar as these can be modeled at the murine aortic root.

Another study reached an apparently contradictory conclusion: that "macrophage-expressed uPA contributes to the progression and complications of atherosclerosis" [38]. This was based on a study in ApoE knockout mice in which macrophages were engineered to overexpress murine uPA. The activity of uPA in

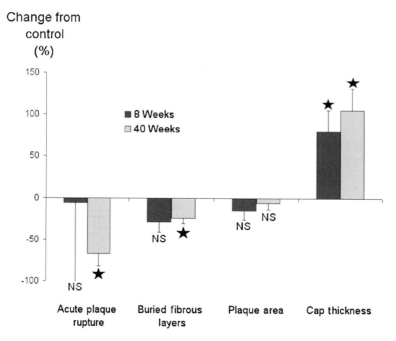

Figure 10.2 Brachiocephalic artery atherosclerosis in ApoE/tPA double knockout mice. Bars represent mean percentage changes from ApoE single knockout controls, with associated standard error bars. Black bars: study conducted for eight weeks of high-fat feeding. Gray bars: study conducted for 40 weeks of high-fat feeding. $*P < 0.05$.

peritoneal macrophages was fully 500-fold higher in the transgenic animals than in ApoE knockout controls. Activity in the aorta was roughly doubled, and there was very significant early mortality with a maximum survival in the transgenic animals of about 30 weeks. Interpreting effects on atherosclerosis in the context of these major physiological changes is difficult, but this study can be said not to rule out the possibility that uPA is involved in lesion expansion in the murine aortic root. Unfortunately, there are good reasons to doubt the suitability of this site of interrogation of atherosclerosis in mice [48].

Because of concerns about the usefulness of studies of atherosclerosis that focus on the murine aortic root, we conducted our own studies on the development of atherosclerosis in the brachiocephalic arteries of ApoE/uPA double knockout mice (K.G.S. Carson, P. Carmeliet, and C.L. Jackson, unpublished observations). The data from these studies are puzzling because different effects were seen at different time-points. As shown in Figure 10.3, plaque area was significantly decreased in the double knockouts after eight weeks of fat-feeding, but was significantly increased after 40 weeks. Despite this, the number of acute plaque ruptures and the number of healed plaque ruptures (seen as buried fibrous caps, [49]) were significantly reduced at both time-points. In some ways this points up the serious limitations inherent in assessing atherosclerosis simply in terms of plaque size:

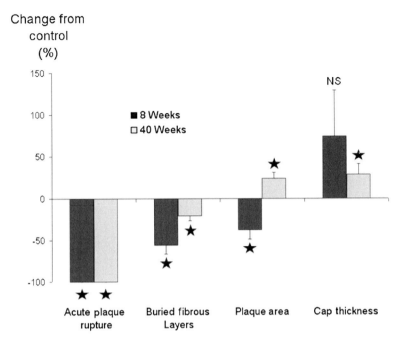

Figure 10.3 Brachiocephalic artery atherosclerosis in ApoE/uPA double knockout mice. Bars represent mean percentage changes from ApoE single knockout controls, with associated standard error bars. Black bars: study conducted for eight weeks of high-fat feeding. Gray bars: study conducted for 40 weeks of high-fat feeding. $*P < 0.05$.

there is no strong correlation between coronary artery plaque size and symptomatology in humans [50], so plaque cross-sectional area is not an appropriate surrogate for disease severity. We therefore conclude from our studies that uPA is an important mediator of plaque destablization but its role in plaque expansion varies across the course of plaque development.

Again, apparently contradictory data have been reported [39]. In the brachiocephalic arteries of chow-fed ApoE/uPA double knockout mice, there were said to be significantly fewer intraplaque hemorrhages and buried fibrous layers. However, the hemorrhages were detected using an antibody that does not distinguish between fibrinogen and fibrin, so may simply reflect increased endothelial permeability with associated accumulation of fibrinogen; these data must be disregarded. The buried fibrous layers appear to have been miscounted, and the incidence of acute plaque rupture did not differ significantly between groups. This appears to be a case of somewhat overenthusiastic reporting of data: it is probably still safe to conclude that uPA plays a destructive role during plaque evolution in hypercholesterolemic mice, favoring destablization and eventual rupture.

10.3.3.4 PAI-1

PAI-1 is expressed at high levels in murine plaques [51], so the finding that uPA is probably a destructive factor in plaques suggests that knocking out PAI-1, its major physiological inhibitor, ought to increase the incidence of plaque rupture. This is indeed what is seen.

ApoE/PAI-1 double knockout mice fed a high-fat diet for 25 weeks had significantly larger plaques in their thoracic and abdominal aortas [40]. Collagen bundles in the plaques were disorganized, suggesting that plaque stability may be compromised. To test this hypothesis, we examined brachiocephalic arteries from the same mice (K.G.S. Carson, A. Luttun, P. Carmeliet, C.L. Jackson, unpublished observations). Although plaque size was unchanged at this site in the double knockouts, the incidence of acute plaque rupture was increased 6.5-fold ($P < 0.05$).

Plaque size is also unaffected by PAI-1 gene deletion in ApoE3-Leiden mice [36], or at the aortic root in ApoE/PAI-1 double knockouts or ApoE mice engineered to overexpress PAI-1 [41]. There is one published report of reduced plaque size in aged chow-fed ApoE/PAI-1 double knockout mice, at the carotid bifurcation [42].

Further support for the idea that PAI-1 is a protective factor comes from studies of rates of cellular apoptosis in plaques of ApoE/PAI-1 double knockout mice. Animals were fed a cholesterol-enriched diet for 25 weeks and plaques were assessed in the thoracic and abdominal aortas: rates of apoptosis were increased about sixfold [43].

However, directed overexpression of murine PAI-1 in the smooth muscle cells of fat-fed ApoE knockout mice resulted in a slight but significant decrease in cellularity in aortic sinus lesions, suggesting that increasing the levels of PAI-1 might reduce plaque stability [44]. Such overexpression studies are hard to interpret, and this result does not necessarily disqualify the notion that PAI-1 is a protective factor in the plaque.

Overall, these data suggest that PAI-1 is protective in terms of the stability, but perhaps not the growth, of plaques, and are consistent with the idea that limiting uPA activity is an effective way of inhibiting plaque rupture.

10.4
Summary and Conclusions

The ubiquitous nature of the involvement in plasmin in conditions where extracellular matrix turnover is increased mean that is no surprise that this system also participates in atherosclerosis. Resolving the precise nature of its involvement has, however, been problematical mainly because of the lack of specific inhibitory drugs.

Studies in genetically altered mice have provided the greatest insights, but many have been compromised by the common assumption that the severity of atherosclerosis can be assessed by measuring plaque size, and furthermore that the aortic sinus is a suitable place for such assessment. It is important to stress that atherosclerosis has clinical significance because plaques rupture, not because they grow large, so changes in the stability of lesions are of much more physiological consequence than changes in size. What is more, none of the published studies utilizing genetically altered mice has involved conditional knockouts, so alterations in atherosclerosis could have been manifested at any stage of development including during embryogenesis. It is not possible in studies of constitutive knockouts to say with certainty that the changes seen are the result of the involvement of the specific gene in a defined process.

The data that are available allow the tentative conclusion that uPA appears to promote plaque rupture, and that inhibition of this factor might be a useful therapeutic avenue. Small molecule uPA inhibitors are beginning to appear [52] and some have undergone successful phase I clinical trials for cancer (http://www.wilex.de/News/2008/250108.htm). Studies of the actions of such drugs in mouse models of atherosclerosis would permit tighter resolution of the involvement of uPA in atherogenesis, plaque development, and plaque rupture.

References

1 Pennica, D., Holmes, W.E., Kohr, W.J., Harkins, R.N., Vehar, G.A., Ward, C.A., Bennett, W.F., Yelverton, E., Seeburg, P.H., Heyneker, H.L., Goeddel, D.V., and Collen, D. (1983) Cloning and expression of human tissue-type plasminogen activator cDNA in *E. coli. Nature*, **301**, 214–221.

2 Tachias, K. and Madison, E.L. (1996) Converting tissue-type plasminogen activator into a zymogen. *J. Biol. Chem.*, **271**, 28749–28752.

3 Myöhänen, H. and Vaheri, A. (2004) Regulation and interactions in the activation of cell-associated plasminogen. *Cell. Mol. Life Sci.*, **61**, 2840–2858.

4 Wun, T.C., Ossowski, L., and Reich, E. (1982) A proenzyme form of human urokinase. *J. Biol. Chem.*, **257**, 7262–7268.

5 Cesarman-Maus, G. and Hajjar, K.A. (2005) Molecular mechanisms of fibrinolysis. *Br. J. Haematol.*, **129**, 307–321.

6 Cubellis, M.V., Nolli, M.L., Cassani, G., and Blasi, F. (1986) Binding of single-chain prourokinase to the urokinase receptor of human U937 cells. *J. Biol. Chem.*, **261**, 15819–15822.

7 Castellino, F.J. and Ploplis, V.A. (2005) Structure and function of the plasminogen/plasmin system. *Thromb. Haemost.*, **93**, 647–654.

8 Andreasen, P.A., Egelund, R., and Petersen, H.H. (2000) The plasminogen activation system in tumor growth, invasion, and metastasis. *Cell. Mol. Life Sci.*, **57**, 25–40.

9 Andreasen, P.A., Nielsen, L.S., Kristensen, P., Grøndahl-Hansen, J., Skriver, L., and Danø, K. (1986) Plasminogen activator inhibitor from human fibrosarcoma cells binds urokinase-type plasminogen activator, but not its proenzyme. *J. Biol. Chem.*, **261**, 7644–7651.

10 Ellis, V., Wun, T.C., Behrendt, N., Rønne, E., and Danø, K. (1990) Inhibition of receptor-bound urokinase by plasminogen-activator inhibitors. *J. Biol. Chem.*, **265**, 9904–9908.

11 Hekman, C.M. and Loskutoff, D.J. (1988) Bovine plasminogen activator inhibitor 1: specificity determinations and comparison of the active, latent, and guanidine-activated forms. *Biochemistry*, **27**, 2911–2918.

12 Lijnen, H.R. (2005) Pleiotropic functions of plasminogen activator inhibitor-1. *J. Thromb. Haemost.*, **3**, 35–45.

13 Andreasen, P.A., Georg, B., Lund, L.R., Riccio, A., and Stacey, S.N. (1990) Plasminogen activator inhibitors: hormonally regulated serpins. *Mol. Cell. Endocrinol.*, **68**, 1–19.

14 Holmes, W.E., Nelles, L., Lijnen, H.R., and Collen, D. (1987) Primary structure of human alpha 2-antiplasmin, a serine protease inhibitor (serpin). *J. Biol. Chem.*, **262**, 1659–1664.

15 Plow, E.F. and Collen, D. (1981) The presence and release of alpha 2-

antiplasmin from human platelets. *Blood*, **58**, 1069–1074.

16 Aoki, N., Moroi, M., and Tachiya, K. (1978) Effects of alpha2-plasmin inhibitor on fibrin clot lysis. Its comparison with alpha2-macroglobulin. *Thromb. Haemost.*, **39**, 22–31.

17 Redlitz, A., Tan, A.K., Eaton, D.L., and Plow, E.F. (1995) Plasma carboxypeptidases as regulators of the plasminogen system. *J. Clin. Invest.*, **96**, 2534–2538.

18 Aoki, N., Moroi, M., Sakata, Y., Yoshida, N., and Matsuda, M. (1978) Abnormal plasminogen: a hereditary molecular abnormality found in a patient with recurrent thrombosis. *J. Clin. Invest.*, **61**, 1186–1195.

19 Fay, W.P., Shapiro, A.D., Shih, J.L., Schleef, R.R., and Ginsburg, D. (1992) Brief report: complete deficiency of plasminogen-activator inhibitor type 1 due to a frame-shift mutation. *N. Engl. J. Med.*, **327**, 1729–1733.

20 Mingers, A.M., Philapitsch, A., Zeitler, P., Schuster, V., Schwarz, H.P., and Kreth, H.W. (1999) Human homozygous type I plasminogen deficiency and ligneous conjunctivitis. *Acta Pathol. Microbiol. Immunol. Scand.*, **107**, 62–72.

21 Salomaa, V., Rasi, V., Kulathinal, S., Vahtera, E., Jauhiainen, M., Ehnholm, C., and Pekkanen, J. (2002) Hemostatic factors as predictors of coronary events and total mortality: the FINRISK '92 Hemostasis Study. *Arterioscler. Thromb. Vasc. Biol.*, **22**, 353–358.

22 Hoffmeister, A., Rothenbacher, D., Khuseyinova, N., Brenner, H., and Koenig, W. (2002) Plasminogen levels and risk of coronary artery disease. *Am. J. Cardiol.*, **90**, 1168–1170.

23 Jansson, J.H., Nilsson, T.K., and Olofsson, B.O. (1991) Tissue plasminogen activator and other risk factors as predictors of cardiovascular events in patients with severe angina pectoris. *Eur. Heart J.*, **12**, 157–161.

24 Geppert, A., Graf, S., Beckmann, R., Hornykewycz, S., Schuster, E., Binder, B.R., and Huber, K. (1998) Concentration of endogenous tPA antigen in coronary artery disease: relation to thrombotic events, aspirin treatment, hyperlipidemia, and multivessel disease.

Arterioscler. Thromb. Vasc. Biol., **18**, 1634–1642.

25 Gils, A. and Declerck, P.J. (2004) The structural basis for the pathophysiological relevance of PAI-1 in cardiovascular diseases and the development of potential PAI-1 inhibitors. *Thromb. Haemost.*, **91**, 425–437.

26 Meade, T.W., Ruddock, V., Stirling, Y., Chakrabarti, R., and Miller, G.J. (1993) Fibrinolytic activity, clotting factors, and long-term incidence of ischaemic heart disease in the Northwick Park Heart Study. *Lancet*, **342**, 1076–1079.

27 Schneiderman, J., Sawdey, M.S., Keeton, M.R., Bordin, G.M., Bernstein, E.F., Dilley, R.B., and Loskutoff, D.J. (1992) Increased type 1 plasminogen activator inhibitor gene expression in atherosclerotic human arteries. *Proc. Natl Acad. Sci. U. S. A.*, **89**, 6998–7002.

28 Lupu, F., Bergonzelli, G.E., Heim, D.A., Cousin, E., Genton, C.Y., Bachmann, F., and Kruithof, E.K. (1993) Localization and production of plasminogen activator inhibitor-1 in human healthy and atherosclerotic arteries. *Arterioscler. Thromb.*, **13**, 1090–1100.

29 Salomaa, V., Stinson, V., Kark, J.D., Folsom, A.R., Davis, C.E., and Wu, K.K. (1995) Association of fibrinolytic parameters with early atherosclerosis. The ARIC Study. Atherosclerosis Risk in Communities Study. *Circulation*, **91**, 284–290.

30 Redondo, M., Carroll, V.A., Mauron, T., Biasiutti, F.D., Binder, B.R., Lammle, B., and Wuillemin, W.A. (2001) Hemostatic and fibrinolytic parameters in survivors of myocardial infarction: a low plasma level of plasmin-alpha2-antiplasmin complex is an independent predictor of coronary re-events. *Blood. Coagul. Fibrinolysis*, **12**, 17–24.

31 Kienast, J., Padro, T., Steins, M., Li, C.X., Schmid, K.W., Hammel, D., Scheld, H.H., and van de Loo, J.C. (1998) Relation of urokinase-type plasminogen activator expression to presence and severity of atherosclerotic lesions in human coronary arteries. *Thromb. Haemost.*, **79**, 579–586.

32 Stone, G.W., Grines, C.L., Browne, K.F., Marco, J., Rothbaum, D., O'Keefe, J.,

Hartzler, G.O., Overlie, P., Donohue, B., Chelliah, N. *et al.* (1995) Implications of recurrent ischemia after reperfusion therapy in acute myocardial infarction: a comparison of thrombolytic therapy and primary angioplasty. *J. Am. Coll. Cardiol.*, **26**, 66–72.

33 Le May, M.R., Labinaz, M., Davies, R.F., Marquis, J.F., Laramee, L.A., O'Brien, E.R., Williams, W.L., Beanlands, R.S., Nichol, G., and Higginson, L.A. (2001) Stenting versus thrombolysis in acute myocardial infarction trial (STAT). *J. Am. Coll. Cardiol.*, **37**, 985–991.

34 Keeley, E.C., Boura, J.A., and Grines, C.L. (2003) Primary angioplasty versus intravenous thrombolytic therapy for acute myocardial infarction: a quantitative review of 23 randomised trials. *Lancet*, **361**, 13–20.

35 Xiao, Q., Danton, M.J., Witte, D.P., Kowala, M.C., Valentine, M.T., Bugge, T.H., and Degen, J.L. (1997) Plasminogen deficiency accelerates vessel wall disease in mice predisposed to atherosclerosis. *Proc. Natl Acad. Sci. U. S. A.*, **94**, 10335–10340.

36 Rezaee, F., Gijbels, M., Offerman, E., and Verheijen, J. (2003) Genetic deletion of tissue-type plasminogen activator (t-PA) in APOE3-Leiden mice reduces progression of cholesterol-induced atherosclerosis. *Thromb. Haemost.*, **90**, 710–716.

37 Carmeliet, P., Moons, L., Lijnen, R., Baes, M., Lemaitre, V., Tipping, P., Drew, A., Eeckhout, Y., Shapiro, S., Lupu, F., and Collen, D. (1997) Urokinase-generated plasmin activates matrix metalloproteinases during aneurysm formation. *Nat. Genet.*, **17**, 439–444.

38 Cozen, A.E., Moriwaki, H., Kremen, M., DeYoung, M.B., Dichek, H.L., Slezicki, K.I., Young, S.G., Veniant, M., and Dichek, D.A. (2004) Macrophage-targeted overexpression of urokinase causes accelerated atherosclerosis, coronary artery occlusions, and premature death. *Circulation*, **109**, 2129–2135.

39 Dellas, C., Schremmer, C., Hasenfuss, G., Konstantinides, S.V., and Schafer, K. (2007) Lack of urokinase plasminogen activator promotes progression and

instability of atherosclerotic lesions in apolipoprotein E-knockout mice. *Thromb. Haemost.*, **98**, 220–227.

40 Luttun, A., Lupu, F., Storkebaum, E., Hoylaerts, M.F., Moons, L., Crawley, J., Bono, F., Poole, A.R., Tipping, P., Herbert, J.M., Collen, D., and Carmeliet, P. (2002) Lack of plasminogen activator inhibitor-1 promotes growth and abnormal matrix remodeling of advanced atherosclerotic plaques in apolipoprotein E-deficient mice. *Arterioscler. Thromb. Vasc. Biol.*, **22**, 499–505.

41 Sjöland, H., Eitzman, D.T., Gordon, D., Westrick, R., Nabel, E.G., and Ginsburg, D. (2000) Atherosclerosis progression in LDL receptor-deficient and apolipoprotein E-deficient mice is independent of genetic alterations in plasminogen activator inhibitor-1. *Arterioscler. Thromb. Vasc. Biol.*, **20**, 846–852.

42 Eitzman, D.T., Westrick, R.J., Xu, Z., Tyson, J., and Ginsburg, D. (2000) Plasminogen activator inhibitor-1 deficiency protects against atherosclerosis progression in the mouse carotid artery. *Blood*, **96**, 4212–4215.

43 Rossignol, P., Luttun, A., Martin-Ventura, J.L., Lupu, F., Carmeliet, P., Collen, D., Angles-Cano, E., and Lijnen, H.R. (2006) Plasminogen activation: a mediator of vascular smooth muscle cell apoptosis in atherosclerotic plaques. *J. Thromb. Haemost.*, **4**, 664–670.

44 Schneider, D.J., Hayes, M., Wadsworth, M., Taatjes, H., Rincon, M., Taatjes, D.J., and Sobel, B.E. (2004) Attenuation of neointimal vascular smooth muscle cellularity in atheroma by plasminogen activator inhibitor type 1 (PAI-1). *J. Histochem. Cytochem.*, **52**, 1091–1099.

45 Ploplis, V.A., Carmeliet, P., Vazirzadeh, S., Van Vlaenderen, I., Moons, L., Plow, E.F., and Collen, D. (1995) Effects of disruption of the plasminogen gene on thrombosis, growth, and health in mice. *Circulation*, **92**, 2585–2593.

46 Bugge, T.H., Flick, M.J., Daugherty, C.C., and Degen, J.L. (1995) Plasminogen deficiency causes severe thrombosis but is compatible with development and reproduction. *Genes Dev.*, **9**, 794–807.

47 Carmeliet, P., Schoonjans, L., Kieckens, L., Ream, B., Degen, J., Bronson, R., De

Vos, R., van den Oord, J.J., Collen, D., and Mulligan, R.C. (1994) Physiological consequences of loss of plasminogen activator gene function in mice. *Nature*, **368**, 419–424.

48 Jackson, C.L., Bennett, M.R., Biessen, E.A., Johnson, J.L., and Krams, R. (2007) Assessment of unstable atherosclerosis in mice. *Arterioscler. Thromb. Vasc. Biol.*, **27**, 714–720.

49 Williams, H., Johnson, J.L., Carson, K.G., and Jackson, C.L. (2002) Characteristics of intact and ruptured atherosclerotic plaques in brachiocephalic arteries of apolipoprotein E knockout mice. *Arterioscler. Thromb. Vasc. Biol.*, **22**, 788–792.

50 Schwartz, S.M. (1995) How vessels narrow. *Z. Kardiol.*, **84** (Suppl. 4), 129–135.

51 Schäfer, K., Müller, K., Hecke, A., Mounier, E., Goebel, J., Loskutoff, D.J., and Konstantinides, S. (2003) Enhanced thrombosis in atherosclerosis-prone mice is associated with increased arterial expression of plasminogen activator inhibitor-1. *Arterioscler. Thromb. Vasc. Biol.*, **23**, 2097–2103.

52 Muehlenweg, B., Sperl, S., Magdolen, V., Schmitt, M., and Harbeck, N. (2001) Interference with the urokinase plasminogen activator system: a promising therapy concept for solid tumours. *Expert Opin. Biol. Ther.*, **1**, 683–691.

11
Mast Cell Proteases and Atherosclerosis

Petri T. Kovanen

11.1
Introduction

The central role of inflammation in atherogenesis has gained wide acceptance and has forced us to re-evaluate our conception of all phases of this long-lasting disease [1]. Three types of inflammatory cells – macrophages, T cells, and mast cells – constitute the inflammatory cell infiltrates typically present in evolving atherosclerotic lesions and in the adventitial layer backing them [2, 3]. In contrast to other chronic inflammatory conditions, the inflammatory component of an atherosclerotic lesion is a response to local retention and ensuing modification of apolipoprotein B–containing lipoproteins [4]. In this complex inflammatory response to the modi-fied lipids, all three types of inflammatory cells become activated and interact with each other. Mast cells were rediscovered in human atherosclerotic plaques in the early 1990s and, using contemporary techniques of mast cell biology, these cells have been assigned a role in the early and late stages of the development of athero-sclerotic plaques and their clinical complications [3, 5].

Approximately 50% of the weight of a mature mast cell consists of various neutral proteases stored in the cell's cytoplasmic secretory granules [6]. Indeed, as a major local source of extracellular neutral serine proteases, mast cells have gained a unique position among the inflammatory cells present in atherosclerotic lesions. Here we discuss the potential role of activated mast cells in the early stage of atherosclerosis, when the clinically silent fatty streaks evolve, and also in the late stages of atherosclerosis when atherosclerotic plaques become unstable and may rupture or erode, as the disease becomes clinically manifest.

11.2
Mast Cell Progenitors Enter the Arterial Intima and Differentiate into Mature Mast Cells Filled with Protease-Containing Cytoplasmic Secretory Granules

Tissue mast cells in humans differentiate from committed circulating progenitor cells that arise in the bone marrow from pluripotent hematopoietic progenitors

Atherosclerosis: Molecular and Cellular Mechanisms. Edited by Sarah Jane George and Jason Johnson
Copyright © 2010 WILEY-VCH Verlag GmbH & Co. KGaA, Weinheim
ISBN: 978-3-527-32448-4

[7]. The chemokine mainly responsible for the immigration of these progenitors into the peripheral tissues is stem cell factor (SCF), which is secreted by tissue stromal cells, such as fibroblasts. Some of the circulating progenitors find their way into the atherosclerotic coronary intima. A role for SCF in this recruitment is suggested by the finding that cultured human aortic endothelial cells and smooth muscle cells express SCF [8]. By immunohistochemical methods, we recently found SCF in the endothelial cells, smooth muscle cells, and mast cells of human coronary atheromas (M. Mäyränpää *et al.*, unpublished data), thus rendering SCF a strong candidate for being responsible for mast cell progenitor influx into the coronary intima. The chemokine eotaxin and its receptor 3 (CCR3) have been found to be prominently expressed in smooth muscle cells, and also in some mast cells and macrophages of human atheromas, suggesting that eotaxin also contributes to the recruitment of mast cell progenitors from the circulation into human atherosclerotic lesions [9].

The main factor that influences mast cell maturation and phenotypic differentiation is SCF, but the complete list of the factors comprises a wide array of cytokines, chemokines, and growth factors [10]. As in other tissues, the mature mast cells in the human aortic, coronary, and carotid artery intima are filled with cytoplasmic secretory granules that contain various effector molecules. Although the granules of arterial intimal mast cells show heterogeneous morphology in electron microscopy [11], they all contain a heparin proteoglycan matrix to which are bound at least two preformed effector molecules, namely histamine and tryptase, the latter being a mast cell–specific neutral serine protease with trypsin-like activity [6, 10]. In addition, a variable mixture of other preformed mediators is contained in the granule compartment. Of these mediators, chymase, a mast cell–specific neutral serine protease with chymotrypsin-like activity, is tightly bound to the negatively charged heparin component of the granular proteoglycans.

Cell culture systems involving human mast cell progenitors have shown that addition of SCF alone induces strong expression of tryptase and only weak expression of chymase, whereas SCF together with interleukin-4 (IL-4) upregulates the expression of chymase and cathepsin G [12]. Because the T cells in the human coronary lesions are of the T-helper 1 (Th1) type rather than of the Th2 type (which express IL-4) [2], we are left with the challenging question of whether other cells in the lesions secrete factors that may modify the phenotype of mast cells from the merely tryptase-expressing phenotype (MC_T) to the tryptase/chymase-expressing phenotype (MC_{TC}) [13].

Because all mast cells, and only mast cells, contain immunohistochemically detectable tryptase, we routinely stain sections of arterial tissue with a commercially available monoclonal antibody directed against tryptase to count all the mast cells and to determine their degree of degranulation (number of extracellularly located granule remnants) [14]. To define the phenotype of the mast cells, we also stain the sections for chymase [11], and, recently also for cathepsin G [15]. Our studies have revealed that in some individuals all of the mast cells in the arterial intima (whether normal or atherosclerotic) contain chymase, in other individuals no mast cells contain chymase, and in yet other individuals a highly variable fraction of the mast cells contains chymase.

11.3
Mast Cells in Fatty Streak Lesions and Mechanisms by Which Their Proteases May Contribute to Foam Cell Formation

The proteolytic enzymes derived from the mast cells have been shown to profoundly modify the composition and function of both low-density lipoprotein (LDL) and high-density lipoprotein (HDL) particles [16]. To gain insight into the mechanisms by which mast cell–dependent proteolytic modification of these lipoproteins might contribute to foam cell formation, we have used experimental cell culture methods employing mast cells derived from the rat peritoneal cavity. Taken together, the cell culture experiments carried out with rodent mast cells have suggested the possibility that stimulated mast cells can accelerate foam cell formation in the intimal areas in which mast cells and macrophages coexist and are surrounded by intimal fluid enriched with plasma-derived LDL and HDL particles (Figure 11.1).

11.3.1
Proteolytic Degradation of LDL Particles by Mast Cell Chymase

Regarding mast cells and LDL, we originally made the intriguing observation that incubation of LDL particles with stimulated peritoneal mast cells leads to rapid and extensive proteolytic degradation of the LDL particles [17]. Further experiments revealed that this degradation is not intracellular, that is, it does not involve particle uptake and lysosomal degradation, but is due to extracellular proteolysis of the apolipoprotein B100 (ApoB100) component of LDL by the neutral serine protease chymase present in the exocytosed cytoplasmic granules (i.e., granule remnants). Of note, the extracellular degradation of the LDL is purely a proteolytic modification without signs of hydrolysis of any of the lipid components of the particles. Neither have we observed any mast cell–dependent oxidation of LDL; rather, by chelating copper ions, histamine effectively blocks the propagation of macrophage-initiated oxidation of LDL.

We also learned that the LDL particles bind to the extracellular granule remnants, and that this binding is due to a strong ionic interaction between the positively charged residues of the amino acids lysine and arginine of the ApoB100 component of LDL, and the negatively charged sulfate groups of the heparin proteoglycans of the granule remnants [5]. Further, the heparin-bound granule remnant chymase most avidly degrades the heparin-bound LDL particles. When proteolyzed, the LDL particles become unstable and fuse into larger lipid droplets. We also found that, due to particle fusion, the capacity of each granule remnant to bind LDL was increased from a full load of 10000 to a full load of 50000 of original LDL particles. Although the ApoB100 component of each individual LDL

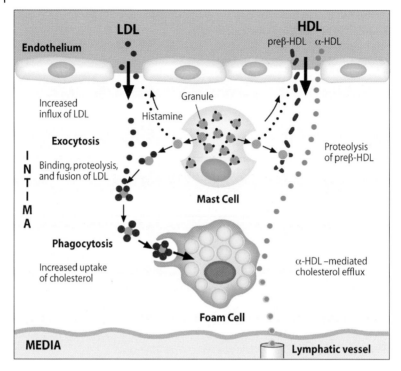

Figure 11.1 Hypothetical model of mast cell–mediated fatty streak formation. A mast cell (MC) is shown to reside in the subendothelial space of the arterial intima. The MC has been activated to exocytose some of its intracellular cytoplasmic granules. In the granules, histamine and two neutral serine peptidases, chymase and tryptase, are bound to heparin proteoglycans. In the extracellular space, histamine is detached from the exocytosed granules, diffuses away, and locally induces enhanced endothelial permeability to plasma lipoproteins, that is, LDLs and HDLs. Left: the heparin proteoglycans of the exocytosed MC granules (i.e., of the "granule remnants") bind LDL particles, and granule remnant chymase proteolyzes them so that the particles become unstable and fuse into larger lipid droplets on remnant surface. The granule remnant, with its load of fused LDL, is phagocytosed by the macrophage, the result being intracellular loading of the macrophage with LDL-derived cholesteryl esters and formation of a macrophage foam cell. Right: HDLs consisting of pre-beta- and alpha-migrating particle subpopulations also enter the intima. Chymase and tryptase of the granule remnants selectively proteolyze the pre-beta particles, so abolishing pre-beta-HDL-dependent high-affinity component of cholesterol efflux from the macrophage foam cell, and leaving alpha-HDL-mediated cholesterol efflux to act alone. The MC-dependent stimulation of cholesterol uptake and partial inhibition of cholesterol efflux operate in concert and may together effectively promote intimal foam cell formation and so convert the arterial intima into a fatty streak lesion. From *Immunological Reviews* [3], with permission.

particle is extensively degraded, the fused particles bind even more strongly to granule remnant heparin than do single native LDL particles. We could calculate that one such lipid particle with a diameter of about 100 nm consists of 100 proteolyzed LDL particles, and accordingly, also contains the proteolyzed residues of

100 ApoB100 molecules. Accordingly, the surface of such fused LDL particle must be fully covered by the residual ApoB100 peptides, thus allowing multivalent ionic interactions with several heparin molecules – an obvious explanation of the increased binding strength.

We next asked whether extracellular binding and degradation of LDL particles by mast cell granule remnants may have implications for foam cell formation [18]. For this purpose, we co-cultured the mast cells with rat or mouse peritoneal macrophages in a medium to which LDL particles had been added. If the mast cells were then stimulated immunologically or non-immunologically to degranulate, the co-cultured macrophages became gradually filled with cholesterol (i.e., foam cells were formed). This conversion of macrophages into foam cells was caused by phagocytosis and ensuing intracellular degradation of the LDL-loaded granule remnants. Cultured rat aortic smooth muscle cells of the synthesizing phenotype (corresponding to the phenotype in atherosclerotic lesions) also ingested these granule remnants and were converted into foam cells. This mast cell–dependent formation of foam cells, whether involving macrophages or smooth muscle cells, critically depended on exocytosis and phagocytosis of mast cell granules, and was termed the "granule-mediated uptake of LDL" [19]. This tightly regulated sequence of events involves stimulation of mast cells, exocytosis of mast cells granules, binding of LDL to granule remnants, proteolysis of the bound LDL with ensuing particle fusion, and, finally, phagocytosis of the LDL-loaded granule remnants by macrophages or smooth muscle cells (Figure 11.1).

We could also trigger a similar granule carrier pathway *in vivo* in the rat. When LDL was injected into the peritoneal cavity of rats, in which mast cells and macrophages coexist, and the mast cells were stimulated to degranulate, uptake of LDL by the macrophages was dramatically increased [20]. With the aid of immunoelectron-microscopic studies we then found evidence of the existence of a similar granule carrier pathway in human atherosclerotic lesions [21]. The electron micrographs revealed that in fatty streaks of human aorta and coronary arteries, exocytosed mast cell granules contained ApoB100-containing lipoproteins. Moreover, such granules appeared to be ingested by the phagocytes residing next to the degranulated mast cell. These pieces of evidence raise the possibility that human aortic and coronary mast cells might contribute to the formation of foam cells when fatty streaks evolve. The ApoB100-containing granule remnants were detected by their tryptase content (double staining), which precluded simultaneous detection of chymase and lipid droplets in the remnants by aid of electron microscopy. Thus, we still miss final proof for the chymase-dependent fusion process on the surface of granule remnants in the human atherosclerotic intima.

It should be noted that human tryptase is not able to induce LDL fusion. Tryptase is a tetrameric enzyme with the active sites of each monomer facing the central pore, which precludes access of large substrates to the enzymatic sites [22]. Thus, tryptase cannot degrade ApoB100, and therefore fails to induce particle fusion (Ken Lindstedt, Jorma Kokkonen, and Petri Kovanen, unpublished observation).

However, the observation of proteolytic fusion on mast cell granules, a physiological "heparin bead," may have wider implications for the study of protease-dependent accumulation of extracellular lipoprotein lipids in atherogenesis. Of particular interest is the observation that the size of the fused LDL particles was of the same magnitude as observed in the extracellular matrix of the human arterial intima [23]. This observation of proteolytic generation of lipid droplets resembling those in early human atherogenesis and possessing enhanced binding strength to glycosaminoglycans, has prompted us to expand our studies to other proteases present in the extracellular space of the arterial wall. The original observation made with mast cell chymase also applies to other extracellular proteases capable of extensively proteolyzing ApoB100 [24]. In fact, the chymase-dependent proteolytic modification of LDL particles with ensuing particle fusion represents the first example of LDL modification leading to enhanced binding strength of LDL to proteoglycans. In addition, the observed "supersaturation" of heparin proteoglycans (binding is enhanced fivefold beyond the level of saturation; see above) allows accumulation of LDL lipids beyond the stage at which all ApoB100-binding sites are occupied with LDL particles. The observed increases in binding strength and maximal binding capacity of LDL have potential implications for the initial phase of LDL retention in atherogenesis, when LDL particles bind to the proteoglycans of the arterial intima [4].

11.3.2
Proteolytic Degradation of HDL Particles by Mast Cell Chymase and Tryptase

Regarding the ability of mast cells to generate dysfunctional HDL particles (i.e., particles with an attenuated ability to induce cholesterol efflux from foam cells), the proteolytic activity of exocytosed mast cell granules has been found to be the only factor of importance [25]. While ApoB100 is the sole apolipoprotein in LDL particles, several apolipoproteins are bound to the lipid monolayer surface of HDL particles, the major apolipoprotein being apolipoprotein AI (ApoAI). This apolipoprotein does not interact with heparin, and, accordingly, HDL particles do not bind to exocytosed mast cell granules. Yet, when present in a specific subtype of HDL particles, ApoAI is exquisitely sensitive to proteolytic degradation by a wide variety of proteolytic enzymes, among them mast cell chymase and tryptase of rat or human origin.

Removal of excess cholesterol from intimal cholesterol-loaded foam cells by HDL particles is the key initial step along the reverse pathway, ultimately resulting in net cholesterol transport from an atherosclerotic arterial intima back to the circulation. ApoAI plays a critical role in the initiation of this pathway, accounting to about half of the total efflux of cholesterol from cholesterol-loaded macrophage foam cells [26]. The HDL particles are heterogeneous in size, shape, and composition, the smallest HDL unit containing only one molecule of ApoAI without associated lipids, or containing a few phospholipid molecules attached to it [27]. These lipid-free and lipid-poor ApoAI molecules induce efflux of phospholipids and cholesterol when they react with ABCA1 transporter molecules on the surface of macrophage foam cells. Acquisition of cellular phospholipids and cholesterol then induces the formation of discoidal nascent HDL particles, which contain two, three, or even four ApoAI molecules per particle [27]. The lipid-free ApoAI, the lipid-poor ApoAI, and the discoidal HDL particles of various sizes all migrate with prebeta-mobility in agarose gel electrophoresis, and are collectively called prebeta-HDL.

The discoidal prebeta particles are progenitors of mature spherical HDL particles that show alpha-mobility (alpha-HDL), and form the major fraction of circulating HDL in human plasma. Thus, over 90% of the ApoAI in plasma is associated with the alpha-HDL particles. The alpha-HDL particles can be separated ultracentrifugally into two subfractions according to their densities: the larger and less dense HDL_2 particles, and the smaller and more dense HDL_3 particles. Notably, in contrast to the HDL_2 fraction, the HDL_3 fraction contains minuscule quantities of prebeta-HDL in addition to the alpha-HDL. The prebeta-particles present in the ultracentrifugally isolated HDL_3 fraction are generated after the isolation procedure, when a fraction of the ApoAI molecules (constituting 4–8% of the ApoAI in the HDL_3 fraction) detaches from the surface monolayer of the spherical HDL_3 particles. The detachment is either spontaneous or caused by the action of the phospholipid transfer protein (PLTP) also present in the HDL_3 fraction [28].

It was originally shown by Kunitake and coworkers [29] that ApoAI in prebeta-HDL particles is very sensitive to proteolytic cleavage. To test whether mast cell chymase would also degrade this minor fraction of HDL particles, we stimulated rat serosal mast cells in an incubation medium containing human HDL_3. It appeared that the smallest prebeta-HDL species containing only one ApoAI (prebeta1-HDL) disappeared within minutes from the medium, reflecting efficient and complete degradation of the ApoAI contained in them [30]. In contrast, ApoAI in the alpha-HDL particles remained relatively intact; only a fraction of ApoAI (28 kDa) was cleaved into peptides, of which two large ones (10 kDa and 26 kDa) remained bound to the monolayer lipid surface of the particles.

When the experiment was repeated with human blood plasma, not only were the prebeta-HDL particles depleted, but other minor species of lipid-poor HDL particles present in human plasma that actively induce cellular cholesterol efflux were also fully depleted. More recently, we found that human recombinant chymase also avidly cleaves ApoAI, and thereby fully depletes prebeta-HDL particles from HDL_3. Human tryptase, the major protease in all human mast cells, also effectively degrades the minor HDL fractions that contain only ApoAI, ApoAIV, or ApoE [25].

The above findings of a specific pattern of HDL_3 proteolysis prompted us to ask whether proteolytic modification of HDL_3 by mast cell proteases would have an impact on the physiological function of HDL_3, in particular the apolipoprotein-mediated high-affinity efflux of cholesterol from macrophage foam cells. Originally, in an experimental cell culture system, rat serosal mast cells were co-cultured with mouse or human macrophage foam cells in a medium to which HDL_3 particles had been added [31]. Provided the mast cells were not stimulated to degranulate and the HDL_3 particles remained intact, the particles efficiently removed cholesterol from the foam cells via a high-affinity process. In sharp contrast, when the mast cells were stimulated to degranulate, the high-affinity efflux of cholesterol efflux was lost as a consequence of the chymase activity present in the exocytosed granules. The above observations were confirmed by treating HDL_3 with human chymase or tryptase, and then adding the proteolyzed HDL_3 to cultures of foam cells. In these experiments, full inhibition of the high-affinity component of cholesterol efflux accounted to about a 50% reduction of the cholesterol efflux-inducing ability of the HDL_3 fraction, that is, it was similar to the reported proportion of total efflux mediated by prebeta-HDL [26] (Figure 11.1). More recently, by the use of cell models that express specific cholesterol efflux mechanisms, we could confirm that the mechanism by which stimulated mast cells block the high-affinity cholesterol efflux from macrophage foam cells is due to extensive proteolysis of the various apolipoproteins, particularly of ApoAI, which remove cholesterol via the ABCA1 pathway [32].

As noted earlier in this chapter, all mast cells in the human arterial intima contain tryptase, and a variable fraction also contains chymase. Both tryptase and chymase are released in heparin-bound active forms from stimulated mast cells [33]. However, a major question about the ability of tryptase and chymase to act on extracellular substrates relates to their abilities to remain active after being secreted into the extracellular fluid, such as the intimal fluid. Once secreted by intimal mast cells, chymase is exposed to its physiological inhibitors, which are filtered from blood plasma into the intimal fluid. Of note, however, chymase of rat serosal mast cells in its natural heparin proteoglycan-bound form remains active and is able to degrade HDL_3 in the presence of its physiological inhibitors, and also can degrade alpha1-antitrypsin, its major inhibitor [34]. In agreement with these findings, we have found that rat granule remnants degrade prebeta-HDL particles present in human aortic intimal fluid, and so block the ability of the intimal fluid to promote cholesterol efflux from human monocyte-derived macrophage foam cells in culture [35].

In contrast to chymase, tryptase is a unique serine protease with no known (major) naturally occurring inhibitors [22, 33]. In order to preserve its tetrameric structure and enzymatic activity, tryptase needs to remain associated with the co-secreted heparin proteoglycans. However, while chymase remains firmly attached to the co-secreted heparin proteoglycans, tryptase is slowly released from them, and subsequently dissociates into inactive monomers. We have found that, in addition to the mast cell–derived heparin proteoglycans, human aortic proteoglycans can also stabilize the tryptase tetramer and maintain its activity [36]. This finding suggests that tryptase secreted by intimal mast cells may associate with the proteoglycans in the proteoglycan-rich extracellular matrix of the intima, enabling tryptase to remain active for prolonged times in human atherosclerotic lesions. Since the activated mast cells present in such coronary lesions also secrete heparin proteoglycans and reside in a proteoglycan-rich environment, tryptase, in addition to chymase, may remain at least partially active for prolonged periods of time in the intimal fluid, allowing HDL particles to become locally proteolyzed.

We have extended the *ex vivo* observations derived from mast cell–macrophage co-culture models, in which rat peritoneal mast cells are stimulated to degranulate by compound 48/80, to *in vivo* conditions in which compound 48/80 is injected intraperitoneally into the rat [25]. For this purpose, peritoneal macrophages were first loaded with radioactively labeled cholesterol, and then the release of the labeled cholesterol from the macrophage foam cells to the surrounding peritoneal fluid was analyzed by isolating the macrophages and measuring the remaining cell-associated amount of radioactive cholesterol. We found that stimulation of peritoneal mast cell during the efflux phase partially blocked cholesterol release from the foam cells. Since the inhibitory effect of activated rat peritoneal mast cells on HDL-dependent cholesterol efflux from foam cells requires chymase activity of exocytosed mast cell granules, the observed inhibitory effect must have reflected proteolytic degradation of the chymase-sensitive cholesterol efflux–inducing apolipoproteins present in the peritoneal fluid. Inasmuch as the peritoneal cavity only acted as a substitute for the actual site of foam cell formation, that is, the arterial intima, complementary approaches to these *in vivo* experimentations are required. However, the above observation paves the way to the definition in living organisms of a novel species of dysfunctional HDL in which the critical apolipoproteins have lost their specific functions in the initiation of reverse cholesterol transport as a result of their proteolytic modification.

11.4
Mast Cells in Advanced Coronary Atherosclerotic Plaques, and Mechanisms by Which They May Promote Plaque Rupture and Erosion

The most important mechanism of sudden onset of coronary syndromes, including unstable angina, acute myocardial infarction, and sudden cardiac death, is rupture or erosion of a coronary atheroma [37, 38]. The risk of atheromatous rupture or erosion appears to depend critically on both the cellular and the extracellular composition of the atheroma. The collagen-rich layer covering the lipid core of an advanced atherosclerotic plaque or atheroma is called the fibrous cap. Typically, a stable lesion with a thick fibrous cap and a small lipid core is stable, whereas a lipid-rich lesion with a thin fibrous cap and a large lipid core is unstable and prone to rupture [39]. If a stable lesion causes significant stenosis of the arterial lumen, and turbulent flow conditions are created, some endothelial cells may detach, that is, the plaque erodes.

Apart from the above described effects directly related to lipid metabolism, novel functions for intimal mast cells are emerging. These functions relate to the hypothesis that coronary atheromas ultimately rupture because of locally increased activities of the enzymes involved in the digestion of the extracellular and pericellular matrices are rendering the plaque unstable [40]. Importantly, in the human arterial intima, mast cells (instead of neutrophilic granulocytes) turned out to be the major local source of neutral serine proteases, providing them with the tools to contribute to the weakening of the atheromatous plaques [15].

In coronary atheromas, we made the observation that mast cells are distributed unevenly; typically they are more abundant in the shoulder region than in the cap and the core regions [41]. Interestingly, the proportion of degranulated mast cells was especially pronounced in the inflamed shoulder region of the atheromas (85%), the site at which the cap merges with normal intima and which is most prone to rupture, compared with the proportion in normal intima (18%). It is noteworthy that in patients who had died of acute myocardial infarction, most of the mast cells were located at the sites of atheromatous erosion or rupture [42]. Importantly, the proportions of degranulated mast cells were 86% at the site of erosion or rupture, 63% in the adjacent atheromatous area, and 27% in the unaffected intima. Thus, the density of degranulated mast cells was 200-fold higher at the eroded or ruptured site than in the unaffected intimal area.

Since our method of discovering mast cell activation in human coronary atherosclerotic lesions is based on detection of exocytosed granules around the mast cells, we have restricted our experimental studies to include only the pre-formed mast cell mediators present in mast cell granules. These include heparin proteoglycans, chymase, tryptase, tumor necrosis factor-α (TNF-α) and transforming growth factor-β (TGF-β) [43]. When selecting targets for these effector molecules, we have focused our studies on the tissue components critically contributing to plaque rupture and erosion. Degradation of the extracellular matrix and loss of its production by the SMCs are critical for plaque rupture, while loss of endothelial cells is critical for plaque erosion. The protease-secreting activity of mast cells turned out to be crucial to the mast cell–dependent effects in the above processes.

11.4.1
Potential Mechanisms by Which Mast Cell Proteases Induce Loss of Extracellular Matrix and Smooth Muscle Cells in Atherosclerotic Plaques

Mast cells can indirectly contribute to the matrix degradation when their secreted neutral proteases activate matrix metalloproteinases (MMPs) (Figure 11.2). We found that human skin chymase effectively activates the proform of the interstitial collagenase, or pro-MMP-1, by cleaving the proenzyme at the Leu83–Thr84 position [44]. Moreover, studies in other laboratories have shown that tryptase can activate prostromelysin (pro-MMP-3), which, again, in addition to being a powerful matrix-degrading enzyme, can activate other MMPs. Finally, gelatinase A (MMP-2), an enzyme highly expressed in human atherosclerotic plaques (like MMP-1 and MMP-3), is activated by cathepsin G, a neutral protease present in a fraction of mast cells in human coronary mast cells [15, 45]. Importantly, an *ex vivo* study in organ culture by Johnson and coworkers [46] has demonstrated that triggering mast cell degranulation in freshly obtained human carotid endarterectomy specimens results in increased extracellular tryptase activity and subsequently also in increased MMP activity in the specimens. Degranulating mast cells may also induce MMP synthesis. Thus, cell culture experiments showed that TNF-α induces cultured human macrophages to synthesize and secrete gelatinase B or MMP-9 [47]. Indeed, two lines of evidence support the possibility that mast cells may actually induce MMP-9 synthesis by intimal cells located in their neighborhood. First, the granules of mast cells in rupture-prone areas of human coronary atheromas

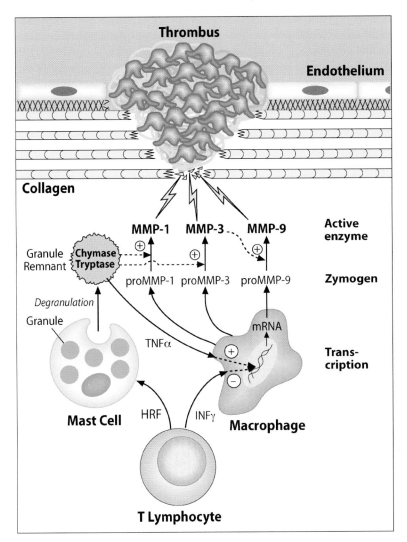

Figure 11.2 Possible contribution of mast cells to plaque rupture. The fibrous cap of a coronary plaque is shown to contain three types of inflammatory cells: a T lymphocyte, a mast cell, and a macrophage. The T lymphocyte secretes mast cell–activating histamine-releasing factors (HRFs) and so induces the mast cells to degranulate. TNF-α released from the extracellular granule remnant stimulates the synthesis of pro-MMP-9 by the macrophage, while interferon-γ secreted by the T lymphocyte counteracts this stimulatory effect. The macrophage also synthesizes and secretes pro-MMP-1 and pro-MMP-3. The neutral serine peptidases, chymase and tryptase, present in the granule remnant activate the inactive zymogen forms (pro-MMPs), as indicated. The active MMP-3, again, activates pro-MMP-9. Together, these MMPs degrade the various components of the extracellular matrix, notably collagen. After initial cleavage of fibrillar collagen by MMP-1 or MMP-3, the activated MMP-9 further degrades the cleaved collagen (gelatin). Breakdown of collagen weakens the fibrous cap and renders the plaque susceptible to rupture. Exposure of the subendothelial prothrombogenic tissue precipitates the formation of an arterial thrombus. From *Immunological Reviews* [3], with permission.

contain TNF-α [48], and second, TNF-α-positive mast cells and MMP-9-positive macrophages are colocalized in coronary atheromas of patients with severe unstable angina [49].

In addition to their matrix-degrading capacity, degranulating mast cells also have the potential to attenuate matrix production by killing collagen-producing SMCs or by inhibiting their collagen synthesis. Thus, coculture experiments revealed that degranulating mast cells can induce apoptotic death of the cocultured SMCs [50]. Notably, the secreted chymase degrades the pericellularly located fibronectin, which results in loss of outside-in signaling with ensuing apoptotic death of the SMCs. Interestingly, caps of ruptured human coronary plaques contain increased numbers of activated mast cells and reduced numbers of SMCs [42]. Moreover, chymase from exocytosed mast cell granules can also inhibit collagen synthesis in cultured SMCs by TGF-β-dependent and -independent mechanisms [51].

Taken together, protease-secreting mast cells may contribute to the weakening of the fibrous cap of advancing atherosclerotic plaques in many ways.

11.4.2
Potential Mechanisms by Which Mast Cell Induce Death or Detachment of Endothelial Cells in Atherosclerotic Plaques

The mechanisms leading to the loss of endothelial cells in human coronary plaques, that is, to plaque erosion, remain enigmatic [39]. Apoptosis of endothelial cells has been suggested as one possible mechanism. By secreting neutral serine proteases, mast cells may degrade the pericellular (or intercellular) matrix of endothelial cells, and thereby induce their apoptotic death. Indeed, we have observed that chymase, cathepsin G, and tryptase are capable of degrading vascular endothelial cadherin, a molecule involved in the outside-in survival signaling of endothelial cells [15]. In addition, *ex vivo* experiments with segments of freshly isolated human coronary arteries demonstrated that proteolytic degradation of the endothelial basement membrane (particularly fibronectin) by mast cell chymase or tryptase loosens the attachment of endothelial cells to the wall of the atherosclerotic plaque, and so predisposes them to acute desquamation in areas of excessive shear stress and turbulent flow. Finally, by secreting the strongly proapoptotic TNF-α, subendothelial mast cells may aggravate the protease-induced apoptotic death of endothelial cells. This hypothesis received support from experimental observations made in cultured cells, which revealed that activation of rat serosal mast cells in the presence of cocultured rat cardiac microvascular endothelial cells resulted in apoptotic death of the endothelial cells, and that the TNF-α activity present in the granule remnants enhanced this effect [52].

In summary, mast cell–dependent desquamation of endothelial cells could involve loosening of vascular endothelial (VE) cadherin–mediated endothelial cell–cell junctions by TNF-α and histamine, or via activation of proteinase-activated receptor-2 (PAR-2) by tryptase. Such loosening of cell–cell contacts between endothelial cells would then provide the mast cell–derived proteases tryptase,

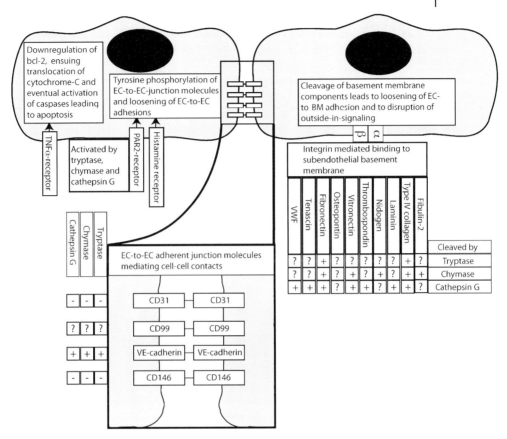

Figure 11.3 Summary of the mechanisms by which mast cells may loosen the adhesions of endothelial cells (ECs). Two neighboring ECs are shown at the top of the figure. Upon activation, subendothelial mast cells release various mediators (histamine, heparin, tumor necrosis factor-α) and the proteases tryptase, chymase, and cathepsin G. These mediators and other proteases may loosen the EC adhesion to basement membrane (BM) and to neighboring endothelial cells by affecting the intracellular signaling cascades and by proteolytically cleaving cellular adhesion molecules and components of BM. The mechanisms are discussed in more detail in the text. +, cleaved by the enzyme; –, not cleaved by the enzyme; ?, cleavage by the enzyme not studied. From *Coronary Artery Disease* [15], with permission.

chymase, and cathepsin G better access to their intercellularly located substrate VE cadherin. Moreover, proteolytic degradation by mast cell proteases of the various components of the endothelial basement membrane, for example, fibronectin, vitronectin, type IV collagen, laminin, nidogen, tenascin, and fibulin-2, which are important for the adhesion of endothelial cells to the basement membrane and/or for the outside-in signaling–mediated endothelial survival, would compromise endothelial integrity and contribute to the onset of plaque erosion [15] (Figure 11.3).

11.5
Mast Cells around Coronary Plaque Neovessels – Additional Mechanisms by Which Mast Cell Proteases May Promote Plaque Vulnerability

As introduced earlier in this chapter, arterial intima is a peculiar tissue in that it lacks a capillary circulation. When atherosclerosis develops, neovascularization may be induced in the deep parts of the lesions in response to hypoxia [53]. These plaque microvessels originating from the adventitial vasa vasorum are accompanied by mast cells [54]. The mast cells located near the neovessels contain basic fibroblast factor (bFGF), a potent pro-angiogenic factor, and are likely to contribute to the growth of the neovascular sprouts [55]. In addition to bFGF, other mast cell constituents, such as histamine, heparin, tryptase, chymase, vascular endothelial growth factor (VEGF), and nerve growth factor (NGF), are pro-angiogenic, thus rendering the mast cell potentially a powerful pro-angiogenic effector cell [56]. The neovessels may also act as a port of entry for circulating mast cell progenitors to the deep areas of plaques normally lacking resident mast cells. If so, the newly recruited mast cells would amplify and sustain the angiogenic process originally triggered by hypoxia present in the advanced atherosclerotic lesions.

Proteolytic activity is required for the physiological growth of neovessels, but under certain conditions of inflammatory angiogenesis, non-targeted and excessive proteolytic activity at the inflamed tissue site may damage the neovessels [53]. Accordingly, mast cells may also injure the microvessels by the action of their neutral proteases chymase and tryptase and by their abilities to induce production of MMPs and to trigger their activation [54, 57, 58]. Inasmuch as the neovessels are fragile [59], proteolytic degradation of their thin basement membrane could lead to endothelial apoptosis and rupture of the microvessels. Indeed, in very advanced human coronary plaques we found local hemorrhages in mast cell–populated areas of neovascularization, indicative of microvascular damage [55]. The resulting intraplaque microhemorrhages, again, would tend to weaken the plaque and render it susceptible to rupture (Figure 11.4).

Importantly, human carotid plaques also demonstrate microfocal neovascularization, which often exceeds that found in coronary atheromas [60, 61]. In contrast to the atherosclerotic lesions of human coronary arteries and aorta, atherosclerotic lesions of human carotid arteries have been shown to contain capillary-like microvessels already in very early stages of lesion development. As the lesions advance, neovascularization becomes more prominent, and in the late-stage plaques it is often very extensive, particularly in the vulnerable shoulder areas of the plaques. Jeziorska and Woolley [61] also provided evidence of local microvascular damage with hemorrhages derived from damaged microvessels in the shoulder areas of carotid arteries. Interestingly, mast cells were found to accompany the microvessels at all stages of carotid lesion development, indicating that they may have participated both in the growth and in the damage of the microvessels [61]. This latter notion is strongly supported by the direct demonstration of MMP activation in explanted human carotid plaques by the two mast cell–derived neutral proteases

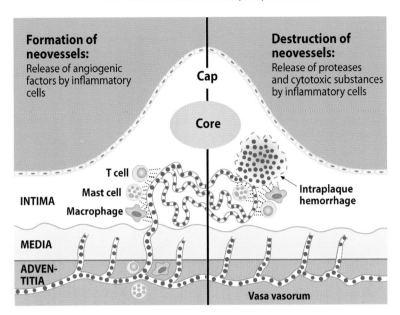

Figure 11.4 Schematic presentation of the two roles of a mast cell–containing infiltrate of proinflammatory cells in an advanced atherosclerotic lesion: left angiogenic and right angiolytic effects. The infiltrate consists of macrophages, T cells, and mast cells. Left: by releasing a variety of angiogenic factors, the cells induce the growth of neovessels which originate from the vasa vasorum in the outer layer of the arterial wall. Right: by releasing a variety of proteases and cytotoxic substances, the cells induce death of endothelial cells and so create local disruption of the microvessels. The ensuing intraplaque hemorrhage tends to weaken the plaque and predispose to plaque rupture with ensuing clinical sequelae, such as myocardial infarction and stroke. From *Annals of Medicine* [53], with permission.

chymase and tryptase [46]. Mast cell protease-dependent damage of endothelial cells and the basal lamina has also been suggested to occur in the scenario leading to brain damage during hypoxic–ischemic insults [62].

Regarding the role of mast cells in ischemic cerebrovascular events of thromboembolic origin, we made the observation that in endarterectomized carotid plaques from patients with ipsilateral stroke, that is, in symptom-causing plaques whose erosion or rupture had caused a thromboembolic complication, the number of mast cells was higher than it was in non-symptom-causing advanced carotid plaques [63]. This finding agrees well with the suggested role of mast cells in the destabilization of advanced carotid atherosclerotic lesions.

Finally, human mast cells have also been shown to produce MMPs, notably MMP-1 and MMP-9 [64–66]. Although mast cells contain tissue inhibitor of metalloproteinases-4 (TIMP-4) [67], mast cell–derived chymase is capable of cleaving and inactivating TIMP-1, and thus has the potential of promoting MMP activity in the vicinity of activated mast cells. Thus, mast cells are not only activators, but also producers of MMPs. Provided the activators of MMP secretion by mast cells are different from those leading to secretion of serine neutral proteases by mast cells (degranulators), these differentially regulated members of the mast cells' proteolytic armament would enable mast cells to perform entirely novel roles in chronic inflammation, particularly in a scenario in which the mast cells actively participate in plaque remodeling, both before and after rupture or erosion.

11.6
Genetically and Pharmacologically Manipulated Mice – An Opportunity to Rigorously Test the Role of Mast Cell Proteases in Atherogenesis

Mast cell biology in atherosclerosis is now entering a new phase, in which the role of mast cells in atherogenesis is being tested in hypercholesterolemic mice susceptible to atherosclerosis [68]. Moreover, the availability of spontaneously mutated and genetically manipulated mice has opened new horizons for rigorous testing of the various hypotheses regarding the postulated roles of mast cells in atherogenesis. Indeed, the availability of $Kit^{W-sh/W-sh}$ mice which lack mast cells have been vital for such hypothesis-testing work. Kit is the receptor for SCF, and it is encoded by the proto-oncogene c-*kit* [69]. In the $Kit^{W-sh/W-sh}$ mouse strain, a spontaneous inversion mutation of the c-*kit* promoter region has occurred. This mutation reduces c-kit tyrosine kinase–dependent signaling in progenitor cells, and consequently, mast cell deficiency results. By cross-breeding $Kit^{W-sh/W-sh}$ mice with $LDLr^{-/-}$ mice, $LDLr^{-/-} Kit^{W-sh/W-sh}$ mice could generated, so allowing to test the effect of mast cells in experimental mouse atherogenesis. These experiments clearly defined adventitial and intimal mast cells as atherosclerosis-promoting cells in the mouse aorta [70, 71].

Mast cell–specific pharmacological manipulation of genetically modified atherosclerosis-prone mice is another approach of *in vivo* testing of the potential role of mast cells in atherogenesis. In the work presented by Bot and coworkers [72], elegant pharmacological experiments were performed with mast cell stabilizers in genetically hyperlipidemic mice ($ApoE^{-/-}$). In these mice, a collar had been placed around the carotid artery to induce advanced atherosclerotic lesions in this area of the arterial tree. An advantage of the collar model is the possibility to locally stimulate the adventitial mast cells to degranulate in the atherosclerotic arterial segment. Based on these findings, the authors concluded that local mast cell stimulation in the adventitia of the carotid arteries results in increased vascular permeability and recruitment of further leukocytes to the plaques, and leads to intraplaque microhemorrhage and macrophage apoptosis. Inasmuch as the mast cells were acutely stimulated to degranulate, the observed effects were likely caused by the preformed mediators released upon degranulation, such as histamine, heparin, and notably, the neutral serine proteases, as discussed in this chapter. Indeed, in terms of experimental testing of the role of mast cells in rupture, erosion, or microhemorrhages, the future looks very promising, since various MMPs, which may also be targets of mast cells, have been assigned a pathophysiological role in atherosclerotic plaque instability in mouse brachiocephalic arteries [73, 74].

However, mouse and human are different, which also holds true for rat and mouse. For example, some of the murine chymases, although phylogenetically closely related to human chymase, have acquired tryptic or elastolytic, rather than chymotryptic activity [6]. Furthermore, although mice, in contrast to rats, possess several chymases in their cytoplasmic secretory granules, only the mouse mast cell protease-4 (mMCP-4) has chymotryptic activity [75]. Again, genetic engineering may help us to determine the role of the individual mouse mast cell proteases in the hyperlipidemia-driven atherosclerotic process in this animal. Thus, mice specifically lacking mMCP-4 are available for testing [76]. However, a caution was expressed by Bischoff [77], who emphasized that mice and rats never suffer from allergic or other mast cell–associated diseases, as humans do. In this regard, experiments with cultured mature human mast cells expressing large amounts of both tryptase and chymase, that is, mast cells resembling those in human arterial intima [12], are necessary intermediary steps when testing the various suggested protease-dependent mechanisms by which mast cells may participate in human atherogenesis.

11.7
Epilogue

We need to appreciate that, when examining and interpreting possible roles of mast cells in atherogenesis, we have one-sidedly selected those hypotheses which assign mast cells the role of the "bad guy." This limitation applies to all stages of atherogenesis, particularly to the advanced stage characterized by generation of vulnerable plaques susceptible to erosion and rupture. This apparently biased approach stems from the general understanding of atherosclerosis as an inflammatory disease combined with the definition of mast cells as proinflammatory cells. However, because the inflammatory reaction of the arterial wall also contains elements that must be considered "good," that is, reparative or even healing, there is also a challenge for mast cells to act as beneficial cells, as has been suggested for mast cells in other chronic inflammatory diseases [78]. Inasmuch as the development of complex advanced atherosclerotic plaques results from repetitive atherothrombotic events in which injury and repair occur alternatively [79], there is a niche for both the "good" and the "bad" mast cell as an effector cell in this disease. Considering the cadre of proteases mast cells may release, either acutely or gradually, into the intimal microenvironment with an already ongoing fine-tuned orchestration of further proteases derived from other types of cells, we may safely adopt the view taken by Newby [80] in his considerate and balanced analysis of the role of MMPs in atherogenesis, and simply state that mast cells are likely to have "a dual role" in this disease.

Acknowledgments

This work has been performed at the Wihuri Research Institute in Helsinki, Finland. The Institute is maintained by the Jenny and Antti Wihuri Foundation. I wish to thank the Foundation for the generous support over the years and for encouragement for the studies on mast cells. Without the ever-lasting devotion and enthusiasm of the entire Wihuri Crew, this work would not have been possible to carry out.

References

1 Hansson, G.K. and Libby, P. (2006) The immune response in atherosclerosis: a double-edged sword. *Nat. Rev. Immunol.*, **6**, 508–519.

2 Hansson, G.K. (2005) Inflammation, atherosclerosis, and coronary artery disease. *N. Engl. J. Med.*, **352**, 1685–1695.

3 Kovanen, P.T. (2007) Mast cells – multipotent local effector cells in atherothrombosis. *Immunol. Rev.*, **217**, 105–122.

4 Tabas, I., Williams, K.J., and Borén, J. (2007) Subendothelial lipoprotein retention as the initiating process in atherosclerosis: update and therapeutic implications. *Circulation*, **116**, 1832–1844.

5 Kovanen, P.T. (1995) Role of mast cells in atherosclerosis. *Chem. Immunol.*, **62**, 132–170.

6 Stevens, R.L. and Adachi, R. (2007) Protease-proteoglycan complexes of mouse and human mast cells and importance of their beta-tryptase-heparin complexes in inflammation and innate immunity. *Immunol. Rev.*, **217**, 155–167.

7 Okayama, Y. and Kawakami, T. (2006) Development, migration, and survival of mast cells. *Immunol. Res.*, **34**, 97–115.

8 Miyamoto, T., Sasaguri, Y., Sasaguri, T., Azakami, S., Yasukawa, H., Kato, S., Arima, N., Sugama, K., and Morimatsu, M. (1997) Expression of stem cell factor in human aortic endothelial and smooth muscle cells. *Atherosclerosis*, **129**, 207–213.

9 Haley, K.J., Lilly, C.M., Yang, J.H., Feng, Y., Kennedy, S.P., Turi, T.G., Thompson, J.F., Sukhova, G.H., Libby, P., and Lee, R.T. (2000) Overexpression of eotaxin and the CCR3 receptor in human atherosclerosis: using genomic technology to identify a potential novel pathway of vascular inflammation. *Circulation*, **102**, 2185–2189.

10 Galli, S.J., Nakae, S., and Tsai, M. (2005) Mast cells in the development of adaptive immune responses. *Nat. Immunol.*, **6**, 135–142.

11 Kaartinen, M., Penttilä, A., and Kovanen, P.T. (1994) Mast cells of two types differing in neutral protease composition in the human aortic intima. Demonstration of tryptase- and tryptase/chymase-containing mast cells in normal intimas, fatty streaks, and the shoulder region of atheromas. *Arterioscler. Thromb.*, **14**, 966–972.

12 Lappalainen, J., Lindstedt, K.A., and Kovanen, P.T. (2007) A protocol for generating high numbers of mature and functional human mast cells from peripheral blood. *Clin. Exp. Allergy*, **37**, 1404–1414.

13 Irani, A.A., Schechter, N.M., Craig, S.S., DeBlois, G., and Schwartz, L. (1986) Two types of human mast cells that have distinct neutral protease compositions. *Proc. Natl Acad. Sci. U. S. A.*, **83**, 4464–4468.

14 Laine, P., Kaartinen, M., Penttilä, A., Panula, P., Paavonen, T., and Kovanen, P.T. (1999) Association between myocardial infarction and the mast cells in the adventitia of the infarct-related coronary artery. *Circulation*, **99**, 361–369.

15 Mäyränpää, M.I., Heikkilä, H.M., Lindstedt, K.A., Walls, A.F., and

Kovanen, P.T. (2006) Desquamation of human coronary artery endothelium by human mast cell proteases: implications for plaque erosion. *Coron. Artery Dis.*, **17**, 611–621.

16 Kovanen, P.T. (1996) Mast cells in human fatty streaks and atheromas: implications for intimal lipid accumulation. *Curr. Opin. Lipidol.*, **7**, 281–286.

17 Kokkonen, J.O. and Kovanen, P.T. (1985) Low density lipoprotein degradation by rat mast cells: demonstration of extracellular proteolysis caused by mast cell granules. *J. Biol. Chem.*, **260**, 14756–14763.

18 Kovanen, P.T. (1991) Mast cell granule-mediated uptake of low density lipoproteins by macrophages: a novel carrier mechanism leading to the formation of foam cells. *Ann. Med.*, **23**, 551–559.

19 Kokkonen, J.O. and Kovanen, P.T. (1987) Stimulation of mast cells leads to cholesterol accumulation in macrophages *in vitro* by a mast cell granule-mediated uptake of low density lipoprotein. *Proc. Natl Acad. Sci. U. S. A.*, **84**, 2287–2291.

20 Kokkonen, J.O. (1989) Stimulation of rat peritoneal mast cells enhances uptake of low density lipoproteins by rat peritoneal macrophages *in vivo*. *Atherosclerosis*, **79**, 213–223.

21 Kaartinen, M., Penttilä, A., and Kovanen, P.T. (1995) Extracellular mast cell granules carry apolipoprotein B-100-containing lipoproteins into phagocytes in human arterial intima: functional coupling of exocytosis and phagocytocis in neighboring cells. *Arterioscler. Thromb. Vasc. Biol.*, **15**, 2047–2054.

22 Sommerhoff, C.P., Bode, W., Matschiner, G., Bergner, A., and Fritz, H. (2000) The human mast cell tryptase tetramer: a fascinating riddle solved by structure. *Biochim. Biophys. Acta*, **1477**, 75–89.

23 Guyton, J.R. and Klemp, K.F. (1993) Transitional features in human atherosclerosis. Intimal thickening, cholesterol clefts, and cell loss in human aortic fatty streaks. *Am. J. Pathol.*, **143**, 1444–1457.

24 Öörni, K., Pentikäinen, M.O., Ala-Korpela, M., and Kovanen, P.T. (2000) Aggregation, fusion, and vesicle formation of modified low density lipoprotein particles: molecular mechanisms and effects on matrix interactions. *J. Lipid Res.*, **41**, 1703–1714.

25 Lee-Rueckert, M. and Kovanen, P.T. (2006) Mast cell proteases: physiological tools to study functional significance of high density lipoproteins in the initiation of reverse cholesterol transport. *Atherosclerosis*, **189**, 8–18.

26 Adorni, M.P., Zimetti, F., Billheimer, J.T., Wang, N., Rader, D.J., Phillips, M.C., and Rothblat, G.H. (2007) The roles of different pathways in the release of cholesterol from macrophages. *J. Lipid Res.*, **48**, 2453–2462.

27 Duong, P.T., Weibel, G.L., Lund-Katz, S., Rothblat, G.H., and Phillips, M.C. (2008) Characterization and properties of pre beta-HDL particles formed by ABCA1-mediated cellular lipid efflux to apoA-I. *J. Lipid Res.*, **49**, 1006–1014.

28 Lee, M., Metso, J., Jauhiainen, M., and Kovanen, P.T. (2003) Degradation of phospholipid transfer protein (PLPT) and PLPT-generated prebeta-high density lipoprotein by mast cell chymase impairs high affinity efflux of cholesterol from macrophage foam cells. *J. Biol. Chem.*, **278**, 13539–13545.

29 Kunitake, S.T., Chen, G.C., Kung, S.F., Schilling, J.W., Hardman, D.A., and Kane, J.P. (1990) Pre-beta high density lipoprotein. Unique disposition of apolipoprotein A-I increases susceptibility to proteolysis. *Arteriosclerosis*, **10**, 25–30.

30 Lee, M., von Eckardstein, A., Lindstedt, L., Assmann, G., and Kovanen, P.T. (1999) Depletion of preß1LpA1 and LpA4 particles by mast cell chymase reduces cholesterol efflux from macrophage foam cells induced by plasma. *Arterioscler. Thromb. Vasc. Biol.*, **19**, 1066–1074.

31 Lee, M., Lindstedt, L.K., and Kovanen, P.T. (1992) Mast cell-mediated inhibition of reverse cholesterol transport. *Arterioscler. Thromb.*, **12**, 1329–1335.

32 Lee, M., Uboldi, P., Giudice, D., Catapano, A.L., and Kovanen, P.T. (2000) Identification of domains in apoA-1 susceptible to proteolysis by mast cell

chymase: implications for HDL function. *J. Lipid Res.*, **41**, 975–984.

33 Caughey, G.H. (2007) Mast cell tryptases and chymases in inflammation and host defense. *Immunol. Rev.*, **217**, 141–154.

34 Lindstedt, L., Lee, M., and Kovanen, P.T. (2001) Chymase bound to heparin is resistant to its natural inhibitors and capable of proteolyzing high density lipoproteins in aortic intimal fluid. *Atherosclerosis*, **155**, 87–97.

35 Lindstedt, L., Lee, M., Castro, G.R., Fruchart, J.-C., and Kovanen, P.T. (1996) Chymase·in exocytosed rat mast cell granules effectively proteolyzes apolipoprotein AI-containing lipoproteins, so reducing the cholesterol efflux-inducing ability of serum and aortic intimal fluid. *J. Clin. Invest.*, **97**, 2174–2182.

36 Lee, M., Sommerhoff, C.P., von Eckardstein, A., Zettl, F., Fritz, H., and Kovanen, P.T. (2002) Mast cell tryptase degrades HDL and blocks its function as an acceptor of cellular cholesterol. *Arterioscler. Thromb. Vasc. Biol.*, **22**, 2086–2091.

37 Stary, H.C., Chandler, A.B., Dinsmore, R.E., Fuster, V., Glagov, S., Insull, W. Jr., Rosenfeld, M.E., Schwartz, C.J., Wagner, W.D., and Wissler, R.W. (1995) A definition of advanced types of athero-sclerotic lesions and a histological classification of atherosclerosis. A report from the Committee on Vascular Lesions of the Council on Arteriosclerosis, American Heart Association. *Circulation*, **92**, 1355–1374.

38 Falk, E. (1992) Why do plaques rupture? *Circulation*, **86** (6 Suppl.), III30–III42.

39 Virmani, R., Burke, A.P., Farb, A., and Kolodgie, F.D. (2006) Pathology of the vulnerable plaque. *J. Am. Coll. Cardiol.*, **47** (8 Suppl.), C13–C18.

40 Libby, P. (1995) Molecular bases of the acute coronary syndromes. *Circulation*, **91**, 2844–2850.

41 Kaartinen, M., Penttilä, A., and Kovanen, P.T. (1994) Accumulation of activated mast cells in the shoulder region of human coronary atheroma, the predilection site of atheromatous rupture. *Circulation*, **90**, 1669–1678.

42 Kovanen, P.T., Kaartinen, M., and Paavonen, T. (1995) Infiltrates of activated mast cells at the site of coronary atheromatous erosion or rupture in myocardial infarction. *Circulation*, **92**, 1084–1088.

43 Kovanen, P.T. (2009) Mast cells in atherogenesis: actions and reactions. *Curr. Atheroscler. Rep.*, **11**, 214–219.

44 Saarinen, J., Kalkkinen, N., Welgus, H.G., and Kovanen, P.T. (1994) Activation of human interstitial procollagenase through direct cleavage of the Leu83-Thr84 bond by mast cell chymase. *J. Biol. Chem.*, **269**, 18134–18140.

45 Shamamian, P., Schwartz, J.D., Pocock, B.J., Monea, S., Whiting, D., Marcus, S.G., and Mignatti, P. (2001) Activation of progelatinase A (MMP-2) by neu-trophil elastase, cathepsin G, and proteinase-3: a role for inflammatory cells in tumor invasion and angio-genesis. *J. Cell Physiol.*, **189**, 197–206.

46 Johnson, J.L., Jackson, C.L., Angelini, G.D., and George, S.J. (1998) Activation of matrix-degrading metalloproteinases by mast cell proteases in atherosclerotic plaques. *Arterioscler. Thromb. Vasc. Biol.*, **18**, 1707–1715.

47 Saren, P., Welgus, H.G., and Kovanen, P.T. (1996) TNF-alpha and IL-1beta selectively induce expression of 92-kDa gelatinase by human macrophages. *J. Immunol.*, **157**, 4159–4165.

48 Kaartinen, M., Penttilä, A., and Kovanen, P.T. (1996) Mast cells in rupture-prone areas of human coronary atheromas produce and store TNF-alpha. *Circula-tion*, **94**, 2787–2792.

49 Kaartinen, M., van der Wal, A.C., van der Loos, C.M., Piek, J.J., Koch, K.T., Becker, A.E., and Kovanen, P.T. (1998) Mast cell infiltration in acute coronary syndromes: implications for plaque rupture. *J. Am. Coll. Cardiol.*, **32**, 606–612.

50 Leskinen, M.J., Kovanen, P.T., and Lindstedt, K.A. (2003) Regulation of smooth muscle cell growth, function and death *in vitro* by activated mast cells – a potential mechanism for the weakening and rupture of atherosclerotic plaques. *Biochem. Pharmacol.*, **66**, 1493–1498.

51 Wang, Y., Shiota, N., Leskinen, M.J., Lindstedt, K.A., and Kovanen, P.T. (2001) Mast cell chymase inhibits smooth muscle cell growth and collagen expression *in vitro*: transforming growth factor-beta1-dependent and -independent effects. *Arterioscler. Thromb. Vasc. Biol.*, **21**, 1928–1933.

52 Lätti, S., Leskinen, M., Shiota, N., Wang, Y., Kovanen, P.T., and Lindstedt, K.A. (2003) Mast cell-mediated apoptosis of endothelial cells *in vitro*: a paracrine mechanism involving TNT-alpha-mediated down-regulation of bcl-2 expression. *J. Cell. Physiol.*, **195**, 130–138.

53 Ribatti, D., Levi-Schaffer, F., and Kovanen, P.T. (2008) Inflammatory angiogenesis in atherogenesis – a double-edged sword. *Ann. Med.*, **40**, 606–621.

54 Kaartinen, M., Penttilä, A., and Kovanen, P.T. (1996) Mast cells accompany microvessels in human coronary atheromas: implications for intimal neovascularization and hemorrhage. *Atherosclerosis*, **123**, 123–131.

55 Lappalainen, H., Laine, P., Pentikäinen, M.O., Sajantila, A., and Kovanen, P.T. (2004) Mast cells in neovascularized human coronary plaques store and secrete basic fibroblast growth factor, a potent angiogenic mediator. *Arterioscler. Thromb. Vasc. Biol.*, **24**, 1880–1885.

56 Norrby, K. (2002) Mast cells and angiogenesis. *APMIS*, **110**, 355–371.

57 Saunders, W.B., Bayless, K.J., and Davis, G.E. (2005) MMP-1 activation by serine proteases and MMP-10 induces human capillary tubular network collapse and regression in 3D collagen matrices. *J. Cell. Sci.*, **118**, 2325–2340.

58 de Nooijer, R., Verkleij, C.J., von der Thüsen, J.H., Jukema, J.W., van der Wall, E.E., van Berkel, T.J., Baker, A.H., and Biessen, E.A. (2006) Lesional overexpression of matrix metalloproteinase-9 promotes intraplaque hemorrhage in advanced lesions but not at earlier stages of atherogenesis. *Arterioscler. Thromb. Vasc. Biol.*, **26**, 340–346.

59 Sluimer, J.C., Kolodgie, F.D., Bijnens, A.P., Maxfield, K., Pacheco, E., Kutys, B., Duimel, H., Frederik, P.M., van Hinsbergh, V.W., Virmani, R., and Daemen, M.J. (2009) Thin-walled microvessels in human coronary atherosclerotic plaques show incomplete endothelial junctions relevance of compromised structural integrity for intraplaque microvascular leakage. *J. Am. Coll. Cardiol.*, **53**, 1517–1527.

60 Jeziorska, M. and Woolley, D.E. (1999) Neovascularization in early atherosclerotic lesions of human carotid arteries: its potential contribution to plaque development. *Hum. Pathol.*, **30**, 919–925.

61 Jeziorska, M. and Woolley, D.E. (1999) Local neovascularization and cellular composition within vulnerable regions of atherosclerotic plaques of human carotid arteries. *J. Pathol.*, **188**, 189–196.

62 Strbian, D., Kovanen, P.T., Karjalainen-Lindsberg, M.L., Tatlisumak, T., and Lindsberg, P.J. (2009) An emerging role of mast cells in cerebral ischemia and hemorrhage. *Ann. Med.*, **41**, 438–450.

63 Lehtonen-Smeds, E.M.P., Mayranpaa, M., Lindsberg, P.J., Soinne, L., Saimanen, E., Jarvinen, A.A., Salonen, O., Carpén, O., Lassila, R., Sarna, S., Kaste, M., and Kovanen, P.T. (2005) Carotid plaque mast cells associated with atherogenetic serum lipids, high grade carotid stenosis and symptomatic carotid artery disease. Results from Helsinki Carotid Endarterectomy Study. *Cerebrovasc. Dis.*, **19**, 291–301.

64 Di Girolamo, N. and Wakefield, D. (2000) *In vitro* and *in vivo* expression of interstitial collagenase/MMP-1 by human mast cells. *Dev. Immunol.*, **7**, 131–142.

65 Baram, D., Vaday, G.G., Salamon, P., Drucker, I., Hershkoviz, R., and Mekori, Y.A. (2001) Human mast cells release metalloproteinase-9 on contact with activated T cells: juxtacrine regulation by TNF-alpha. *J. Immunol.*, **167**, 4008–4016.

66 Di Girolamo, N., Indoh, I., Jackson, N., Wakefield, D., McNeil, H.P., Yan, W., Geczy, C., Arm, J.P., and Tedla, N. (2006) Human mast cell-derived gelatinase B (matrix metalloproteinase-9) is regulated by inflammatory cytokines: role in cell migration. *J. Immunol.*, **177**, 2638–2650.

67 Koskivirta, I., Rahkonen, O., Mäyränpää, M., Pakkanen, S., Husheem, M., Sainio,

A., Hakovirta, H., Laine, J., Jokinen, E., Vuorio, E., Kovanen, P., and Järveläinen, H. (2006) Tissue inhibitor of metalloproteinases 4 (TIMP4) is involved in inflammatory processes of human cardiovascular pathology. *Histochem. Cell Biol.*, **126**, 335–342.

68 Libby, P. and Shi, G.P. (2007) Mast cells as mediators and modulators of atherogenesis. *Circulation*, **115**, 2471–2473.

69 Grimbaldeston, M.A., Chen, C.C., Piliponsky, A.M., Tsai, M., Tam, S.Y., and Galli, S.J. (2005) Mast cell-deficient W-sash c-kit mutant Kit W-sh/W-sh mice as a model for investigating mast cell biology *in vivo. Am. J. Pathol.*, **167**, 835–848.

70 Sun, J., Sukhova, G.K., Wolters, P.J., Yang, M., Kitamoto, S., Libby, P., MacFarlane, L.A., Mallen-St Clair, J., and Shi, G.P. (2007) Mast cells promote atherosclerosis by releasing proinflammatory cytokines. *Nat. Med.*, **13**, 719–724.

71 Lindstedt, K.A., Mäyränpää, M.I., and Kovanen, P.T. (2007) Mast cells in vulnerable atherosclerotic plaques – a view to a kill. *J. Cell. Mol. Med.*, **11**, 739–758.

72 Bot, I., de Jager, S.C., Zernecke, A., Lindstedt, K.A., van Berkel, T.J., Weber, C., and Biessen, E.A. (2007) Perivascular mast cells promote atherogenesis and induce plaque destabilization in apolipoprotein E-deficient mice. *Circulation*, **15** (115), 2516–2525.

73 Johnson, J.L., George, S.J., Newby, A.C., and Jackson, C.L. (2005) Divergent effects of matrix metalloproteinases 3, 7, 9, and 12 on atherosclerotic plaque

stability in mouse brachiocephalic arteries. *Proc. Natl Acad. Sci. U. S. A.*, **102**, 15575–15580.

74 Gough, P.J., Gomez, I.G., Wille, P.T., and Raines, E.W. (2006) Macrophage expression of active MMP-9 induces acute plaque disruption in apoE-deficient mice. *J. Clin. Invest.*, **116**, 59–69.

75 Pejler, G., Abrink, M., Ringvall, M., and Wernersson, S. (2007) Mast cell proteases. *Adv. Immunol.*, **95**, 167–255.

76 Tchougounova, E., Pejler, G., and Abrink, M. (2003) The chymase, mouse mast cell protease 4, constitutes the major chymotrypsin-like activity in peritoneum and ear tissue. A role for mouse mast cell protease 4 in thrombin regulation and fibronectin turnover. *J. Exp. Med.*, **198**, 423–431.

77 Bischoff, S.C. (2007) Role of mast cells in allergic and non-allergic immune responses: comparison of human and murine data. *Nat. Rev. Immunol.*, **7**, 93–104.

78 Metz, M., Grimbaldeston, M.A., Nakae, S., Piliponsky, A.M., Tsai, M., and Galli, S.J. (2007) Mast cells in the promotion and limitation of chronic inflammation. *Immunol. Rev.*, **217**, 304–328.

79 Burke, A.P., Kolodgie, F.D., Farb, A., Weber, D.K., Malcom, G.T., Smialek, J., and Virmani, R. (2001) Healed plaque ruptures and sudden coronary death: evidence that subclinical rupture has a role in plaque progression. *Circulation*, **103**, 934–940.

80 Newby, A.C. (2005) Dual role of matrix metalloproteinases (matrixins) in intimal thickening and atherosclerotic plaque rupture. *Physiol. Rev.*, **85**, 1–31.

PART IV
Hyperlipidemia

Atherosclerosis: Molecular and Cellular Mechanisms. Edited by Sarah Jane George and Jason Johnson
Copyright © 2010 WILEY-VCH Verlag GmbH & Co. KGaA, Weinheim
ISBN: 978-3-527-32448-4

12

Macrophage Foam Cell Formation: The Pathways to Cholesterol Engorgement

Kathryn J. Moore and Katey Rayner

12.1
Introduction

The subendothelial retention of cholesterol-rich, atherogenic lipoproteins is thought to be the central pathogenic event that incites atherosclerotic lesion formation. This intramural retention of apolipoprotein B (ApoB)–rich lipoproteins in susceptible areas of the vasculature has been linked to the interaction of positively charged domains on ApoB to negatively charged elements of arterial matrix, particularly proteoglycans [1]. Lipoproteins sequestered in this microenvironment are then susceptible to various modifications including oxidation, enzymatic cleavage, and aggregation, which render these particles proinflammatory (summarized in Figure 12.1). The ensuing immune response to these modified lipoproteins, a maladaptive macrophage-dominated inflammation, both establishes macrophage foam cell formation and drives the pathogenic evolution of the atherosclerotic plaque.

Atherosclerotic inflammation of the artery wall is notable for the preponderance of macrophages relative to other immune cells. Monocytes are rapidly recruited to regions of lipoprotein retention in the intima, presumably in an effort to clear these inflammatory lipids. While macrophage uptake of these lipoproteins is likely to be beneficial at the outset of this immune response, these cells become grossly engorged with lipid. Termed "foam cells" because of the foamy appearance of lipid droplets in their cytoplasm, these cholesterol-laden macrophages persist in the intima and are the foundation of the fatty streak lesion. Electron micrographs of such early atherosclerotic lesions in animal models have documented lipid-laden foam cells emerging from the intima into the arterial lumen [2–5], suggesting that, at least initially, macrophages ferry lipid out of the artery wall. However, the forces that promote foam cell accumulation appear to prevail over those that allow their egress from the artery wall, establishing a chronic inflammation and lesion formation. As the atherosclerotic plaque evolves, macrophages are joined by other inflammatory cell subsets (T cells, dendritic cells, neutrophils, mast cells) and local smooth muscle cells that contribute to the cellular environ-

Atherosclerosis: Molecular and Cellular Mechanisms. Edited by Sarah Jane George and Jason Johnson
Copyright © 2010 WILEY-VCH Verlag GmbH & Co. KGaA, Weinheim
ISBN: 978-3-527-32448-4

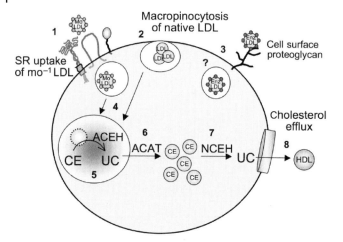

Figure 12.1 Macrophage cholesterol uptake and storage leading to foam cell formation. The macrophage internalization of atherogenic lipoproteins is presumed to occur via: (1) endocytosis by lipoprotein receptors that are not downregulated by increased cholesterol content, such as the scavenger receptors, (2) by macropinocytic uptake of native LDL, or (3) uptake of sPLA$_2$ modified LDL (mo-LDL) by cell surface proteoglycans, such as syndecan-4. While the metabolic fate of lipoproteins in this third category is not yet known, lipoproteins that enter the endocytic pathway, such as ox-LDL and pinocytosed native LDL, are internalized into vesicles that subsequently fuse with lysosomes and become acidified (4). Internalized cholesterol esters (CE) are then cleaved to unesterified cholesterol (UC) by acid cholesterol ester hydrolase (ACEH) (5). This UC is transported out of lysosomes and re-esterified by acyl-coenzyme A:cholesterol acyltransferase (ACAT) to form cholesterol esters that are stored in intracellular lipid droplets (6). Cholesterol esters are cleaved to cholesterol by neutral cholesterol ester hydrolase (NCEH) (7), which is available to be effluxed to the extracellular space by several potential transporters (including ABCA1, ABCG1, and SR-B1) to acceptor molecules such as HDL and ApoAI (8). Imbalance in this lipid metabolism pathway results in excessive intracellular cholesterol accumulation and macrophage foam cell formation.

ment of the plaque. In these more advanced lesions, macrophage foam cells remain key regulators of plaque behavior through their secretion of inflammatory mediators (cytokines, reactive oxygen species) and matrix-degrading proteases. Furthermore, the eventual demise of the foam cell and release of its lipid contents leads to the formation of a key component of the vulnerable plaque, the necrotic core. Thus, in both life and death, the macrophage foam cell profoundly influences atherogenesis.

As a result of its central role in the pathobiology of atherosclerosis, the mechanisms underlying macrophage foam cell formation have been intensely studied during the last 40 years. Multiple means of low-density lipoprotein (LDL) modification have been identified that facilitate the conversion of macrophage into foam cells *in vitro* (Figure 12.1); however, the physiologically relevant *in vivo* pathways remain a matter of speculation. A widely accepted paradigm for much of the last two decades has been the oxidative modification hypothesis proposed by Steinberg and colleagues, in which heightened oxidative stress in the vascular wall gives rise to oxidized LDL (ox-LDL) that is readily taken up by macrophages [6, 7]. However, other credible hypotheses have emerged in recent years which suggest that modification of LDL by lipolytic enzymes such as secretory phospholipase A_2, [8, 9] as well as the uptake of native LDL by novel mechanisms also contribute to macrophage cholesterol loading in the artery wall. With the appreciation that macrophage foam cell formation likely arises from multiple, simultaneous pathways, new questions have arisen regarding the different effects of these lipid loading pathways on macrophage behavior. It has become clear that the cellular pathways by which lipoprotein particles are internalized and subsequently metabolized may not only influence the amount of lipid taken up, but also its availability storage and/or efflux, as well as the inflammatory state of the macrophage. This complex interaction is likely to impact strategies to modify the development of foam cells by manipulating receptor-mediated uptake or efflux of lipids.

12.2
ox-LDL, the Scavenger Receptors and Foam Cell Formation

Oxidized forms of LDL have been shown to be abundant in human and mouse atheroma and are a major target of the humoral immune response [6]. The purification of these lipids has identified biologically active oxidation epitopes that bind with high affinity to pattern recognition receptors, including those of the scavenger receptor family [10–13]. The prevailing model for much of the last two decades indicated that macrophage foam cell formation occurs through endocytosis of oxidized lipoproteins via these receptors, which unlike the LDL receptor, are not downregulated by increased cellular cholesterol content. Consequently, blocking scavenger receptor–mediated uptake of oxidized lipids was proposed to be a mechanism to abrogate foam cell formation and atherosclerosis. However, disappointing outcomes of antioxidant trials and inconsistent findings upon genetic manipulation of the scavenger receptors have led some investigators to question this widely accepted paradigm. Nevertheless, these studies have revealed the breadth of scavenger receptor functions in atherogenesis, including the activation of signal transduction pathways regulating inflammation, apoptotic cell clearance, chemoattraction, and angiogenesis. It is now clear that in addition to promoting macrophage foam cell formation these receptors contribute to both to pro- and anti-inflammatory forces in the artery wall.

12.2.1
Scavenger Receptor A

Scavenger receptor A (SR-A), originally termed the macrophage scavenger receptor (gene name *Msr1*), was the first identified member of the scavenger receptor family. This receptor was found by Brown and Goldstein to mediate the unregulated uptake and degradation of acetylated LDL, leading to massive cholesterol loading of macrophages [14]. The identification of this receptor spawned a new field of macrophage biology in which multiple similar receptors were identified that can endocytose acetylated or oxidized lipoproteins, as well as other polyanionic molecules and pathogen components [15].

The principal hypothesis surrounding the role for SR-A in atherogenesis is that this receptor responds to increased modified (i.e., oxidized) lipid deposits within the vessel wall and attempts to clear these unwanted deposits by engulfing them. Using receptor-mediated endocytosis, the lipoproteins are swallowed into clathrin-coated pits that are sorted to the endosome and subsequently the lysosome, where, because of the acidic pH of this compartment, ligands are released from the receptor and receptors can be recycled back to the plasma membrane (reviewed in [16]). The modified lipid taken up via SR-A is processed by the lysosome and broken up into its building blocks, which includes free cholesterol (FC) and fatty acids (FAs). The FC accumulates in lipid droplets, becoming larger and larger as the macrophage continues to engulf lipids. Normally this FC can be recycled back to high-density lipoprotein (HDL) particles through reverse cholesterol transport via another member of the scavenger receptor family, SR-B1, and the ATP-binding cassette (ABC) family of transporters, but this transport appears to become impaired due to the overwhelming amount of modified lipid engulfed.

These macrophages become loaded with lipid and appear trapped in the vessel wall, where they readily secrete proinflammatory cytokines (i.e., interleukin-1β (IL-1β), tumor necrosis factor-α (TNF-α)), chemokines (i.e., monocyte chemotactic protein-1 (MCP-1)), and reactive oxygen species, recruiting additional inflammatory cells and ultimately leading to the demise of the surrounding healthy tissue.

If SR-A is indeed responsible for this inflammatory propagation of the atherosclerotic process, then it stands to reason that deletion of this receptor would lead to decreased foam cell formation and smaller and less inflammatory lesions. Indeed, the initial reports of SR-AI/II knockout mice ($Msr^{-/-}$) crossed onto an $Apoe^{-/-}$ hybrid (ICR/129) background demonstrated a greater than 50% reduction in atherosclerotic lesion after feeding with a normal chow diet [17]. Similarly, decreased foam cell formation in response to acetylated LDL (ac-LDL) was observed in macrophages from these SR-A-null mice, confirming that this receptor is indeed responsible for the majority of the uptake of this modified lipid. Subsequent studies also demonstrated a 20–30% reduction in lesion area in $Msr^{-/-}$ mice on the $Ldlr^{-/-}$ background compared to $Ldlr^{-/-}$ control mice [18].

These early studies solidified the role for SR-A in the propagation of foam cell formation and its contribution to atherosclerosis. However, disparate results were obtained when SR-A was deleted on the ApoE3-Leiden background with more extensive hyperlipidemia, where there was no observed protection from atherosclerosis and lesions were even slightly (though not statistically significantly) larger when SR-A was absent [19]. The plot continued to thicken when a >80% decrease in lesion size was observed in $Msr^{-/-}$ mice backcrossed six generations on the $Apoe^{-/-}$ C57BL6 background fed a moderately high-fat diet [20]; however, a 40%

increase in lesion size was observed in male *Msr*$^{-/-}$ mice backcrossed seven generations onto the *Apoe*$^{-/-}$ C57BL6 background and no change at all was observed in their female counterparts [21].

An overexpression model was employed by independent groups and showed that high levels of SR-A in the vessel wall does not exacerbate atherosclerosis as it would be predicted to do, and macrophage-specific overexpression of SR-A even protected from atherosclerosis in the arch by >70% [22–24]. More recently, Kuchibholta *et al.* showed that *Msr*$^{-/-}$ mice backcrossed seven generations onto the *Apoe*$^{-/-}$ C57BL6 background showed no difference in lesion size throughout the aorta in male mice after 12 weeks of high-fat feeding, although female mice showed a 30% decrease in overall lesion area with the absence of SR-A [25]. The overwhelming conclusion from these mouse models is that the role for SR-A in atherosclerosis development is not as straightforward as we had once believed. Experimental differences in each model such as the background strain used (i.e., the C57BL6 strain is considered atherosclerosis-susceptible whereas the ICR/129 strain used in the earlier studies is considered atherosclerosis-resistant) and the severity of hyperlipidemia are likely to have profound effects on the conclusions reached by each model and the resulting confusion surrounding the role for this receptor in the disease process.

There also appears to be gender-specific effects and region-specific effects on lesion development in the absence of SR-A that are, as of yet, incompletely understood. Moreover, we now appreciate that SR-A plays a role in a variety of cellular processes (i.e., the clearance of apoptotic cells and regulation of cell migration) that may be extremely important for the maintenance of a normal healthy vessel wall and become perturbed upon deletion of this receptor [15]. An unambiguous role for SR-A in the development of atherosclerosis has yet to be elucidated.

12.2.2
The Class B Scavenger Receptor CD36

Another member of the scavenger receptor family with important roles in foam cell formation is CD36, a B-class scavenger receptor responsible for the majority of ox-LDL uptake in macrophages [26, 27]. Like SR-A, CD36 engulfs modified lipids within the vessel wall in an attempt to clear them; however this receptor is believed to set off an inflammatory cascade leading to recruitment of additional macrophages and propagation of disease. In the absence of CD36, macrophages *in vitro* have reduced accumulation of lipid and overall reduced foam cell formation [27]. Based on these findings, it was predicted that in the absence of CD36 *in vivo*, fewer foam cells would accumulate within the vessel wall and as such atherosclerotic lesion development would be impaired. The first of such studies using the *Apoe*$^{-/-}$ mouse model as a background reported a >70% decrease in *en face* aortic lesion area in mice lacking CD36 [28]. Interestingly, however, only male *Cd36*$^{-/-}$ mice had reduced lesions at the aortic sinus (45% reduction) whereas female mice showed no difference, suggesting a possible gender-specific and site-specific role for CD36 in disease progression.

In mice fed a normal chow diet (and thus with only modest hyperlipidemia) the absence of CD36 reduced sinus lesion size in male mice, whereas female mice had only a slight and non-statistically significant difference in lesion area. However, like the studies of SR-A deletion, additional studies of *Cd36*[-/-] mice provided inconsistent protection against atherosclerosis. In *Cd36*[-/-]/*Apoe*[-/-] mice backcrossed an additional three generations into the C57BL6 strain, the absence of CD36 resulted in an increase in aortic sinus lesion area in both male and female mice fed a Western diet for eight weeks, despite the fact that macrophages from these mice showed reduced cholesterol ester content [21]. Moreover, while lesion area in the descending aorta of mice lacking CD36 was decreased, the benefit was much less than previously seen at 12 weeks. If the primary role of CD36 was indeed in macrophage foam cell formation, the earlier time-point might have been predicted to show more dramatic effects of CD36 deletion.

Most recently, an additional CD36-mutant mouse created in the C57BL6 background (thus eliminating even 1% differences in background strain) was crossed onto *Apoe*[-/-] and fed a high-fat diet for 12 weeks. A 45% and 70% reduction in lesion area was found in the descending aorta of male and female mice, respectively [25].

Transplantation studies have shed more light on the role of macrophage-specific expression of CD36, *Apoe*[-/-] mice reconstituted with *Cd36*[-/-] bone marrow prior to high-fat diet show significant reductions in *en face* aortic lesion area [29]. Together, these studies suggest that targeted deletion of CD36 in mouse models of atherosclerosis is protective, albeit in a site-specific manner in the descending aorta. CD36 also likely plays a temporal role in atheroma development, which may differ at early versus later stages of the disease process. While CD36 knockout mouse studies have underscored the importance of this receptor in the pathogenesis of atherosclerosis, clearly many questions remain to be addressed.

Given the confusing and often conflicting results in the development of atherosclerosis in mice deficient in either of the two "primary" receptors for modified LDL uptake (SR-A and CD36), it was hypothesized that one of these receptors might be compensating for the lack of the other one, allowing foam cell formation to continue and as such atherosclerosis to ensue. Studies were performed in mice deficient for both CD36 and SR-A, backcrossed seven generations onto the *Apoe*[-/-] background and fed a high-fat diet. Two groups demonstrated that in the absence of both scavenger receptors, foam cell accumulation within the atherosclerotic plaque was unimpaired, and lesions in the aortic sinus were not reduced in size [25, 30]. These results indicate that other receptors, or potentially non-receptor-mediated uptake pathways, participate in the conversion of macrophages to foam cells *in vivo*.

In the absence of a change in lesion size, however, Tobin-Manning *et al.* showed significant reductions in the degree of apoptosis and necrotic core formation within the plaque of mice deficient for both CD36 and SR-A [30]. Moreover, both groups reported a reduction in inflammatory markers in macrophages lacking CD36 and SR-A [25, 30], suggesting that these receptors promote proinflammatory processes that may lead to more complex atherosclerotic plaques.

12.2.3
Lectin-Like ox-LDL Receptor LOX-1

The lectin-like ox-LDL receptor (LOX-1) is a transmembrane receptor capable of taking up ox-LDL and other SR ligands (reviewed in [31]). Unlike SR-A and CD36, LOX-1 appears to be primarily expressed on endothelial cells, and to a lesser extent on smooth muscle cells, platelets and macrophages [32]. Given its endothelial localization, LOX-1 is thought to be the primary receptor responsible for ox-LDL uptake by endothelial cells, and consequently a mediator of endothelial dysfunction and apoptosis [33]. The expression of LOX-1 is induced by ox-LDL as well as cytokines, shear stress, and hypertension, and LOX-1 is highly expressed within atherosclerotic plaques in both mice and humans (reviewed in [31]). In a mouse model of atherosclerosis, overexpression of LOX-1 increased the uptake of ox-LDL and caused intramyocardial vasculopathy, implying that this receptor does indeed play a role in foam cell formation *in vivo* [34]. Similarly, deletion of LOX-1 in $Ldlr^{-/-}$ mice reduced atherosclerotic lesion size by 40% and also preserved normal endothelial function [31].

   Interestingly, these mice had reduced collagen content within the plaque, as well as decreased expression of extracellular matrix proteases such as MMP-2 and MMP-9, which have been implicated in increasing plaque vulnerability [35]. These studies indicate that LOX-1 deletion may protect from atherosclerosis by maintaining endothelial function and reducing accumulation of extracellular matrix (ECM) proteins, resulting in increased plaque stability. Recently, using a hepatic-specific overexpression model, LOX-1 was demonstrated to reduce circulating ox-LDL, resulting in a 38% decrease in lesion area in mice overexpressing LOX-1 in the liver [36]. These results suggest that LOX-1 may function to remove ox-LDL from the circulation, making it less available to tissue macrophages to induce inflammatory lesion formation. However, it is not known what, if any, contribution LOX-1 makes to macrophage foam cell formation and further studies in this area are warranted.

12.2.4
CXCL16/SR-PSOX: A Scavenger Receptor and a Chemokine

With dual functions as a chemokine and a scavenger receptor, CXCL16, also known as SR-PSOX, is a unique member of the scavenger receptor family. On the cell surface, CXCL16 can engulf ox-LDL and hence incite foam cell formation [37], whereas in its cleaved secreted form it can lead to the recruitment of T cells that express the CXCR6 receptor, making it a potent inducer of cell migration and hence inflammatory plaque progression [38, 39]. Upon deletion of this receptor, macrophages showed >30% reduction in ox-LDL uptake, suggesting that this receptor accounts for a significant proportion of modified lipid uptake in macrophages and could potentially account for the significant foam cell formation *in vivo* in SR-A and CD36-null mice. Surprisingly however, targeted deletion of CXCL16 in $Apoe^{-/-}$ mice increased atherosclerotic lesion formation compared with controls, indicating that the observed reduction of macrophage *in vitro* did not correlate with protection of atherosclerosis *in vivo* [40]. The reasons for this discrepancy remain unclear, and given the complexity of this scavenger receptor family member, CXCL16 may protect against atherosclerosis through any number of mechanisms.

12.2.5
New Insights into Scavenger Receptor Function in Atherosclerosis

The paradigm for the role of scavenger receptors as simply cell surface molecules intended to mop up excessive lipid products within the vessel wall is changing. We now understand that in the absence of these receptors, foam cell formation is unimpaired and lesions can progress without reserve. However, we are also beginning to understand more about the complex role that scavenger receptors play in a variety of other cellular processes that are central to the pathogenesis of atherosclerosis, including cell migration, apoptosis, and inflammatory signaling [15].

12.2.5.1 Cell Migration
There is a new theory emerging regarding the effects of modified lipoproteins on macrophage function in the atherosclerotic plaque: that ox-LDL affects the migration of foam cells in and out of the vessel wall via interactions with scavenger receptors at the surface. Recently Park *et al.* showed that macrophages exposed to ox-LDL had reduced migratory ability in response to an inflammatory stimulus compared to naïve macrophages [41]. This occurred due to increased cell attachment and spreading, and sustained polymerization of actin via activation of focal adhesion kinase (FAK). Surprisingly, in the absence of CD36, the migratory ability of these cells was restored, indicating that activation of CD36 via ox-LDL has profound effects on cell migration. These results suggest that in the presence of elevated levels of ox-LDL (as is the case in human atheroma) macrophage foam cells have reduced ability to emigrate from the lesions to the lymph nodes, resulting in increased cell number, impaired clearance of lipid and subsequent increased inflammatory signaling.

Similarly, Stuart *et al.* reported that CD36 signals through the actin cytoskeleton via a p130Cas-PYK2 focal adhesion complex to control migration of macrophage microglial cells in response to β-amyloid, a potent inflammatory stimulus in Alzheimer's disease that acts in an analogous fashion to ox-LDL in the vessel wall [42]. This new role for CD36 cross-talking with the actin cytoskeleton to regulate cell motility could have profound effects on lesion development, and further insight into these mechanisms will certainly lead to new ways to target this pathway therapeutically.

12.2.5.2 Apoptosis
In addition to the traditional role for scavenger receptors in ox-LDL-induced inflammatory gene expression, these receptors appear to participate in both the regulation of macrophage foam cell apoptosis and the clearance of these dying cells. Recently it was demonstrated that in the absence of both CD36 and SR-A,

both macrophage apoptosis and necrotic core formation were reduced within the atherosclerotic plaque [30]. This is consistent with *in vitro* observations that free cholesterol can trigger apoptosis in macrophages downstream of SR-A via activation of ER stress pathways [43]. Although free cholesterol accumulation within the foam cell itself causes initiation of the unfolded protein response (a stress pathway activated by the ER), it is only following a "second hit" initiated by SR-A (and perhaps CD36) that apoptosis progresses. If apoptosis is excessive or if apoptotic cells are not cleared effectively, this results in the formation of a necrotic core and weakened plaque stability [44]. Notably, targeted inhibition of both SR-A and CD36 reduced necrotic core size, as well as markers of inflammation, suggesting that loss of these pathways retards the progression to more advanced necrotic lesions.

12.2.5.3 Novel Ligands

As well as having a high affinity for the oxidized and acetylated forms of LDL, scavenger receptors are known to bind bacterial pathogens, apoptotic cells, and negatively charged phospholipids [15]. Although foam cell formation proceeds after modified lipids are taken up via SR engagement, the accumulation of these other SR ligands within plaques could influence this process. For example, extracellular heat shock protein 27 (HSP27) was recently shown to bind SR-A, and prevent uptake of ac-LDL and inflammatory foam cell development [45]. This may be responsible for the observed decrease in atherosclerotic lesion development in mice overexpressing HSP27, suggesting that a competitive ligand for SR-A could assist in the prevention of disease progression [45]. Recently, a novel such peptide inhibitor of SR-A was identified using a phage display library that prevents uptake of ac-LDL *in vitro* [46]. However, further studies are required to assess its efficacy *in vivo*. In addition, various forms of amyloid fibrils found within atherosclerotic plaques bind and activate the SRs, including those found within oxidized lipoproteins, β-amyloid and α_1-antitrypsin [47]. At present, it is not known what role these alternative ligands may be playing in the pathobiology of atherosclerosis.

12.2.5.4 Other Mechanisms of SRs in Atherosclerosis

Although we now know SR-mediated uptake of modified lipids is not essential for *in vivo* foam cell formation [25, 30], the expression of these receptors throughout the development of disease can affect lesion development. A key molecule involved in antioxidant defense, Nrf2, was recently shown to regulate the expression of CD36 [48]. Deletion of Nrf2 reduced expression of CD36 in both macrophages and within the atherosclerotic plaque of *Nrf2$^{-/-}$/Apoe$^{-/-}$* mice, which resulted in decreased modified lipid uptake and reduced lesion formation. This study demonstrates a new role for an antioxidant protein in the regulation of foam cell formation, not via alterations in the oxidative state of the lipoproteins as it might have been predicted, but instead through the regulation of the expression of the receptors that modify their uptake. Similarly, in human macrophages *in vitro*, inhibition of protein kinase C-β (PKCβ) reduced expression of SR-A and subsequent uptake of ox-LDL and ac-LDL [49]. These represent novel mechanisms of manipulating foam cell formation via specific regulation of SR expression, which could have potential future therapeutic benefits.

SRs are also emerging as potential biomarkers of disease, and may be useful in the diagnosis of atherosclerosis and type 2 diabetes. Soluble, circulating forms of CD36 and LOX-1 are found in human plasma and can be detected using standard immunoassay techniques. A soluble form of CD36 (sCD36) was found to be elevated in patients with type 2 diabetes compared to healthy controls [50]. More recently, levels of this protein in the circulation were also found to correlate with symptomatic carotid atherosclerosis [51]. sCD36 levels were highest in patients with unstable plaques, and its expression was found to be enriched in foam cells within these plaques [51]. The authors speculate that CD36 is released from macrophage foam cells, perhaps due to increased proteolytic activity within the plaque as it becomes more unstable, resulting in the observed positive correlation with symptomatic events. Likewise, soluble LOX-1 was found to be elevated in patients with acute coronary syndrome [52].

Variations within the LOX-1 gene were recently shown to be associated with variations in the soluble form of this receptor, although the physiological effects of this gene variant and its relationship with atherosclerosis remain to be investigated.

Clearly scavenger receptors play a fundamental role in the pathogenesis of atherosclerosis. Whether through the excessive uptake of modified lipids or through signaling events leading to apoptosis and cell migration, these receptors are dynamic and multifaceted. As our understanding improves of the intricate ways in which these receptors can function within cells, we will no doubt gain additional insight into how these receptors can be targeted therapeutically to prevent or treat atherosclerosis.

12.3
Enzymatic Modifications of LDL Leading to Foam Cell Formation

Phospholipolytic enzymes, such as secretory phospholipase A_2 (sPLA$_2$) and secretory sphingomylenase (s-SMase), have been implicated in atherogenesis through their ability to modify LDL and promote its retention in the subendothelial space. Hydrolysis by either sPLA$_2$ or s-SMase brings about changes in the structure of the LDL particle that enhance its atherogenicity. These enzymatically modified lipoprotein particles have increased association with ECM proteoglycans, which not only mediate the retention of LDL, but also appear to modulate the activity of various enzymes towards LDL.

The sPLA$_2$ family comprises a group of enzymes that hydrolyze glycerophospholipids at the *sn-2* position to generate lysophospholipids and free fatty acids. Accumulating evidence suggests that at least four members of this family (groups IIA, III, V, and X) are upregulated in murine or human atherosclerotic lesions [53–57]. Hydrolysis of native LDL by these sPLA$_2$ family members alters the conformation of ApoB100 on the phospholipid-depleted particle, leading to exposure of cryptic proteoglycan-binding sites and an increased affinity for arterial proteoglycans [58, 59]. Furthermore, oxidative modification of LDL phospholipids increases its susceptibility to sPLA$_2$ transforming ox-LDL to a more atherogenic form [60]. sPLA$_2$-modified proteoglycan-bound lipoproteins also show a greater tendency to aggregation and fusion. The combined result of these sPLA$_2$ actions is a progressive deposition of lipids within the extracellular matrices of the arterial intima and an increased susceptibility to atherosclerotic lesion formation [61].

Early interest in this family of enzymes centered on sPLA$_2$-IIa, as it was highly regulated by inflammatory processes and found to be an independent risk factor for cardiovascular events in humans [62]. Subsequent evidence from gain-of-function studies in C57BL6 mice indicated that sPLA2-IIa was not just a marker for atherosclerosis, but was likely to also have a causal role in this disease. Mice globally overexpressing sPLA2-IIA were found to have spontaneous atherosclerotic lipid deposition in the absence of hypercholesterolemia [63]. Furthermore, bone marrow–restricted transgenic expression of human sPLA$_2$-IIA also promoted atherosclerosis in *Ldlr*$^{-/-}$ mice, suggesting that sPLA$_2$ can exert local effects in the artery wall [64–66]. Subsequent studies of sPLA$_2$-III and sPLA$_2$-V overexpression have also revealed enhanced atherosclerotic lesion formation in transgenic mice, indicating that several sPLA$_2$ subtypes may contribute to atherosclerotic processes [67, 68].

In addition to enhancing LDL trapping in the vessel wall, there is evidence that sPLA$_2$ modification also promotes macrophage LDL uptake and foam cell formation [57, 58, 67]. Mouse peritoneal macrophages avidly take up LDL hydrolyzed by sPLA$_2$-III, sPLA$_2$-V, and sPLA$_2$-X to form foam cells [53, 57, 69]. Studies using peritoneal macrophages deficient in CD36 and SR-A showed that the binding, internalization, and intracellular accumulation of sPLA$_2$-V-hydrolyzed LDL are independent of these two scavenger receptors [70]. In contrast to oxidative modification, sPLA$_2$-V hydrolysis does not increase the negative charge of the LDL particle which may account for the lack of interaction of hydrolyzed LDL with CD36 or SR-A, which are known to bind polyanionic molecules [70]. Rather, sPLA$_2$-V-modified LDL was found to interact with syndecan-4, a cell surface proteoglycan that mediates macropinocytic uptake via a PI3-kinase dependent mechanism [71]. Thus, sPLA$_2$-V hydrolysis appears to bring about structural changes in LDL particles that promote interactions with both cellular and ECM proteoglycans.

Recent results from a bone marrow transplantation study using the *Ldlr*$^{-/-}$ mouse model provide evidence that sPLA$_2$-V is indeed pro-atherogenic *in vivo* [72]. *Ldlr*$^{-/-}$ mice reconstituted with sPLA$_2$-V-deficient bone marrow showed a 36% reduction in atherosclerotic lesion area in the aortic arch/thoracic aorta compared to control *Ldlr*$^{-/-}$ mice after 14 weeks on an atherogenic diet. These data suggest that targeted inhibition of sPLA$_2$-V, and potentially other sPLA$_2$ family members, may have therapeutic value.

Recent results from studies using the sPLA$_2$ inhibitor Varespladib (A-002), which simultaneously inhibits groups IIa, V, and X, suggest such a prediction is true. Male *Apoe*$^{-/-}$ mice showed a dramatic decrease in atherosclerotic lesion formation when supplemented with Varespladib during 12 weeks high-fat diet feeding [73]. Compared with control mice, Varespladib-treated mice showed a 75% decrease in atherosclerotic lesion area, while also exhibiting a 200% increase in fibrous cap size. Interestingly, the macrophage content of the lesions did not significantly decrease when measured as a percent of lesion area [73]. The athero-protective effects of this sPLA$_2$ inhibitor were further enhanced by simultaneous treatment with the HMGCoA reductase inhibitor Pravastatin. While Pravastatin monotherapy of *Apoe*$^{-/-}$ mice showed no effect on atherosclerotic lesion size, when administered with Varespladib, the combined treatment led to additional decreases in lesion area and total cholesterol levels of 50% and 18%, respectively [73].

Similar protective effects of Varespladib have also been reported in a study using an accelerated *Apoe*$^{-/-}$ atherosclerosis model [74]. When angiotensin-II was used to induce aortic lesions and aneurysms in *Apoe*$^{-/-}$ mice, Varespladib reduced atherosclerosis by 40% and also attenuated aneurysm formation. Furthermore, as in the study by Shaposhnik *et al.*, treatment with Varespladib dramatically reduced atherosclerosis in *Apoe*$^{-/-}$ mice fed a high fat diet in the absence of angiotensin-II; however, the authors also reported a reduction in total plasma cholesterol not seen in the previous study. Together, these studies suggest that inhibiting sPLA$_2$ in human disease holds promise as a strategy to retard atherogenesis, particularly when combined with statin therapy.

Another lipolytic enzyme well-characterized for its effects in the artery wall is s-SMase. This enzyme has been detected by immunohistochemistry in human atherosclerotic lesions in association with the extracellular matrix, and its expression by arterial endothelial cells and macrophages is induced by inflammatory cytokines [75]. Exposure of lipoproteins (LDL, intermediate-density lipoprotein (IDL), or very low-density lipoprotein (VLDL)) to SMase *in vitro* induces aggregation and fusion of particles, likely due to increased ceramide content [9, 76–78]. Such SMase-induced aggregated/fused LDL shows enhanced binding to human aortic proteoglycans, and is a potent inducer of macrophage foam cell formation *in vitro* [9, 79]. Notably, while plasma LDL is not hydrolyzed by s-SMase at neutral pH, LDL extracted from human atherosclerotic lesions is hydrolyzed at pH 7.4 [80].

LDL oxidation or pretreatment with sPLA$_2$ confer susceptibility to s-SMase at neutral pH, and thus these different modifications of LDL may act in concert to promote subendothelial LDL retention, aggregation, and foam cell formation. Macrophage internalization of s-SMase-aggregated LDL occurs via a pathway that is distinct from receptor-mediated endocytosis of monomeric, soluble LDL [81]. Whereas the internalization of s-SMase-modified LDL is blocked by compounds that disrupt actin polymerization (cytochalasin D and latrunculin A), the uptake and degradation of soluble LDL is unaffected. However whether lipid loading of macrophages by these alternative pathways results in differences in inflammatory responses by the macrophage has not been investigated.

12.4
Native Lipoprotein Receptors and Foam Cell Formation

In addition to the modified lipoproteins discussed above, VLDL and VLDL remnants have been shown to induce macrophage foam cell formation in their native states [82–85]. Interaction with cell surface lipoprotein lipase hydrolyzes the core triglyceride of VLDL to free fatty acids that are taken up by the cell and re-esterified into triglyceride. The cholesterol ester (CE)-rich remnants are then readily endocytosed via receptors, including the LDL-R [86–88]. C57BL/6 mice placed on an atherogenic diet develop modest hypertriglyceridemia as a result of β-VLDL accumulation and reconstitution of these mice with LDLR-deficient bone marrow significantly decreases diet-induced atherosclerosis, supporting a role for this foam cell formation pathway *in vivo* [89, 90]. Residual accumulation of cholesterol ester in β-VLDL treated *Ldlr*$^{-/-}$ macrophages suggests that additional pathways can contribute to macrophage foam cell formation by this lipoprotein [88]. The LDL-R-related protein (LRP), VLDL-R, and SR-BI have all been proposed to mediate this residual VLDL uptake, however *in vivo* evidence for these receptors has not yet been obtained. LRP, which exhibits structural homology with the LDL-R and recognizes ApoE-containing lipoproteins, can endocytose VLDL and chylomicron remnants and has been suggested to interact with proteoglycans to allow lipoprotein uptake [91, 92]. However, inhibition of LRP in *Ldlr*$^{-/-}$ macrophages did not abrogate VLDL-induced foam cell formation [88].

Another promising candidate, the VLDL-R, also binds ApoE-containing lipoproteins such as VLDL and IDL, is not downregulated by intracellular cholesterol content and facilitates macrophage foam cell formation *in vitro* [93]. Expression of this receptor is markedly induced in rabbit and human atherosclerotic lesions, however targeted deletion of the VLDL-R in *Ldlr*$^{-/-}$ mice failed to establish a role for this receptor in atherogenesis [94, 95]. Studies in *Apoe*$^{-/-}$ macrophages indicate that VLDL and IDL also readily induce macrophage cholesterol ester accumulation in an ApoE-independent manner suggesting a role for additional unknown receptors [96]. The large body of evidence implicating triglyceride-rich lipoproteins in human atherosclerotic disease merits further study of these potential pathways for macrophage foam cell formation by native lipoproteins.

12.5
Receptor-Independent Macrophage Cholesterol Accumulation

Native LDL has traditionally not been considered capable of generating foam cells because cells downregulate their LDL receptor number with increasing cellular cholesterol content, precluding lipid loading via this pathway [97]. Early studies showing that native LDL, at concentrations that were saturating for the LDL receptor, did not trigger macrophage cholesterol loading established the paradigm that LDL modification in the intima was required to incite foam cell formation. However, recent findings suggest that when LDL levels reach a concentration range comparable to that achieved in the arterial intima under hyperlipidemic conditions (0.5–2 mg/ml), fluid-phase endocytosis by macrophages can mediate substantial uptake of native LDL [98]. This receptor-independent pathway of LDL internalization, known as macropinocytosis, has been demonstrated in human monocyte-derived macrophages treated with the protein kinase C activator PMA or macrophage colony-stimulating factor [99, 100].

Macropinocytosis is a non-saturable endocytic pathway that leads to LDL accumulation in endolysosomes and the stimulation of cholesterol esterification [99]. Thus, it appears that the fate of native LDL that enters macrophages by macropinocytosis is similar to that observed for scavenger receptor internalized modified LDL–entry into an endocytic pathway that leads to hydrolysis of cholesterol esters and trafficking to the endoplasmic reticulum for re-esterification and storage.

While macrophage macropinocytosis of native LDL is a plausible mechanism of cellular cholesterol ester accumulation, the contribution of this pathway to *in vivo* foam cell formation may be difficult to establish. Whereas inhibitors of actin polymerization and microtubule function (that disrupt macropinocytosis) blocked macrophage foam cell formation by native LDL *in vitro* [99], the essential nature of these cellular functions precludes their targeted disruption *in vivo*. Notably, a recent study reported that liver X receptor (LXR) ligands which promote reverse cholesterol transport also inhibit fluid-phase pinocytosis of native LDL by approximated 70%. By contrast, these ligands do not affect receptor-mediated uptake of ac-LDL, suggesting a specific downregulation by LXR agonists of macrophage pinocytosis of LDL and cholesterol accumulation. Thus, inhibition of this macrophage cholesterol loading pathway by LXR agonists may contribute to their atheroprotective effects.

12.6
Macrophage Cholesterol Efflux Pathways

In addition to lipoprotein-uptake pathways, macrophage foam cells are also modulated by pathways that promote the removal of excess cellular cholesterol. Although not the focus of this review, the transport mechanisms that promote the efflux of excess cholesterol from macrophages to extracellular acceptors (the first step of "reverse cholesterol transport") are of major importance in protecting against atherosclerotic lipid accumulation. HDL and HDL apolipoproteins play a key role in this reverse cholesterol transport by serving as the primary acceptors for macrophage cholesterol efflux (reviewed in [101, 102]). A major advance in our understanding of macrophage cholesterol efflux pathways has been the identification of ABC transporters that facilitate the transfer of cellular cholesterol to extracellular acceptors. It is now recognized that two members of the ABC superfamily of transmembrane transporters, ABCA1 and ABCG1, play critical roles in preventing cholesterol lipid accumulation in macrophages (see recent reviews on this topic [102–104]).

Extensive studies have shown that ABCA1 promotes the efflux of both cholesterol and choline-containing phospholipids to HDL apolipoproteins, such as ApoAI [105, 106]. Mutations in the ABCA1 transporter result in Tangiers disease, which is characterized by low HDL levels and the accumulation of cholesterol in macrophages [107–110]. In addition, ABCG1 plays a critical role in maintaining lipid homeostasis in a variety of tissues and cells, most notably macrophages, through its ability to stimulate cholesterol efflux to HDL, but not to lipid-poor ApoAI [111].

In vivo studies addressing the roles of these transporters in macrophage cholesterol efflux suggest that ABCA1 and ABCG1 act cooperatively to promote macrophage reverse cholesterol transport [112]. Thus, strategies to increase ABCA1 and ABCG1 activity in macrophages may also provide attractive targets for ameliorating foam cell formation and atherosclerosis.

12.7
Summary

Thirty years have passed since the original observation that macrophages take up and degrade modified forms of LDL. Despite the abundance of research directed at understanding macrophage foam cell biology during the intervening years, many questions remain to be addressed. Clearly, a large number of molecules contribute to LDL retention and uptake, and unraveling the nature in which they interact will provide challenges for future investigations. The selective inhibition of individual pathways involved in LDL uptake or modification in mouse models of atherosclerosis has revealed the complexity of foam cell formation in vivo. As might be expected, these in vivo studies have uncovered a redundancy in the lipid uptake pathways leading to foam cell formation. However, the identification of various alternative pathways of macrophage cholesterol loading has also led to a greater appreciation of the notion that the cellular pathways by which particles are internalized and subsequently metabolized may not only influence the amount of lipid taken up by macrophages, but also the inflammatory state of the macrophage.

Although scavenger receptors were long postulated to represent the main pathways for the internalization of modified lipoproteins and cholesterol loading of macrophages, studies of scavenger receptor inhibition in vivo have not uniformly shown a clear benefit. While these studies have yielded several surprising results that have opened new lines of inquiry, such as the contribution of these pathways to inflammation and tissue homeostasis, experiments in mouse models of atherosclerosis, particularly those using Western diet supplementation, should be interpreted with some caution. Whereas Western diet feeding conveniently accelerates atherosclerosis development in the mouse and reduces the length of these studies, the circulating cholesterol levels achieved in $Apoe^{-/-}$ or $Ldlr^{-/-}$ mice under these conditions do not approximate those observed in humans. In the presence of such marked hypercholesterolemia, the local concentration of LDL in the aortic intima may lead to saturating conditions that overwhelm specific receptors, triggering backup receptor pathways not previously implicated or promoting additional modifications of lipoproteins that result in uptake by non-receptor-mediated pathways.

Rather than the originally envisioned "single modification model" in which a single change in LDL would render it atherogenic, it appears more likely that several different modifications of LDL may act in concert to promote its suben-dothelial retention, aggregation and its unrestricted uptake by macrophages. It has also become appreciated that native and modified lipoprotein receptors are not the only portal of macrophage entry for lipoproteins accumulating in the artery wall. A number of studies support a role for non-receptor-mediated lipid uptake path-ways, however, the relevance of these non-receptor-mediated mechanisms for macrophage cholesterol loading *in vivo* may be difficult to determine. Classic loss-of-function studies to address the role of such pathways in mice are unlikely to be successful, as such strategies would also target functions essential to basic mac-rophage homeostatic responses.

Many questions remain regarding how the different mechanisms of cholesterol loading alter the inflammatory status of the macrophage and the availability of cholesterol ester pools to undergo efflux. Lipoproteins or their products have the potential to activate key signaling pathways that influence inflammatory responses such as nuclear hormone receptors, scavenger receptors, and Toll-like receptors [15]. Depending on the presentation of these lipids, macrophage responses to cholesterol loading may be varied. Clarification of how intervention of specific pathways affects these parameters will be required in order to facilitate the devel-opment of therapeutic agents to redirect lipid uptake and alter macrophage behavior.

References

1 Skalen, K., Gustafsson, M., Rydberg, E.K., Hulten, L.M., Wiklund, O., Innerarity, T.L., and Boren, J. (2002) Subendothelial retention of atherogenic lipoproteins in early atherosclerosis. *Nature*, **417** (6890), 750–754.

2 Gerrity, R.G. (1981) The role of the monocyte in atherogenesis: II. Migration of foam cells from atheroscle-rotic lesions. *Am. J. Pathol.*, **103** (2), 191–200.

3 Gerrity, R.G. and Naito, H.K. (1980) Lipid clearance from fatty streak lesions by foam cell migration. *Artery*, **8** (3), 215–219.

4 Kling, D., Holzschuh, T., and Betz, E. (1993) Recruitment and dynamics of leukocytes in the formation of arterial intimal thickening – a comparative study with normo- and hypercholesterolemic rabbits. *Atherosclerosis*, **101** (1), 79–96.

5 Landers, S.C., Gupta, M., and Lewis, J.C. (1994) Ultrastructural localization of tissue factor on monocyte-derived macrophages and macrophage foam

cells associated with atherosclerotic lesions. *Virchows Arch.*, **425** (1), 49–54.

6 Stocker, R. and Keaney, J.F., Jr. (2004) Role of oxidative modifications in atherosclerosis. *Physiol. Rev.*, **84** (4), 1381–1478.

7 Steinberg, D., Parthasarathy, S., Carew, T.E., Khoo, J.C., and Witztum, J.L. (1989) Beyond cholesterol. Modifica-tions of low-density lipoprotein that increase its atherogenicity. *N. Engl. J. Med.*, **320** (14), 915–924.

8 Gesquiere, L., Cho, W., and Subbaiah, P.V. (2002) Role of group IIa and group V secretory phospholipases A(2) in the metabolism of lipoproteins. Substrate specificities of the enzymes and the regulation of their activities by sphingomyelin. *Biochemistry*, **41** (15), 4911–4920.

9 Xu, X.X. and Tabas, I. (1991) Sphingo-myelinase enhances low density lipoprotein uptake and ability to induce cholesteryl ester accumulation in

macrophages. *J. Biol. Chem.*, **266** (36), 24849–24858.

10 Boullier, A., Gillotte, K.L., Horkko, S., Green, S.R., Friedman, P., Dennis, E.A., Witztum, J.L., Steinberg, D., and Quehenberger, O. (2000) The binding of oxidized low density lipoprotein to mouse CD36 is mediated in part by oxidized phospholipids that are associated with both the lipid and protein moieties of the lipoprotein. *J. Biol. Chem.*, **275** (13), 9163–9169.

11 Podrez, E.A., Poliakov, E., Shen, Z., Zhang, R., Deng, Y., Sun, M., Finton, P.J., Shan, L., Febbraio, M., Hajjar, D.P., Silverstein, R.L., Hoff, H.F., Salomon, R.G., and Hazen, S.L. (2002) A novel family of atherogenic oxidized phospholipids promotes macrophage foam cell formation via the scavenger receptor CD36 and is enriched in atherosclerotic lesions. *J. Biol. Chem.*, **277** (41), 38517–38523.

12 Podrez, E.A., Poliakov, E., Shen, Z., Zhang, R., Deng, Y., Sun, M., Finton, P.J., Shan, L., Gugiu, B., Fox, P.L., Hoff, H.F., Salomon, R.G., and Hazen, S.L. (2002) Identification of a novel family of oxidized phospholipids that serve as ligands for the macrophage scavenger receptor CD36. *J. Biol. Chem.*, **277** (41), 38503–38516.

13 Watson, A.D., Leitinger, N., Navab, M., Faull, K.F., Horkko, S., Witztum, J.L., Palinski, W., Schwenke, D., Salomon, R.G., Sha, W., Subbanagounder, G., Fogelman, A.M., and Berliner, J.A. (1997) Structural identification by mass spectrometry of oxidized phospholipids in minimally oxidized low density lipoprotein that induce monocyte/ endothelial interactions and evidence for their presence *in vivo*. *J. Biol. Chem.*, **272** (21), 13597–13607.

14 Brown, M.S., Goldstein, J.L., Krieger, M., Ho, Y.K., and Anderson, R.G. (1979) Reversible accumulation of cholesteryl esters in macrophages incubated with acetylated lipoproteins. *J. Cell. Biol.*, **82** (3), 597–613.

15 Moore, K.J. and Freeman, M.W. (2006) Scavenger receptors in atherosclerosis: beyond lipid uptake. *Arterioscler. Thromb. Vasc. Biol.*, **26** (8), 1702–1711.

16 Murphy, J.E., Tedbury, P.R., Homer-Vanniasinkam, S., Walker, J.H., and Ponnambalam, S. (2005) Biochemistry and cell biology of mammalian scavenger receptors. *Atherosclerosis*, **182** (1), 1–15.

17 Suzuki, H., Kurihara, Y., Takeya, M., Kamada, N., Kataoka, M., Jishage, K., Ueda, O., Sakaguchi, H., Higashi, T., Suzuki, T., Takashima, Y., Kawabe, Y., Cynshi, O., Wada, Y., Honda, M., Kurihara, H., Aburatani, H., Doi, T., Matsumoto, A., Azuma, S., Noda, T., Toyoda, Y., Itakura, H., Yazaki, Y., Kodama, T. (1997) A role for macro-phage scavenger receptors in atheroscle-rosis and susceptibility to infection. *Nature*, **386** (6622), 292–296.

18 Sakaguchi, H., Takeya, M., Suzuki, H., Hakamata, H., Kodama, T., Horiuchi, S., Gordon, S., van der Laan, L.J., Kraal, G., Ishibashi, S., Kitamura, N., and Takahashi, K. (1998) Role of macro-phage scavenger receptors in diet-induced atherosclerosis in mice. *Lab. Invest.*, **78** (4), 423–434.

19 de Winther, M.P., Gijbels, M.J., van Dijk, K.W., van Gorp, P.J., Suzuki, H., Kodama, T., Frants, R.R., Havekes, L.M., and Hofker, M.H. (1999) Scavenger receptor deficiency leads to more complex atherosclerotic lesions in APOE3Leiden transgenic mice. *Atherosclerosis*, **144** (2), 315–321.

20 Babaev, V.R., Gleaves, L.A., Carter, K.J., Suzuki, H., Kodama, T., Fazio, S., and Linton, M.F. (2000) Reduced atheroscle-rotic lesions in mice deficient for total or macrophage-specific expression of scavenger receptor-A. *Arterioscler. Thromb. Vasc. Biol.*, **20** (12), 2593–2599.

21 Moore, K.J., Kunjathoor, V.V., Koehn, S.L., Manning, J.J., Tseng, A.A., Silver, J.M., McKee, M., and Freeman, M.W. (2005) Loss of receptor-mediated lipid uptake via scavenger receptor A or CD36 pathways does not ameliorate atherosclerosis in hyperlipidemic mice. *J. Clin. Invest.*, **115** (8), 2192–2201.

22 Herijgers, N., de Winther, M.P., Van Eck, M., Havekes, L.M., Hofker, M.H., Hoogerbrugge, P.M., and Van Berkel, T.J. (2000) Effect of human scavenger

receptor class A overexpression in bone marrow-derived cells on lipoprotein metabolism and atherosclerosis in low density lipoprotein receptor knockout mice. *J. Lipid Res.*, **41** (9), 1402–1409.

23 Van Eck, M., De Winther, M.P., Herijgers, N., Havekes, L.M., Hofker, M.H., Groot, P.H., and Van Berkel, T.J. (2000) Effect of human scavenger receptor class A overexpression in bone marrow-derived cells on cholesterol levels and atherosclerosis in ApoE-deficient mice. *Arterioscler. Thromb. Vasc. Biol.*, **20** (12), 2600–2606.

24 Whitman, S.C., Rateri, D.L., Szilvassy, S.J., Cornicelli, J.A., and Daugherty, A. (2002) Macrophage-specific expression of class A scavenger receptors in LDL receptor$^{-/-}$ mice decreases atherosclerosis and changes spleen morphology. *J. Lipid Res.*, **43** (8), 1201–1208.

25 Kuchibhotla, S., Vanegas, D., Kennedy, D.J., Guy, E., Nimako, G., Morton, R.E., and Febbraio, M. (2008) Absence of CD36 protects against atherosclerosis in ApoE knock-out mice with no additional protection provided by absence of scavenger receptor A I/II. *Cardiovasc. Res.*, **78** (1), 185–196.

26 Kunjathoor, V.V., Febbraio, M., Podrez, E.A., Moore, K.J., Andersson, L., Koehn, S., Rhee, J.S., Silverstein, R., Hoff, H.F., and Freeman, M.W. (2002) Scavenger receptors class A-I/II and CD36 are the principal receptors responsible for the uptake of modified low density lipoprotein leading to lipid loading in macrophages. *J. Biol. Chem.*, **277** (51), 49982–49988.

27 Podrez, E.A., Febbraio, M., Sheibani, N., Schmitt, D., Silverstein, R.L., Hajjar, D.P., Cohen, P.A., Frazier, W.A., Hoff, H.F., and Hazen, S.L. (2000) Macrophage scavenger receptor CD36 is the major receptor for LDL modified by monocyte-generated reactive nitrogen species. *J. Clin. Invest.*, **105** (8), 1095–1108.

28 Febbraio, M., Podrez, E.A., Smith, J.D., Hajjar, D.P., Hazen, S.L., Hoff, H.F., Sharma, K., and Silverstein, R.L. (2000) Targeted disruption of the class B scavenger receptor CD36 protects against atherosclerotic lesion develop-

ment in mice. *J. Clin. Invest.*, **105** (8), 1049–1056.

29 Febbraio, M., Guy, E., and Silverstein, R.L. (2004) Stem cell transplantation reveals that absence of macrophage CD36 is protective against atherosclerosis. *Arterioscler. Thromb. Vasc. Biol.*, **24**, 2333–2338.

30 Manning-Tobin, J.J., Moore, K.J., Seimon, T.A., Bell, S.A., Sharuk, M., Alvarez-Leite, J.I., de Winther, M.P., Tabas, I., and Freeman, M.W. (2009) Loss of SR-A and CD36 activity reduces atherosclerotic lesion complexity without abrogating foam cell formation in hyperlipidemic mice. *Arterioscler. Thromb. Vasc. Biol.*, **29** (1), 19–26.

31 Mehta, J.L., Chen, J., Hermonat, P.L., Romeo, F., and Novelli, G. (2006) Lectin-like, oxidized low-density lipoprotein receptor-1 (LOX-1): a critical player in the development of atherosclerosis and related disorders. *Cardiovasc. Res.*, **69** (1), 36–45.

32 Sawamura, T., Kume, N., Aoyama, T., Moriwaki, H., Hoshikawa, H., Aiba, Y., Tanaka, T., Miwa, S., Katsura, Y., Kita, T., and Masaki, T. (1997) An endothelial receptor for oxidized low-density lipoprotein. *Nature*, **386** (6620), 73–77.

33 Li, D., and Mehta, J.L. (2000) Upregulation of endothelial receptor for oxidized LDL (LOX-1) by oxidized LDL and implications in apoptosis of human coronary artery endothelial cells: evidence from use of antisense LOX-1 mRNA and chemical inhibitors. *Arterioscler. Thromb. Vasc. Biol.*, **20** (4), 1116–1122.

34 Inoue, K., Arai, Y., Kurihara, H., Kita, T., and Sawamura, T. (2005) Overexpression of lectin-like oxidized low-density lipoprotein receptor-1 induces intramyocardial vasculopathy in apolipoprotein E-null mice. *Circ. Res.*, **97** (2), 176–184.

35 Hu, C., Dandapat, A., Sun, L., Chen, J., Marwali, M.R., Romeo, F., Sawamura, T., and Mehta, J.L. (2008) LOX-1 deletion decreases collagen accumulation in atherosclerotic plaque in low-density lipoprotein receptor knockout mice fed a high-cholesterol diet. *Cardiovasc. Res.*, **79** (2), 287–293.

36 Ishigaki, Y., Katagiri, H., Gao, J., Yamada, T., Imai, J., Uno, K., Hasegawa, Y., Kaneko, K., Ogihara, T., Ishihara, H., Sato, Y., Takikawa, K., Nishimichi, N., Matsuda, H., Sawamura, T., and Oka, Y. (2008) Impact of plasma oxidized low-density lipoprotein removal on atherosclerosis. *Circulation*, **118** (1), 75–83.

37 Ludwig, A., and Weber, C. (2007) Transmembrane chemokines: versatile "special agents" in vascular inflammation. *Thromb. Haemost.*, **97** (5), 694–703.

38 Gough, P.J., Garton, K.J., Wille, P.T., Rychlewski, M., Dempsey, P.J., and Raines, E.W. (2004) A disintegrin and metalloproteinase 10-mediated cleavage and shedding regulates the cell surface expression of CXC chemokine ligand 16. *J. Immunol.*, **172** (6), 3678–3685.

39 Kim, C.H., Kunkel, E.J., Boisvert, J., Johnston, B., Campbell, J.J., Genovese, M.C., Greenberg, H.B., and Butcher, E.C. (2001) Bonzo/CXCR6 expression defines type 1-polarized T-cell subsets with extralymphoid tissue homing potential. *J. Clin. Invest.*, **107** (5), 595–601.

40 Aslanian, A.M. and Charo, I.F. (2006) Targeted disruption of the scavenger receptor and chemokine CXCL16 accelerates atherosclerosis. *Circulation*, **114** (6), 583–590.

41 Park, Y.M., Febbraio, M., and Silverstein, R.L. (2009) CD36 modulates migration of mouse and human macrophages in response to oxidized LDL and may contribute to macrophage trapping in the arterial intima. *J. Clin. Invest.*, **119**, 136–145.

42 Stuart, L.M., Bell, S.A., Stewart, C.R., Silver, J.M., Richard, J., Goss, J.L., Tseng, A.A., Zhang, A., El Khoury, J.B., and Moore, K.J. (2007) CD36 signals to the actin cytoskeleton and regulates microglial migration via a p130Cas complex. *J. Biol. Chem.*, **282** (37), 27392–27401.

43 Devries-Seimon, T., Yao, Y., Li, P.M., Stone, E., Wang, Y., Davis, R.J., Flavell, R., and Tabas, I. (2005) Cholesterol-induced macrophage apoptosis requires ER stress pathways and engagement of the type A scavenger receptor. *J. Cell. Biol.*, **171** (1), 61–73.

44 Seimon, T. and Tabas, I. (2009) Mechanisms and consequences of macrophage apoptosis in atherosclerosis. *J. Lipid Res.*, **50**, S382–S387.

45 Rayner, K., Chen, Y.X., McNulty, M., Simard, T., Zhao, X., Wells, D.J., de Belleroche, J., and O'Brien, E.R. (2008) Extracellular release of the atheroprotective heat shock protein 27 is mediated by estrogen and competitively inhibits acLDL binding to scavenger receptor-A. *Circ. Res.*, **103** (2), 133–141.

46 Wang, X., Zheng, Y., Xu, Y., Ben, J., Gao, S., Zhu, X., Zhuang, Y., Yue, S., Bai, H., Chen, Y., Jiang, L., Ji, Y., Xu, Y., Fan, L., Sha, J., He, Z., and Chen, Q. (2009) A novel peptide binding to the cytoplasmic domain of class A scavenger receptor reduces lipid uptake in THP-1 macrophages. *Biochim. Biophys. Acta*, **1791** (1), 76–83.

47 Howlett, G.J. and Moore, K.J. (2006) Untangling the role of amyloid in atherosclerosis. *Curr. Opin. Lipidol.*, **17** (5), 541–547.

48 Sussan, T.E., Jun, J., Thimmulappa, R., Bedja, D., Antero, M., Gabrielson, K.L., Polotsky, V.Y., and Biswal, S. (2008) Disruption of Nrf2, a key inducer of antioxidant defenses, attenuates ApoE-mediated atherosclerosis in mice. *PLoS ONE*, **3** (11), e3791.

49 Osto, E., Kouroedov, A., Mocharla, P., Akhmedov, A., Besler, C., Rohrer, L., von Eckardstein, A., Iliceto, S., Volpe, M., Luscher, T.F., and Cosentino, F. (2008) Inhibition of protein kinase Cbeta prevents foam cell formation by reducing scavenger receptor A expression in human macrophages. *Circulation*, **118** (21), 2174–2182.

50 Handberg, A., Levin, K., Hojlund, K., and Beck-Nielsen, H. (2006) Identification of the oxidized low-density lipoprotein scavenger receptor CD36 in plasma: a novel marker of insulin resistance. *Circulation*, **114** (11), 1169–1176.

51 Handberg, A., Skjelland, M., Michelsen, A.E., Sagen, E.L., Krohg-Sorensen, K., Russell, D., Dahl, A., Ueland, T., Oie, E., Aukrust, P., and Halvorsen, B.

(2008) Soluble CD36 in plasma is increased in patients with symptomatic atherosclerotic carotid plaques and is related to plaque instability. *Stroke*, **39** (11), 3092–3095.

52 Hayashida, K., Kume, N., Murase, T., Minami, M., Nakagawa, D., Inada, T., Tanaka, M., Ueda, A., Kominami, G., Kambara, H., Kimura, T., and Kita, T. (2005) Serum soluble lectin-like oxidized low-density lipoprotein receptor-1 levels are elevated in acute coronary syndrome: a novel marker for early diagnosis. *Circulation*, **112** (6), 812–818.

53 Hanasaki, K., Yamada, K., Yamamoto, S., Ishimoto, Y., Saiga, A., Ono, T., Ikeda, M., Notoya, M., Kamitani, S., and Arita, H. (2002) Potent modification of low density lipoprotein by group X secretory phospholipase A2 is linked to macrophage foam cell formation. *J. Biol. Chem.*, **277** (32), 29116–29124.

54 Karabina, S.A., Brocheriou, I., Le Naour, G., Agrapart, M., Durand, H., Gelb, M., Lambeau, G., and Ninio, E. (2006) Atherogenic properties of LDL particles modified by human group X secreted phospholipase A2 on human endothelial cell function. *FASEB J.*, **20** (14), 2547–2549.

55 Menschikowski, M., Kasper, M., Lattke, P., Schiering, A., Schiefer, S., Stockinger, H., and Jaross, W. (1995) Secretory group II phospholipase A2 in human atherosclerotic plaques. *Atherosclerosis*, **118** (2), 173–181.

56 Romano, M., Romano, E., Bjorkerud, S., and Hurt-Camejo, E. (1998) Ultrastructural localization of secretory type II phospholipase A2 in atherosclerotic and nonatherosclerotic regions of human arteries. *Arterioscler. Thromb. Vasc. Biol.*, **18** (4), 519–525.

57 Wooton-Kee, C.R., Boyanovsky, B.B., Nasser, M.S., de Villiers, W.J., and Webb, N.R. (2004) Group V sPLA2 hydrolysis of low-density lipoprotein results in spontaneous particle aggregation and promotes macrophage foam cell formation. *Arterioscler. Thromb. Vasc. Biol.*, **24** (4), 762–767.

58 Rosengren, B., Peilot, H., Umaerus, M., Jonsson-Rylander, A.C., Mattsson-Hulten, L., Hallberg, C., Cronet, P., Rodriguez-Lee, M., and Hurt-Camejo, E. (2006) Secretory phospholipase A2 group V: lesion distribution, activation by arterial proteoglycans, and induction in aorta by a Western diet. *Arterioscler. Thromb. Vasc. Biol.*, **26** (7), 1579–1585.

59 Flood, C., Gustafsson, M., Pitas, R.E., Arnaboldi, L., Walzem, R.L., and Boren, J. (2004) Molecular mechanism for changes in proteoglycan binding on compositional changes of the core and the surface of low-density lipoprotein-containing human apolipoprotein B100. *Arterioscler. Thromb. Vasc. Biol.*, **24** (3), 564–570.

60 Eckey, R., Menschikowski, M., Lattke, P., and Jaross, W. (1997) Minimal oxidation and storage of low density lipoproteins result in an increased susceptibility to phospholipid hydrolysis by phospholipase A2. *Atherosclerosis*, **132** (2), 165–176.

61 Hakala, J.K., Oorni, K., Pentikainen, M.O., Hurt-Camejo, E., and Kovanen, P.T. (2001) Lipolysis of LDL by human secretory phospholipase A(2) induces particle fusion and enhances the retention of LDL to human aortic proteoglycans. *Arterioscler. Thromb. Vasc. Biol.*, **21** (6), 1053–1058.

62 Kugiyama, K., Ota, Y., Takazoe, K., Moriyama, Y., Kawano, H., Miyao, Y., Sakamoto, T., Soejima, H., Ogawa, H., Doi, H., Sugiyama, S., and Yasue, H. (1999) Circulating levels of secretory type II phospholipase A(2) predict coronary events in patients with coronary artery disease. *Circulation*, **100** (12), 1280–1284.

63 Ivandic, B., Castellani, L.W., Wang, X.P., Qiao, J.H., Mehrabian, M., Navab, M., Fogelman, A.M., Grass, D.S., Swanson, M.E., de Beer, M.C., de Beer, F., and Lusis, A.J. (1999) Role of group II secretory phospholipase A2 in atherosclerosis: 1. Increased atherogenesis and altered lipoproteins in transgenic mice expressing group IIa phospholipase A2. *Arterioscler. Thromb. Vasc. Biol.*, **19** (5), 1284–1290.

64 Ghesquiere, S.A., Gijbels, M.J., Anthonsen, M., van Gorp, P.J., van der Made, I., Johansen, B., Hofker, M.H.,

and de Winther, M.P. (2005) Macro-phage-specific overexpression of group IIa sPLA2 increases atherosclerosis and enhances collagen deposition. *J. Lipid Res.*, **46** (2), 201–210.

65 Tietge, U.J., Pratico, D., Ding, T., Funk, C.D., Hildebrand, R.B., Van Berkel, T., and Van Eck, M. (2005) Macrophage-specific expression of group IIA sPLA2 results in accelerated atherogenesis by increasing oxidative stress. *J. Lipid Res.*, **46** (8), 1604–1614.

66 Webb, N.R., Bostrom, M.A., Szilvassy, S.J., van der Westhuyzen, D.R., Daugherty, A., and de Beer, F.C. (2003) Macrophage-expressed group IIA secretory phospholipase A2 increases atherosclerotic lesion formation in LDL receptor-deficient mice. *Arterioscler. Thromb. Vasc. Biol.*, **23** (2), 263–268.

67 Murakami, M. and Kudo, I. (2003) New phospholipase A_2 isozymes with a potential role in atherosclerosis. *Curr. Opin. Lipidol.*, **14** (5), 431–436.

68 Webb, N.R. (2005) Secretory phospholipase A2 enzymes in atherogenesis. *Curr. Opin. Lipidol.*, **16** (3), 341–344.

69 Curfs, D.M., Ghesquiere, S.A., Vergouwe, M.N., van der Made, I., Gijbels, M.J., Greaves, D.R., Verbeek, J.S., Hofker, M.H., and de Winther, M.P. (2008) Macrophage secretory phospholipase A2 group X enhances anti-inflammatory responses, promotes lipid accumulation, and contributes to aberrant lung pathology. *J. Biol. Chem.*, **283** (31), 21640–21648.

70 Boyanovsky, B.B., van der Westhuyzen, D.R., and Webb, N.R. (2005) Group V secretory phospholipase A2-modified low density lipoprotein promotes foam cell formation by a SR-A- and CD36-independent process that involves cellular proteoglycans. *J. Biol. Chem.*, **280** (38), 32746–32752.

71 Boyanovsky, B.B., Shridas, P., Simons, M., van der Westhuyzen, D.R., and Webb, N.R. (2009) Syndecan-4 mediates macrophage uptake of group V secretory phospholipase A2-modified low density lipoprotein. *J. Lipid Res.*, **50**, 641–650.

72 Bostrom, M.A., Boyanovsky, B.B., Jordan, C.T., Wadsworth, M.P., Taatjes,

D.J., de Beer, F.C., and Webb, N.R. (2007) Group v secretory phospholipase A2 promotes atherosclerosis: evidence from genetically altered mice. *Arterioscler. Thromb. Vasc. Biol.*, **27** (3), 600–606.

73 Shaposhnik, Z., Wang, X., Trias, J., Fraser, H., and Lusis, A.J. (2009) The synergistic inhibition of atherogenesis in ApoE-/- mice between pravastatin and the sPLA2 inhibitor varespladib (A-002). *J. Lipid Res.*, **50**, 623–629.

74 Fraser, H., Hislop, C., Christie, R.M., Rick, H.L., Reidy, C.A., Chouinard, M.L., Eacho, P.I., Gould, K.E., and Trias, J. (2009) Varespladib (A-002), a secretory phospholipase A2 inhibitor, reduces atherosclerosis and aneurysm formation in ApoE–/– mice. *J. Cardio-vasc. Pharmacol.*, **53**, 60–65.

75 Marathe, S., Kuriakose, G., Williams, K.J., and Tabas, I. (1999) Sphingomyeli-nase, an enzyme implicated in atherogenesis, is present in atheroscle-rotic lesions and binds to specific components of the subendothelial extracellular matrix. *Arterioscler. Thromb. Vasc. Biol.*, **19** (11), 2648–2658.

76 Oorni, K., Hakala, J.K., Annila, A., Ala-Korpela, M., and Kovanen, P.T. (1998) Sphingomyelinase induces aggregation and fusion, but phospholi-pase A2 only aggregation, of low density lipoprotein (LDL) particles. Two distinct mechanisms leading to increased binding strength of LDL to human aortic proteoglycans. *J. Biol. Chem.*, **273** (44), 29127–29134.

77 Oorni, K., Posio, P., Ala-Korpela, M., Jauhiainen, M., and Kovanen, P.T. (2005) Sphingomyelinase induces aggregation and fusion of small very low-density lipoprotein and intermedi-ate-density lipoprotein particles and increases their retention to human arterial proteoglycans. *Arterioscler. Thromb. Vasc. Biol.*, **25** (8), 1678–1683.

78 Schissel, S.L., Tweedie-Hardman, J., Rapp, J.H., Graham, G., Williams, K.J., and Tabas, I. (1996) Rabbit aorta and human atherosclerotic lesions hydrolyze the sphingomyelin of retained low-density lipoprotein. Proposed role

for arterial-wall sphingomyelinase in subendothelial retention and aggregation of atherogenic lipoproteins. *J. Clin. Invest.*, **98** (6), 1455–1464.

79 Tabas, I., Li, Y., Brocia, R.W., Xu, S.W., Swenson, T.L., and Williams, K.J. (1993) Lipoprotein lipase and sphingomyelinase synergistically enhance the association of atherogenic lipoproteins with smooth muscle cells and extracellular matrix. A possible mechanism for low density lipoprotein and lipoprotein(a) retention and macrophage foam cell formation. *J. Biol. Chem.*, **268** (27), 20419–20432.

80 Schissel, S.L., Jiang, X., Tweedie-Hardman, J., Jeong, T., Camejo, E.H., Najib, J., Rapp, J.H., Williams, K.J., and Tabas, I. (1998) Secretory sphingomyelinase, a product of the acid sphingomyelinase gene, can hydrolyze atherogenic lipoproteins at neutral pH. Implications for atherosclerotic lesion development. *J. Biol. Chem.*, **273** (5), 2738–2746.

81 Sakr, S.W., Eddy, R.J., Barth, H., Wang, F., Greenberg, S., Maxfield, F.R., and Tabas, I. (2001) The uptake and degradation of matrix-bound lipoproteins by macrophages require an intact actin Cytoskeleton, Rho family GTPases, and myosin ATPase activity. *J. Biol. Chem.*, **276** (40), 37649–37658.

82 Goldstein, J.L., Ho, Y.K., Brown, M.S., Innerarity, T.L., and Mahley, R.W. (1980) Cholesteryl ester accumulation in macrophages resulting from receptor-mediated uptake and degradation of hypercholesterolemic canine beta-very low density lipoproteins. *J. Biol. Chem.*, **255** (5), 1839–1848.

83 Mahley, R.W., Innerarity, T.L., Brown, M.S., Ho, Y.K., and Goldstein, J.L. (1980) Cholesteryl ester synthesis in macrophages: stimulation by beta-very low density lipoproteins from cholesterol-fed animals of several species. *J. Lipid Res.*, **21** (8), 970–980.

84 Huff, M.W., Evans, A.J., Sawyez, C.G., Wolfe, B.M., and Nestel, P.J. (1991) Cholesterol accumulation in J774 macrophages induced by triglyceride-rich lipoproteins. Comparison of very low density lipoprotein from subjects

with type III, IV, and V hyperlipoproteinemias. *Arterioscler. Thromb.*, **11** (2), 221–233.

85 Evans, A.J., Sawyez, C.G., Wolfe, B.M., Connelly, P.W., Maguire, G.F., and Huff, M.W. (1993) Evidence that cholesteryl ester and triglyceride accumulation in J774 macrophages induced by very low density lipoprotein subfractions occurs by different mechanisms. *J. Lipid Res.*, **34** (5), 703–717.

86 Koo, C., Wernette-Hammond, M.E., Garcia, Z., Malloy, M.J., Uauy, R., East, C., Bilheimer, D.W., Mahley, R.W., and Innerarity, T.L. (1988) Uptake of cholesterol-rich remnant lipoproteins by human monocyte-derived macrophages is mediated by low density lipoprotein receptors. *J. Clin. Invest.*, **81** (5), 1332–1340.

87 Ellsworth, J.L., Kraemer, F.B., and Cooper, A.D. (1987) Transport of beta-very low density lipoproteins and chylomicron remnants by macrophages is mediated by the low density lipoprotein receptor pathway. *J. Biol. Chem.*, **262** (5), 2316–2325.

88 Herijgers, N., Van Eck, M., Korporaal, S.J., Hoogerbrugge, P.M., and Van Berkel, T.J. (2000) Relative importance of the LDL receptor and scavenger receptor class B in the beta-VLDL-induced uptake and accumulation of cholesteryl esters by peritoneal macrophages. *J. Lipid Res.*, **41** (7), 1163–1171.

89 Herijgers, N., Van Eck, M., Groot, P.H., Hoogerbrugge, P.M., and Van Berkel, T.J. (2000) Low density lipoprotein receptor of macrophages facilitates atherosclerotic lesion formation in C57Bl/6 mice. *Arterioscler. Thromb. Vasc. Biol.*, **20** (8), 1961–1967.

90 Linton, M.F., Babaev, V.R., Gleaves, L.A., and Fazio, S. (1999) A direct role for the macrophage low density lipoprotein receptor in atherosclerotic lesion formation. *J. Biol. Chem.*, **274** (27), 19204–19210.·

91 Krieger, M. and Herz, J. (1994) Structures and functions of multiligand lipoprotein receptors: macrophage scavenger receptors and LDL receptor-

related protein (LRP). *Annu. Rev. Biochem.*, **63**, 601–637.

92 Hussain, M.M., Maxfield, F.R., Mas-Oliva, J., Tabas, I., Ji, Z.S., Innerarity, T.L., and Mahley, R.W. (1991) Clearance of chylomicron remnants by the low density lipoprotein receptor-related protein/alpha 2-macroglobulin receptor. *J. Biol. Chem.*, **266** (21), 13936–13940.

93 Kosaka, S., Takahashi, S., Masamura, K., Kanehara, H., Sakai, J., Tohda, G., Okada, E., Oida, K., Iwasaki, T., Hattori, H., Kodama, T., Yamamoto, T., and Miyamori, I. (2001) Evidence of macrophage foam cell formation by very low-density lipoprotein receptor: interferon-gamma inhibition of very low-density lipoprotein receptor expression and foam cell formation in macrophages. *Circulation*, **103** (8), 1142–1147.

94 Hiltunen, T.P., Luoma, J.S., Nikkari, T., and Yla-Herttuala, S. (1998) Expression of LDL receptor, VLDL receptor, LDL receptor-related protein, and scavenger receptor in rabbit atherosclerotic lesions: marked induction of scavenger receptor and VLDL receptor expression during lesion development. *Circulation*, **97** (11), 1079–1086.

95 Tacken, P.J., Delsing, D.J., Gijbels, M.J., Quax, P.H., Havekes, L.M., Hofker, M.H., and van Dijk, K.W. (2002) VLDL receptor deficiency enhances intimal thickening after vascular injury but does not affect atherosclerotic lesion area. *Atherosclerosis*, **162** (1), 103–110.

96 Hakamata, H., Sakaguchi, H., Zhang, C., Sakashita, N., Suzuki, H., Miyazaki, A., Takeya, M., Takahashi, K., Kitamura, N., and Horiuchi, S. (1998) The very low- and intermediate-density lipoprotein fraction isolated from apolipoprotein E-knockout mice transforms macrophages to foam cells through an apolipoprotein E-independent pathway. *Biochemistry*, **37** (39), 13720–13727.

97 Brown, M.S. and Goldstein, J.L. (1975) Regulation of the activity of the low density lipoprotein receptor in human fibroblasts. *Cell*, **6** (3), 307–316.

98 Kruth, H.S., Huang, W., Ishii, I., and Zhang, W.Y. (2002) Macrophage foam cell formation with native low density lipoprotein. *J. Biol. Chem.*, **277** (37), 34573–34580.

99 Kruth, H.S., Jones, N.L., Huang, W., Zhao, B., Ishii, I., Chang, J., Combs, C.A., Malide, D., and Zhang, W.Y. (2005) Macropinocytosis is the endocytic pathway that mediates macrophage foam cell formation with native LDL. *J. Biol. Chem.*, **280**, 2352–2360.

100 Zhao, B., Li, Y., Buono, C., Waldo, S.W., Jones, N.L., Mori, M., and Kruth, H.S. (2006) Constitutive receptor-independent low density lipoprotein uptake and cholesterol accumulation by macrophages differentiated from human monocytes with macrophage-colony-stimulating factor (M-CSF). *J. Biol. Chem.*, **281** (23), 15757–15762.

101 Yancey, P.G., Bortnick, A.E., Kellner-Weibel, G., de la Llera-Moya, M., Phillips, M.C., and Rothblat, G.H. (2003) Importance of different pathways of cellular cholesterol efflux. *Arterioscler. Thromb. Vasc. Biol.*, **23** (5), 712–719.

102 Yokoyama, S. (2006) Assembly of high-density lipoprotein. *Arterioscler. Thromb. Vasc. Biol.*, **26** (1), 20–27.

103 Jessup, W., Gelissen, I.C., Gaus, K., and Kritharides, L. (2006) Roles of ATP binding cassette transporters A1 and G1, scavenger receptor BI and membrane lipid domains in cholesterol export from macrophages. *Curr. Opin. Lipidol.*, **17** (3), 247–257.

104 Oram, J.F. and Vaughan, A.M. (2006) ATP-Binding cassette cholesterol transporters and cardiovascular disease. *Circ. Res.*, **99** (10), 1031–1043.

105 Oram, J.F. and Vaughan, A.M. (2000) ABCA1-mediated transport of cellular cholesterol and phospholipids to HDL apolipoproteins. *Curr. Opin. Lipidol.*, **11** (3), 253–260.

106 Remaley, A.T., Stonik, J.A., Demosky, S.J., Neufeld, E.B., Bocharov, A.V., Vishnyakova, T.G., Eggerman, T.L., Patterson, A.P., Duverger, N.J., Santamarina-Fojo, S., and Brewer, H.B., Jr. (2001) Apolipoprotein specificity for lipid efflux by the human ABCAI

transporter. *Biochem. Biophys. Res. Commun.*, **280** (3), 818–823.

107 Bodzioch, M., Orso, E., Klucken, J., Langmann, T., Bottcher, A., Diederich, W., Drobnik, W., Barlage, S., Buchler, C., Porsch-Ozcurumez, M., Kaminski, W.E., Hahmann, H.W., Oette, K., Rothe, G., Aslanidis, C., Lackner, K.J., and Schmitz, G. (1999) The gene encoding ATP-binding cassette transporter 1 is mutated in Tangier disease. *Nat. Genet.*, **22** (4), 347–351.

108 Brooks-Wilson, A., Marcil, M., Clee, S.M., Zhang, L.H., Roomp, K., van Dam, M., Brewer, L., Yu, C., Collins, J.A., Molhuizen, H.O., Loubser, O., Ouelette, B.F., Fichter, K., Ashbourne-Excoffon, K.J., Sensen, C.W., Scherer, S., Mott, S., Denis, M., Martindale, D., Frohlich, J., Morgan, K., Koop, B., Pimstone, S., Kastelein, J.J., Genest, J., Jr., and Hayden, M.R. (1999) Mutations in ABC1 in Tangier disease and familial high-density lipoprotein deficiency. *Nat. Genet.*, **22** (4), 336–345.

109 Lawn, R.M., Wade, D.P., Garvin, M.R., Wang, X., Schwartz, K., Porter, J.G., Seilhamer, J.J., Vaughan, A.M., and Oram, J.F. (1999) The Tangier disease gene product ABC1 controls the cellular apolipoprotein-mediated lipid removal pathway. *J. Clin. Invest.*, **104** (8), R25–R31.

110 Rust, S., Rosier, M., Funke, H., Real, J., Amoura, Z., Piette, J.C., Deleuze, J.F., Brewer, H.B., Duverger, N., Denefle, P., and Assmann, G. (1999) Tangier disease is caused by mutations in the gene encoding ATP-binding cassette transporter 1. *Nat. Genet.*, **22** (4), 352–355.

111 Wang, N., Lan, D., Chen, W., Matsuura, F., and Tall, A.R. (2004) ATP-binding cassette transporters G1 and G4 mediate cellular cholesterol efflux to high-density lipoproteins. *Proc. Natl Acad. Sci. U. S. A.*, **101** (26), 9774–9779.

112 Wang, X., Collins, H.L., Ranalletta, M., Fuki, I.V., Billheimer, J.T., Rothblat, G.H., Tall, A.R., and Rader, D.J. (2007) Macrophage ABCA1 and ABCG1, but not SR-BI, promote macrophage reverse cholesterol transport *in vivo*. *J. Clin. Invest.*, **117** (8), 2216–2224.

PART V
Oxidative Stress

Atherosclerosis: Molecular and Cellular Mechanisms. Edited by Sarah Jane George and Jason Johnson
Copyright © 2010 WILEY-VCH Verlag GmbH & Co. KGaA, Weinheim
ISBN: 978-3-527-32448-4

13
NADPH Oxidase and Atherosclerosis

Jamie Y. Jeremy, Saima Muzaffar, Carina Mill, and Nilima Shukla

13.1
Introduction

Reactive oxygen species (ROS), in particular, superoxide (O_2^-) and hydrogen peroxide (H_2O_2) have been implicated in every aspect of atherosclerosis from its origins as fatty streaks to plaque instability and rupture. A major inducible source of ROS is NADPH oxidase (NOX), which is expressed by neutrophils, monocytes, vascular smooth muscle cells (VSMCs), endothelial cells (ECs), fibroblasts, macrophages, and mast cells [1–3]. In relation to plaque biology, NOX has been linked to the replication and migration of cells, apoptosis, necrosis, lipid oxidation, angiogenesis, and matrix deposition and turnover. In this review, therefore, the role of NOX will be considered in relation to these functions as well as possible therapeutic interventions.

13.2
Biology of NAPDH Oxidases

There are seven members of the NOX family: NOX1, NOX2 (formerly termed gp91^{phox}), NOX3, NOX4, NOX5, Duox1, and Duox2 [1]. Neutrophils and monocytes contain NOX2 which is composed of two subunits, p22^{phox} and gp91^{phox}, and reside in the plasma membrane and together form a flavocytochrome (cytochrome b558) (Figure 13.1). On activation, regulatory subunits resident in the cytosol (p47^{phox}, p67^{phox}, p40^{phox}, and Rac-1) translocate to the membrane where they associate with cytochrome b558, resulting in electron transfer from NADPH to molecular oxygen and the generation of superoxide which then destroys bacteria according to the equation: $2O_2 + NAD(P)H > 2O_2^- + NAD(P)^+ + H^+$ [2, 3]. The binding of p67^{phox} to an activation site on NOX2 initiates catalytic activity, p47^{phox} and p22^{phox} being required to facilitate this process. Phosphorylation of p47^{phox} is essential for its translocation to the membrane cytochrome b558, as is isoprenylation of Rac-1 [2, 3].

Atherosclerosis: Molecular and Cellular Mechanisms. Edited by Sarah Jane George and Jason Johnson
Copyright © 2010 WILEY-VCH Verlag GmbH & Co. KGaA, Weinheim
ISBN: 978-3-527-32448-4

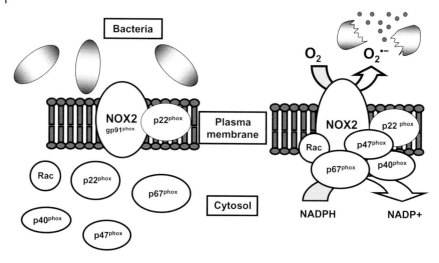

Figure 13.1 Acute activation of NADPH oxidase (NOX) in monocytes and neutrophils. Sensing the presence of bacteria, components of the NOX complex residing in the cytosol rapidly move and combine at the plasma membrane level. Superoxide released in a respiratory burst destroys bacteria. Similar mechanisms apply to the NOX isoforms expressed in vascular cells and fibroblasts.

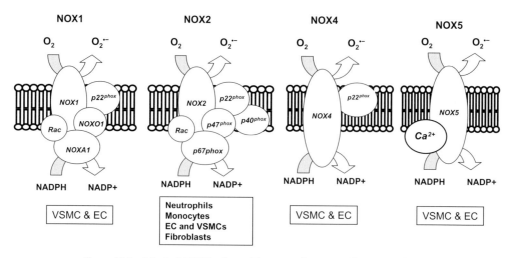

Figure 13.2 Principal NOX isoforms. These vary between cells types and are often located in quite distinctly different intracellular locations. Nevertheless, all NOXs generate superoxide, which elicits a plethora of pro-atherogenic effects.

ECs, VSMCs, and fibroblasts express NOX1, NOX2, NOX4, and NOX5, all requiring p22^{phox} for full activity [1–5]. NOX1 activation is similar to NOX2 except that it interacts with NOX organizer 1 (NOXO1) and NOX activator 1 (NOXA1), which are homologs of p47^{phox} and p67^{phox}, respectively (Figure 13.2). The organization and activation of NOX2 in these cells is similar to that of NOX2 in leukocytes. NOX1–4 all contain six transmembrane domains and have NOX and FAD-binding domains at the cytoplasmic C-terminus. NOX5 has a similar basic structure but with an additional N-terminal calmodulin-like Ca^{2+}-binding domain. Duox1 and Duox 2 include a further N-terminal extension with a peroxidase homology domain that is separated from the Ca^{2+}-binding domain by an additional transmembrane segment [3]. NOX4-containing NADPH oxidase does not require any of the conventional cytosolic subunits required for activation of other NOX isoforms (Figure 13.2).

There are two cardinal facets of NOXs found in plaque tissues/cells:

i) their expression is upregulated by a myriad of factors associated with plaque biology and
ii) acute activation, once upregulated, is elicited by these same factors [5, 6].

These factors include growth factors (platelet-derived growth factor (PDGF), epidermal growth factor (EGF), and transforming growth factor-β (TGF-β)), cytokines (tumor necrosis factor-α, interleukin-1, and platelet aggregation factor), mechanical forces (cyclic stretch, laminar, and oscillatory shear stress), metabolic factors (hyperglycemia, hyperinsulinemia, free fatty acids, advanced glycation end-products) and G protein–coupled receptor agonists (angiotensin II, serotonin, thrombin, bradykinin, and endothelin). The assembly of NOX is controlled by signaling pathways that include c-Src, p21^{Ras}, protein kinase C, phospholipase D, and phospholipase A_2 [5] (Figure 13.3).

The subcellular localization of the NOX family varies markedly between each isoform and cell type, which is indicative of specific functions. For example, NOX2 is present at the plasma membrane, in caveolae, and in perinuclear membranes of vascular cells, while NOX4 is present in the endoplasmic reticulum and nucleus [7]. This aspect of NOX distribution will be expanded upon below.

ROS derived from NOX activate an array of downstream intracellular signaling systems that include protein kinases, phosphatases, phospholipases, and calcium mobilization [1–7, 9], which in turn exert profound effects on cell behavior related to atherogenesis and plaque fragility. A full description of all the signaling systems influenced by ROS is beyond the scope of this chapter. However, there are a number of definitive and erudite reviews devoted to this topic [1–7, 9]. In subsequent sections, however, the relevant intracellular systems will be expanded upon in more detail. O_2^- and H_2O_2 also elicit other effects due to their highly reactive nature.

13.3
Studies in Human Tissues and Knockout Mice

In clinical samples, p22^{phox}, p67^{phox}, and p47^{phox} protein were significantly increased in diseased coronary arteries, as were mRNA levels for p22^{phox} and NOX2 [10].

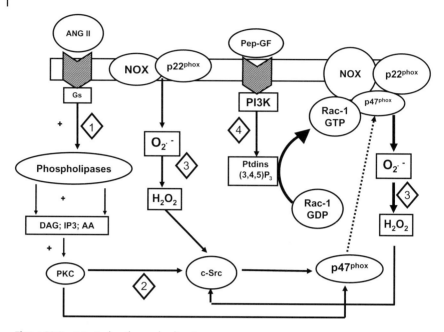

Figure 13.3 Principal pathways leading to activation of NOX1 in VSMCs. (1) Angiotensin II and other vasoconstrictors such as endothelin and thromboxane A$_2$ activate phospholipases A$_2$, C, and D to generate diacyl glycerol (DAG), inositol trisphosphate (IP3), and arachidonic acid (AA). (2) These activate protein kinase C (PKC) and cSrc which phosphorylates and triggers the translocation of p47^{phox} and Rac-1 from the cytosol to the plasmalemma. Once assembled the enzyme then converts oxygen to superoxide (O$_2^{\cdot-}$) and by dismutation to hydrogen peroxide (H$_2$O$_2$). (3) ROS feedback into the pathway to augment activation of the above pathways amplifying action of NOX by positive feedback. (4) Peptide growth factors (Pep-GF) activate phosphinositide 3 kinase (PI3K), which through IP$_3$ activates and triggers translocation of NOX components in particular Rac-1 to the membrane. For more detailed description of these pathways see [1, 5–8].

More recently it has been reported that NOX5 is present in diseased coronary arteries collected from heart transplant patients [11].

In mice, atherosclerotic (aortic) lesions, as opposed to those localized at the aortic sinus, were found to be dramatically decreased in p47^{phox}-null, NADPH oxidase–deficient animals [12]. More recently the specific role of NOX was studied in the different cells types that make up the plaque [13, 14]. Cell-specific NOX inhibition was achieved via allogenic, sex-mismatched bone marrow transplantation. Aortic atherosclerosis and superoxide production with functional NOX in monocytes, macrophages, and vascular wall cells was compared with nonfunctional monocyte/macrophage NOX (BMO) or nonfunctional vessel wall NOX (VWO) in apolipoprotein E–deficient (ApoE$^{\prime}$) mice after arterial injury. A significant decrease in O$_2^{\cdot-}$ production and atherosclerotic lesions, attenuated neointimal hyperplasia, and mitogenic proteins was observed in both BMO and VWO mice compared with control mice. This elegant study confirmed that atherosclerosis increased VSMC replication and migration which involves an equal contribution of NOX-derived ROS from monocyte/macrophage and VSMCs, at least in the mouse.

Plaque development, acceleration, and rupture are associated with clinical conditions, classically: dyslipidemia, hypercholesterolemia, diabetes mellitus, obesity, hyperhomocysteinemia, cigarette smoking, and hypertension [15, 16]. In turn, increased intravascular expression and activity of the NOX proteins have been widely documented in both animal models and clinical samples taken from patients with these conditions [15, 16], further implicating NOX in atherogenesis.

13.4
Apoptosis

Apoptosis is now firmly associated with atherogenesis, plaque fragility, and rupture [17–19]. In experimental models, chronic apoptosis of VSMCs accelerates atherogenesis and augments the progression of atherosclerosis in established plaques [17–19]. Fibrous cap thinning, enlargement of the necrotic core, plaque calcification, medial expansion and degeneration, elastin breaks, and failure of outward remodeling are all consequences of VSMC apoptosis [17–19]. Apoptosis of ECs lining diseased arteries can render plaques more susceptible to thrombosis and inflammation and plays a key role in retarding angiogenesis.

Little is known of the role of NOX in the regulation of apoptosis in relation to atherosclerosis. It has been established in other cell types, however, that the overproduction of ROS promotes apoptosis through damage to DNA, lipids, and proteins or by activation of signaling molecules that include PI3 kinase, MAP kinases, SAPK/JNK, JAK/STAT, ERK1/2, and p38 as well as inhibition of tyrosine phosphatase [20–23]. These in turn activate the orchestrators of apoptosis, principally, calpain, caspases, the Bcl-2 family, and cytochrome c [20–23].

In direct contrast, there are reports that ROS augment survival, an effect that is dependent on the magnitude and duration of the ROS signal and the subcellular localization of the NOX isoform [1]. Indeed, low levels of ROS derived from NOX have a pro-survival effect through activation of the NF-κB and/or the Akt/ASK1 pathways and inhibition of Fas-mediated cell death [23]. It has been proposed that the subcellular localization of NOX, the signaling targets (e.g., transcription factors, kinases, phosphatases, caspases) and the metabolism of superoxide by catalase/SOD status determine whether ROS are pro- or anti-apoptotic.

In ECs, $O_2^{\cdot-}$ derived from monocyte NOX2 induces apoptosis, an effect dependent on long-lasting ROS release, direct cell contact between ECs and monocytes, and mediation by β_2-integrins and their corresponding counterligand [21]. ROS also promote apoptosis in ECs via elevation of intracellular calcium concentration, leading to membrane depolarization of mitochondria and caspase activation [24]. There is also evidence that ROS derived from NOX promotes endoplasmic reticular (ER) stress through disruption of calcium homeostasis [23–27]. ER stress results in an accumulation and aggregation of unfolded protein (the unfolded protein response (UPR)) [28, 29], which, if not alleviated, will switch the cell from pro-survival to pro-apoptotic status. UPR is mediated by ER-resident transmembrane proteins including protein kinase activated by double-stranded RNA (PKR) and PKR-like ER kinase (PERK), which phosphorylate eukaryotic initiation factor (eIF) [28, 29]. UPR also promotes the expression of the growth inhibitory transcription factor, GADD153 (now known as CHOP, a phosphatase upregulated by the PERK pathway) and the glucose response protein 78 kDa/immunoglobulin heavy chain–binding protein (BiP) [28, 29].

Disruption of Ca^{2+} homeostasis triggers the UPR and ER stress [27]. NOX promotes apoptosis in rat endothelial cells through an elevation of intracellular calcium concentrations, membrane depolarization of mitochondria, and caspase activation [24]. Peroxynitrite (ONOO), which is formed by interaction of O_2^- with NO, induces ERS in ECs, depletes ER Ca^{2+} and induces apoptosis, indicating either an overactivation of or an inhibition of SERCA pumps [26]. Notably, NOX4 is present on ER of coronary artery VSMCs which in turn regulates Ca^{2+} release from the sarcoplasmic reticulum [26]. In a key study it was demonstrated that ER stress in ECs is mediated by an upregulation of NOX4 [27].

13.5
Cell Proliferation and Migration

VSMC replication and migration is central to arterial remodeling that characterizes atherosclerotic plaques, and as such the role of the NOX family in relation to these key events has been the subject of intense research [4, 5, 7]. *In vivo*, the replication and migration of VSMCs is manifest by neointima formation, which involves the migration of VSMCs from the media of the artery to the intima where they continue to proliferate [30]. Matrix-degrading metalloproteinase (MMP) secretion is crucial to this process (see later). Neointima formation is studied in animal models by inducing arterial injury (e.g., internal wires, cuff, and ligation) [30]. In turn, several studies have now demonstrated that NOX inhibition attenuates neointima formation in these animal models [31–33].

In vitro, mitogenic factors, including PDGF, angiotensin II, thromboxane A₂, and cytokines, increase ROS formation though NOX activation [34–38]. Conversely, ROS scavengers block VSMC replication and migration. Studies using antisense or siRNA suppression have established a predominant role for NOX1 in VSMC proliferation and a role for p22phox in proliferation of ECs [5]. Notably, migration of VSMCs derived from p22phox-overexpressing mouse aorta is increased [32].

Growth factors, such as PDGF and angiotensin II, insulin-like growth factor-1 (IGF-1), thrombin, vascular endothelial growth factor (VEGF), and lysophosphatidic acid, activate phosphinositide 3 kinase (PI3K), phospholipases A, D, and C, and PKC, which in turn activates c-Src and Rac-1 which then activate NOX (Figure

13.3). The resultant generation of ROS then facilitates replication and/or migration of vascular cells through a number of interrelated pathways, including:

i) activation of mitogen-activated protein kinases (MAPKs);

ii) activation of p21 protein-activated kinase 1 (PAK1) through tyrosine phosphorylation of pyruvate dehydrogenase kinase isozyme 1 (PDK1);

iii) activation of Janus tyrosine kinases (JAKs) which in turn triggers phosphorylation and translocation to the nucleus of eukaryotic replication kinase (ERK)1/2, STAT1, STAT3, and increased expression of heat shock protein-70;

iv) inhibition of phosphatases, for example, protein tyrosine phosphatases (PTPs), of which at least seven cytoplasmic PTPs are expressed in VSMCs. These include low M(r) protein tyrosine phosphatase (LMW-PTP) which tonically inhibits VSMC proliferation and migration. Another important phosphatase inactivated by ROS is phosphatase and tensin homolog (PTEN);

v) transactivation of tyrosine kinase receptors by G protein–coupled receptors;

vi) *S*-glutathiolation and activation of Ras; and

vii) modulation of transcription. One of the best-studied ROS-sensitive transcription factors is activator protein-1 (AP-1), a heterodimer of Fos and Jun. Antisense p22^{phox} oligonucleotides inhibit AP-1 binding to DNA in response to mitogens. Other ROS-sensitive transcription factors include cyclic AMP response element–binding protein (CREB), hypoxia-inducible factor-1α (HIF-1α), and the growth arrest homeobox gene Gax.

The amount of published work on this topic is considerable and detailed coverage beyond the scope of this review. However, there are several excellent reviews that fully cover this area [1–7, 9].

13.6
Nitric Oxide

Nitric oxide (NO), which is generated by vascular endothelial NO synthase (eNOS), elicits a battery of vasculoprotective effects which include inhibition of VSMC replication and migration and neointima formation, platelet and leukocyte activity, adhesion molecule expression, lipid oxidation, and angiogenesis [9]. Impairment of NO formation increases the risk of atherosclerotic disease and conversely the administration of drugs that augment NO release or augment NO formation attenuate atherosclerotic lesion progression [9]. NO status has therefore been widely implicated in atherogenesis and plaque stability.

Because of its innate chemical reactivity, NO bioactivity is negated by ROS to form peroxynitrite (ONOO) and other reactive nitrogen species (RNS) [9, 39]

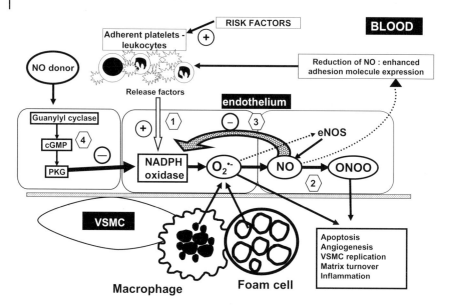

Figure 13.4 Interactions between NOX and nitric oxide. (1) Risk and release factors (including cytokines and eicosanoids) upregulate the expression of NADPH oxidase, thereby increasing the endogenous formation of superoxide ($O_2^{\cdot-}$). Resident macrophages and foam cells also contribute to local $O_2^{\cdot-}$ formation. (2) $O_2^{\cdot-}$ reduces NO derived from eNOS to form reactive nitrogen species (RNS), including peroxynitrite (ONOO) and uncouples eNOS. (3) Since NO also suppresses the expression and activity of NOX, a reduction of NO results in further augmentation of NOX expression and activity. $O_2^{\cdot-}$ itself also induces the expression of NOX, thereby increasing endogenous $O_2^{\cdot-}$ formation by a self-amplifying loop.

(Figure 13.4). $O_2^{\cdot-}$ derived from NOX generates large quantities of ONOO and downregulates the expression of eNOS [37, 38]. ONOO also oxidizes tetrahydrobiopterin (BH_4) an obligatory cofactor for eNOS activity which shifts eNOS from an NO-generating enzyme to an ROS-producing enzyme [40]. ONOO is notably highly chemically reactive and readily nitrates tyrosine residues, which disrupts a plethora of cellular functions that would promote atherogenesis [9]. As mentioned above, ONOO induces ER stress in human vascular ECs and induces programmed cell death.

By contrast, NO derived from eNOS inhibits the upregulation and acute activity of NOX in VSMCs through inhibition of Rac-1 [35, 37]. NO donors also inhibit NOX expression and activity in ECs and VSMCs, an effect mediated by the cyclic GMP-PKG signaling axis [37]. Thus, NO appears to be in dynamic balance with NOX, such that a reduction of NO formation would result in a greater expression and activity of NOX proteins (Figure 13.4).

The NO-cGMP-PKG axis is controlled in part by type 5 phosphodiesterase (PDE5), which hydrolyzes cGMP to inactive GMP [41] (Figure 13.5). $O_2^{\cdot-}$ derived

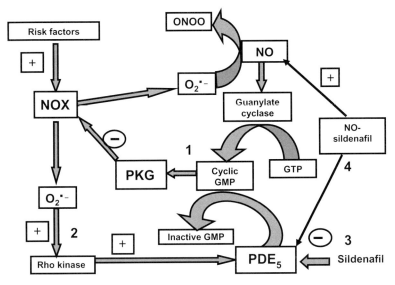

Figure 13.5 Mechanism underlying the antioxidant effect of sildenafil. (1) NO inhibits NOX expression and activity through elevation of cGMP and activation of protein kinase G (PKG). $O_2^{\cdot-}$ negates this protective effect of NO. (2) Type 5 phosphodiesterase (PDE_5) is upregulated by $O_2^{\cdot-}$ derived from NOX (via Rho kinase activation) and hydrolyzes cGMP, which in turn reduces that inhibitory action of the NO-PKG axis on NOX expression and activity. (3) Sildenafil, a PDE_5 inhibitor increases cGMP levels and therefore augments inhibition of NOX. (4) NO-donating sildenafil (NCX 911) would promote even greater relaxation, since the drug instrinsically provides NO drive.

from NOX upregulates PDE5 thereby blunting the vasculoprotective effects of NO, including inhibition of NOX [42]. In turn, sildenafil, a specific inhibitor of PDE5 activity, is a potent inhibitor of NOX expression and activity, an effect mediated by augmentation of cGMP levels [41] (Figure 13.5). Furthermore, a NO-donating derivative of sildenafil (NCX 911) was even more potent at inhibiting NOX expression in ECs, a facet attributable to the simultaneous provision of NO drive by the drug [41].

13.7
Eicosanoids

Prostaglandins (PGs), thromboxane A_2, and leukotrienes (LTs) are axiomatic in inflammation and have been implicated in every facet of atherosclerosis [43, 44]. Eicosanoid formation is triggered by a diverse range of factors associated with atherosclerosis that activate phospholipase A_2 (PLA_2) which releases arachidonic acid (AA) from phospholipid stores (Figure 13.6) [45]. AA is then converted to endoperoxides (PGH_2 and PGG_2) by cyclooxygenase (COX)-1 or COX-2 which are then converted to prostanoids by specific synthases or isomerases [43, 44].

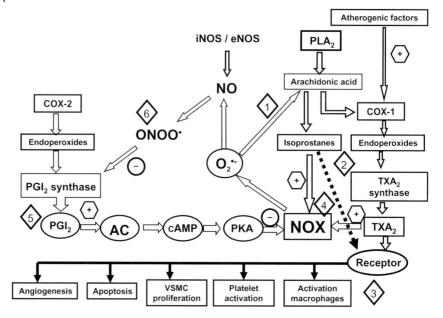

Figure 13.6 Eicosanoids and NADPH oxidase. (1) O_2^- from NOX activates phospholipase A_2 which generates arachidonic acid. (2) Arachidonic acid is converted to either isoprostanes (again instigated by O_2^- from NOX) or thromboxane A_2 (TXA_2) by cyclooxygenase 1 (COX-1) and TXA_2 synthase, both of which are upregulated by O_2^-. (3) Isoprostanes and TXA_2 activate the same receptor which triggers a plethora of event associated with atherosclerosis and plaque rupture. (4) These effects include the upregulation and activation of NOX thereby creating a self-perpetuating cascade since O_2^- would result in increased formation of isoprostanes and TXA_2. (5) In direct contrast, prostacyclin (PGI_2) through activation of adenylyl cyclase and cAMP activation of protein kinase A (PKA) inhibits both the activity and upregulation of NOX. This axis therefore represents an innate protective system that blocks the vasculopathic impact of isoprostanes and TXA_2. (6) Peroxynitrite ($ONOO^-$) formed by the reaction between NO and O_2^- from NOX inactivates PGI_2 synthase by nitration. This would reduce the inhibitory effect of PGI_2 and augment TXA_2-driven increase in NOX expression and activity.

COX-1 is constitutive and COX-2 is inducible by similar factors that upregulate NOX [43, 44].

In VSMCS and ECs, the predominant prostanoid generated is PGI_2 although other PGs are formed (namely PGE_2, PGD_2, and PGJ_2). TXA_2 is the most abundant product of AA in platelets (which only express COX-1) and is also a prominent product of monocyte and macrophages. Expression of COX-1 and COX-2 is augmented in endothelium, VSMCs, and macrophages in human atherosclerotic lesions [46, 47]. In case–control studies COX-2-derived PGE_2 has been implicated in the activation of MMPs and consequent destabilization of human atherosclerotic plaques [48, 49]. PGI_2 and PGE_2 inhibit the activity and adhesion of platelets and leukocytes, dilate blood vessels, and modulate MMP activity and VSMC replication. In direct contrast, TXA_2 is a potent vasoconstrictor, activator of platelets and leukocytes, promoter of VSMC replication, and is powerfully pro-angiogenic. Thus, considerable attention has been paid to the changes in the ratio of these prostanoids in relation to the etiology of atherogenesis.

TXA_2 upregulates NOX protein expression and also activates established NOX activity, in part, through activation of Rac-1 in VSMCs and ECs [50]. In direct contrast, stable analogs of PGI_2 inhibit the upregulation and activation of NOX induced by TXA_2 via activation of adenylyl cyclase and the cyclic AMP-protein kinase axis [50] (Figure 13.6). ROS derived from NOX also promotes the formation of isoprostanes [50], which have identical properties to TXA_2 through activation of TXA_2 receptors [50]. Plasma levels of isoprostane are increased in atherosclerotic patients, and in experimental mice their generation increases during atherogenesis [51]. In turn, inhibition of their formation retards atherogenesis [51]. Isoprostanes upregulate the expression NOX, an effect again blocked by PGI_2 [50]. However, as a consequence of increased 8-isoprostane formation there is a reduction of PGI_2 formation which would favor upregulation of NOX expression and activity (Figure 13.6).

The inhibitory effects of PGI_2 on NOX are mediated by activation of adenylyl cyclase that generates cyclic AMP which in turn activates protein kinase A (PKA) [52]. PKA then phosphorylates enzymes that mediate the biological impact of PGI_2. cAMP is hydrolyzed to inactive AMP by type 4 phosphodiesterases (PDE_4) thereby reducing the bio-impact of PKA. A recent study has demonstrated that upregulation of NOX1 in VSMCs elicits an upregulation of PDE_4, an effect mediated by activation of Rho kinase by O_2^- [52]. In turn, the PDE_4 inhibitor roflumilast augments the inhibitory action of PGI_2 on both expression and activity of NOX by TXA_2 [53], consolidating the potential importance of PDE_4 status in plaque biology.

In clinical trials, specific COX-2 inhibitors (designed to reduce proinflammatory PGI_2 and PGE_2) elicited a two- to fourfold increase in myocardial infarction and stroke that has led to withdrawal of the drug from the market [43, 44]. It was suggested that the inhibition of COX-2-derived PGI_2 removes a constraint on platelet COX-1-derived TXA_2 that augments the thrombotic response to plaque rupture. It does point to the importance of the PGI_2–TXA_2 balance in plaque biology, both at the progression and rupture level since myocardial infarction and stoke are direct consequences of plaque rupture. Thus, the anti-atherogenic effects of TXA_2 receptor antagonists have been investigated in animal models [43, 44]. In mice, TXA_2 receptor antagonism delays both spontaneous atherogenesis lesion formation and that accelerated by a high-fat diet [43, 44]. However, a COX-2 inhibitor failed to augment the impact of TXA_2 receptor antagonism. Most notably, lesions contained more necrotic cores and fibrotic caps were thinner, indicating that combined treatment with COX-2 inhibitor and TXA_2 receptor antagonist results in plaque destabilization [43, 44]. TXA_2 may therefore play a dual role: (i) augmentation of the progression of atherosclerosis and (ii) stabilization of the atherosclerotic plaque, close to terminal stages, possibly through the promotion of VSMC replication and matrix deposition [43, 44].

13.8
Angiogenesis and Neovascularization

All conduit arteries possess a vasa vasorum, a complex of microvessels housed in the adventitia that provides oxygen and nutrients to the artery. Experimental disruption of the vasa vasorum promotes atherogenesis, an effect ascribed to hypoxia [54–56]. Microvessel density is increased in human ruptured aortic plaques and is increased in lesions with overt inflammation, intraplaque hemorrhage, and in thin-cap fibroatheromas [57, 58]. Ectopic neovascularization in the intima and media is considered a hallmark of advanced, vulnerable atherosclerotic lesions. Although there is abundant evidence that NOX plays an axiomatic role in angiogenesis [59, 60], a direct link between NOX, angiogenesis, and atherogenesis is far less clear.

During angiogenesis, ECs proliferate, secrete MMPs, and migrate and organize themselves into tubules (sprouting). Maturation and stabilization of the new vessels occurs through the recruitment of pericytes and VSMCs, the beginning of arteriogenesis. Once a microvessel has formed, ECs become resistent to exogenous factors and survive for years. However, in the dynamic and ever-changing environment of the plaque, microvessels are likely to be modulated on an ongoing basis.

In wound-healing, angiogenesis is orchestrated by macrophages which release angiogenic factors that include cytokines, thromboxane A_2, leukotrienes, and peptide growth factors. Since macrophages are in constant residence within plaques it is reasonable to surmise that macrophages would augment neovascularization. Hypoxia is also a primary trigger for angiogenesis but the atherosclerotic plaque is intrinsically hypoxic [61–63].

Principal among the factors that control angiogenesis are HIF, VEGF, AP-1, early growth response 1 (EGR-1), nuclear factor interleukin-6 (NF-IL6), and NF-κB [59–66]. In turn, ROS derived from NOX have been shown to trigger the activation and effects of these key signal transduction pathways [59].

HIF is an α,β heterodimer. HIF-1β is constitutively expressed whereas HIF-1α subunits are inducible by hypoxia [63]. HIF-1 binds to a consensus region of the gene termed the hypoxia response element (HRE). HIF-1α is subject to rapid ubiquination and proteosomal degradation, a process that is inhibited under normoxic conditions. Hypoxia, cytokines, peptide growth factors, insulin, and basic fibroblast growth factor all upregulate HIF-1α [63] and a number of target genes for HIF-1α, including VEGF and VEGF receptor [63]. Intracellular signal transduction systems involved in the effect of HIF include Ras/Raf/MAP kinase and receptor tyrosine kinase, PI3 kinase and PTEN Akt kinase pathways [63]. NOX4 siRNA decreased HIF-1α and VEGF levels and inhibited angiogenesis, indicating that NOX4 is required for the induction of angiogenesis. In a key study, VEGF was shown to trigger phosphorylation of p47^{phox} and colocalized with WAVE1, Rac-1, and the Rac-1 effector PAK1 [65]. It was suggested that WAVE1 may act as a scaffold to recruit the NOX to a complex involved with both cytoskeletal regulation and downstream JNK activation [65].

Angiopoietins are also axiomatic in mediating angiogenesis. Angiopoietin 1 (Ang1), via phosphorylation of Tie-2 receptors, is chemotactic for ECs and induces sprouting and stimulates EC–pericyte interactions. Angiopoietin 2 (Ang2) is a natural antagonist of Ang1 and is involved in detaching VSMCs and loosening underlying matrix, thereby allowing ECs to migrate. Ang2 in concert with VEGF promotes angiogenesis but alone can be an inhibitor. Inhibition of NOX significantly suppresses endothelial cell migration and sprouting [67]. In ECs from $p47^{phox-/-}$ knockout mice, intracellular ROS, phosphorylation of both Akt and p42/44 MAPK, and cell migration was reduced [68].

13.9
MMPs, Wnt-β Catenin and Forkhead Box (FOX) Signaling Pathway

MMPs are enormously important in plaque pathobiology since they are axiomatic in controlling cell proliferation and migration, apoptosis, angiogenesis, and plaque fragility [69]. MMPs include interstitial collagenase (MMP-1), gelatinase A (MMP-2), stromelysin-1 (MMP-3), matrilysin (MMP-7) and gelatinase B (MMP-9) [69]. MMP activity, is regulated by tissue inhibitors of MMPs (TIMPs) [69]. Of the MMP family, MMPs-1,2,3 and 9 have been shown to be modulated by NOX in cell typess present the plaque. In ECs, increased MMP-2 secretion is mediated by NOX [70]. In VSMCs, the latent form of MMP-2 (pro-MMP-2) is released after mechanical stretch-stimulated production of ROS by NOX [71]. NOX activates MMP-1 and MMP-9 expression in VSMCs through activation of MAPK by ROS [72]. In cell-free systems, low concentrations of ROS can activate pro-MMPs by oxidation of the sulfide bond in the pro-domain of the MMP followed by release of this pro-domain by autocatalytic cleavage [73]. Although little is known about TIMPs and NOX, a recent study demonstrated that NOX1 deficiency protects from aortic dissection in response to angiotensin II in mice, and TIMP-1 mRNA and protein was markedly increased in aortas from NOX1-deficient mice, suggestive of an association [74]. In a related study Hu *et al.* found that deletion of LDL receptor-1 (LOX-1) in LDL receptor (LDLR) knockout mice fed a high-cholesterol diet decreased collagen accumulation in plaques and was accompanied by reduced expression/activity of osteopontin, fibronectin, MMP-2 and MMP-9, and levels of $p47^{phox}$, $p22^{phox}$, $gp91^{phox}$, and NOX4 subunits [75]. These data confirm that there may be a close relationship between NOX and matrix proteins [75]

Another system influenced by ROS is the Wnt pathway which regulates cell proliferation, differentiation, polarity, migration, invasion, and survival and as such has been recently implicated in plaque development and rupture [76]. A relationship between NOX and the Wnt pathway remains to be established, however. Closely associated with the Wnt signaling/β-catenin pathways are the Forkhead box (FOX) proteins, a family of transcription factors that play roles in regulating cell growth, proliferation, differentiation, and survival [77, 78]. A subgroup of the FOX family are the FoxO proteins which are emerging as critical transcriptional integrators of cell fate [77, 78]. FoxO-mediated transcription requires binding of β-catenin, which appears to fulfill a critical function in balancing positive and negative regulation of cell cycle and apoptosis [77, 78]. The FoxO subfamily has a well-established role in induction of apoptosis in various cell types through regulation of pro-apoptotic genes, such as Bim, Bcl-6, and FasL. Furthermore, in VSMCs FoxO3a overexpression induces apoptosis *in vitro* and *in vivo* [77, 78]. The role of NOX in mediating these emerging systems especially in relation to apoptosis and/or survival is in its infancy and warrants further investigation.

13.10
Therapeutic Strategies Involving Inhibition of NOX

Given the possible involvement of NOX in atherosclerosis, NOX may represent an interventional target for drugs. Drugs that inhibit NOX include apocynin and diphenyliodonium chloride [79, 80]. The administration of apocynin in animal models reduced intravascular oxidative stress and atherosclerosis. However, recent evidence has emerged that these are not as specific as hitherto assumed [81]. For example, apocynin inhibits Rho kinase and thromboxane synthase [81].

Alternatively, the inhibition of O_2^- formation or its negation (quenching) constitutes a potentially valid approach to the prevention and treatment of atherosclerosis. Drugs that do so are termed "antioxidants." The diversity and number of antioxidants that have been shown to ameliorate cardiovascular disease (CVD) in animal models are considerable and include: vitamins C, D, and E, pomegranate juice, selenate, metal (iron and copper) chelators, probucol, and SOD mimetics [15, 16]. By contrast, although the therapeutic use of antioxidant vitamins, including vitamins C, D, and E, for treating CVD has been widely studied in humans the results have been disappointing [15, 16]. A possible reason for the discrepancy between improvements by antioxidants in animal models and human studies is that the disease was treated at a more advanced stage in clinical studies [15, 16]. Put another way, antioxidant therapy may be a case of "shutting the therapeutic door when the pathological horse has bolted." Furthermore, the synergistic effect of multiple risk factors or other dietary or lifestyle confounding variables may be significant. It is also possible that these antioxidants do not elicit direct inhibition of O_2^- formation in relevant tissues (i.e., plaque). Thus, agents specifically reducing intracellular O_2^- may be more useful than classical antioxidants such as vitamins C and E. In this context, it has been demonstrated that folic acid inhibits the expression and activity of NOX in animal models of diabetes, hyperhomocysteinemia, and copper overload [82, 83].

As has been stressed in this chapter, NOX is activated and upregulated by a bewildering array of factors (cytokines, peptide growth factors, eicosanoids, shear stress, vasoconstrictors, ROS, and risk factors such as cholesterol and glucose). It follows, therefore, that drugs that inhibit the action of these effectors would inhibit NOX. For example, angiotensin II (Ang-II) receptor antagonists reduces NOX activity which is as one would expect since Ang-II is a classic activator of the complex. Most notable, perhaps, is that 3-hydroxy-3-methylglutaryl coenzyme A (HMG-CoA) reductase inhibitors (statins) inhibit both the activity and expression of NOX1 and NOX2 in vascular cells [80]. Membrane association of the GTPase Rac-1 is dependent on geranylgeranylation, which is prevented by HMG-CoA reductase inhibitors [80]. Spontaneously hypertensive rats treated with atorvastatin reduced aortic ROS production, p22^{phox} expression, and NOX1 expression and Rac-1 translocation [80].

In addition, as has been discussed above, it has been shown that endogenous vasculoprotective systems (NO, PGI$_2$, H$_2$S) and donors and mimetics thereof, inhibit NOX activity and upregulation [35–38]. In turn, drugs that augment the action of these protective factors, namely PDE inhibitors, may also be effective through modulation of these pathways and that the principal mode of action of these drugs may involve, at least in part, effects on the influential NOX family.

Furthermore, NO-donating aspirins but not aspirin alone are potent inhibitors of NOX expression elicited by cytokines and endotoxin [37]. Although aspirin is undoubtedly beneficial in preventing cardiovascular episodes, it does have limitations. These include gastric erosion and a lack of effect on platelet and leukocyte adhesion molecule expression and VSMC proliferation and migration [37]. By contrast, NO is a potent inhibitor of these functions [37]. Thus, the NO-donating moiety of NO–aspirin may compensate for the innate limitations of aspirin, while retaining the positive effects of aspirin alone.

13.11
Concluding Remarks

It is clear that NOX expression and activity is associated with the pathobiology of atherosclerosis since NOX and the overproduction of ROS modulates inflammation, apoptosis, VSMC replication, angiogenesis, and matrix turnover. However, as with the etiology of atherosclerosis and plaque rupture, exactly how prominent NOX is in this scenario will require further experimentation. From an interventional perspective, inhibition of NOX may be a double-edged sword. NOX inhibition may be effective in retarding the rate of progression of plaque development. However, at end-stage, NOX inhibition may be deleterious since it appears to influence VSMC replication and MMP activity.

References

1 Bedard, K. and Krause, K.H. (2007) *Physiol. Rev.*, **87**, 245–313.

2 Babior, B.M. (2004) NADPH oxidase. *Curr. Opin. Immunol.*, **16**, 42–47.

3 Lambeth, J.D. (2004) *Nat. Rev. Immunol.*, **4**, 181–189.

4 Griendling, K.K., Sorescu, D., Lassegue, B., *et al.* (2000) *Arterioscler. Thromb. Vasc. Biol.*, **20**, 2175–2183.

5 Clempus, R.E. and Griendling, K.K. (2006) *Cardiovasc. Res.*, **71**, 216–225.

6 Muzaffar, S., Shukla, N., and Jeremy, J.Y. (2005) *Trends. Cardiovasc. Med.*, **15**, 278–282.

7 Ushio-Fukai, M. (2009) *Antioxid. Redox Signal.*, **11**, 1289–1299.

8 Adachi, T., Pimentel, D.R., Heibeck, T., *et al.* (2004) *J. Biol. Chem.*, **279**, 29857–29862.

9 Jeremy, J.Y., Rowe, D., Emsley, A.M., and Newby, A.C. (1999) *Cardiovasc. Res.*, **43**, 580–594.

10 Guzik, T.J., Sadowski, J., Guzik, B., *et al.* (2006) *Arterioscler. Thromb. Vasc. Biol.*, **26**, 333–339.

11 Guzik, T.J., Chen, W., Gongora, M.C., *et al.* (2008) *J. Am. Coll. Cardiol.*, **52**, 1803–1809.

12 Barry-Lane, P.A., Patterson, C., van der Merwe, M., *et al.* (2001) *J. Clin. Invest.*, **108**, 1513–1522.

13 Vendrov, A.E., Hakim, Z.S., Madamanchi, N.R., *et al.* (2007) *Arterioscler. Thrombos. Vasc. Biol.*, **27**, 2714.

14 Kalinina, N., Agrotis, A., Tararak, E., *et al.* (2002) *Arterioscler. Thromb. Vasc. Biol.*, **22**, 2037–2043.

15 Jeremy, J.Y., Shukla, N., Muzaffar, S., and Angelin, G.D. (2004) *Curr. Vasc. Pharmacol.*, **2**, 229–236.

16 Jeremy, J.Y., Yim, A.P., Wan, S., and Angelini, G.D. (2002) *J. Cardiovasc. Surg.*, **17**, 324–327.

17 Clarke, M. and Bennett, M. (2006) *Cell Cycle*, **5**, 2329.

18 Clarke, M., Figg, N., Maguire, J., *et al.* (2007) *Nat. Med.*, **12**, 1075–1080.

19 Clarke, M.C.H., Littlewood, T.D., Figg, N., *et al.* (2008) *Circ. Res.*, **102**, 1529–1538.

20 Ryter, S.W., Kim, H.P., Hoetzel, A., *et al.* (2007) *Antioxid. Redox Signal.*, **9**, 49–89.

21 Warren, M.C., Bump, E.A., Medeiros, D., and Braunhut, S.J. (2000) *Free. Radic. Biol. Med.*, **29**, 537–547.

22 Martindale, J.L. and Holbrook, N.J. (2002) *J. Cell. Physiol.*, **192**, 1–15.

23 Clement, M.V. and Stamenkovic, I. (1996) *EMBO J.*, **15**, 216–225.

24 Madesh, M., Hawkins, B.J., Milovanova, T., *et al.* (2005) *J. Cell Biol.*, **170**, 1079–1090.

25 Pervaiz, S. and Clement, M.V. (2002) *Biochem. Biophys. Res. Commun.*, **290**, 1145–1150.

26 Dickhout, J.G., Hossain, G.S., Pozza, L.M., *et al.* (2005) *Arterioscler. Thrombos. Vasc. Biol.*, **25**, 2623.

27 Pedruzzi, E., Guichard, C., Ollivier, V., *et al.* (2004) *Mol. Cell. Biol.*, **24**, 10703–10717.

28 Yokouchi, M., Iramatsu, N., Hayakawa, K., *et al.* (2008) *J. Biol. Chem.*, **283**, 4252.–4260

29 Boyce, M. and Yuan, J. (2006) *Cell Death Diff.*, **13**, 363–373.

30 Jeremy, J.Y. and Thomas, A.C. (2010) *Curr. Vasc. Pharmacol.*, in press.

31 Menshikov, M., Plekhanova, O., Cai, *et al.* (2006) *Arterioscler. Thromb. Vasc. Biol.*, **26**, 801–807.

32 Jacobson, G.M., Dourron, H.M., Liu, J., *et al.* (2003) *Circ. Res.*, **92**, 637–643.

33 Paravicni, T.M., Gulluyan, L.M., Dusting, G.J., *et al.* (2002) *Circ. Res.*, **91**, 54–61.

34 Muzaffar, S., Jeremy, J.Y., Sparatore, A., *et al.* (2008) *Br. J. Pharmacol.*, **155**, 984–994.

35 Muzaffar, S., Shukla, N., Bond, M., *et al.* (2008) *Prostaglandins Leukot. Essent. Fatty Acids*, **78**, 247–255.

36 Muzaffar, S., Jeremy, J.Y., Sparatore, A., *et al.* (2008) *J. Vasc. Res.*, **45**, 521–528.

37 Muzaffar, S., Shukla, N., Angelini, G.D., and Jeremy, J.Y. (2004) *Circulation*, **110**, 1140–1147.

38 Muzaffar, S., Jeremy, J.Y., Angelini, G.D., Stuart-Smith, K., and Shukla, N. (2003) *Thorax*, **58**, 598–604.

39 Chatterjee, A., Black, S.M., and Catravas, J.D. (2008) *Vasc. Pharmacol.*, **49**, 134–140.

40 Xu, J., Xie, Z., Reece, R., *et al.* (2006) *Arterioscler. Thromb. Vasc. Biol.*, **26**, 2688–2695.

41 Muzaffar, S., Shukla, N., Angelini, G.D., and Jeremy, J.Y. (2005) *Br. J. Pharmacol.*, **146**, 109–117.

42 Muzaffar, S., Shukla, N., Bond, M., *et al.* (2008) *Br. J. Pharmacol.*, **155**, 847–856.

43 Fitzgerald, G.A. (2004) *N. Engl. J. Med.*, **351**, 1709–1711.

44 Funk, C.D. and FitzGerald, G.A. (2007) *J. Cardiovasc. Pharmacol.*, **50**, 470–479.

45 Jeremy, J.Y., Jackson, C.L., Bryan, A.J., and Angelini, G.D. (1996) *Prostaglandins Leukot. Essent. Fatty Acids*, **54**, 385–402.

46 Schonbeck, U., Sukhova, G.K., Graber, P., *et al.* (1999) *Am. J. Pathol.*, **155**, 1281–1291.

47 Baker, C.S., Hall, R.J., Evans, T.J., *et al.* (1999) *Arterioscler. Thromb. Vasc. Biol.*, **19**, 646–655.

48 Shankavaram, U.T., Lai, W.C., Netzel-Arnett, S., *et al.* (2001) *J. Biol. Chem.*, **276**, 19027–19032.

49 Cipollone, F., Fazia, M., Iezzi, A., *et al.* (2003) *Circulation*, **107**, 1479–1485.

50 Muzaffar, S., Shukla, N., Lobo, C., *et al.* (2004) *Br. J. Pharmacol.*, **141**, 488–496.

51 Pratico, D., Tangirala, R.K., Rader, D.J., *et al.* (1998) *Nat. Med.*, **4**, 1189–1192.

52 Houslay, M.D. (2005) *Mol. Pharmacol.*, **68**, 563–567.

53 Muzaffar, S., Shukla, N., and Jeremy, J.Y. (2009) *Proc. Br. Pharm. Soc.* (abstract), **161**, p 27.

54 Martin, J.F., Booth, R.F.G., and Moncada, S. (1991) *Eur. J. Clin. Invest.*, **21**, 355–359.

55 McGeachie, J.K., Campbell, P.A., and Predergast, F.J. (1981) *Ann. Surg.*, **194**, 100–107.

56 Barker, S.G., Talbert, A., Cottam, S., *et al.* (1993) *Arterioscler. Thromb.*, **13**, 70–77.

57 Moreno, P.R., Purushothaman, K.R., Zias, E., Sanz, J., and Fuster, V. (2006) *Curr. Mol. Med.*, **6**, 457–477.

58 Moreno, P.R., Purushothamam, K.R., Sirol, M., *et al.* (2006) *Circulation*, **113**, 2245–2252.

59 Ushio-Fukai, M. (2006) *Cardiovasc. Res.*, **71**, 226–235.

60 Irani, K. (2000) *Circ. Res.*, **87**, 179–183.

61 Lugue, A., Turu, M., Jean Babot, A., *et al.* (2008) *Front. Biosci.*, **13**, 6483–6490.

62 Vink, A., Schoneveld, A.H., Lamers, D., *et al.* (2007) *Atherosclerosis*, **195**, 69–75.

63 Ruas, J.L., Lendahl, U., and Poellinger, L. (2007) *Curr. Opinin. Lipidol.*, **18**, 508–514.

64 Xia, C., Meng, Q., Liu, L.-Z., *et al.* (2007) *Am. J. Cancer Res.*, **67**, 10823–10830.

65 Wu, RF, Gu, Y., Xu, Y.C., *et al.* (2003) *J. Biol. Chem.*, **278**, 36830–36840.

66 Forsythe, J.A., Jiang, B.H., Iyer, N.V., *et al.* (1996) *Mol. Cell. Biol.*, **16**, 4604–4613.

67 Chen, J.X., Zeng, H., Lawrence, M.L., *et al.* (2006) *Am. Heart J.*, **291**, H1563–72.

68 Harfouche, R., Malak, N.A., Brandes, R.P., *et al.* (2005) *FASEB J.*, **19**, 1728–1730.

69 Newby, A.C. (2008) *Arterioscler. Thrombos. Vasc. Biol.*, **28**, 2108–2114.

70 Inoue, N., Takeshita, S., Gao, D., *et al.* (2001) *Atherosclerosis*, **155**, 45–52.

71 Grote, K., Flach, I., Luchtefeld, M., *et al.* (2003) *Circ. Res.*, **92**, e80–e86.

72 Shin, M.H., Moon, Y.J., Seo, J.-E., *et al.* (2008) *Free Rad. Biol. Med.*, **44**, 635–645.

73 Rajagopalan, S., Meng, X.P., Ramasamy, S., *et al.* (1996) *J. Clin. Invest.*, **98**, 2572–2579.

74 Gavazzi, G., Deffert, C., Trocme, C., *et al.* (2007) *Hypertension*, **50**, 189–196.

75 Hu, C., Dandapat, A., Sun, L., *et al.* (2008) *Cardiovasc. Res.*, **79**, 287–293.

76 George, S.J. and Beeching, C.A. (2006) *Atherosclerosis*, **188**, 1–11.

77 Sedding, D.G. (2008) *Biol. Chem.*, **389**, 279–283.

78 Essers, M.A.G., de Vries-Smits, L.M.M., Barker, N., *et al.*. (2005) *Science*, **308**, 1181–1184.

79 Guzik, T.J. and Harrison, D.G. (2006) *Drug Discov. Today*, **11**, 524–533.

80 Jiang, F., Drummond, G.R., and Dusting, G.J. (2004) *Endothelium*, **111**, 79–88.

81 Aldieri, E., Rignati, C., Polimeni, M., *et al.* (2008) *Curr. Drug Metab.*, **9**, 686–696.

82 Shukla, N., Angelini, G.D., and Jeremy, J.Y. (2008) *Metabolism.*, **57**, 774–781.

83 Shukla, N., Hotston, M., Persad, R., Angelini, G.D., and Jeremy, J.Y. (2009) *BJU Int.*, **103**, 98–103.

14
Uncoupling of Endothelial Nitric Oxide Synthase in Atherosclerosis

Colin Cunnington and Keith M. Channon

14.1
Introduction: Nitric Oxide and the Endothelium

The influence of endothelial cells on vascular homeostasis was first described by Furchgott and Zawadzki [1], who proposed in 1980 that endothelial cells release an endothelial-derived relaxing factor (EDRF) in response to acetylcholine. EDRF has since been shown to be nitric oxide (NO) [2]. In addition to its vasodilatory capacity, NO has a number of anti-atherogenic properties, including downregulation of leukocyte adhesion molecule expression [3], inhibition of platelet aggregation [4], and suppression of smooth muscle proliferation [5]. It is now widely accepted that endothelial dysfunction, defined as decreased NO bioavailability, is a critical early step in the development of atherosclerosis, being associated with risk factors for atherosclerotic vascular disease [6–9] and predictive of adverse cardiovascular events in prospective studies [10, 11].

Numerous pharmacological and genetic experimental models indicate that reduced endothelial NO production accelerates atherosclerosis [12]. Furthermore, under certain pathophysiological conditions, endothelial nitric oxide synthase (eNOS), the enzymatic source of NO, can instead become a producer of superoxide anion ($O_2^{\cdot-}$) [13, 14]. This process, termed "eNOS uncoupling," can give rise to increased vascular oxidative stress and further loss of NO bioavailability. In this chapter we aim to review the evidence underlying eNOS uncoupling and its importance in the pathogenesis of atherosclerosis.

14.2
Production of NO in the Endothelium

In addition to its role in endothelial biology, NO has been shown to function as an important signaling molecule in other physiological systems, mediating the cytotoxic effect of macrophages [15] and acting as a messenger molecule in plate-

Atherosclerosis: Molecular and Cellular Mechanisms. Edited by Sarah Jane George and Jason Johnson
Copyright © 2010 WILEY-VCH Verlag GmbH & Co. KGaA, Weinheim
ISBN: 978-3-527-32448-4

lets [4] and the brain [16]. Characterizing the enzymatic source of NO production in these tissues was the focus of much research in the late 1980s and early 1990s.

14.2.1
NOS Structure and Function

It was established that the amino acid L-arginine is the substrate for NO production, and more specifically that the terminal guanidino-nitrogen atom of L-arginine is used to synthesize NO [17, 18], via an N^ω-hydroxy-L-arginine intermediate [19], with L-citrulline as a further product [20]. The oxygen atom in NO was shown to be derived from molecular oxygen [21].

Three isoforms of NOS have been defined: neuronal (nNOS, NOS I), inducible (iNOS, NOS II), and endothelial (eNOS, NOS III). eNOS and nNOS are expressed constitutively, whereas iNOS expression is induced by cytokines and other inflammatory stimuli [22]. In addition to requiring L-arginine as a substrate for NO production, all isoforms require the cofactors 6R-5,6,7,8-tetrahydrobiopterin (BH4, discussed below), nicotinamide adenine dinucleotide phosphate (NADPH) [19, 23, 24], and the flavins flavin adenine dinucleotide (FAD) and flavin mononucleotide (FMN) [22, 24, 25]. All three isoforms are activated by the presence of intracellular Ca^{2+} ions [23, 24, 26, 27], however only eNOS and nNOS are regulated by agonist stimulation through calcium-calmodulin activation [27, 28], whereas iNOS remains tonically active because calmodulin remains bound at intracellular calcium concentrations, independent of agonist stimulation [24]. In contrast to nNOS and iNOS, which are cytosolic enzymes, eNOS is mainly membrane-associated, localized in plasma membrane caveolae and other intracellular membranes through posttranslational lipid modifications and protein–protein interactions with caveolin and other caveolae-associated proteins [26, 28, 29].

All three NOS isoforms function as catalytically active homodimers, comprising an oxidoreductase bidomain structure, with calmodulin facilitating the transfer of electrons from the C-terminal reductase domain to the heme-containing oxygenase domain of the opposite monomer within the dimer pair [30]. Upon binding of Ca^{2+} to constitutive NOS, oxidation of the electron donor NADPH results in electron transfer via FAD and FMN of the reductase domain to the heme center of the oxygenase domain where molecular oxygen and L-arginine are bound; NO catalysis can then proceed via the stepwise oxidation of L-arginine (Figure 14.1a) [30].

Figure 14.1 Mechanism of endothelial nitric oxide synthase (eNOS) regulation by tetrahydrobiopterin (BH4). Under optimal conditions (a), the eNOS enzyme is a homodimer with each monomer comprising a flavin-containing reductase domain and a heme-containing oxygenase domain. Upon binding of calcium (Ca^{2+})/calmodulin (CaM), electron flow (e^-) generated from NADPH is transferred via the flavins FAD and FMN to the oxygenase domain of the other monomer. BH4 is bound close to the heme group (H) in the active site and also interacts with residues in both monomers, contributing to eNOS dimer formation.

(a)

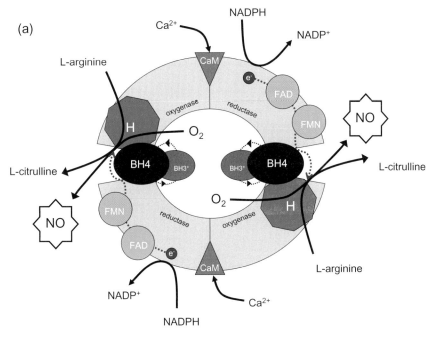

(b)

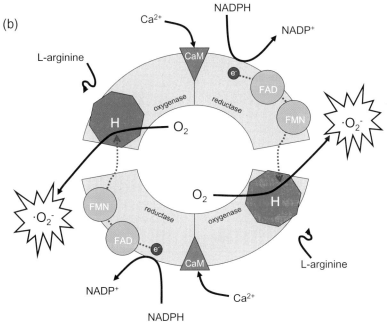

Figure 14.1 *Continued* Nitric oxide (NO) is generated by the oxidation of the guanidino-nitrogen of L-arginine using molecular oxygen, together with the production of L-citrulline. BH4 undergoes temporary oxidation to the BH3+ radical but is rapidly recycled back to BH4 in a continuous process. In conditions of BH4 deficiency (b), electron flow from flavins to L-arginine becomes uncoupled from L-arginine oxidation, resulting in formation of superoxide (O_2^-) from molecular oxygen.

14.2.2
Requirement of BH4 as a Cofactor for NOS

Prior to the discovery of NO, the pteridine cofactor BH4 was known to catalyze the conversion of phenylalanine to tyrosine by phenylalanine hydroxylase [31], as well as acting as a cofactor for other aromatic amino acid hydroxylases. Deficiency of BH4 has been shown to lead to hyperphenylalaninemia [32]. Other pterin species, particularly neopterin, had been shown to be released by cytokine-activated macrophages [33]. Thus, it was speculated that BH4 may have cofactor activity in other enzyme systems. The requirement for BH4 as a NOS cofactor was first demonstrated for iNOS in macrophages [34, 35]; it was shown that both BH4 and NADPH were required for enzyme activity, as defined by the consumption of L-arginine and the production of citrulline and NO_2^-/NO_3^-, the breakdown products of NO. Evidence for BH4 cofactor activity was later demonstrated for NOS in brain [27], neutrophils [24], fibroblasts [36], and endothelial cells [37]. Studies of purified NOS suggest that BH4 functions both via allosteric regulation of L-arginine binding [38] and by promoting dimer formation and stability [30, 39].

14.2.3
BH4 Synthesis

Intracellular BH4 levels are regulated principally by the activity of the *de novo* synthetic pathway. Guanosine triphosphate cyclohydrolase I (GTP-CH) catalyzes the conversion of GTP to dihydroneopterin triphosphate. Two further enzymatic steps, catalyzed by 6-pyruvoyltetrahydropterin synthase (PTPS) and sepiapterin reductase (SR), are required to synthesize BH4, although under normal circumstances GTP-CH is the rate-limiting enzyme [40]. GTP-CH, but not PTPS or SR, is upregulated in some cell types by inflammatory cytokines such as interferon-γ and tumor necrosis factor-α [40, 41]. Additionally, BH4 may be synthesized via a "salvage pathway" involving conversion of exogenous sepiapterin to 7,8-dihydro-biopterin (BH2), and then via NADPH-dependent dihydrofolate reductase (DHFR) to BH4 [42]. DHFR may also play a role in "recycling" BH4 from BH2 produced within cells by oxidation of BH4.

14.3
NOS Uncoupling

It was observed that the amino acid sequence of NOS bore close resemblance to the cytochrome P450 reductases, enzymes known to generate O_2^-. It was therefore hypothesized that NOS may also be able to produce reactive oxygen species (ROS) as well as NO. Indeed, in 1992 two groups showed that under certain conditions brain NOS could catalyze the formation of reactive oxygen species instead of NO [43, 44]. In the absence of BH4 and L-arginine, purified porcine cerebellar NOS consumed molecular oxygen at the expense of NADPH, with generation of hydrogen peroxide (H_2O_2, the dismutation product of O_2^- by superoxide dismutase

(SOD)) [43], a process only partially inhibited by supplementary L-arginine. The addition of BH4 inhibited H_2O_2 formation and restored arginine-to-citrulline conversion [43]. Using electron paramagnetic resonance (EPR) spin-trapping techniques it was then demonstrated that NOS produces $O_2^{\cdot-}$ rather than H_2O_2 in the absence of L-arginine [44]. The addition of the arginine analog N^G-nitro-L-arginine methyl ester (L-NAME), a specific inhibitor of NOS, inhibited both the production of $O_2^{\cdot-}$ and the consumption of NADPH.

These findings from purified NOS were later confirmed in intact cells transfected with nNOS [45], with $O_2^{\cdot-}$ being produced only in the absence of L-arginine. Although these investigators did not address the role of BH4 in this study, they demonstrated for the first time that $O_2^{\cdot-}$ produced by NOS can combine with NO to form peroxynitrite ($ONOO^-$), a potent oxidant which is significantly more reactive than $O_2^{\cdot-}$ itself. Furthermore, $ONOO^-$ is able to rapidly oxidize BH4 to inactive pterin species, including BH2, 7,8-dihydropterin, and dihydroxanthopterin [46] which have no NOS cofactor activity.

14.3.1
eNOS Uncoupling

A number of studies over the last 10 years have shown that the endothelium is a source of $O_2^{\cdot-}$ under pro-atherosclerotic conditions, including hypercholesterolemia [47]. It was hypothesized that the enzymatic "uncoupling" of L-arginine oxidation observed in purified brain NOS, leading to the unopposed reduction of molecular oxygen to $O_2^{\cdot-}$, may implicate the endothelial isoform of NOS as a source of $O_2^{\cdot-}$ in the vascular wall. A fundamental observation was made by Pritchard *et al.*, who noted that incubation of cultured human umbilical vein endothelial cells (HUVECs) with native low-density lipoprotein (LDL) resulted in increased $O_2^{\cdot-}$ production, which could be inhibited by L-NAME, implying that eNOS was the source of $O_2^{\cdot-}$ [48]. Conversely, L-NAME increased the production of $O_2^{\cdot-}$ in control cells due to decreased scavenging of $O_2^{\cdot-}$ by NO. LDL also increased levels of nitrotyrosine, a stable oxidation product of $ONOO^-$ [48].

14.3.1.1 Studies of Purified eNOS Enzyme
In cell-free systems, three separate groups using recombinant human [13, 14] or bovine [49] eNOS demonstrated that BH4, and not L-arginine, was critical in preventing $O_2^{\cdot-}$ generation by eNOS; addition of L-arginine in the absence of BH4 was insufficient to inhibit $O_2^{\cdot-}$ formation [13], whereas BH4 could abolish the formation of $O_2^{\cdot-}$ in the absence of L-arginine [50]. $O_2^{\cdot-}$ production, inhibitable by L-NAME, was reported to occur at the heme center, as $O_2^{\cdot-}$ was shown to be inhibited by the heme iron ligands cyanide and imidazole [13, 14]. In addition, BH4 was able to scavenge $O_2^{\cdot-}$ produced from xanthine/xanthine oxidase *in vitro*, indicating a direct antioxidant effect in addition to regulating eNOS coupling [49]. Finally, BH4 increased the eNOS dimer:monomer ratio in purified enzyme heated to 55°C, suggesting a role in maintaining dimerization [49].

The term "eNOS uncoupling," therefore, describes the process whereby, in the absence of BH4, electron flow from the reductase domain of eNOS is diverted towards molecular oxygen rather than L-arginine, resulting in $O_2^{\cdot-}$ production rather than NO synthesis (Figure 14.1b). While changes in eNOS dimer structure may be associated with eNOS uncoupling, it is important to note that uncoupling refers specifically to enzymatic function rather than protein structure.

14.3.1.2 Studies of Pharmacological BH4 Supplementation and Depletion

To support these findings in purified eNOS and further investigate the role of BH4 in regulating eNOS function, a number of experimental strategies have been used to examine the effect of altered BH4 levels in cellular or animal models.

Exogenous BH4 administration caused L-NAME-inhibitable relaxations in pre-constricted rat aorta, associated with elevated cGMP levels [51]. A series of studies from the Katusic laboratory investigated the regulation of eNOS by BH4 in canine arteries *ex vivo* [52–55], first demonstrating increased H_2O_2 production in canine coronary arteries depleted of BH4 by incubation with 2,4-diamino-6-hydroxypyrimidine (DAHP, an inhibitor of GTP-CH) [52]. BH4 depletion by DAHP, however, resulted in increased endothelium-dependent relaxations (EDR) to calcium iono-phore A23187, an effect inhibited by catalase and L-NAME, implying that H_2O_2 generated by eNOS mediates vasorelaxation in these arteries. Second, EDR in canine middle cerebral arteries could be augmented by exogenous sepiapterin (a BH4 analog converted to BH4 via the salvage pathway), but only in the presence of adequate levels of SOD to prevent scavenging of NO by $O_2^{\cdot-}$ [53]. This finding was explained by the fact that BH4 is rapidly oxidized *in vitro* in Krebs buffer, leading to the spontaneous generation of $O_2^{\cdot-}$ [53]. This hypothesis was explored further in a third study using canine basilar arteries, where exogenous BH4 pro-duced strong endothelium-dependent contractions inhibitable by SOD [54]. In contrast, endothelium-denuded vessels showed only minor contraction in response to exogenous BH4, with vasorelaxation at higher doses. Lastly, in contrast to coro-nary arteries [52], canine basilar arteries incubated with DAHP exhibited decreased EDR which were restored by sepiapterin or SOD [55]. The differences observed between coronary and cerebral arteries were thought be due to differential sensitiv-ity of these arteries to $O_2^{\cdot-}$ and H_2O_2.

14.3.1.3 Alteration of BH4 Levels via Genetic Manipulation of GTP-CH

Further evidence for a causative role of BH4 in eNOS uncoupling was provided by the hyperphenylalaninemic mouse mutant (hph-1), which displays a 90% defi-ciency in GTP-CH activity [56], the rate-limiting enzyme in BH4 synthesis. In comparison to wild-type (WT) mice, the hph-1 mouse had significantly lower aortic BH4 levels, associated with decreased basal arginine-to-citrulline conversion (NOS activity) which could be augmented by exogenous BH4 supplementation [57]. Basal $O_2^{\cdot-}$ levels were slightly higher compared with WT; importantly however, the addition of L-NAME increased $O_2^{\cdot-}$ in WT mice (due to decreased NO scavenging), whereas $O_2^{\cdot-}$ levels in hph-1 were decreased, implying uncoupled eNOS as a prin-cipal source of $O_2^{\cdot-}$. Similar to the findings in canine coronary arteries [52], these BH4-deficient mice exhibited increased EDR compared with WT. EDR to acetyl-choline were inhibited by catalase and L-NAME, and augmented by SOD, implying that endothelium-dependent vasomotion in these mice is mediated by eNOS derived O_2^- and H_2O_2. Impaired EDR in the presence of catalase could be restored by exogenous BH4. In contrast, EDR in WT mice were augmented by catalase, whereas SOD had no effect on EDR [57].

Cai *et al.* used adenoviral gene transfer of GTP-CH (AdGCH) to augment intracellular BH4 levels in human endothelial cells [58]. AdGCH significantly increased both BH4 and total biopterin levels (the sum of BH4, BH2, and biopterin) compared to control cells. AdGCH cells exhibited a parallel increased in eNOS activity, which was inhibited by DAHP or the NOS inhibitor *N*-monomethyl-L-arginine (L-NMMA), despite these agents having no effect on eNOS protein levels. Despite a similar increase in BH4, sepiapterin incubation resulted in a lesser increase in eNOS activity. Furthermore, AdGCH transfection resulted in both increased eNOS dimer : monomer ratio, providing further evidence of the role of dimerization regulating eNOS function [58].

14.3.1.4 Paradoxical Findings in Chronic eNOS Overexpression

Studies using animals overexpressing eNOS have also proved useful in clarifying the regulatory function of BH4 *in vivo*. Mice with a deficiency in eNOS have been shown to develop accelerated atherosclerosis [59, 60], whereas short-term gene transfer of NOS to augment vascular NO production has been demonstrated to ameliorate endothelial dysfunction and improve markers of atherosclerosis [61, 62]. It was therefore expected that mice with transgenic overexpression of eNOS in the setting of atherosclerosis would exhibit decreased lesion formation. However, ApoE-KO/eNOS-Tg mice demonstrated a paradoxical increase in atherosclerosis compared to ApoE-KO controls, despite slightly lower blood pressure [63]. These animals were shown to have both increased vascular NO production and increased O_2^- levels (with a larger decrease in O_2^- following endothelial denudation) than matched ApoE-KO mice. Despite significantly higher levels of eNOS protein in ApoE-KO/eNOS-Tg mice, both strains had similar aortic BH4 levels (although tending to be reduced in ApoE-KO/eNOS-Tg animals), implying that decreased BH4 : eNOS ratio was contributory to the increased O_2^- observed in eNOS transgenic animals. Furthermore, oral BH4 supplementation resulted in a significantly greater reduction of atherosclerotic plaque and O_2^- levels in ApoE-KO/eNOS-Tg mice. Thus, the inhibitory effect of BH4 feeding on atherosclerosis is greatest when eNOS is overexpressed, implying that the underlying mechanism is via increased eNOS coupling, rather than a direct antioxidant effect [63].

Bendall *et al.* combined mouse models of eNOS and GTP-CH overexpression to further define the quantitative relationship between eNOS and BH4, in the absence of atherosclerosis [64]. First, eNOS-Tg mice had decreased eNOS dimer : monomer ratio compared to WT mice which was normalized in eNOS-Tg/GCH-Tg mice. Second, the aortic BH4 content of eNOS-Tg mice was reduced by 50% compared to WT, whereas both GCH-Tg and eNOS-Tg/GCH-Tg animals exhibited a twofold increase in aortic BH4. Third, despite an eightfold increase in eNOS protein, eNOS-Tg mice had similar levels of NO production compared to GCH-Tg mice; eNOS-Tg/GCH-Tg animals had a further significant increase in

NO production above eNOS-Tg mice despite similar levels of eNOS protein. Finally, L-NAME-inhibitable O_2^- was increased in eNOS-Tg mice and ameliorated by the GCH transgene [64]. Taken together, these data provide compelling evidence to support the notion that eNOS activity is dependent on BH4 availability, even in healthy animals.

14.3.2
Oxidation of Tetrahydrobiopterin: The Role of Peroxynitrite

As BH4 is in itself highly redox sensitive, being easily oxidized in buffer alone to other inactive pterin species with no NOS cofactor activity, it was hypothesized that BH4 levels may become limiting in the setting of oxidative stress via oxidative scavenging of BH4. O_2^- is known to react rapidly with NO to produce the potent biological oxidant $ONOO^-$. It was demonstrated *in vitro* that $ONOO^-$ can rapidly oxidize BH4 [46, 65], and that $ONOO^-$ is a far more potent scavenger of BH4 than O_2^- itself, with H_2O_2 having no reaction with BH4 [66]. In this study, Laursen *et al.* demonstrated firstly increased production of O_2^- from the aortas of ApoE-KO mice compared to WT, inhibitable by sepiapterin. Second, $ONOO^-$ production was elevated in ApoE-KO in comparison to WT, and uric acid, a $ONOO^-$ scavenger, restored the impaired EDR in ApoE-KO aorta. Lastly, incubation of WT aorta with $ONOO^-$ increased O_2^- production in vessels with intact endothelium, but not vessels incubated with L-NAME or vessels from eNOS knockout mice [66].

In a later study, $ONOO^-$ was shown to oxidize BH4 via the $BH3^+$ radical, and ascorbic acid was able to reduce $BH3^+$ back to BH4 [65]. Ascorbic acid is known to ameliorate impaired endothelium-dependent vasomotion in subjects with endothelial dysfunction [67–69], thought to be due to its ability to scavenge O_2^- [70]. Ascorbate, however, was only a weak scavenger of $ONOO^-$ [65]. Incubation with $ONOO^-$ gave rise to L-NAME-inhibitable O_2^- generation in bovine aortic endothelial cells (BAEC), indicating eNOS uncoupling, however O_2^- was unable to uncouple eNOS directly. Furthermore, the $ONOO^-$ donor 3-morpholinosydnon-imine (SIN-1) reduced NO production in BAEC pre-incubated with BH4 and vitamin C, however co-incubation with BH4 (during SIN-1 exposure) partially restored NO production, which was further increased by the addition of ascorbic acid [65].

Taken together, these data suggest that $ONOO^-$, resulting from the interaction of NO and O_2^-, can induce eNOS uncoupling via oxidation of BH4, and that some of the observed beneficial effects of ascorbic acid on endothelial function may be due to increased recycling of $BH3^+$ back to BH4, thus promoting eNOS coupling. Furthermore, 5-methyltetrahydrofolic acid (5-MTHF), the circulating form of folic acid, has also been shown to be a potent scavenger of $ONOO^-$ and can promote eNOS coupling and increased BH4:BH2 ratio in human blood vessels [71].

Finally, it has also been observed that $ONOO^-$ may disrupt eNOS dimerization via direct oxidation of the zinc–thiolate complex, independently of BH4, thus providing an alternative mechanism underlying $ONOO^-$-mediated endothelial dysfunction [72].

14.3.3
The Ratio of Reduced to Oxidized Biopterins in eNOS Regulation

Considerable debate exists as to whether it is solely the absolute levels of BH4 that are critical in determining eNOS catalytic activity, or alternatively whether the levels of oxidized biopterin species, for example, BH2, play an important inhibitory role. Total biopterins (the sum of BH4, BH2 and biopterin) represent a measure of overall GTP-CH synthetic activity, whereas the ratio of reduced biopterins to oxidized biopterins (expressed either as the BH4:BH2 ratio or BH4:total biopterins ratio) functions as a measure of biopterin redox status. Studies of purified eNOS in a cell-free system have provided robust evidence that BH2 can both bind to eNOS with a similar affinity to that of BH4, and can also promote O_2^{-} generation [50], and thus the ratio of BH4:BH2 is important in regulating eNOS in this model. It is less clear, however, whether such findings can be applied *in vivo*. Under conditions of oxidative stress, BH4 is oxidized to BH2, thereby resulting in both a diminished BH4 level coupled with a decreased BH4:BH2 ratio; consequently, it has been challenging to determine which process underlies eNOS uncoupling, making data interpretation difficult [73]. Recently published work by Crabtree and colleagues provides the best evidence to date on the role of BH2 *in vivo*. In cells with a fixed BH4 concentration, supplementation of cells with BH2 (in the presence of the DHFR inhibitor methotrexate to prevent reduction to BH4) to decrease the BH4:BH2 ratio was sufficient to promote eNOS-dependent O_2^{-} formation [74]. Furthermore, in a model of glucose-induced oxidative stress, changes in BH4:BH2 ratio were shown to be more closely correlated to uncoupled eNOS activity than changes in BH4 alone [75]. In the study by Cai *et al.* [58] the rise in BH4 achieved by GTP-CH gene transfer was similar to that achieved by incubation with sepiapterin, although sepiapterin incubation resulted in much higher levels of BH2, and thus a lower BH4:total biopterins ratio (<10% versus 30%); this was associated with decreased eNOS activity. In contrast to these findings, the low but detectable concentrations of BH4 in diabetic GCH-Tg mice were sufficient to maintain normal eNOS function in comparison to the eNOS uncoupling observed in diabetic WT mice, despite a comparable reduction in BH4:BH2 ratio following the onset of experimental diabetes in both groups, suggesting that the absolute availability of BH4 may be more important in this model [76]. It is likely to be a combination of both the eNOS:BH4 ratio and BH4:BH2 ratio that is important for regulating eNOS function *in vivo* [74].

14.4
eNOS Uncoupling in Experimental Disease Models

In addition to the studies described in detail above, a number of others have provided evidence for the role of eNOS uncoupling, and specifically strategies to augment BH4 levels, in experimental vascular disease states. Table 14.1 provides an overview of these studies.

Table 14.1 Summary of studies utilizing increased BH4 to modulate vascular function in experimental models of disease.

First author	Species	Tissue	Disease model	Intervention		Effect	Ref
Cai	Human	Endothelial cells	High glucose	GTP-CH gene transfer	In vitro	↑eNOS dimer:monomer, ↑eNOS coupling	[58]
Meininger	Rat	Endothelial cells	Spontaneous diabetes	Sepiapterin incubation	In vitro	↑NO, ↑intracellular BH4	[77]
Pieper	Rat	Aorta	STZ diabetes	Methyltetrahydropterin incubation	Ex vivo	↑EDR	[78]
Shinozaki	Rat	Aorta	Insulin resistance	BH4 incubation	Ex vivo	↑eNOS coupling	[79]
Shinozaki	Rat	Aorta	Insulin resistance	Oral BH4	In vivo	↑eNOS coupling, ↑EDR, ↓lipid peroxidation	[80]
Alp	Mouse	Aorta	STZ diabetes	Transgenic GTP-CH overexpression	In vivo	↑eNOS coupling, ↑EDR	[76]
Cosentino	Rat	Aorta	Spontaneous hypertension	BH4 incubation	Ex vivo	↑eNOS coupling, ↑EDR	[81]
Hong	Rat	Aorta	Spontaneous hypertension	Intraperitoneal BH4	In vivo	↓O_2^-, ↑EDR, ↓BP, ↓iNOS expression	[82]
Kase	Rat	Aorta	Angiotensin-II hypertension	Oral BH4	In vivo	↑EDR, ↓NO production by iNOS, ↓NADPH oxidase subunit expression	[83]

Author	Species	Tissue	Model	Treatment		Effects	Ref.
Landmesser	Mouse	Aorta	DOCA-salt hypertension	Oral BH4	In vivo	↑eNOS coupling, ↑EDR	[84]
Zheng	Rat	Carotid artery	DOCA-salt hypertension	GTP-CH gene transfer	Ex vivo	↑EDR, ↑NO, ↓O_2^-	[85]
Mitchell	Rat	Aorta	Glucocorticoid hypertension	Sepiapterin incubation	Ex vivo	↑EDR	[86]
Laursen	Mouse	Aorta	ApoE-KO hypercholesterolemia	Sepiapterin incubation	Ex vivo	↑EDR, ↓O_2^-	[66]
Ozaki	Mouse	Aorta	ApoE-KO hypercholesterolemia + eNOS overexpression	Oral BH4	In vivo	↓atherosclerosis, ↑eNOS coupling	[63]
Alp	Mouse	Aorta	ApoE-KO hypercholesterolemia	Transgenic GTP-CH overexpression	In vivo	↓atherosclerosis, ↑eNOS coupling, ↑EDR	[87]
Ali	Mouse	Vein graft	ApoE-KO hypercholesterolemia	Transgenic GTP-CH overexpression	In vivo	↓atherosclerosis, ↑eNOS coupling, ↓macrophage content, ↓MCP-1	[88]
Hattori	Mouse	Aorta	ApoE-KO hypercholesterolemia	Oral BH4	In vivo	↓atherosclerosis, ↑eNCS coupling, ↑EDR, ↓eNOS & iNOS protein, ↓NADPH subunit expression, ↓adhesion molecules ↓MCP-1	[89]

EDR, endothelium-dependent relaxations; STZ, streptozotocin; DOCA, deoxycorticosterone acetate; BP, blood pressure; MCP-1, monocyte chemotactic protein-1.

14.4.1
Diabetes and Insulin Resistance

Despite the fact that NO-mediated vasodilatation had been shown to be impaired in patients with diabetes [67], stimulation of human endothelial cells with high concentrations of glucose was shown to produce a paradoxical increase in eNOS gene expression and a 40% increase in NO production; O_2^- production, however was increased threefold [90], resulting in a net alteration in redox balance towards pro-oxidation in the setting of hyperglycemia. Cai *et al.* extended this work by demonstrating decreased eNOS dimer : monomer ratio and decreased intracellular BH4 and total biopterin content, implying decreased biopterin synthesis by GTP-CH [91]. Furthermore adenoviral gene transfer of GTP-CH restored eNOS dimer : monomer ratio, increased NO production, and decreased eNOS-derived O_2^-.

The Meininger group isolated coronary endothelial cells from the spontaneously diabetic BioBreeding (BB) rat, a model of type 1 diabetes mellitus, and found significant BH4 deficiency compared to control cells [77]. Impaired NO production was restored by incubation with sepiapterin. Similarly, impaired EDR in a rat model of streptozotocin-induced (STZ) diabetes were ameliorated by the BH4 analog 6-methyl-5,6,7,8-tetrahydropterin [78]. eNOS uncoupling was also demonstrated in the fructose-fed rat, a model of insulin resistance, as evidenced by L-NAME-inhibitable O_2^- production which was inhibited by *ex vivo* incubation with BH4 [79]. In comparison with control rats, fructose-fed rats had decreased GTP-CH activity and aortic BH4, and increased levels of oxidized biopterins, associated with impaired EDR. Treatment with oral BH4 in these animals resulted in augmented aortic BH4, restored EDR and NO production, and decreased eNOS-derived O_2^- [80]. Furthermore, oral BH4 decreased both lipid peroxidation and the binding of redox-sensitive transcription factors including NF-κB [80].

Alp *et al.* used a model of endothelial-specific overexpression of human GTP-CH to produce a constitutive threefold increase in aortic BH4 levels in mice [76]. Importantly, systemic levels of BH4, that is, in plasma and liver, were not affected, thus excluding the potential confounding antioxidant effects of high-dose exogenous BH4 treatment used in other models. STZ-induced diabetes had no effect on GTP-CH mRNA or total biopterins in WT mice, however all biopterins were present in an oxidized form, implying that the loss of BH4 in this model of diabetes is due to oxidation alone, rather than decreased synthesis. The resulting impaired EDR and increased eNOS-derived O_2^- were restored to control levels by the GTP-CH transgene [76].

14.4.2
Hypertension

Studies in the spontaneously hypertensive rat (SHR) [81, 82] revealed increased aortic O_2^- (inhibitable by L-NMMA) and H_2O_2, and impaired EDR to acetylcholine and A23187 compared to WT animals. One study reported similar endogenous levels of vascular biopterins in SHR and WT [81], while the other demonstrated reduced biopterin levels in the plasma of SHR [82], although later studies in humans demonstrate that biopterin levels in plasma may not act as a reliable surrogate marker of endothelial biopterin availability [92]. Both these studies assayed biopterins via a high-performance liquid chromatography (HPLC) technique that oxidizes BH4 and BH2 to fully oxidized biopterin, therefore it is not possible to ascertain the relative levels of reduced (BH4) and partially oxidized (BH2) biopterins in these models. However, *ex vivo* incubation with BH4 restored eNOS coupling [81] and chronic intraperitoneal supplementation of BH4 suppressed the development of hypertension in these animals [82]. Interestingly, BH4 supplementation was also seen to suppress the increased basal NO production observed in SHR via decreased expression of iNOS [82], an observation replicated using oral BH4 in a rat model of angiotensin-II-induced oxidative stress [83].

Animals with deoxycorticosterone acetate–salt (DOCA-salt) hypertension have yielded further evidence of eNOS uncoupling as a potential mechanism underlying the impaired vascular function observed in hypertension. By using mice deficient in eNOS or nNOS, Landmesser *et al.* demonstrated that the endothelial isoform was the principal source of NOS-derived $O_2^{.-}$, and importantly that inactivation of NADPH oxidase was able to restore the decreased BH4 : total biopterin ratio observed in these animals [84]. These data provide strong evidence to support $O_2^{.-}$-mediated BH4 oxidation as an initial step in eNOS uncoupling in hypertension. Furthermore, oral BH4 supplementation resulted in reduced L-NAME-inhibitable $O_2^{.-}$ release, whereas there was no effect of tetrahydroneopterin, which is reported to have similar direct antioxidant properties to BH4 but no effect on eNOS coupling [93]. Additionally, adenoviral gene transfer of GTP-CH has been shown to restore decreased GTP-CH activity in DOCA-salt rats, resulting in restored eNOS coupling and EDR [85], and sepiapterin has been shown to improve endothelial function in glucocorticoid hypertension [86].

14.4.3
Hypercholesterolemia

In addition to the work by Ozaki *et al.* [63] described previously, further studies have investigated the effects of BH4 on atherosclerosis progression using the ApoE-KO mouse model of hypercholesterolemia [87–89]. Endothelium-specific GTP-CH transgenic overexpression was shown to significantly increase the diminished BH4 : total biopterins ratio observed in ApoE-KO mice (71% versus 38%), resulting in improved eNOS coupling as evidenced by decreased endothelial and L-NAME-inhibitable $O_2^{.-}$, and augmented EDR. Aortic root plaques in GTP-CH transgenic mice were less advanced at 16 weeks compared with ApoE-KO controls, demonstrating that atherosclerosis in the ApoE-KO mouse is modulated by NO-mediated endothelial function through endothelial BH4 availability [87]. Recently, the GTP-CH transgenic mouse has been extended to a venous bypass graft model of accelerated atherosclerosis [88]. Experimental venous bypass grafting increases ROS formation that is implicated in inflammation and smooth muscle cell proliferation that leads to vein graft neointimal hyperplasia and accelerated atherosclerosis [94]. In addition to reducing both total and eNOS-derived $O_2^{\cdot-}$, GTP-CH overexpression significantly reduced vein graft neointimal hyperplasia and atherosclerosis at both 28- and 56-day time-points. Furthermore, the inflammatory cell content of these lesions was decreased, specifically via attenuated CCR2-mediated cell signaling, in turn likely mediated through decreased oxidative stress [88].

Reductions in atherosclerosis have also been observed using oral BH4 supplementation [89]. Consistent with observations in hypertensive mice [82, 83], BH4 was also shown to decrease both eNOS and iNOS protein levels, in addition to reduced expression of NADPH oxidase subunits, adhesion molecules and CC chemokines [89].

14.5
Human Studies

With a plethora of evidence from experimental models implicating eNOS uncoupling as an important mediator of endothelial dysfunction and oxidative stress in conditions predisposing to atherosclerosis, it is not surprising that there has been significant attention given to the possibility that targeting BH4 in the endothelium might represent a realistic treatment strategy to prevent or treat vascular disease. Accordingly, many investigators have sought to elucidate the role of BH4 in human blood vessels. However, human studies have been limited by difficulties in obtaining vascular tissue, and by an inability to directly measure important biological processes *in vivo*, for example, ROS production. Nonetheless, important evidence exists regarding the role of BH4 in regulating eNOS function in humans, which we review in the following sections.

14.5.1
Studies of Isolated Blood Vessels

Vasomotor studies of coronary arterioles dissected from atrial myocardial samples obtained during cardiac surgery demonstrated impaired EDR in subjects with atherosclerosis compared to those without; furthermore, *ex vivo* incubation with sepiapterin augmented EDR in these vessels [95]. The first demonstration of eNOS uncoupling in human endothelium, however, was in saphenous vein and internal mammary artery conduit samples obtained from patients undergoing coronary artery bypass graft (CABG) surgery [96]. In this study, Guzik *et al.* reported not only increased vascular O_2^- in patients with diabetes compared to those without, but more specifically an increased eNOS-derived component inhibited by L-NAME or endothelial denudation. Similar to L-NAME, sepiapterin incubation inhibited endothelial O_2^- production, but there was no effect on O_2^- production in the media or adventitia, implying that the action of sepiapterin was mediated via restored eNOS coupling [96].

Recently published work by Antoniades *et al.* [92, 97] provides further robust evidence to support the role of BH4 in human endothelium. Their first study examined vascular tissue and plasma from 219 patients with atherosclerosis undergoing CABG surgery [92]. 80% of vascular BH4 was present within the endothelium. Patients with the highest levels of vascular BH4 demonstrated the greatest EDR to acetylcholine and the lowest eNOS-derived O_2^- production. Furthermore, there remained significant differences in O_2^- between patients in the highest and middle tertiles of vascular BH4 content, despite a similar BH4:total biopterins ratio in these groups, suggesting that absolute BH4 levels may be more important than BH4:total biopterin ratio *in vivo*. Surprisingly, there was no association between plasma and vascular BH levels, and furthermore there was an inverse relationship between plasma BH4 and EDR, with patients with high plasma BH4 having impaired NO-mediated endothelial function [92]. These data suggest that plasma and vascular compartments of biopterin availability may be regulated separately. Indeed, plasma levels of total biopterins correlated with levels of C-reactive protein, suggesting that the chronic inflammation associated with atherosclerosis may augment systemic biopterin synthesis through cytokine-mediated activation of GTP-CH [41], whereas inflammation within the vascular wall leads to loss of BH4, perhaps through oxidation.

The same investigators then used the concept of Mendelian randomization [98] to determine the effect of genetic variation in GTP-CH activity on BH4 levels and endothelial function [97]. A haplotype of the *GCH1* gene (defined by three single nucleotide polymorphisms (SNPs) and denoted as "X"), which encodes GTP-CH, was previously demonstrated to confer reduced induction of *GCH1* expression in leukocytes following inflammatory stimulation [99]. Among patients undergoing CABG surgery, homozygote X haplotype carriers exhibited significantly lower *GCH1* expression and vascular BH4 content, associated with increased eNOS-derived O_2^- and impaired EDR [97]. These data provide further evidence to implicate GTP-CH and BH4 levels as important regulators of eNOS function in the human vasculature.

14.5.2
Clinical Studies

In addition to the small number of *ex vivo* human studies described in the previous section, there are a larger number of *in vivo* clinical studies using pharmacological supplementation of BH4 to modulate endothelial function. These studies are summarized in Table 14.2. The majority of these studies involved either acute intra-arterial administration of BH4 into either the brachial or coronary arteries or single-dose oral treatment. These studies have frequently reported an improvement of NO-mediated endothelial function in a number of disease states linked with the development of atherosclerosis, although the chief limitation of the methodology is the high dose of BH4 used. Forearm plethysmography protocols using typical infusion rates of 400–500 µg/min or more have demonstrated elevations of plasma BH4 concentrations to between 50 and 100 µM [112, 113]; in contrast,

Table 14.2 Summary of clinical BH4 studies.

First author	Clinical setting	Number of subjects	Route of administration	Duration	Effect	Ref.
Stroes	Hypercholesterolemia	26	Brachial artery	Acute infusion	Restores FBF responses to L-NMMA and 5-HT	[100]
Fukuda	Hypercholesterolemia	18	Coronary artery	Acute infusion	Restores CBF response to ACh and L-NMMA	[101]
Cosentino	Hypercholesterolemia	22	Oral	4 weeks	Restores FBF response to ACh; ↓plasma F_2-isoprostanes	[102]
Higashi	Hypertension	16	Brachial artery	Acute infusion	Restores FBF response to ACh	[103]
Heitzer	Smokers	58	Brachial artery	Acute infusion	Restores FBF response to ACh	[93]
Ueda	Smokers	17	Oral	Single dose	Improves brachial FMD	[104]
Ihlemann	Glucose challenge (healthy subjects)	39	Brachial artery	Acute infusion	Prevents impairment of FBF response to 5-HT	[105]
Heitzer	Type 2 diabetes	35	Brachial artery	Acute infusion	Improves FBF response to ACh	[106]
Nystrom	Type 2 diabetes + CAD	32	Brachial artery	Acute infusion	Increases insulin sensitivity; no effect on brachial FMD	[107]
Maier	CAD	19	Coronary artery	Acute infusion	Prevents vasoconstrictor response to ACh	[108]
Worthley	CAD	47	Coronary artery	Acute infusion	No change in CBF	[109]
Fukuda	Vasospastic angina	28	Coronary artery	Acute infusion	Prevents vasoconstrictor response to low-dose ACh; does not prevent coronary spasm in response to high-dose ACh	[110]
Setoguchi	Coronary endothelial dysfunction	15	Coronary artery	Acute infusion	Restores CBF response to ACh	[111]
Setoguchi	Heart failure	21	Brachial artery	Acute infusion	Improves FBF response to ACh	[112]
Mayahi	Ischemia-reperfusion injury	48	Brachial artery	Acute infusion	Restores FBF response to ACh following ischemia-reperfusion	[113]
Higashi	Aging	37	Brachial artery	Acute infusion	Augments FBF response to ACh	[114]
Eskurza	Exercise/aging	31	Oral	Single dose	Restores brachial FMD in older, sedentary subjects	[115]

CAD, coronary artery disease; FBF, forearm blood flow; CBF, coronary blood flow; FMD, flow-mediated dilatation; ACh, acetylcholine; 5-HT, serotonin; L-NMMA, N-monomethyl-L-arginine.

plasma BH4 levels in healthy subjects [116] and patients with atherosclerosis [92] have been reported to lie within a range of approximately 10–50 nM. Thus the use of supraphysiological BH4 doses may be confounded by nonspecific effects, including a direct antioxidant effect on the vessel wall. Some investigators have attempted to account for these effects by comparing the effects of BH4 to other known antioxidants. For example, BH4 restored endothelial function in chronic smokers, whereas tetrahydroneopterin (NH4, which has equipotent antioxidant capability *in vitro*) had no effect, suggesting restored eNOS coupling as the underlying mechanism [93]. Similarly, the 6*S*-stereoisomer of BH4 showed no effect on endothelial function following a glucose challenge [105]. While these elegantly controlled pharmacologic studies have suggested a specific role for BH4, others do not. For example the amelioration of forearm ischemia-reperfusion injury by 6*R*-BH4 was also observed with 6*S*-BH4 and NH4, implying a more nonspecific antioxidant effect [113].

To date, there is only one published randomized, placebo-controlled clinical trial of chronic oral BH4 supplementation [102]. This study reported an improvement in NO-mediated endothelial function after four weeks of treatment, although it is not possible to be sure whether this effect is due to alteration in eNOS coupling or an alternative mechanism. A reduction in plasma F_2-isoprostanes, a marker of oxidative stress, implies a favorable effect on systemic antioxidant status; nonetheless, further clinical trials are warranted to establish the utility of oral BH4 as a realistic therapeutic strategy. The OXBIO study (NCT00423280) has completed recruitment of patients randomized to receive either placebo or BH4 at low or high dose for four weeks prior to CABG surgery, and will evaluate the effects of oral BH4 therapy on vascular biopterin levels, eNOS coupling, and endothelial function.

Other treatment strategies, beyond direct BH4 supplementation, may influence vascular disease pathogenesis through effects on BH4 and eNOS coupling. For example, folates improve endothelial function in conditions such as hypercholesterolemia [117]. More recent studies suggest that 5-methyltetrahydrofolate (5-MTHF) may significantly increase vascular BH4 levels, most likely by protecting BH4 from oxidation by peroxynitrite, leading to improvements in eNOS coupling and endothelial function [71]. Co-administration of agents that "protect" BH4 either in plasma or within endothelial cells, such as 5-MTHF or ascorbic acid [65], may be an important aspect to realizing the potential of therapies to target BH4 and eNOS coupling.

14.6
Conclusion

Data from studies of purified eNOS enzyme support the notion that dysfunctional eNOS may contribute to ROS production under certain conditions, specifically when BH4 availability is limiting, or when increased levels of oxidized biopterins compete with BH4 for eNOS binding. These data are supported by observations in experimental cellular and animal models of pro-oxidant vascular disease states, under which circumstances BH4 can be oxidized by ROS, particularly $ONOO^-$, thus leading to further eNOS uncoupling and a self-perpetuating cycle of increased ROS production. Furthermore, eNOS protein levels can be elevated in atherosclerosis, thus providing further apparatus for O_2^- generation when BH4 levels are already limiting. Findings from studies examining the effects of BH4 supplementation or depletion, either by means of pharmacological intervention or genetic manipulation, have driven speculation that BH4 may represent a rational therapeutic target in the human vasculature. However, although data from observational human studies supports an important role of BH4 in preserving endothelial NO bioavailability, we still await answers from rigorously conducted clinical trials to establish the utility of this therapeutic strategy. The BH4-eNOS coupling pathway is a rational and promising therapeutic target to intervene in the pathogenesis of atherosclerosis. Future therapeutic success must first identify the appropriate timing, cellular targets and specific molecular aspects of BH4-eNOS coupling that will intervene in atherosclerosis most effectively.

Acknowledgments

Work in the authors' laboratory is supported by the British Heart Foundation (grants RG/07/003/23133 and PG/08/119/26263) and by the National Institutes of Health Oxford Biomedical Research Center.

References

1 Furchgott, R.F. and Zawadzki, J.V. (1980) The obligatory role of endothelial cells in the relaxation of arterial smooth muscle by acetylcholine. *Nature*, **288**, 373–376.

2 Palmer, R.M.J., Ferrige, A.G., and Moncada, S. (1987) Nitric oxide release accounts for the biological activity of endothelium-derived relaxing factor. *Nature*, **327**, 524–526.

3 De Caterina, R., Libby, P., Peng, H.-B., et al. (1995) Nitric oxide decreases cytokine-induced endothelial activation: nitric oxide selectively reduces endothelial expression of adhesion molecules and proinflammatory cytokines. *J. Clin. Invest.*, **96**, 60–68.

4 Radomski, M.W., Palmer, R.M., and Moncada, S. (1990) An L-arginine/nitric oxide pathway present in human platelets regulates aggregation. *Proc. Natl Acad. Sci. U. S. A.*, **87**, 5193–5197.

5 Garg, U.C. and Hassid, A. (1989) Nitric oxide-generating vasodilators and 8-bromo cyclic guanosine monophosphate inhibit mitogenesis and proliferation of cultured rat vascular smooth muscle cells. *J. Clin. Invest.*, **83**, 1774–1777.

6 Heitzer, T., Yla-Herttuala, S., Luoma, J., et al. (1996) Cigarette smoking potentiates endothelial dysfunction of forearm resistance vessels in patients with hypercholesterolemia. Role of oxidized LDL. *Circulation*, **93**, 1346–1353.

7 Panza, J.A., García, C.E., Kilcoyne, C.M., Quyyumi, A.A., and Cannon R.O., III (1995) Impaired endothelium-dependent vasodilation in patients with essential hypertension: evidence that nitric oxide abnormality is not localized to a single signal transduction pathway. *Circulation*, **91**, 1732–1738.

8 Stroes, E.S., Koomans, H.A., de Bruin, T.W., and Rabelink, T.J. (1995) Vascular function in the forearm of hypercholesterolaemic patients off and on lipid-lowering medication. *Lancet*, **346**, 467–471.

9 Williams, S.B., Cusco, J.A., Roddy, M.A., Johnstone, M.T., and Creager, M.A. (1996) Impaired nitric oxide-mediated vasodilation in patients with non-insulin-dependent diabetes mellitus. *J. Am. Coll. Cardiol.*, **27**, 567–574.

10 Gokce, N., Keaney, J.F., Jr., Hunter L.M., *et al.* (2003) Predictive value of noninvasively determined endothelial dysfunction for long-term cardiovascular events in patients with peripheral vascular disease. *J. Am. Coll. Cardiol.*, **41**, 1769–1775.

11 Schachinger, V., Britten, M.B., and Zeiher, A.M. (2000) Prognostic impact of coronary vasodilator dysfunction on adverse long-term outcome of coronary heart disease. *Circulation*, **101**, 1899–1906.

12 Cayatte, A.J., Palacino, J.J., Horten, K., and Cohen, R.A. (1994) Chronic inhibition of nitric oxide production accelerates neointima formation and impairs endothelial function in hypercholesterolemic rabbits. *Arterioscler. Thromb.*, **14**, 753–759.

13 Vasquez-Vivar, J., Kalyanaraman, B., Martasek, P., *et al.* (1998) Superoxide generation by endothelial nitric oxide synthase: the influence of cofactors. *Proc. Natl Acad. Sci. U. S. A.*, **95**, 9220–9225.

14 Xia, Y., Tsai, A.L., Berka, V., and Zweier, J.L. (1998) Superoxide generation from endothelial nitric-oxide synthase. A Ca^{2+}/calmodulin-dependent and tetrahydrobiopterin regulatory process. *J. Biol. Chem.*, **273**, 25804–25808.

15 Hibbs, J.B., Jr., Taintor R.R., Vavrin, Z., and Rachlin, E.M. (1988) Nitric oxide: a cytotoxic activated macrophage effector molecule. *Biochem. Biophys. Res. Commun.*, **157**, 87–94.

16 Bult, H., Boeckxstaens, G.E., Pelckmans, P.A., Jordaens, F.H., Van Maercke, Y.M., and Herman, A.G. (1990) Nitric oxide as an inhibitory non-adrenergic non-cholinergic neurotransmitter. *Nature*, **345**, 346–347.

17 Palmer, R.M.J., Ashton, D.S., and Moncada, S. (1988) Vascular endothelial cells synthesize nitric oxide from L-arginine. *Nature*, **333**, 664–666.

18 Schmidt, H.H., Nau, H., Wittfoht, W., *et al.* (1988) Arginine is a physiological precursor of endothelium-derived nitric oxide. *Eur. J. Pharmacol.*, **154**, 213–216.

19 Stuehr, D.J., Kwon, N.S., Nathan, C.F., Griffith, O.W., Feldman, P.L., and Wiseman, J. (1991) N omega-hydroxy-L-arginine is an intermediate in the biosynthesis of nitric oxide from L-arginine. *J. Biol. Chem.*, **266**, 6259–6263.

20 Palmer, R.M. and Moncada, S. (1989) A novel citrulline-forming enzyme implicated in the formation of nitric oxide by vascular endothelial cells. *Biochem. Biophys. Res. Commun.*, **158**, 348–352.

21 Kwon, N.S., Nathan, C.F., Gilker, C., Griffith, O.W., Matthews, D.E., and Stuehr, D.J. (1990) L-Citrulline production from L-arginine by macrophage nitric oxide synthase. The ureido oxygen derives from dioxygen. *J. Biol. Chem.*, **265**, 13442–13445.

22 Stuehr, D.J., Cho, H.J., Kwon, N.S., Weise, M.F., and Nathan, C.F. (1991) Purification and characterization of the cytokine-induced macrophage nitric oxide synthase: an FAD- and FMN-containing flavoprotein. *Proc. Natl Acad. Sci. U. S. A.*, **88**, 7773–7777.

23 Mayer, B., Schmidt, K., Humbert, P., and Bohme, E. (1989) Biosynthesis of endothelium-derived relaxing factor: a cytosolic enzyme in porcine aortic endothelial cells Ca^{2+}-dependently converts L-arginine into an activator of soluble guanylyl cyclase. *Biochem. Biophys. Res. Commun.*, **164**, 678–685.

24 Yui, Y., Hattori, R., Kosuga, K., *et al.* (1991) Calmodulin-independent nitric oxide synthase from rat polymorphonuclear neutrophils. *J. Biol. Chem.*, **266**, 3369–3371.

25 Mayer, B., John, M., Heinzel, B., *et al.* (1991) Brain nitric oxide synthase is a biopterin- and flavin-containing

multi-functional oxido-reductase. *FEBS Lett.*, **288**, 187–191.

26 Pollock, J.S., Forstermann, U., Mitchell, J.A., *et al.* (1991) Purification and characterization of particulate endothelium-derived relaxing factor synthetase from cultured and native bovine aortic endothelial cells. *Proc. Natl Acad. Sci. U. S. A.*, **88**, 10480–10484.

27 Mayer, B., John, M., and Bohme, E. (1990) Purification of a Ca^{2+}/calmodulin-dependent nitric oxide synthase from porcine cerebellum. Cofactor-role of tetrahydrobiopterin. *FEBS Lett.*, **277**, 215–219.

28 Förstermann, U., Pollock, J.S., Schmidt, H.H.H.W., Heller, M., and Murad, F. (1991) Calmodulin-dependent endothelium-derived relaxing factor/nitric oxide synthase activity is present in the particulate and cytosolic fractions of bovine aortic endothelial cells. *Proc. Natl Acad. Sci. U. S. A.*, **88**, 1788–1792.

29 Mitchell, J.A., Forstermann, U., Warner, T.D., *et al.* (1991) Endothelial cells have a particulate enzyme system responsible for EDRF formation: measurement by vascular relaxation. *Biochem. Biophys. Res. Commun.*, **176**, 1417–1423.

30 Stuehr, D.J. (1997) Structure-function aspects in the nitric oxide synthases. *Annu. Rev. Pharmacol. Toxicol.*, **37**, 339–359.

31 Kaufman, S. (1987) Enzymology of the phenylalanine-hydroxylating system. *Enzyme*, **38**, 286–295.

32 Kaufman, S., Berlow, S., Summer, G.K., *et al.* (1978) Hyperphenylalaninemia due to a deficiency of biopterin. A variant form of phenylketonuria. *N. Engl. J. Med.*, **299**, 673–679.

33 Schoedon, G., Troppmair, J., Fontana, A., Huber, C., Curtius, H., and Niederwieser, A. (1987) Biosynthesis and metabolism of pterins in peripheral blood mononuclear cells and leukemia lines of man and mouse. *Eur. J. Biochem.*, **166**, 303–310.

34 Tayeh, M.A. and Marletta, M.A. (1989) Macrophage oxidation of L-arginine to nitric oxide, nitrite, and nitrate. Tetrahydrobiopterin is required as a cofactor. *J. Biol. Chem.*, **264**, 19654–19658.

35 Kwon, N.S., Nathan, C.F., and Stuehr, D.J. (1989) Reduced biopterin as a cofactor in the generation of nitrogen oxides by murine macrophages. *J. Biol. Chem.*, **264**, 20496–20501.

36 Werner-Felmayer, G., Werner, E.R., Fuchs, D., Hausen, A., Reibnegger, G., and Wachter, H. (1990) Tetrahydrobiopterin-dependent formation of nitrite and nitrate in murine fibroblasts. *J. Exp. Med.*, **172**, 1599–1607.

37 Schmidt, K., Werner, E.R., Mayer, B., Wachter, H., and Kukovetz, W.R. (1992) Tetrahydrobiopterin-dependent formation of endothelium-derived relaxing factor (nitric oxide) in aortic endothelial cells. *Biochem. J.*, **281**, 297–300.

38 Klatt, P., Schmid, M., Leopold, E., Schmidt, K., Werner, E.R., and Mayer, B. (1994) The pteridine binding site of brain nitric oxide synthase. Tetrahydrobiopterin binding kinetics, specificity, and allosteric interaction with the substrate domain. *J. Biol. Chem.*, **269**, 13861–13866.

39 Tzeng, E., Billiar, T.R., Robbins, P.D., Loftus, M., and Stuehr, D.J. (1995) Expression of human inducible nitric oxide synthase in a tetrahydrobiopterin (H4B)-deficient cell line – H4B promotes assembly of enzyme subunits into an active enzyme. *Proc. Natl Acad. Sci. U. S. A.*, **92**, 11771–11775.

40 Werner-Felmayer, G., Werner, E.R., Fuchs, D., *et al.* (1993) Pteridine biosynthesis in human endothelial cells. Impact on nitric oxide-mediated formation of cyclic GMP. *J. Biol. Chem.*, **268**, 1842–1846.

41 Rosenkranz-Weiss, P., Sessa, W.C., Milstien, S., Kaufman, S., Watson, C.A., and Pober, J.S. (1993) Regulation of nitric oxide synthesis by proinflammatory cytokines in human umbilical vein endothelial cells. *J. Clin. Invest.*, **93**, 2236–2243.

42 Nichol, C.A., Lee, C.L., Edelstein, M.P., Chao, J.Y., and Duch, D.S. (1983) Biosynthesis of tetrahydrobiopterin by de novo and salvage pathways in adrenal medulla extracts, mammalian

cell cultures, and rat brain *in vivo*. *Proc. Natl Acad. Sci. U. S. A.*, **80**, 1546–1550.

43 Heinzel, B., John, M., Klatt, P., Bohme, E., and Mayer, B. (1992) Ca2+/ calmodulin-dependent formation of hydrogen peroxide by brain nitric oxide synthase. *Biochem. J.*, **281** (Pt 3), 627–630.

44 Pou, S., Pou, W.S., Bredt, D.S., Snyder, S.H., and Rosen, G.M. (1992) Generation of superoxide by purified brain nitric oxide synthase. *J. Biol. Chem.*, **267**, 24173–24176.

45 Xia, Y., Dawson, V.L., Dawson, T.M., Snyder, S.H., and Zweier, J.L. (1996) Nitric oxide synthase generates superoxide and nitric oxide in arginine-depleted cells leading to peroxynitrite-mediated cellular injury. *Proc. Natl Acad. Sci. U. S. A.*, **93**, 6770–6774.

46 Milstien, S. and Katusic, Z. (1999) Oxidation of tetrahydrobiopterin by peroxynitrite: implications for vascular endothelial function. *Biochem. Biophys. Res. Commun.*, **263**, 681–684.

47 Ohara, Y., Peterson, T.E., and Harrison, D.G. (1993) Hypercholesterolemia increases endothelial superoxide anion production. *J. Clin. Invest.*, **91**, 2546–2551.

48 Pritchard, K.A., Jr., Groszek L., Smalley, D.M., *et al.* (1995) Native low-density lipoprotein increases endothelial cell nitric oxide synthase generation of superoxide anion. *Circ. Res.*, **77**, 510–518.

49 Wever, R.M.F., van Dam, T., van Rijn, H.J., de Groot, F., and Rabelink, T.J. (1997) Tetrahydrobiopterin regulates superoxide and nitric oxide generation by recombinant endothelial nitric oxide synthase. *Biochem. Biophys. Res. Commun.*, **237**, 340–344.

50 Vasquez-Vivar, J., Martasek, P., Whitsett, J., Joseph, J., and Kalyanaraman, B. (2002) The ratio between tetrahydrobiopterin and oxidized tetrahydrobiopterin analogues controls superoxide release from endothelial nitric oxide synthase: an EPR spin trapping study. *Biochem. J.*, **362**, 733–739.

51 Van Amsterdam, J.G. and Wemer, J. (1992) Tetrahydrobiopterin induces vasodilation via enhancement of cGMP level. *Eur. J. Pharmacol.*, **215**, 349–350.

52 Cosentino, F. and Katusic, Z.S. (1995) Tetrahydrobiopterin and dysfunction of endothelial nitric oxide synthase in coronary arteries. *Circulation*, **91**, 139–144.

53 Tsutsui, M., Milstien, S., and Katusic, Z.S. (1996) Effect of tetrahydrobiopterin on endothelial function in canine middle cerebral arteries. *Circ. Res.*, **79**, 336–342.

54 Kinoshita, H. and Katusic, Z.S. (1996) Exogenous tetrahydrobiopterin causes endothelium-dependent contractions in isolated canine basilar artery. *Am. J. Physiol.*, **271**, H738–43.

55 Kinoshita, H., Milstien, S., Wambi, C., and Katusic, Z.S. (1997) Inhibition of tetrahydrobiopterin biosynthesis impairs endothelium-dependent relaxations in canine basilar artery. *Am. J. Physiol.*, **273**, H718–24.

56 McDonald, J.D., Cotton, R.G., Jennings, I., Ledley, F.D., Woo, S.L., and Bode, V.C. (1988) Biochemical defect of the hph-1 mouse mutant is a deficiency in GTP- cyclohydrolase activity. *J. Neurochem.*, **50**, 655–657.

57 Cosentino, F., Barker, J.E., Brand, M.P., *et al.* (2001) Reactive oxygen species mediate endothelium-dependent relaxations in tetrahydrobiopterin-deficient mice. *Arterioscler. Thromb. Vasc. Biol.*, **21**, 496–502.

58 Cai, S., Alp, N.J., McDonald, D., Canevari, L., Heales, S., and Channon, K.M. (2002) GTP cyclohydrolase I gene transfer augments intracellular tetrahydrobiopterin in human endothelial cells: effects on nitric oxide synthase activity, protein levels and dimerization. *Cardiovasc. Res.*, **55**, 838–849.

59 Knowles, J.W., Reddick, R.L., Jennette, J.C., Shesely, E.G., Smithies, O., and Maeda, N. (2000) Enhanced athero-sclerosis and kidney dysfunction in eNOS$^{-/-}$Apoe$^{-/-}$ mice are ameliorated by enalapril treatment. *J. Clin. Invest.*, **105**, 451–458.

60 Kuhlencordt, P.J., Gyurko, R., Han, F., et al. (2001) Accelerated atherosclerosis, aortic aneurysm formation, and ischemic heart disease in apolipoprotein E/endothelial nitric oxide synthase double-knockout mice. *Circulation*, **104**, 448–454.

61 Channon, K.M., Qian, H.S., Neplioueva, V., et al. (1998) *In vivo* gene transfer of nitric oxide synthase enhances vasomotor function in carotid arteries from normal and cholesterol-fed rabbits. *Circulation*, **98**, 1905–1911.

62 Qian, H.S., Neplioueva, V., Shetty, G.A., Channon, K.M., and George, S.E. (1999) Nitric oxide synthase gene therapy rapidly reduces adhesion molecule expression and inflammatory cell infiltration in carotid arteries of cholesterol-fed rabbits. *Circulation*, **99**, 2979–2982.

63 Ozaki, M., Kawashima, S., Yamashita, T., et al. (2002) Overexpression of endothelial nitric oxide synthase accelerates atherosclerotic lesion formation in apoE-deficient mice. *J. Clin. Invest.*, **110**, 331–340.

64 Bendall, J.K., Alp, N.J., Warrick, N., et al. (2005) Stoichiometric relationships between endothelial tetrahydrobiopterin, eNOS activity and eNOS coupling *in vivo*: insights from transgenic mice with endothelial-targeted GTPCH and eNOS over-expression. *Circ. Res.*, **97**, 864–871.

65 Kuzkaya, N., Weissmann, N., Harrison, D.G., and Dikalov, S. (2003) Interactions of peroxynitrite, tetrahydrobiopterin, ascorbic acid, and thiols: implications for uncoupling endothelial nitric-oxide synthase. *J. Biol. Chem.*, **278**, 22546–22554.

66 Laursen, J.B., Somers, M., Kurz, S., et al. (2001) Endothelial regulation of vasomotion in apoE-deficient mice: implications for interactions between peroxynitrite and tetrahydrobiopterin. *Circulation*, **103**, 1282–1288.

67 Ting, H.H., Timimi, F.K., Boles, K.S., Creager, S.J., Ganz, P., and Creager, M.A. (1996) Vitamin C improves endothelium-dependent vasodilation in patients with non-insulin-dependent diabetes mellitus. *J. Clin. Invest.*, **97**, 22–28.

68 Heitzer, T., Just, H., and Munzel, T. (1996) Antioxidant vitamin C improves endothelial dysfunction in chronic smokers. *Circulation*, **94**, 6–9.

69 Ting, H.H., Timimi, F.K., Haley, E.A., Roddy, M.A., Ganz, P., and Creager, M.A. (1997) Vitamin C improves endothelium-dependent vasodilation in forearm resistance vessels of humans with hypercholesterolemia. *Circulation*, **95**, 2617–2622.

70 Gotoh, N. and Niki, E. (1992) Rates of interactions of superoxide with vitamin E, vitamin C and related compounds as measured by chemiluminescence. *Biochim. Biophys. Acta*, **1115**, 201–207.

71 Antoniades, C., Shirodaria, C., Warrick, N., et al. (2006) 5-methyltetrahydrofolate rapidly improves endothelial function and decreases superoxide production in human vessels: effects on vascular tetrahydrobiopterin availability and endothelial nitric oxide synthase coupling. *Circulation*, **114**, 1193–1201.

72 Zou, M.H., Shi, C., and Cohen, R.A. (2002) Oxidation of the zinc-thiolate complex and uncoupling of endothelial nitric oxide synthase by peroxynitrite. *J. Clin. Invest.*, **109**, 817–826.

73 Taylor, N.E., Maier, K.G., Roman, R.J., and Cowley, A.W., Jr. (2006) NO synthase uncoupling in the kidney of Dahl S rats: role of dihydrobiopterin. *Hypertension*, **48**, 1066–1071.

74 Crabtree, M.J., Tatham, A.L., Al-Wakeel, Y., et al. (2009) Quantitative regulation of intracellular endothelial nitric oxide synthase (eNOS) coupling by both tetrahydrobiopterin-eNOS stoichiometry and biopterin redox status: insights from cells with tet-regulated GTP cyclohydrolase I expression. *J. Biol. Chem.*, **284**, 1136–1144.

75 Crabtree, M.J., Smith, C.L., Lam, G., Goligorsky, M.S., and Gross, S.S. (2008) Ratio of 5,6,7,8-tetrahydrobiopterin to 7,8-dihydrobiopterin in endothelial cells determines glucose-elicited changes in NO vs. superoxide production by eNOS. *Am. J. Physiol. Heart Circ. Physiol.*, **294**, H1530–H1540.

76 Alp, N.J., Mussa, S., Khoo, J., *et al.* (2003) Tetrahydrobiopterin-dependent preservation of nitric oxide-mediated endothelial function in diabetes by targeted transgenic GTP-cyclohydrolase I overexpression. *J. Clin. Invest.*, **112**, 725–735.

77 Meininger, C.J., Marinos, R.S., Hatakeyama, K., *et al.* (2000) Impaired nitric oxide production in coronary endothelial cells of the spontaneously diabetic BB rat is due to tetrahydrobiopterin deficiency. *Biochem. J.*, **349**, 353–356.

78 Pieper, G.M. (1997) Acute amelioration of diabetic endothelial dysfunction with a derivative of the nitric oxide synthase cofactor, tetrahydrobiopterin. *J. Cardiovasc. Pharmacol.*, **29**, 8–15.

79 Shinozaki, K., Kashiwagi, A., Nishio, Y., *et al.* (1999) Abnormal biopterin metabolism is a major cause of impaired endothelium- dependent relaxation through nitric oxide/O^{2-} imbalance in insulin- resistant rat aorta. *Diabetes*, **48**, 2437–2445.

80 Shinozaki, K., Nishio, Y., Okamura, T., *et al.* (2000) Oral administration of tetrahydrobiopterin prevents endothelial dysfunction and vascular oxidative stress in the aortas of insulin-resistant rats. *Circ. Res.*, **87**, 566–573.

81 Cosentino, F., Patton, S., d'Uscio, L.V., *et al.* (1998) Tetrahydrobiopterin alters superoxide and nitric oxide release in prehypertensive rats. *J. Clin. Invest.*, **101**, 1530–1537.

82 Hong, H.J., Hsiao, G., Cheng, T.H., and Yen, M.H. (2001) Supplementation with tetrahydrobiopterin suppresses the development of hypertension in spontaneously hypertensive rats. *Hypertension*, **38**, 1044–1048.

83 Kase, H., Hashikabe, Y., Uchida, K., Nakanishi, N., and Hattori, Y. (2005) Supplementation with tetrahydrobiopterin prevents the cardiovascular effects of angiotensin II-induced oxidative and nitrosative stress. *J. Hypertens.*, **23**, 1375–1382.

84 Landmesser, U., Dikalov, S., Price, S.R., *et al.* (2003) Oxidation of tetrahydrobiopterin leads to uncoupling of endothelial cell nitric oxide synthase in hypertension. *J. Clin. Invest.*, **111**, 1201–1209.

85 Zheng, J.-S., Yang, X.-Q., Lookingland, K.J., *et al.* (2003) Gene transfer of human guanosine 5′-triphosphate cyclohydrolase I restores vascular tetrahydrobiopterin level and endothelial function in low renin hypertension. *Circulation*, **108**, 1238–1245.

86 Mitchell, B.M., Dorrance, A.M., and Webb, R.C. (2003) GTP cyclohydrolase 1 downregulation contributes to glucocorticoid hypertension in rats. *Hypertension*, **41**, 669–674.

87 Alp, N.J., McAteer, M.A., Khoo, J., Choudhury, R.P., and Channon, K.M. (2004) Increased endothelial tetrahydrobiopterin synthesis by targeted transgenic GTP-cyclohydrolase I overexpression reduces endothelial dysfunction and atherosclerosis in ApoE-knockout mice. *Arterioscler. Thromb. Vasc. Biol.*, **24**, 445–450.

88 Ali, Z.A., Bursill, C.A., Douglas, G., *et al.* (2008) CCR2-mediated antiinflammatory effects of endothelial tetrahydrobiopterin inhibit vascular injury-induced accelerated atherosclerosis. *Circulation*, **118**, S71–7.

89 Hattori, Y., Hattori, S., Wang, X., Satoh, H., Nakanishi, N., and Kasai, K. (2007) Oral administration of tetrahydrobiopterin slows the progression of atherosclerosis in apolipoprotein E-knockout mice. *Arterioscler. Thromb. Vasc. Biol.*, **27**, 865–870.

90 Cosentino, F., Hishikawa, K., Katusic, Z.S., and Luscher, T.F. (1997) High glucose increases nitric oxide synthase expression and superoxide anion generation in human aortic endothelial cells. *Circulation*, **96**, 25–28.

91 Cai, S., Khoo, J., and Channon, K.M. (2005) Augmented BH4 by gene transfer restores nitric oxide synthase function in hyperglycemic human endothelial cells. *Cardiovasc. Res.*, **65**, 823–831.

92 Antoniades, C., Shirodaria, C., Crabtree, M., *et al.* (2007) Altered plasma versus vascular biopterins in human atherosclerosis reveal relationships between endothelial nitric oxide synthase coupling, endothelial function, and inflammation. *Circulation*, **116**, 2851–2859.

93 Heitzer, T., Brockhoff, C., Mayer, B., *et al.* (2000) Tetrahydrobiopterin improves endothelium-dependent vasodilation in chronic smokers: evidence for a dysfunctional nitric oxide synthase. *Circ. Res.*, **86**, E36–E41.

94 West, N.E.J., Guzik, T.J., Black, E., and Channon, K.M. (2001) Enhanced superoxide production in experimental venous bypass graft intimal hyperplasia: role of NAD(P)H oxidase. *Arterioscler. Thromb. Vasc. Biol.*, **21**, 189–194.

95 Tiefenbacher, C.P., Bleeke, T., Vahl, C., Amann, K., Vogt, A., and Kubler, W. (2000) Endothelial dysfunction of coronary resistance arteries is improved by tetrahydrobiopterin in atherosclerosis. *Circulation*, **102**, 2172–2179.

96 Guzik, T.J., Mussa, S., Gastaldi, D., *et al.* (2002) Mechanisms of increased vascular superoxide production in human diabetes mellitus: role of NAD(P)H oxidase and endothelial nitric oxide synthase. *Circulation*, **105**, 1656–1662.

97 Antoniades, C., Shirodaria, C., Van Assche, T., *et al.* (2008) GCH1 haplotype determines vascular and plasma biopterin availability in coronary artery disease effects on vascular superoxide production and endothelial function. *J. Am. Coll. Cardiol.*, **52**, 158–165.

98 Hingorani, A. and Humphries, S. (2005) Nature's randomised trials. *Lancet*, **366**, 1906–1908.

99 Tegeder, I., Costigan, M., Griffin, R.S., *et al.* (2006) GTP cyclohydrolase and tetrahydrobiopterin regulate pain sensitivity and persistence. *Nat. Med.*, **12**, 1269–1277.

100 Stroes, E., Kastelein, J., Cosentino, F., *et al.* (1997) Tetrahydrobiopterin restores endothelial function in hypercholesterolemia. *J. Clin. Invest.*, **99**, 41–46.

101 Fukuda, Y., Teragawa, H., Matsuda, K., Yamagata, T., Matsuura, H., and Chayama, K. (2002) Tetrahydrobiopterin restores endothelial function of coronary arteries in patients with hypercholesterolaemia. *Heart*, **87**, 264–269.

102 Cosentino, F., Hurlimann, D., Delli Gatti, C., *et al.* (2008) Chronic treatment with tetrahydrobiopterin reverses endothelial dysfunction and oxidative stress in hypercholesterolaemia. *Heart*, **94**, 487–492.

103 Higashi, Y., Sasaki, S., Nakagawa, K., *et al.* (2002) Tetrahydrobiopterin enhances forearm vascular response to acetylcholine in both normotensive and hypertensive individuals. *Am. J. Hypertens.*, **15**, 326–332.

104 Ueda, S., Matsuoka, H., Miyazaki, H., Usui, M., Okuda, S., and Imaizumi, T. (2000) Tetrahydrobiopterin restores endothelial function in long-term smokers. *J. Am. Coll. Cardiol.*, **35**, 71–75.

105 Ihlemann, N., Rask-Madsen, C., Perner, A., *et al.* (2003) Tetrahydrobiopterin restores endothelial dysfunction induced by an oral glucose challenge in healthy subjects. *Am. J. Physiol. Heart Circ. Physiol.*, **285**, H875–82.

106 Heitzer, T., Krohn, K., Albers, S., and Meinertz, T. (2000) Tetrahydrobiopterin improves endothelium-dependent vasodilation by increasing nitric oxide activity in patients with type II diabetes mellitus. *Diabetologia*, **43**, 1435–1438.

107 Nystrom, T., Nygren, A., and Sjoholm, A. (2004) Tetrahydrobiopterin increases insulin sensitivity in patients with type 2 diabetes and coronary heart disease. *Am. J. Physiol. Endocrinol. Metab.*, **287**, E919–E925.

108 Maier, W., Cosentino, F., Lutolf, R.B., *et al.* (2000) Tetrahydrobiopterin improves endothelial function in patients with coronary artery disease. *J. Cardiovasc. Pharmacol.*, **35**, 173–178.

109 Worthley, M.I., Kanani, R.S., Sun, Y.H., *et al.* (2007) Effects of tetrahydrobiopterin on coronary vascular reactivity in atherosclerotic human coronary arteries. *Cardiovasc. Res.*, **76**, 539–546.

110 Fukuda, Y., Teragawa, H., Matsuda, K., Yamagata, T., Matsuura, H., and Chayama, K. (2002) Tetrahydrobiopterin improves coronary endothelial function, but does not prevent coronary spasm in patients with vasospastic angina. *Circ. J.*, **66**, 58–62.

111 Setoguchi, S., Mohri, M., Shimokawa, H., and Takeshita, A. (2001) Tetrahydrobiopterin improves endothelial dysfunction in coronary microcirculation in patients without epicardial coronary artery disease. *J. Am. Coll. Cardiol.*, **38**, 493–498.

112 Setoguchi, S., Hirooka, Y., Eshima, K., Shimokawa, H., and Takeshita, A. (2002) Tetrahydrobiopterin improves impaired endothelium-dependent forearm vasodilation in patients with heart failure. *J. Cardiovasc. Pharmacol.*, **39**, 363–368.

113 Mayahi, L., Heales, S., Owen, D., *et al.* (2007) 6*R*-5,6,7,8-tetrahydro-L-biopterin and its stereoisomer prevent ischemia reperfusion injury in human forearm. *Arterioscler. Thromb. Vasc. Biol.*, **27**, 1334–1339.

114 Higashi, Y., Sasaki, S., Nakagawa, K., *et al.* (2006) Tetrahydrobiopterin improves aging-related impairment of endothelium-dependent vasodilation through increase in nitric oxide production. *Atherosclerosis*, **186**, 390–395.

115 Eskurza, I., Myerburgh, L.A., Kahn, Z.D., and Seals, D.R. (2005) Tetrahydrobiopterin augments endothelium-dependent dilatation in sedentary but not in habitually exercising older adults. *J. Physiol.*, **568**, 1057–1065.

116 Fekkes, D. and Voskuilen-Kooijman, A. (2007) Quantitation of total biopterin and tetrahydrobiopterin in plasma. *Clin. Biochem.*, **40**, 411–413.

117 Verhaar, M.C., Wever, R.M., Kastelein, J.J., van Dam, T., Koomans, H.A., and Rabelink, T.J. (1998) 5-methyltetrahydrofolate, the active form of folic acid, restores endothelial function in familial hypercholesterolemia. *Circulation*, **97**, 237–241.

15
Heme Oxygenase-1 and Atherosclerosis

Justin C. Mason and Faisal Ali

15.1
Background

Atherogenesis is an inflammatory disease process confined to the arterial wall. It is the leading cause of mortality in the Western world and of increasing importance in developing countries. Despite this, treatment is still largely confined to those presenting with symptomatic disease, typically in the form of angina or myocardial infarction, leaving a vast pool of patients with asymptomatic disease untreated. This group of patients is likely to increase in light of the "obesity epidemic" and the greater prevalence of metabolic syndrome and type 2 diabetes mellitus (DM). However, there is also increasing public awareness of atherosclerosis, its associated risk factors, and the relationship with ischemic heart disease and stroke. In light of the fact that we can readily identify groups at significantly increased risk of atherosclerotic disease, such as those suffering from DM, hyperlipidemia, the metabolic syndrome, chronic renal failure, and systemic inflammatory diseases such as rheumatoid arthritis and systemic lupus erythematosus, interest in the development of preventative therapies above and beyond statins is increasing. Among the novel targets being considered is heme oxygenase-1 (HO-1), an important cytoprotective, anti-inflammatory, antioxidant enzyme [1].

15.2
Heme Oxygenases

The inducible form of heme oxygenase, HO-1, and the constitutive isoform HO-2 are the products of separate genes, while HO-3 is a pseudogene confined to the rat and derived from an HO-2 transcript (reviewed in [2, 3]). The role of heme oxygenases in the catalysis of the first and rate-limiting step in the oxidative degradation of heme (Fe-protoporphrin-IX) was first reported by Tenhunen and colleagues in 1968 [4]. Although hemoglobin is the principal source of heme, alternatives include myoglobin, cytochromes, peroxidases, and respiratory burst enzymes. The reaction involves consumption of 3 moles of molecular oxygen for

Atherosclerosis: Molecular and Cellular Mechanisms. Edited by Sarah Jane George and Jason Johnson
Copyright © 2010 WILEY-VCH Verlag GmbH & Co. KGaA, Weinheim
ISBN: 978-3-527-32448-4

each heme molecule and the NADPH:cytochrome P450 reductase system as the predominant electron source. The outcome is the release of ferrous iron (Fe^{2+}), carbon monoxide, and the generation of biliverdin-IXα. The latter is rapidly metabolized to bilirubin-IXα by biliverdin reductase [4]. The increase in intracellular Fe^{2+} induces expression of the iron-binding protein heavy chain ferritin and the opening of Fe^{2+} export channels [5] (Figure 15.1). Thus, HO-1 plays a central role in iron

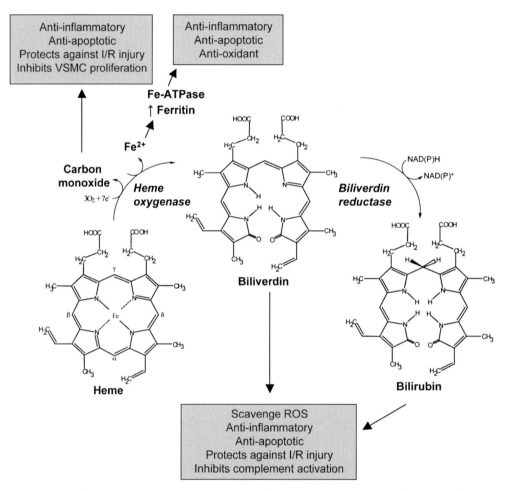

Figure 15.1 Heme oxygenase-1–mediated degradation of heme. The interaction of HO-1 with heme generates equimolar amounts of carbon monoxide (CO), biliverdin, and free iron (Fe^{2+}). Biliverdin reductase subsequently catalyzes the conversion of biliverdin to bilirubin. The increase in intracellular Fe^{2+} induces expression of the iron-binding protein heavy-chain ferritin and the opening of Fe^{2+} export channels. The products of heme degradation exert a variety of effects on endothelial cells (ECs) and vascular smooth muscle cells (VSMCs) which are protective against atherosclerosis. ROS, reactive oxygen species; I/R, ischemia-reperfusion.

homeostasis and iron generated as a consequence of HO-1 activity is predominantly returned to the circulation, transported to the marrow by transferrin and utilized in hemoglobin synthesis. The importance of HO-1 in iron handling at sites of atherosclerosis remains to be determined. However, elevated iron deposition in atherosclerotic plaques has been linked to increased oxidative stress and accelerated atherogenesis [3].

HO-1, a 32 kDa stress protein also known as heat shock protein 32, is highly expressed in reticuloendothelial tissues involved in erythrocyte breakdown, whereas it exhibits low basal expression and activity, typically undetectable, in the majority of normal tissues where it is readily transcriptionally induced. HO-2 (36 kDa) is constitutively expressed in the vascular endothelium and found in the liver, testes, intestine, brain, and central nervous system [2, 3]. The relationship between HO-1 and HO-2 remains to be determined, however recent experiments targeting HO-2 with short-interfering RNA (siRNA) suggest a functional interaction in which HO-2 may affect HO-1 activity by modulating levels of intracellular heme [6].

15.3
Epidemiology

Oxidative stress contributes significantly to atherogenesis and epidemiologic studies were the first to imply that HO-1 antioxidant activity may be protective. An initial report suggested low serum concentrations of bilirubin are related to an increase in the risk of developing ischemic heart disease [7], and subsequent studies are strongly supportive. Gilbert's syndrome leads to a mild unconjugated bilirubinemia and a >fivefold reduced risk of coronary artery disease (CAD) [8], while population studies also show that a high plasma bilirubin is protective (Table 15.1) [9–11].

15.3.1
HO-1 Promoter Polymorphisms

The demonstration that the *Hmox1* gene is polymorphic, leading to variation in the degree to which HO-1 activity can be regulated, led to studies exploring the relevance of this in a variety of disease states including atherosclerosis [16]. *Hmox1* polymorphisms relate predominantly to variability in the 5′-flanking sequence and specifically in the number of $(GT)_n$ repeats, with >30 repeats resulting in reduced HO-1 induction. Moreover, individuals with short repeats (median 23 pairs) have significantly increased HO-1 induction [16].

Long $(GT)_n$ repeats (≥ 32) have been associated with an almost fivefold increase in CAD in Chinese patients with type 2 DM [12]. Likewise, angiographic CAD was reduced in Japanese patients with short repeats (<27) [13], and in those with a single nucleotide polymorphism in the HO-1 promoter [14]. In contrast to these studies from the Asia, a report from Germany revealed increased high-density

Table 15.1 Epidemiologic study of heme oxygenase-1 and atherosclerosis.

Protective effects of bilirubin	Low serum bilirubin is associated with ↑ CAD [7]
	Gilbert's syndrome/mild bilirubinemia ↓ CAD [8]
	Raised plasma bilirubin is protective against CAD [9–11]
Hmox1 polymorphisms	Long (GT)$_n$ repeats (≥32) increase risk of CAD [12]
	Short (GT)$_n$ repeats (<27) protects against CAD [13]
	Single nucleotide polymorphism protects against CAD [14]
	Short (GT)$_n$ repeats ↑bilirubin and ↑HDL, without reducing CAD [15]

The epidemiologic evidence for the protective actions of bilirubin against coronary artery disease (CAD) is strong. A number of studies also suggest that *Hmox1* polymorphisms may influence the risk of CAD. However, further data are required in order to determine the clinical implications of these studies for the management of atherosclerosis.

lipoprotein and bilirubin in those with short (GT)$_n$ repeats, but no difference in incidence of CAD or myocardial infarction (Table 15.1) [15]. Although intriguing and largely supportive of a protective effect associated with short repeats, further larger prospective studies of *Hmox1* polymorphisms are needed in a variety of disease states.

15.4
Expression of HO-1 in Atherosclerotic Plaques

The role of inflammation in atherogenesis is now well recognized [17]. Experimental models suggest endothelial injury is the earliest detectable event in atherogenesis and that chronic endothelial dysfunction precedes structural lesions. Endothelial injury induces an inflammatory response with reduced nitric oxide (NO) biosynthesis, local generation of reactive oxygen species (ROS) and increased permeability to lipids. Moreover, endothelial apoptosis occurs preferentially at sites of endothelial injury and atherosclerosis, where denudation of vascular endothelium enhances the risk of thrombosis. Oxidative modification of low-density lipoproteins (LDLs) by ROS in the subintimal space increases lipid retention. Oxidized LDL (ox-LDL) predisposes endothelial cells (ECs) to injury and may induce proinflammatory genes including vascular cell adhesion molecule-1 (VCAM-1), monocyte chemotactic protein-1 (MCP-1), and chemokine receptor 2 (CCR2), which facilitate further monocyte migration. Ingestion of lipid droplets by macrophages leads to the generation of foam cells in the core region and these are surrounded by migrating vascular smooth muscle cells (VSMCs) and a collagen-rich matrix forming a fibrous cap [17].

Detailed immunohistochemical analysis of human atherosclerotic plaques and those found in mice with accelerated atherogenesis has revealed significant HO-1 expression. Although HO-1 was not detected in unaffected arteries of 4- to

6-month-old apolipoprotein E–deficient (ApoE$^{-/-}$) mice, it was clearly expressed by vascular endothelium overlying plaques in which foam cells also expressed HO-1 [18]. In more complex lesions VSMCs also expressed HO-1. Likewise, HO-1 was undetectable in normal human arteries, while prominent expression was seen in foam cells, VSMCs and at high power in the endothelium over the plaque [18]. These results were confirmed at the mRNA level using *in situ* hybridization. Kinetic analysis revealed a rise in the aortic expression of antioxidant enzymes including HO-1 in ApoE$^{-/-}$ mice up to 12 weeks [19]. Further studies have demonstrated HO-1 expression in advanced human atherosclerosis [20] and in cholesterol-fed LDL receptor–deficient (LDLR$^{-/-}$) mice [21] and rabbits [22]. However, not all results are consistent and HO-1 was not detected in aortae of 26- to 30-week-old male ApoE$^{-/-}$ mice [23].

The precise role of HO-1 at these sites remains to be determined and its expression does not necessarily imply that it exerts a protective influence. Indeed direct evidence that HO-1 catalyzes the breakdown of heme in atherosclerotic lesions is lacking. However, expression of HO-1 in foam cells combined with immunohistochemical detection of bilirubin IXα suggests that heme is degraded in atherosclerotic plaques [22]. Furthermore, ox-LDL is able to induce HO-1 expression by macrophages, human aortic ECs and VSMCs [18, 21]. HO-1 induction reduced monocyte migration induced by ox-LDL, an effect reproduced by biliverdin and bilirubin [21]. Taken together these data suggest a protective role for HO-1 and its products during atherogenesis, a hypothesis supported by models of HO-1 deficiency and overexpression (see below).

15.5
Genetic Modification of HO-1

A variety of studies of natural and induced genetic modification of HO-1 have been performed to explore the role of the enzyme and its products in atherogenesis. In the main, these support a protective role which may help to slow disease progress, although expression of the HO-1 in advanced and unstable plaques suggest that the beneficial effects may be overwhelmed. However, perhaps most importantly these observations raise the possibility of HO-1 as a realistic therapeutic target, particularly for early disease [1].

Human HO-1 deficiency has been reported in a child who died at the age of 6 years having suffered growth retardation, anemia, thrombocytosis, hyperlipidemia and coagulation abnormalilties. The patient was asplenic and suffered from intravascular hemolysis leading to a marked rise in circulating heme and haptoglobin [24]. Endothelial cell damage was severe and multifactorial, with both direct injury and an inadequate cytoprotective response. Importantly, in addition to the toxic effects of heme and iron, reactive oxygen species oxidized hemoglobin to methemoglobin, which in turn led to enhanced oxidative modification of LDL and endothelial injury [25]. Post-mortem findings were extensive and included evidence for accelerated atherosclerosis with fatty streaks and fibrous plaques in the aorta [26].

Yet and colleagues crossed $Hmox1^{-/-}$ mice with $ApoE^{-/-}$ mice and observed no difference in total plasma cholesterol compared to $ApoE^{-/-}$ animals. However, $Hmox1^{-/-}/ApoE^{-/-}$ mice exhibited accelerated atherogenesis with more extensive and complex atherosclerotic plaques [27]. In addition, HO-1 may protect against thrombosis as evidenced by the study of photochemical-induced vascular injury in $Hmox1^{-/-}$ mice, which developed accelerated occlusive arterial thrombosis when compared to $Hmox1^{+/+}$ animals [28]. Potential mechanisms identified included enhanced EC injury, apoptosis, and oxidative stress in $Hmox1^{-/-}$ mice, along with platelet-rich microthrombi in the subendothelium. In addition, the authors reproduced the phenotype in $Hmox1^{+/+}$ mice transplanted with $Hmox1^{-/-}$ bone marrow, and demonstrated that carbon monoxide (CO) and biliverdin rescued the phenotype in $Hmox1^{-/-}$ mice [28]. Macrophages from $Hmox1^{+/-}$ and $Hmox1^{-/-}$ mice generated more ROS, monocyte chemotactic protein-1 (MCP-1), and interleukin-6 (IL-6) than wild-type (WT) macrophages, and demonstrated increased foam cell formation in response to ox-LDL [29]. Furthermore, reconstitution of irradiated $LDLR^{-/-}$ mice with $Hmox1^{-/-}$ marrow resulted in an increased macrophage content in atherosclerotic plaques, although the extent of disease was not increased when compared to $LDLR^{-/-}$ mice with $Hmox1^{+/+}$ marrow [29].

The protective potential of HO-1 has been confirmed by overexpression using adenoviral (Ad) vectors [30]. Initial experiments revealed reduced hemin-induced iron overload in VSMCs overexpressing HO-1. Following intraventricular delivery of Ad-HO-1, $ApoE^{-/-}$ mice were significantly protected against atherogenesis compared to animals infected with control virus. Furthermore, iron metabolism was improved in the Ad-HO-1 mice with reduced iron deposition, a known cause of oxidative stress and a contributor to atherogenesis [30].

15.6
Protection Against Ischemia-Reperfusion Injury

In addition to its ability to modulate atherogenesis, HO-1 activity is protective against ischemia-reperfusion injury in a variety of organs including the heart, where it protects against myocardial infarction [2]. $Hmox1^{-/-}$ mice, in response to chronic hypoxia (10% oxygen), develop right ventricular infarcts and exhibit enhanced lipid peroxidation and oxidative damage in right ventricular cardiomyocytes when compared to WT animals [31]. Likewise, diabetic $Hmox1^{-/-}$ mice exhibit increased oxidative stress and develop larger myocardial infarcts following coronary artery occlusion than controls [32], while $Hmox1^{+/-}$ animals also demonstrate increased susceptibility [33]. Induction of HO-1 in cardiomyocytes has been reported following ischemia-reperfusion and cardiac-specific expression of HO-1 limited myocardial infarct size, with evidence of reduced inflammatory infiltration, oxidative damage [34], and associated long-term protection [35]. A similar beneficial effect was seen in diabetic mice [32].

The mechanism underlying HO-1 protection against ischemia-reperfusion injury remains to be fully determined. It is likely that the combined anti-inflammatory, antioxidant and anti-apoptotic actions of CO, bilverdin, and bilirubin are involved. However, protective mechanisms downstream of these mediators are the subject of current research (see below). Thus, while CO and biliverdin minimize ischemia-reperfusion injury, they influence different aspects of the pathophysiology and act synergistically in combination. Exogenous CO can substitute for HO-1, conferring protection against ischemia-reperfusion, while biliverdin and bilirubin may exert similar effects [36, 37].

15.7
How Does HO-1 Exert Its Effects

The efficacy of HO-1 is thought to reflect the complementary actions of its products biliverdin, free Fe and CO, which may utilize distinct mechanisms [38]. Microarray analysis of arterial tissue from $Hmox1^{+/+}$ and $Hmox1^{-/-}$ mice demonstrates the role of HO-1 in the transcriptional response to injury through regulation of the cell cycle, redox homeostasis, and coagulation [28]. One of the current challenges in this field is to determine the molecular mechanisms utilized by the products of HO-1 to exert their protective effects. Although a comprehensive discussion is beyond the scope of this review, an overview of the mechanisms relevant to protection against atherosclerosis is included (see also [2, 3]).

15.7.1
Carbon Monoxide

CO is able to activate soluble guanylyl cyclase, generating cyclic GMP (cGMP), and may act as a vasodilator. In addition, CO exerts cGMP-independent actions via activation of p38 mitogen-activated protein kinase (MAPK) and calcium-dependent potassium channels. In oxidative conditions, depletion of tetrahydrobiopterin (BH4) or L-arginine leads to nitric oxide synthase (eNOS) switching from an NO- to a superoxide-generating state. The interaction of superoxide with NO results in vascular dysfunction and peroxynitrite generation, which further depletes BH4 resulting in pathogenic uncoupling of eNOS. However, peroxynitrite may induce HO-1 activity and CO release and, under conditions of oxidative stress, CO may substitute for the loss of NO [39].

Activation of p38 MAPK has been implicated in the anti-inflammatory, anti-apoptotic actions of CO. Thus, CO inhibits release of tumor necrosis factor-α (TNF-α) and interleukin-1β (IL-1β) by lipopolysaccharide (LPS)-treated macrophages while also inducing the anti-inflammatory cytokine IL-10 [40]. Protection of ECs against apoptosis is an important action of CO in the vasculature. Reduction in TNF-α-induced EC apoptosis is dependent upon p38 MAPK [41], with both p38α and β implicated [42]. Inhibition of NF-κB is also important for

CO's ability to attenuate the proinflammatory and pro-apoptotic actions of TNF-α [43]. In particular, the ability of HO-1 to reduce TNF-α-mediated induction of VCAM-1 and E-selectin is predominantly dependent upon inhibition of NF-κB by CO [38].

Signal transducer and activator of transcription-3 (STAT3) has also been implicated in the cytoprotective actions of CO in vascular endothelium. Zhang and colleagues initially demonstrated that CO differentially modulated activity of STAT1 and STAT3 and that the latter was involved in CO-mediated protection of EC against apoptosis. CO increased STAT3 activity via activation of phosphoinositide 3-kinase (PI3K)/Akt and p38 MAPK and exerted its anti-apoptotic actions via inhibition of Fas expression and caspase-3 activity. They subsequently confirmed the role of STAT3 in HO-1 and CO-mediated protection against hyperoxia-induced lung injury in a murine model, demonstrating that STAT3 regulates both HO-1-dependent and independent protective effects [44].

15.7.2
Bilirubin and Biliverdin

Bilirubin also exerts antioxidant, anti-apoptotic, and anti-inflammatory actions, although the underlying mechanisms remain to be fully established. *In vivo*, bilirubin is able to inhibit LPS-mediated induction of P- and E-selectin and abrogate leukocyte trafficking [45]. Interestingly, biliverdin may increase vascular endothelial growth factor (VEGF) expression via an extracellular signal–regulated kinases 1 and 2 (ERK1/2), Sp-1-dependent pathway. This may represent an important vasculoprotective action [3, 46]. Bilirubin is also a potent antioxidant [47] and *in vivo* is able to suppress NADPH oxidase–dependent superoxide production stimulated by angiotensin II (Ang-II) in VSMCs and neutrophils [23]. Biliverdin reductase plays a critical role in the antioxidant properties of bile pigments. Biliverdin is reduced by biliverdin reductase to form water-soluble bilirubin. This reaction is part of a catalytic amplification cycle in which bilirubin, while acting as an antioxidant, is oxidized to biliverdin and recycled by biliverdin reductase back to bilirubin [47]. In atherosclerotic lesions, bilirubin may inhibit monocyte migration in addition to suppressing ROS generation. Likewise, Morita *et al.* demonstrated that HO-1 induction in monocytes reversed Ang-II-mediated increases in CCR2 and chemotactic activity [48].

15.7.3
Ferritin

The release of Fe^{2+} from heme by HO-1 leads to increased expression of intracellular ferritin heavy chain (ferritin H), which catalyzes oxidation of Fe^{2+} to ferric iron so allowing iron storage [5]. This is turn reduces generation of oxygen free radicals by limiting free iron for the Fenton reaction. Ferritin expression is increased atherosclerotic aortas [49] and protects vascular ECs against oxidant injury [50].

15.7.4
HO-1 and the Activation of the Complement Pathway

Complement activation has been implicated in the pathogenesis of atherosclerosis, myocardial infarction and posttransplant vasculopathy. Biliverdin and bilirubin have been reported to inhibit the classical pathway of complement activation at the level of C1, and to reduce complement-mediated hemolysis and inhibit the Forsmann reaction [51]. We have recently reported that HO-1 and its products induce expression of the complement-inhibitory protein decay-accelerating factor (DAF) and protect human ECs against complement-mediated injury. We found that ECs derived from $Hmox1^{-/-}$ mice have reduced DAF expression and increased susceptibility to complement-mediated injury [52] (Figure 15.2). These data suggest that interference with the classical complement pathway at the level of C1, and the C3 and C5 convertases of both alternative and classical pathways, represents an important additional mechanism by which HO-1 and its products mediate their potent cytoprotective actions and may have direct relevance to atherogenesis.

15.7.5
Modulation of Cell Proliferation and Angiogenesis

The use of pharmacological agonists and antagonists for HO-1 has revealed differential effects of HO-1 activity on the proliferation of VSMCs and ECs, with HO-1 increasing EC cycle progression while inhibiting that of VSMCs [53], actions that may be important in atherogenesis. Overexpression of HO-1 increases EC proliferation and formation of capillary-like structures in a two-dimensional Matrigel assay [54], while retroviral expression of antisense HO-1 has the opposite effect [55]. In contrast, the same approach induced apoptosis in VSMCs [56]. These responses were significantly reversed by CO but not by bilirubin. Furthermore, murine aortic ECs isolated from $Hmox1^{-/-}$ mice demonstrated reduced proliferation when compared with ECs from matched $Hmox1^{+/+}$ littermates [3]. In VSMCs CO activates p38β MAPK and upregulates caveolin-1, while in ECs CO inhibits synthesis of endothelin 1, VEGF and platelet-derived growth factor with the combined actions leading to a reduction in VSMC proliferation [57, 58].

The role of angiogenesis in atherosclerosis is complex and remains to be fully determined. On the one hand angiogenesis and arteriogenesis may help revascularize ischemic tissue, while on the other hand neovascularization at the base of the plaque may predispose to intraplaque hemorrhage and plaque rupture. HO-1 has an established and complex role in angiogenesis [3]. However, the protective actions of HO-1 in models of atherogenesis suggest any influence on angiogenesis is likely to be protective. The combined vasculoprotective actions of HO-1 in the endothelium, combined with its ability to inhibit VSMC proliferation [53] may retard atherogenesis.

Based on data obtained from two murine models: (1) angiogenesis initiated by VEGF or an anti-CD40 agonist mAb and (2) an LPS-induced model of inflammatory angiogenesis in which angiogenesis is secondary to leukocyte invasion, we

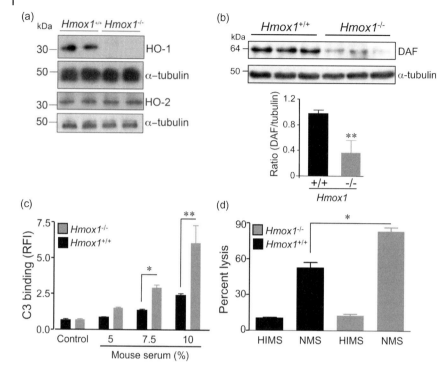

Figure 15.2 Heme oxygenase-1 (HO-1) regulates human and murine decay-accelerating factor (DAF) expression on vascular endothelial cells. (a) Aortic endothelial cells (ECs) from *Hmox1*$^{-/-}$ and *Hmox1*$^{+/+}$ mice were lyzed and immunoblotted for expression of HO-1 and HO-2. (b) Murine cardiac ECs isolated from *Hmox1*$^{-/-}$ and *Hmox1*$^{+/+}$ mice were lyzed, transblotted to polyvinylidene difluoride membranes, and immunoblotted with rat anti-mouse DAF mAb MD1. Equal loading was confirmed by re-probing with anti-α-tubulin. (c) Murine cardiac ECs isolated from *Hmox1*$^{-/-}$ and *Hmox1*$^{+/+}$ mice were opsonized with anti-endoglin mAb MJ7/18 and incubated with 5–10% normal mouse serum (NMS) or heat-inactivated NMS (HIMS) for 90 min at 37 °C, prior to analysis of C3 deposition by flow cytometry using FITC-labeled anti-mouse C3. Results expressed as relative fluorescence intensity ± SEM (*n* = 3 experiments). *$P < 0.05$. (d) To estimate complement-mediated cell lysis, murine ECs were opsonized as above and incubated with NMS or HIMS respectively for 2 h at 37 °C. Following washing, ECs were resuspended in veronal buffered saline and propidium iodide added (50 μg/ml). ECs were analyzed by flow cytometry and percentage EC lysis calculated as above (mean ± SEM, *n* = 3). *$P < 0.05$. Reproduced with permission from [52]. This research was originally published in Blood. Kinderlerer *et al*. Heme oxygenase-1 expression enhances vascular endothelial resistance to complement-mediated injury through induction of decay-accelerating factor. A role for increased bilirubin and ferritin. *Blood*. 2009; 113:1598–1607. © The American Society of Hematology.

have proposed that during chronic inflammation HO-1 has two roles, first an anti-inflammatory action inhibiting leukocyte infiltration and second, promotion of VEGF-driven non-inflammatory angiogenesis that facilitates tissue repair [59, 60]. Although further studies are required, such an effect might be protective against atherosclerosis, particularly during the early phase of the disease. Moreover, the emerging role of HO-1 activity in mobilization and recruitment of endothelial progenitor cells (EPCs) to the neovasculature [61, 62] may also contribute to protection against atherosclerosis (see below).

15.8
Regulation of HO-1 Expression and Activity

The *Hmox1* gene is regulated predominantly at the level of transcription, with expression and enzyme activity induced by a wide variety of stimuli including heme, heavy metals, ROS, growth factors including VEGF and platelet-derived growth factor (PDGF), ox-LDL, cytokines such as IL-10 and IL-6, ultraviolet A, CO and NO [3]. At the molecular level, the diverse stimuli reported to regulate HO-1 activity are reflected in the capacity of *Hmox1* to be induced by multiple transcription factors including Nrf2, AP-1, Sp-1, Ets, CREB, and NF-κB binding to sites in the 5′-flanking region (for detailed review see [2]). Protein phosphorylation-dependent signaling cascades, including those involving PI3K/Akt, protein kinase C, Jun kinase (JNK), ERK1/2, and p38 MAPKs, may influence HO-1 expression [2]. The majority of studies have focused on the MAPKs, identifying agonist, cell type and species-specific pathways, which may include a combination of MAPK isoforms [63].

15.8.1
Nuclear Factor Erythroid 2–Related Factor (Nrf2)

The Nrf2 transcription factor, a member of the Cap'n'Collar-related basic leucine zipper family of proteins, regulates antioxidant response element (ARE)–mediated expression of phase II detoxification enzymes and antioxidant proteins including HO-1. Nrf2 is sequestered in the cytoplasm by Kelch-like ECH-associated protein 1 (Keap 1). Upon activation Keap 1 is released, allowing translocation of Nrf2 to the nucleus where it associates with small Maf proteins and the complex binds the ARE. HO-1 and other phase II detoxification enzymes are not inducible in Nrf2$^{-/-}$mice [64], while activation of Nrf2 by oxidized phospholipids results in induction of antioxidant genes which may in turn limit vascular injury [65]. This led to the hypothesis that Nrf2 deficiency would lead to accelerated atherosclerosis. Intriguingly, however, the opposite was the case when Nrf2$^{-/-}$ mice were crossed with ApoE$^{-/-}$ animals. A significantly reduced plaque area was found, along with decreased uptake of modified LDL by ApoE$^{-/-}$/Nrf2$^{-/-}$ macrophages, which may reflect reduced expression of the CD36 scavenger receptor in these mice [66].

15.8.2
Shear Stress

Vascular endothelium exposed to unidirectional, pulsatile laminar shear stress (LSS) >10 dynes/cm^2 is protected against atherogenesis. LSS increases nitric oxide (NO) biosynthesis, prolongs EC survival and generates an anticoagulant, antiadhesive cell surface. In contrast, endothelium exposed to disturbed blood flow (DF), with low shear reversing or oscillatory flow patterns, such as that located at arterial branch points and curvatures, exhibits reduced eNOS, increased apoptosis, oxidative stress, permeability to LDL, and leukocyte adhesion and is atheroprone.

Nrf2 activity is central to LSS-mediated regulation of antioxidant genes including HO-1, thioredoxin reductase 1, and glutathione reductase [67, 68]. Nrf2 is preferentially activated in ECs exposed to unidirectional LSS and, although it translocates to the nucleus in ECs exposed to oscillatory shear stress, Nrf2 fails to bind the ARE [68, 69]. The mechanisms underlying these responses remain to be established. LSS activates PI3K upstream of Nrf2 [68] and both PI3K and ERK1/2 have been implicated in LSS induction of HO-1 [70]. In addition, the demonstration that NO is a potent inducer of HO-1 [71] and LSS of eNOS/NO has led investigators to explore the relationship between LSS, NO, and HO-1 expression and activity with contrasting results. A recent study has suggested that following exposure to LSS for 6h, bovine aortic EC mitochondria-derived hydrogen peroxide induces HO-1 upregulation, a response reversed by both the nitric oxide synthase antagonist, L-NAME and the ROS scavenger n-acetyl cysteine [70]. In contrast, studies in HUVECs have not found a role for NO in LSS induction of HO-1 [68, 72, 73], and this may reflect species variability or perhaps technical differences such as the duration of exposure to LSS.

15.9
Therapeutic Manipulation of HO-1

The diverse cytoprotective actions of HO-1 and its inducibility have focused attention on the role of HO-1 and its products as therapeutic targets (Table 15.2) [1, 74]. In particular, induction and inhibition of HO-1 activity in murine disease models suggested that modulation of HO-1 may be of benefit in human diseases. Thus, treatment with hemin inhibited atherogenesis in LDLR$^{-/-}$ mice on a high-fat diet, while inhibition of HO-1 with SnPPIX led to larger atherosclerotic plaques [89]. Likewise, Watanabe rabbits on a 1% cholesterol diet developed significantly larger atherosclerotic lesions when treated with SnPPIX [90].

15.9.1
Lipid-Lowering Drugs

Probucol, a lipid-lowering drug, inhibits lipid oxidation and protects against carotid artery atherosclerosis. Using three separate in vivo models and the HO inhibitor SnPPIX, Wu and colleagues demonstrated that HO-1 is an important molecular target of probucol-mediated protection against atherogenesis and restenosis injury [75]. Pharmacological agonists of the peroxisome proliferator–activated receptors (PPARs) PPARα (fibrates) and PPARγ (thiazolidinediones) are in current clinical use for the treatment of hyperlipidemia and type 2 diabetes mel-

Table 15.2 Therapeutic induction of heme oxygenase-1 (HO-1) [1, 74].

Lipid-lowering drugs	
Probucol	Probucol treatment induces expression of HO-1 and inhibition of HO-1 abrogates protective effects of probucol against atherogenesis and restenosis [75]
Peroxisome proliferator–activated receptor (PPAR) agonists	PPARα agonists (fibrates) and PPARγ agonists (thiazolidinediones) induce HO-1
	PPAR agonists exert HO-1-dependent effects on VSMC proliferation and COX-2 expression [76]
Statins	HO-1 induction at supra-therapeutic concentrations, raising questions regarding clinical relevance [1, 3]
	Simvastatin upregulates HO-1 in VSMCs and ECs *in vitro* and in VSMCs *in vivo* [77]
	Atorvastatin and rosuvastatin increased HO-1 activity and generation of CO and bilirubin in mice [78]
	Simvastatin at low concentration (500 nM) increased HO-1 protein expression in HAECs [79]
	Atorvastatin and simvastatin induce HO-1 mRNA, protein and activity in HUVECs and HAECs via KLF2 and protect against oxidative stress [80]
	Atorvastatin induced HO-1 mRNA, but not protein, in HUVECs and human microvascular ECs [3, 81]
	Simvastatin induces HO-1 expression in macrophages and protects against ischemia-reperfusion [82]
	Laminar shear stress enhances statin-mediated induction of HO-1 *in vitro* and *in vivo* [83]
Anti-inflammatory and immunomodulatory drugs	Aspirin induces HO-1 activity in ECV304 cells via an NO-dependent pathway [84]
	Celecoxib upregulates expression of HO-1 in glomerular mesangial cells [85]
	Rapamycin induces HO-1 expression in pulmonary VSMCs and ECs [86]
	HO-1 is critical for the vasculoprotective actions of rapamycin in a rat model of PAH [87]
	Synergistic induction of DAF by rapamycin and statins is HO-1-dependent [88]

The diverse cytoprotective actions of HO-1 have focused attention on the role of HO-1 and its products as therapeutic targets. Moreover, these have been demonstrated in murine models of atherosclerosis. However, clinical trials are now needed to explore the ability of drugs such as rapamycin and the statins to induce HO-1 following therapeutic dosing and to determine whether increased HO-1 activity affects clinical outcomes.
EC, endothelial cell; VSMC, vascular smooth muscle cell; HAEC, human aortic ECs; HUVEC, human umbilical vein endothelial cell; PAH, pulmonary arterial hypertension.

litus. These agents have been shown to inhibit the development of atherosclerosis and to exert lipid-lowering, anti-inflammatory vasculoprotective effects. It has been proposed that increased HO-1 activity may contribute to the pleiotropic effects of PPARα and PPARγ agonists. Thus, human VSMCs and ECs treated with fenofibrate or thiazolidinediones exhibit increased HO-1 expression and an HO-1-dependent effect on cell proliferation and cycloxygenase-2 (COX-2) protein levels [76].

Statins inhibit cholesterol synthesis and reduce serum LDL-cholesterol through inhibition of 3-hydroxy-3-methylglutaryl coenzyme A (HMG-CoA) reductase, the rate-limiting enzyme of the mevalonate pathway. In addition, wide-ranging lipid-independent effects of statins including anti-inflammatory, antioxidant, and immunomodulatory actions have been reported [91]. The regulation of HO-1 expression by statins remains to be fully understood, with differences in experimental conditions, statin preparation, and cell-type specificity contributing to the variability seen. Simvastatin, atorvastatin, and rosuvastatin have been shown to increase HO-1 promoter activity and mRNA levels, to induce HO-1 enzyme activity and increase its antioxidant capacity in human ECs [77–81]. In addition, statins induce HO-1 expression in VSMCs and macrophages [77, 82]. However, the statin concentration used in these experiments often exceeds that measured in the plasma during pharmacokinetic studies, raising the question of therapeutic relevance [74].

We have recently demonstrated that induction of HO-1 by atorvastatin requires inhibition of HMG-CoA reductase and geranylgeranylation. Moreover, we demonstrated that the ability of statins to increase HO-1 activity and protect against oxidative stress is dependent upon the transcription factor Kruppel-like factor 2 (KLF2), the generation of bile pigments and increased intracellular ferritin [80]. In HUVECs preconditioned by 24 h LSS, the concentration of statin required to induce HO-1 is significantly reduced and approaches that achieved by therapeutic dosing. We were able to demonstrate a synergistic relationship between LSS and atorvastatin and defined a signaling pathway dependent upon PI3K/Akt, activation of both KLF2 and Nrf2, and eNOS [83]. Other signaling pathways implicated in HO-1 induction by statins include p38 MAPK, ERK1/2, and protein kinase G [77]. Following treatment of mice with atorvastatin we identified increased HO-1 expression at sites of the murine aorta exposed to atheroprotective LSS, a response that was significantly attenuated in atheroprone regions [83]. In separate studies, atorvastatin and rosuvastatin increased HO-1 activity and generation of CO and bilirubin in mice [78], while induction of HO-1 in infiltrating macrophages may protect against renal ischemia-reperfusion injury in a rat model [82].

15.9.2
Anti-Inflammatory and Immunomodulatory Drugs

Non-steroidal anti-inflammatory drugs and COX-2 selective agents (COXIBs) are important therapies, recognized to have a small but significant pro-thrombotic risk. Aspirin, an antagonist of COX-1 and COX-2, is widely used in patients with CAD and may induce HO-1 activity in ECV304 cells via an NO-dependent pathway

[84]. In contrast, indomethacin failed to induce HO-1 [84], while the COX-2-selective agent celecoxib, which may improve vascular function in CAD, increased HO-1 expression in mesangial cells [85]. Further work is required to determine the clinical relevance of HO-1 induction by these agents.

HO-1 induction is an interesting and important contributor to the actions of the immunosuppressive drug rapamycin (Sirolimus). Rapamycin, a macrolide antibiotic which targets mTOR, inhibits proliferation of lymphocytes, VSMCs, and ECs and is particularly effective in prevention of vasculopathy following cardiac transplantation [92]. The drug induces HO-1 expression in pulmonary VSMCs and ECs and its ability to inhibit their proliferation is reversed by SnPPIX [86]. Furthermore, induction of HO-1 is critical for the vasculoprotective actions of rapamycin in a rat model of pulmonary arterial hypertension [87]. In recent studies we have described an HO-1-dependent synergistic interaction between rapamycin and statins, which results in maximal induction of the complement inhibitor DAF, a response not seen with the combination of statins and cyclosporin A [88].

The potent vasculoprotective actions of HO-1 suggest that modulation of its expression or delivery of its products may have therapeutic potential in patients with atherosclerosis and posttransplant vasculopathy. The epidemiologic studies cited above suggesting protective roles for a moderately raised bilrubin and the short $(GT)_n$ repeats support this idea. However, such an approach is not straightforward in light of the potential toxicity of CO, free iron, and bilirubin. Exposure to exogenous CO for as little as 1 h can confer protection against ischemia-reperfusion [93], restenosis injury, and allograft rejection [36]. Likewise, biliverdin and bilirubin exert similar effects [36, 37]. Amplifying positive feedback loops offer one explanation for the prolonged efficacy of HO-1 products [94]. This suggests that the frequency of CO, biliverdin, or bilirubin treatment required may be less than anticipated. If so, this might reduce any toxicity associated with the therapeutic use of these HO-1 products. However, it must be borne in mind that HO-1 activity has on occasion been reported to have deleterious effects on vascular endothelium and its role in tumor propagation is currently under investigation [95]. Thus clinical trials investigating the manipulation of HO-1 and its products are awaited with interest.

15.10
Stem cell and Gene Therapy

The role of HO-1 in the mobilization and recruitment of EPCs to sites of injury and the developing neovasculature is intriguing [96]. HO-1 activity may be required to establish a stromal cell-derived factor-1 (SDF-1) gradient between an ischemic limb and the bone marrow [62]. Furthermore, SDF-1 itself induces HO-1 expression in ECs, and the subsequent release of CO induces redistribution of vasodilator-stimulated phosphoprotein at the leading edge of EPCs, which in turn directs their migration and facilitates neovascularization and repair [61]. These data suggest that a moderately raised bilirubin, or the short $(GT)_n$ polymorphism genotype, might increase circulating EPCs and optimize their delivery to sites of vascular injury, so protecting against endothelial dysfunction and atherogenesis. Likewise, treatment with EPCs in which HO-1 expression has been optimized may be an option in high-risk patients, such as those with atherosclerosis and *Hmox1* polymorphisms associated with attenuated transcriptional induction. Moreover, the fact that statins induce HO-1 [80] and raise the number of circulating EPCs [97] is of particular interest. However, further studies are required to demonstrate the ability of statins to induce HO-1 following therapeutic dosing in humans, and to establish a link between HO-1 and EPC numbers in this setting.

HO-1 deficiency impairs the function of EPCs [61]. Two studies have investigated the therapeutic potential of genetically manipulating stem cells. In a study of murine bone marrow mesenchymal stem cells (MSCs) transfected with a plasmid expressing hypoxia-inducible HO-1, the overexpression of HO-1 enhanced MSC tolerance to hypoxia-reoxygenation injury and improved their viability in the ischemic myocardium [98]. Contrasting results were seen with rabbit EPCs transduced with HO-1 or eNOS. Treatment with eNOS-transfected cells enhanced re-endothelialization following arterial injury, while no beneficial effect of HO-1 overexpression was apparent [99]. One possible explanation put forward was that the level of HO-1 activity in the transfected EPCs was insufficient to generate CO.

A number of features favor HO-1 as a target for gene therapy approaches. These include its small size (32 kDa), the availability of heme, and the ability of CO to diffuse to surrounding cells [100]. Following intraventricular delivery of an adenoviral vector expressing HO-1, the transgene was expressed in the endothelium and atherosclerotic plaques of ApoE$^{-/-}$ mice leading to protection against atherogenesis when compared to control animals [30]. Furthermore, generation of a transgenic mouse with cardiac-specific expression of HO-1 limited myocardial infarct size [34], and this was associated with increased Bcl-2 expression, reduced apoptosis, oxidative stress, and inflammation, so conferring long-term myocardial protection [35].

15.11
Conclusions

The potent diverse cytoprotective actions of HO-1 are now well established, and the weight of evidence suggests HO-1 activity is protective against atherosclerosis. However, the question remains as to whether HO-1 can be manipulated therapeutically and if so, how is this best achieved? One option is activation by drugs, while alternatives include delivery of HO-1 products including CO either as a gas or in the form of carbon monoxide–releasing molecules (CORMs), and gene therapy approaches such as direct viral delivery of HO-1 to the vasculature or via manipulation of EPCs. We need to bear in mind that such approaches might be hazardous, and that induction of HO-1 might increase neovascularization at the base of the atherosclerotic plaque, predisposing to instability, or even enhance vascularization of occult tumours. Clinical trials are needed to explore the ability of drugs such as statins to induce HO-1 following therapeutic dosing. Likewise, the potential influence of *Hmox1* polymorphisms needs to be defined in more detail, as these may have direct effects on the efficacy of HO-1-inducing therapies, and indeed on which therapy is selected. For example, the prolonged beneficial effects of CO may depend upon positive feedback cycles involving HO-1 itself, and these may be attenuated in individuals with low-expression polymorphisms. In this instance, direct gene therapy approaches may be of particular benefit if they prove to be technically feasible and within health system budgets. The next decade should provide the answers to these questions and hopefully the introduction of HO-1 targeting as an important therapy for atherosclerosis.

References

1 Stocker, R. and Perrella, M.A. (2006) Heme oxygenase-1: a novel drug target for atherosclerotic diseases? *Circulation*, **114**, 2178–2189.

2 Ryter, S.W., Alam, J., and Choi, A.M. (2006) Heme oxygenase-1/carbon monoxide: from basic science to therapeutic applications. *Physiol. Rev.*, **86**, 583–650.

3 Loboda, A., Jazwa, A., Grochot-Przeczek, A., Rutkowski, A.J., Cisowski, J., Agarwal, A., Jozkowicz, A., and Dulak, J. (2008) Heme oxygenase-1 and the vascular bed: from molecular mechanisms to therapeutic opportunities. *Antioxid. Redox. Signal.*, **10**, 1767–1812.

4 Tenhunen, R., Marver, H.S., and Schmid, R. (1968) The enzymatic conversion of heme to bilirubin by microsomal heme oxygenase. *Proc. Natl Acad. Sci. U. S. A.*, **61**, 748–755.

5 Ferris, C.D., Jaffrey, S.R., Sawa, A., Takahashi, M., Brady, S.D., Barrow, R.K., Tysoe, S.A., Wolosker, H., Baranano, D.E., Dore, S., Poss, K.D., and Snyder, S.H. (1999) Haem oxygenase-1 prevents cell death by regulating cellular iron. *Nat. Cell Biol.*, **1**, 152–157.

6 Ding, Y., Zhang, Y.Z., Furuyama, K., Ogawa, K., Igarashi, K., and Shibahara, S. (2006) Down-regulation of heme oxygenase-2 is associated with the increased expression of heme oxygenase-1 in human cell lines. *FEBS J.*, **273**, 5333–5346.

7 Schwertner, H.A., Jackson, W.G., and Tolan, G. (1994) Association of low serum concentration of bilirubin with increased risk of coronary artery disease. *Clin. Chem.*, **40**, 18–23.

8 Vitek, L., Jirsa, M., Brodanova, M., Kalab, M., Marecek, Z., Danzig, V., Novotny, L., and Kotal, P. (2002) Gilbert syndrome and ischemic heart disease: a protective effect of elevated bilirubin levels. *Atherosclerosis*, **160**, 449–456.

9 Djousse, L., Levy, D., Cupples, L.A., Evans, J.C., D'Agostino, R.B., and Ellison, R.C. (2001) Total serum bilirubin and risk of cardiovascular disease in the Framingham offspring study. *Am. J. Cardiol.*, **87**, 1196–12000; A1194, 1197.

10 Hopkins, P.N., Wu, L.L., Hunt, S.C., James, B.C., Vincent, G.M., and Williams, R.R. (1996) Higher serum bilirubin is associated with decreased risk for early familial coronary artery disease. *Arterioscler. Thromb. Vasc. Biol.*, **16**, 250–255.

11 Novotny, L. and Vitek, L. (2003) Inverse relationship between serum bilirubin and atherosclerosis in men: a meta-analysis of published studies. *Exp. Biol. Med. (Maywood)*, **228**, 568–571.

12 Chen, Y.H., Lin, S.J., Lin, M.W., Tsai, H.L., Kuo, S.S., Chen, J.W., Charng, M.J., Wu, T.C., Chen, L.C., Ding, Y.A., Pan, W.H., Jou, Y.S., and Chau, L.Y. (2002) Microsatellite polymorphism in promoter of heme oxygenase-1 gene is associated with susceptibility to

coronary artery disease in type 2 diabetic patients. *Hum. Genet.*, **111**, 1–8.

13 Kaneda, H., Ohno, M., Taguchi, J., Togo, M., Hashimoto, H., Ogasawara, K., Aizawa, T., Ishizaka, N., and Nagai, R. (2002) Heme oxygenase-1 gene promoter polymorphism is associated with coronary artery disease in Japanese patients with coronary risk factors. *Arterioscler. Thromb. Vasc. Biol.*, **22**, 1680–1685.

14 Ono, K., Goto, Y., Takagi, S., Baba, S., Tago, N., Nonogi, H., and Iwai, N. (2004) A promoter variant of the heme oxygenase-1 gene may reduce the incidence of ischemic heart disease in Japanese. *Atherosclerosis*, **173**, 315–319.

15 Endler, G., Exner, M., Schillinger, M., Marculescu, R., Sunder-Plassmann, R., Raith, M., Jordanova, N., Wojta, J., Mannhalter, C., Wagner, O.F., and Huber, K. (2004) A microsatellite polymorphism in the heme oxygenase-1 gene promoter is associated with increased bilirubin and HDL levels but not with coronary artery disease. *Thromb. Haemost.*, **91**, 155–161.

16 Exner, M., Minar, E., Wagner, O., and Schillinger, M. (2004) The role of heme oxygenase-1 promoter polymorphisms in human disease. *Free Radic. Biol. Med.*, **37**, 1097–1104.

17 Ross, R. (1999) Atherosclerosis-an inflammatory disease. *N. Engl. J. Med.*, **340**, 115–126.

18 Wang, L.J., Lee, T.S., Lee, F.Y., Pai, R.C., and Chau, L.Y. (1998) Expression of heme oxygenase-1 in atherosclerotic lesions. *Am. J. Pathol.*, **152**, 711–720.

19 t Hoen, P.A., Van der Lans, C.A., Van Eck, M., Bijsterbosch, M.K., Van Berkel, T.J., Twisk, J., and Aorta of Apo (2003) E-deficient mice responds to atherogenic stimuli by a prelesional increase and subsequent decrease in the expression of antioxidant enzymes. *Circ. Res.*, **93**, 262–269.

20 Morsi, W.G., Shaker, O.G., Ismail, E.F., Ahmed, H.H., El-Serafi, T.I., Maklady, F.A., Abdel-Aziz, M.T., El-Asmar, M.F., and Atta, H.M. (2006) HO-1 and VGEF gene expression in human arteries with advanced atherosclerosis. *Clin. Biochem.*, **39**, 1057–1062.

21 Ishikawa, K., Navab, M., Leitinger, N., Fogelman, A.M., and Lusis, A.J. (1997) Induction of heme oxygenase-1 inhibits the monocyte transmigration induced by mildly oxidized LDL. *J. Clin. Invest.*, **100**, 1209–1216.

22 Nakayama, M., Takahashi, K., Komaru, T., Fukuchi, M., Shioiri, H., Sato, K., Kitamuro, T., Shirato, K., Yamaguchi, T., Suematsu, M., and Shibahara, S. (2001) Increased expression of heme oxygenase-1 and bilirubin accumulation in foam cells of rabbit atherosclerotic lesions. *Arterioscler. Thromb. Vasc. Biol.*, **21**, 1373–1377.

23 Datla, S.R., Dusting, G.J., Mori, T.A., Taylor, C.J., Croft, K.D., and Jiang, F. (2007) Induction of heme oxygenase-1 *in vivo* suppresses NADPH oxidase derived oxidative stress. *Hypertension*, **50**, 636–642.

24 Yachie, A., Niida, Y., Wada, T., Igarashi, N., Kaneda, H., Toma, T., Ohta, K., Kasahara, Y., and Koizumi, S. (1999) Oxidative stress causes enhanced endothelial cell injury in human heme oxygenase-1 deficiency. *J. Clin. Invest.*, **103**, 129–135.

25 Jeney, V., Balla, J., Yachie, A., Varga, Z., Vercellotti, G.M., Eaton, J.W., and Balla, G. (2002) Pro-oxidant and cytotoxic effects of circulating heme. *Blood*, **100**, 879–887.

26 Kawashima, A., Oda, Y., Yachie, A., Koizumi, S., and Nakanishi, I. (2002) Heme oxygenase-1 deficiency: the first autopsy case. *Hum. Pathol.*, **33**, 125–130.

27 Yet, S.F., Layne, M.D., Liu, X., Chen, Y.H., Ith, B., Sibinga, N.E., and Perrella, M.A. (2003) Absence of heme oxygenase-1 exacerbates atherosclerotic lesion formation and vascular remodeling. *FASEB J.*, **17**, 1759–1761.

28 True, A.L., Olive, M., Boehm, M., San, H., Westrick, R.J., Raghavachari, N., Xu, X., Lynn, E.G., Sack, M.N., Munson, P.J., Gladwin, M.T., and Nabel, E.G. (2007) Heme oxygenase-1 deficiency accelerates formation of arterial thrombosis through oxidative damage to the endothelium, which is rescued by inhaled carbon monoxide. *Circ. Res.*, **101**, 893–901.

29 Orozco, L.D., Kapturczak, M.H., Barajas, B., Wang, X., Weinstein, M.M., Wong, J., Deshane, J., Bolisetty, S., Shaposhnik, Z., Shih, D.M., Agarwal, A., Lusis, A.J., and Araujo, J.A. (2007) Heme oxygenase-1 expression in macrophages plays a beneficial role in atherosclerosis. *Circ. Res.*, **100**, 1703–1711.

30 Juan, S.H., Lee, T.S., Tseng, K.W., Liou, J.Y., Shyue, S.K., Wu, K.K., and Chau, L.Y. (2001) Adenovirus-mediated heme oxygenase-1 gene transfer inhibits the development of atherosclerosis in apolipoprotein E-deficient mice. *Circulation*, **104**, 1519–1525.

31 Yet, S.F., Perrella, M.A., Layne, M.D., Hsieh, C.M., Maemura, K., Kobzik, L., Wiesel, P., Christou, H., Kourembanas, S., and Lee, M. (1999) E., Hypoxia induces severe right ventricular dilatation and infarction in heme oxygenase-1 null mice. *J. Clin. Invest.*, **103**, R23–R99.

32 Liu, X., Wei, J., Peng, D.H., Layne, M.D., and Yet, S.F. (2005) Absence of heme oxygenase-1 exacerbates myocardial ischemia/reperfusion injury in diabetic mice. *Diabetes*, **54**, 778–784.

33 Yoshida, T., Maulik, N., Ho, Y.S., Alam, J., and Das, D.K. (2001) H(mox-1) constitutes an adaptive response to effect antioxidant cardioprotection: a study with transgenic mice heterozygous for targeted disruption of the Heme oxygenase-1 gene. *Circulation*, **103**, 1695–1701.

34 Yet, S.-F., Tian, R., Layne, M.D., Wang, Z.Y., Maemura, K., Solovyeva, M., Ith, B., Melo, L.G., Zhang, L., Ingwall, J.S., Dzau, V.J., Lee, M.-E., and Perrella, M.A. (2001) Cardiac-specific expression of heme oxygenase-1 protects against ischemia and reperfusion injury in transgenic mice. *Circ. Res.*, **89**, 168–173.

35 Melo, L.G., Agrawal, R., Zhang, L., Rezvani, M., Mangi, A.A., Ehsan, A., Griese, D.P., Dell'Acqua, G., Mann, M.J., Oyama, J., Yet, S.F., Layne, M.D., Perrella, M.A., and Dzau, V.J. (2002) Gene therapy strategy for long-term myocardial protection using adeno-associated virus-mediated delivery of heme oxygenase gene. *Circulation*, **105**, 602–607.

36 Otterbein, L.E., Zuckerbraun, B.S., Haga, M., Liu, F., Song, R., Usheva, A., Stachulak, C., Bodyak, N., Smith, R.N., Csizmadia, E., Tyagi, S., Akamatsu, Y., Flavell, R.J., Billiar, T.R., Tzeng, E., Bach, F.H., Choi, A.M., and Soares, M.P. (2003) Carbon monoxide suppresses arteriosclerotic lesions associated with chronic graft rejection and with balloon injury. *Nat. Med.*, **9**, 183–190.

37 Ollinger, R., Bilban, M., Erat, A., Froio, A., McDaid, J., Tyagi, S., Csizmadia, E., Graca-Souza, A.V., Liloia, A., Soares, M.P., Otterbein, L.E., Usheva, A., Yamashita, K., and Bach, F.H. (2005) Bilirubin. A natural inhibitor of vascular smooth muscle cell proliferation. *Circulation*, **112**, 1030–1039.

38 Soares, M.P., Seldon, M.P., Gregoire, I.P., Vassilevskaia, T., Berberat, P.O., Yu, J., Tsui, T.Y., and Bach, F.H. (2004) Heme oxygenase-1 modulates the expression of adhesion molecules associated with endothelial cell activation. *J. Immunol.*, **172**, 3553–3563.

39 Motterlini, R., Foresti, R., Intaglietta, M., and Winslow, R.M. (1996) NO-mediated activation of heme oxygenase: endogenous cytoprotection against oxidative stress to endothelium. *Am. J. Physiol.*, **270**, H107–H114.

40 Otterbein, L.E., Bach, F.H., Alam, J., Soares, M., Tao Lu, H., Wysk, M., Davis, R.J., Flavell, R.A., and Choi, A.M. (2000) Carbon monoxide has anti-inflammatory effects involving the mitogen-activated protein kinase pathway. *Nat. Med.*, **6**, 422–428.

41 Brouard, S., Otterbein, L.E., Anrather, J., Tobiasch, E., Bach, F.H., Choi, A.M.K., and Soares, M.P. (2000) Carbon monoxide generated by heme oxygenase 1 suppresses endothelial cell apoptosis. *J. Exp. Med.*, **192**, 1015–1026.

42 Silva, G., Cunha, A., Gregoire, I.P., Seldon, M.P., and Soares, M.P. (2006) The antiapoptotic effect of heme oxygenase-1 in endothelial cells involves the degradation of p38 alpha MAPK isoform. *J. Immunol.*, **177**, 1894–1903.

43 Brouard, S., Berberat, P.O., Tobiasch, E., Seldon, M.P., Bach, F.H., and Soares, M.P. (2002) Heme oxygenase-1-derived carbon monoxide requires the activation of transcription factor NF-kappa B to protect endothelial cells from tumor necrosis factor-alpha-mediated apoptosis. *J. Biol. Chem.*, **277**, 17950–17961.

44 Zhang, X., Shan, P., Jiang, G., Zhang, S.S., Otterbein, L.E., Fu, X.Y., and Lee, P.J. (2006) Endothelial STAT3 is essential for the protective effects of HO-1 in oxidant-induced lung injury. *FASEB J.*, **20**, 2156–2158.

45 Hayashi, S., Takamiya, R., Yamaguchi, T., Matsumoto, K., Tojo, S.J., Tamatani, T., Kitajima, M., Makino, N., Ishimura, Y., and Suematsu, M. (1999) Induction of heme oxygenase-1 suppresses venular leukocyte adhesion elicited by oxidative stress: role of bilirubin generated by the enzyme. *Circ. Res.*, **85**, 663–671.

46 Jazwa, A., Loboda, A., Golda, S., Cisowski, J., Szelag, M., Zagorska, A., Sroczynska, P., Drukala, J., Jozkowicz, A., and Dulak, J. (2006) Effect of heme and heme oxygenase-1 on vascular endothelial growth factor synthesis and angiogenic potency of human keratinocytes. *Free Radic. Biol. Med.*, **40**, 1250–1263.

47 Baranano, D.E., Rao, M., Ferris, C.D., and Snyder, S.H. (2002) Biliverdin reductase: a major physiologic cytoprotectant. *Proc. Natl Acad. Sci. U. S. A.*, **99**, 16093–16098.

48 Morita, T., Imai, T., Yamaguchi, T., Sugiyama, T., Katayama, S., and Yoshino, G. (2003) Induction of heme oxygenase-1 in monocytes suppresses angiotensin II-elicited chemotactic activity through inhibition of CCR2: role of bilirubin and carbon monoxide generated by the enzyme. *Antioxid. Redox. Signal.*, **5**, 439–447.

49 Pang, J.H., Jiang, M.J., Chen, Y.L., Wang, F.W., Wang, D.L., Chu, S.H., and Chau, L.Y. (1996) Increased ferritin gene expression in atherosclerotic lesions. *J. Clin. Invest.*, **97**, 2204–2212.

50 Balla, G., Jacob, H.S., Balla, J., Rosenberg, M., Nath, K., Apple, F., Eaton, J.W., and Vercellotti, G.M. (1992) Ferritin: a cytoprotective antioxidant strategem of endothelium. *J. Biol. Chem.*, **267**, 18148–18153.

51 Nakagami, T., Toyomura, K., Kinoshita, T., and Morisawa, S. (1993) A beneficial role of bile pigments as an endogenous tissue protector: anti-complement effects of biliverdin and conjugated bilirubin. *Biochim. Biophys. Acta*, **1158**, 189–193.

52 Kinderlerer, A.R., Gregoire, I.P., Hamdulay, S.S., Steinberg, R., Ali, F., Silva, G., Ali, N., Haskard, D.O., Soares, M.P., and Mason, J.C. (2009) Heme oxygenase-1 expression enhances vascular endothelial resistance to complement-mediated injury through induction of decay-accelerating factor. A role for bilirubin and ferritin. *Blood*, **113**, 1598–1607.

53 Li Volti, G., Wang, J., Traganos, F., Kappas, A., and Abraham, N.G. (2002) Differential effect of heme-oxygenase-1 in endothelial and smooth muscle cell cycle progression. *Biochem. Biophys. Res. Commun.*, **296**, 1077–1082.

54 Deramaudt, B.M., Braunstein, S., Remy, P., and Abraham, N.G. (1998) Gene transfer of human heme oxygenase into coronary endothelial cells potentially promotes angiogenesis. *J. Cell. Biochem.*, **68**, 121–127.

55 Li Volti, G., Sacerdoti, D., Sangras, B., Vanella, A., Mezentsev, A., Scapagnini, G., Falck, J.R., and Abraham, N.G. (2005) Carbon monoxide signaling in promoting angiogenesis in human microvessel endothelial cells. *Antioxid. Redox. Signal.*, **7**, 704–710.

56 Liu, X.-M., Chapman, G.B., Wang, H., and Durante, W. (2002) Adenovirus-mediated heme oxygenase-1 gene expression stimulates apoptosis in vascular smooth muscle cells. *Circulation*, **105**, 70–84.

57 Kim, H.P., Wang, X., Nakao, A., Kim, S.I., Murase, N., Choi, M.E., Ryter, S.W., and Choi, A.M.K. (2005) Caveolin-1 expression by means of p38 mitogen-activated protein kinase mediates the antiproliferative effect of carbon monoxide. *Proc. Natl Acad. Sci. U. S. A.*, **102**, 11319–11324.

58 Morita, T., Mitsialis, S.A., Koike, H., Liu, Y., and Kourembanas, S. (1997) Carbon monoxide controls the proliferation of hypoxic vascular smooth muscle cells. *J. Biol. Chem.*, **272**, 32804–32809.

59 Bussolati, B., Ahmed, A., Pemberton, H., Landis, R.C., Di Carlo, F., Haskard, D.O., and Mason, J.C. (2004) Bifunctional role for VEGF-induced heme oxygenase-1 *in vivo*: induction of angiogenesis and inhibition of leukocytic infiltration. *Blood*, **103**, 761–766.

60 Bussolati, B. and Mason, J.C. (2006) Dual role of VEGF-induced heme-oxygenase-1 in angiogenesis. *Antioxid. Redox. Signal.*, **8**, 1153–1163.

61 Deshane, J., Chen, S., Caballero, S., Grochot-Przeczek, A., Was, H., Li Calzi, S., Lach, R., Hock, T.D., Chen, B., Hill-Kapturczak, N., Siegal, G.P., Dulak, J., Jozkowicz, A., Grant, M.B., and Agarwal, A. (2007) Stromal cell-derived factor 1 promotes angiogenesis via a heme oxygenase 1-dependent mechanism. *J. Exp. Med.*, **204**, 615–618.

62 Tongers, J., Knapp, J.-M., Korf, M., Kempf, T., Limbourg, A., Limbourg, F.P., Li, Z., Fraccarollo, D., Bauersachs, J., Han, X., Drexler, H., Fiedler, B., and Wollert, K.C. (2008) Haeme oxygenase promotes progenitor cell mobilization, neovascularization, and functional recovery after critical hind-limb ischaemia in mice. *Cardiovasc. Res.*, **78**, 294–300.

63 Zhang, X., Bedard, E.L., Potter, R., Zhong, R., Alam, J., Choi, A.M., and Lee, P.J. (2002) Mitogen-activated protein kinases regulate HO-1 gene transcription after ischemia-reperfusion lung injury. *Am. J. Physiol. Lung. Cell. Mol. Physiol.*, **283**, L815–L829.

64 Chan, K. and Kan, Y.W. (1999) Nrf2 is essential for protection against acute pulmonary injury in mice. *Proc. Natl Acad. Sci. U. S. A.*, **96**, 12731–12736.

65 Jyrkkanen, H.K., Kansanen, E., Inkala, M., Kivela, A.M., Hurttila, H., Heinonen, S.E., Goldsteins, G., Jauhiainen, S., Tiainen, S., Makkonen, H., Oskolkova, O., Afonyushkin, T., Koistinaho, J., Yamamoto, M., Bochkov, V.N., Yla-Herttuala, S., and Levonen, A.L. (2008) Nrf2 regulates antioxidant gene expression evoked by oxidized phospholipids in endothelial cells and murine arteries in vivo. *Circ. Res.*, **103**, e1–e9.

66 Sussan, T.E., Jun, J., Thimmulappa, R., Bedja, D., Antero, M., Gabrielson, K.L., Polotsky, V.Y., and Biswal, S. (2008) Disruption of Nrf2, a key inducer of antioxidant defenses, attenuates ApoE-mediated atherosclerosis in mice. *PLoS ONE*, **3**, e3791.

67 Chen, X.-L., Varner, S.E., Rao, A.S., Grey, J.Y., Thomas, S., Cook, C.K., Wasserman, M.A., Medford, R.M., Jaiswal, A.K., and Kunsch, C. (2003) Laminar flow induction of antioxidant response element-mediated genes in endothelial cells. A novel anti-inflammatory mechanism. *J. Biol. Chem.*, **278**, 703–711.

68 Dai, G., Vaughn, S., Zhang, Y., Wang, E.T., Garcia-Cardena, G., and Gimbrone, M.A., Jr. (2007) Biomechanical forces in atherosclerosis-resistant vascular regions regulate endothelial redox balance via phosphoinositol 3-kinase/Akt-dependent activation of Nrf2. *Circ. Res.*, **101**, 723–733.

69 Hosoya, T., Maruyama, A., Kang, M.I., Kawatani, Y., Shibata, T., Uchida, K., Warabi, E., Noguchi, N., Itoh, K., and Yamamoto, M. (2005) Differential responses of the Nrf2-Keap1 system to laminar and oscillatory shear stresses in endothelial cells. *J. Biol. Chem.*, **280**, 27244–27250.

70 Han, Z., Varadharaj, S., Giedt, R.J., Zweier, J.L., Szeto, H.H., and Alevriadou, B.R. (2009) Mitochondria-derived reactive oxygen species mediate heme oxygenase-1 expression in sheared endothelial cells. *J. Pharmacol. Exp. Ther.*, **329**, 94–101.

71 Motterlini, R., Green, C.J., and Foresti, R. (2002) Regulation of heme oxygenase-1 by redox signals involving nitric oxide. *Antioxid. Redox. Signal.*, **4**, 615–624.

72 Braam, B., de Roos, R., Bluyssen, H., Kemmeren, P., Holstege, F., Joles, J.A., and Koomans, H. (2005) Nitric oxide-dependent and nitric oxide-independent transcriptional responses to high shear stress in endothelial cells. *Hypertension*, **45**, 672–680.

73 Warabi, E., Takabe, W., Minami, T., Inoue, K., Itoh, K., Yamamoto, M., Ishii, T., Kodama, T., and Noguchi, N. (2007) Shear stress stabilizes NF-E2-related factor 2 and induces antioxidant genes in endothelial cells: role of reactive oxygen/nitrogen species. *Free Radic. Biol. Med.*, **42**, 260–269.

74 Li, C., Hossieny, P., Wu, B.J., Qawas-meh, A., Beck, K., and Stocker, R. (2007) Pharmacologic induction of heme oxygenase-1. *Antioxid. Redox. Signal.*, **9**, 2227–2239.

75 Wu, B.J., Kathir, K., Witting, P.K., Beck, K., Choy, K., Li, C., Croft, K.D., Mori, T.A., Tanous, D., Adams, M.R., Lau, A.K., and Stocker, R. (2006) Antioxidants protect from atherosclerosis by a heme oxygenase-1 pathway that is independent of free radical scavenging. *J. Exp. Med.*, **203**, 1117–1127.

76 Kronke, G., Kadl, A., Ikonomu, E., Bluml, S., Furnkranz, A., Sarembock, I.J., Bochkov, V.N., Exner, M., Binder, B.R., and Leitinger, N. (2007) Expression of heme oxygenase-1 in human vascular cells is regulated by peroxisome proliferator-activated receptors. *Arterioscler. Thromb. Vasc. Biol.*, **27**, 1276–1282.

77 Lee, T.S., Chang, C.C., Zhu, Y., and Shyy, J.Y. (2004) Simvastatin induces heme oxygenase-1: a novel mechanism of vessel protection. *Circulation*, **110**, 1296–1302.

78 Muchova, L., Wong, R.J., Hsu, M., Morioka, I., Vitek, L., Zelenka, J., Schroder, H., and Stevenson, D.K. (2007) Statin treatment increases formation of carbon monoxide and bilirubin in mice: a novel mechanism of *in vivo* antioxidant protection. *Can. J. Physiol. Pharmacol.*, **85**, 800–810.

79 Uchiyama, T., Atsuta, H., Utsugi, T., Ohyama, Y., Nakamura, T., Nakai, A., Nakata, M., Maruyama, I., Tomura, H., Okajima, F., Tomono, S., Kawazu, S., Nagai, R., and Kurabayashi, M. (2006) Simvastatin induces heat shock factor 1 in vascular endothelial cells. *Atherosclerosis*, **188**, 265–273.

80 Ali, F., Hamdulay, S.S., Kinderlerer, A.R., Boyle, J.J., Lidington, E.A., Yamaguchi, T., Soares, M.P., Haskard, D.O., Randi, A.M., and Mason, J.C. (2007) Statin-mediated cytoprotection of human vascular endothelial cells: a role for Kruppel-like factor 2-dependent induction of heme oxygenase-1. *J. Thromb. Haemost.*, **5**, 2537–2546.

81 Dulak, J., Loboda, A., Jazwa, A., Zagorska, A., Dorler, J., Alber, H., Dichtl, W., Weidinger, F., Frick, M., and Jozkowicz, A. (2005) Atorvastatin affects several angiogenic mediators in human endothelial cells. *Endothelium*, **12**, 233–241.

82 Gueler, F., Park, J.K., Rong, S., Kirsch, T., Lindschau, C., Zheng, W., Elger, M., Fiebeler, A., Fliser, D., Luft, F.C., and Haller, H. (2007) Statins attenuate ischemia-reperfusion injury by inducing heme oxygenase-1 in infiltrating macrophages. *Am. J. Pathol.*, **170**, 1192–1199.

83 Ali, F., Zakkar, M., Lidington, E.A., Hamdulay, S.S., Boyle, J.J., Haskard, D.O., Evans, P.C., and Mason, J.C. (2009) Induction of the cytoprotective enzyme heme oxygenase-1 by statins is enhanced in vascular endothelium exposed to laminar shear stress. *J. Biol. Chem.*, **284**, 18882–18892.

84 Grosser, N., Abate, A., Oberle, S., Vreman, H.J., Dennery, P.A., Becker, J.C., Pohle, T., Seidman, D.S., and Schroder, H. (2003) Heme oxygenase-1 induction may explain the antioxidant profile of aspirin. *Biochem. Biophys. Res. Commun.*, **308**, 956–960.

85 Hou, C.C., Hung, S.L., Kao, S.H., Chen, T.H., and Lee, H.M. (2005) Celecoxib induces heme-oxygenase expression in glomerular mesangial cells. *Ann. N. Y. Acad. Sci.*, **1042**, 235–245.

86 Visner, G.A., Lu, F., Zhou, H., Liu, J., Kazemfar, K., and Agarwal, A. (2003) Rapamycin induces heme oxygenase-1 in human pulmonary vascular cells: implications in the antiproliferative response to rapamycin. *Circulation*, **107**, 911–916.

87 Zhou, H., Liu, H., Porvasnik, S.L., Terada, N., Agarwal, A., Cheng, Y., and

Visner, G.A. (2006) Heme oxygenase-1 mediates the protective effects of rapamycin in monocrotaline-induced pulmonary hypertension. *Lab. Invest.*, **86**, 62–71.

88 Hamdulay, S.S., Ali, F., Ali, N., Steinberg, R., Lidington, E.A., Haskard, D.O., and Mason, J.C. (2007) Statins and rapamycin: therapeutic synergy in vascular protection. *Rheumatology (Oxford)*, 46, i9.

89 Ishikawa, K., Sugawara, D., Wang, X.-P., Suzuki, K., Itabe, H., Maruyama, Y., and Lusis, A.J. (2001) Heme oxygenase-1 inhibits atherosclerotic lesion formation in LDL-receptor knockout mice. *Circ. Res.*, **88**, 506–512.

90 Ishikawa, K., Sugawara, D., Goto, J., Watanabe, Y., Kawamura, K., Shiomi, M., Itabe, H., and Maruyama, Y. (2001) Heme oxygenase-1 inhibits atherogenesis in Watanabe heritable hyperlipidemic rabbits. *Circulation*, **104**, 1831–1836.

91 Greenwood, J. and Mason, J.C. (2007) Statins and the vascular endothelial inflammatory response. *Trends Immunol.*, **28**, 88–98.

92 Keogh, A., Richardson, M., Ruygrok, P., Spratt, P., Galbraith, A., O'Driscoll, G., Macdonald, P., Esmore, D., Muller, D., and Faddy, S. (2004) Sirolimus in de novo heart transplant recipients reduces acute rejection and prevents coronary artery disease at 2 years: a randomized clinical trial. *Circulation*, **110**, 2694–2700.

93 Fujita, T., Toda, K., Karimova, A., Yan, S.F., Naka, Y., Yet, S.F., and Pinsky, D.J. (2001) Paradoxical rescue from ischemic lung injury by inhaled carbon monoxide driven by derepression of fibrinolysis. *Nat. Med.*, **7**, 598–604.

94 Bach, F.H. (2005) Heme oxygenase-1: a therapeutic amplification funnel. *FASEB J.*, **19**, 1216–1219.

95 Jozkowicz, A., Was, H., and Dulak, J. (2007) Heme oxygenase-1 in tumors: is it a false friend? *Antioxid. Redox. Signal.*, **9**, 2099–2117.

96 Dulak, J., Deshane, J., Jozkowicz, A., and Agarwal, A. (2008) Heme oxygenase-1 and carbon monoxide in vascular pathobiology: focus on angiogenesis. *Circulation*, **117**, 231–241.

97 Dimmeler, S., Aicher, A., Vasa, M., Mildner-Rihm, C., Adler, K., Tiemann, M., Rutten, H., Fichtlscherer, S., Martin, H., and Zeiher, A. (2001) M., HMG-CoA reductase inhibitors (statins) increase endothelial progenitor cells via the PI 3-kinase/Akt pathway. *J. Clin. Invest.*, **108**, 391–397.

98 Tang, Y.L., Tang, Y., Zhang, Y.C., Qian, K., Shen, L., and Phillips, M.I. (2005) Improved graft mesenchymal stem cell survival in ischemic heart with a hypoxia-regulated heme oxygenase-1 vector. *J. Am. Coll. Cardiol.*, **46**, 1339–1350.

99 Kong, D., Melo, L.G., Mangi, A.A., Zhang, L., Lopez-Ilasaca, M., Perrella, M.A., Liew, C.C., Pratt, R.E., and Dzau, V.J. (2004) Enhanced inhibition of neointimal hyperplasia by genetically engineered endothelial progenitor cells. *Circulation*, **109**, 1769–1775.

100 Levonen, A.L., Vahakangas, E., Koponen, J.K., and Yla-Herttuala, S. (2008) Antioxidant gene therapy for cardiovascular disease: current status and future perspectives. *Circulation*, **117**, 2142–2150.

PART VI
Cell Growth and Phenotype

16
Phenotypic Heterogeneity of Smooth Muscle Cells – Implications for Atherosclerosis

Matteo Coen and Marie-Luce Bochaton-Piallat

16.1
The Concept of Smooth Muscle Cell Heterogeneity

During atherosclerosis, local secretion of chemokines and growth factors following endothelial injury, adhesion of platelets, and infiltration of inflammatory cells into the intima, as well as the presence of oxidized low-density lipoproteins (ox-LDLs), lead to migration of smooth muscle cells (SMCs) from the media into the intima and proliferation [1]. Albeit over a shorter period of time, similar phenomena occur during restenosis following angioplasty or stent implantation in humans, as well as in experimentally induced intimal thickening [1]. During these processes, SMCs undergo very complex changes: they acquire a de-differentiated state characterized by increased matrix synthesis, production of various proteases, altered contractility, and downregulation of contractile proteins that have been referred to as phenotypic modulation. The predominant concept called the "response-to-injury" hypothesis [1] implies that any SMCs in the media can undergo phenotypic modulation. However, the question remains open as to whether a pre-existing SMC subpopulation accumulates in the intimal thickening. In this respect, Benditt and Benditt [2] suggested that the origin of SMC accumulation in the atheromatous plaque is monoclonal or oligoclonal. More recently, microdissection of different portions of human plaques followed by PCR amplification of the DNA of an X-inactivated gene has confirmed that SMCs of the fibrous cap are monoclonal [3]. All together, the SMC phenotypic changes and the monoclonal hypothesis support the concept of SMC heterogeneity (i.e., the existence of an atheroma-prone phenotype).

Atherosclerosis: Molecular and Cellular Mechanisms. Edited by Sarah Jane George and Jason Johnson
Copyright © 2010 WILEY-VCH Verlag GmbH & Co. KGaA, Weinheim
ISBN: 978-3-527-32448-4

16.2
In Vivo Establishment of SMC Phenotypic Switch

16.2.1
Morphological Features

The SMC phenotypic switch was defined in the 1970s by means of electron microscopy studies (for review see [4, 5]). The contractile phenotype is typical of SMCs
in a healthy artery (i.e., the differentiated artery); these SMCs contain many microfilament bundles. The synthetic phenotype is typical of developing and pathological arteries (i.e., barely differentiated or de-differentiated arteries) and is
characterized by a cytoplasm with a predominance of rough endoplasmic reticulum and a well-developed Golgi apparatus. Electron micrographs of experimentally
induced intimal thickening in rat (Figure 16.1), rabbit, and pig after endothelial
injury and/or hypercholesterolemic diet show SMC phenotypic changes similar to
that of human situation.

16.2.2
Cytoskeletal Features

SMCs contain a specific repertoire of cytoskeletal proteins that form a filamentous
structure. Their expression varies according to the phenotypic changes (Figure
16.2). The SMC cytoskeleton is composed of three types of filaments: (1) microfila-
ments mainly consisting of actin and myosin which form the thin and thick filaments, respectively, (2) intermediate filament proteins, and (3) microtubules.
Actin and myosin form the contractile machinery of the cell whereas intermediate
filament proteins belong more specifically to the cell skeleton. Several isoforms of

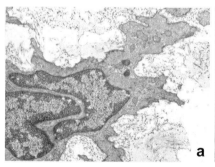 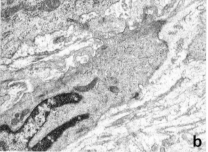

Figure 16.1 Electron micrographs showing
morphological features of (a) a contractile
(i.e., differentiated) smooth muscle cell in the
rat aortic media and (b) a synthetic (i.e.,
de-differentiated) smooth muscle cell in the
thickened intima 15 days after balloon
catheter–induced endothelial injury. (Courtesy
of Professor G. Gabbiani, Department of
Pathology and Immunology, Faculty of
Medicine, University of Geneva, Geneva,
Switzerland.)

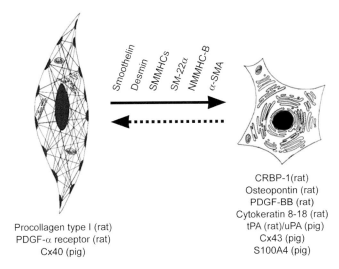

Procollagen type I (rat)
PDGF-α receptor (rat)
Cx40 (pig)

CRBP-1(rat)
Osteopontin (rat)
PDGF-BB (rat)
Cytokeratin 8-18 (rat)
tPA (rat)/uPA (pig)
Cx43 (pig)
S100A4 (pig)

Figure 16.2 Schematic representation of smooth muscle cell phenotypic switch related to the appearance of widely used cytoskeletal protein markers. Relevant biomarkers specific for each smooth muscle cell phenotype are depicted. The species in which they were identified are indicated in brackets.

these proteins as well as cytoskeletal-associated proteins, most of them related to the presence of microfilament bundles observed in the contractile phenotype, are useful differentiation SMC markers (for review see [6–8]). The most studied markers are α-smooth muscle actin (α-SMA), the actin isoform typical of vascular SMCs, desmin, an intermediate filament protein, and smooth muscle myosin heavy chain (SMMHC) isoforms SM1 and SM2. Their expression is dependent on different cis-elements in their promoter region that are activated by specific binding factors; in particular the CArG (i.e., CC[A/T-rich]$_6$GG) sequence motif binds serum response factor and myocardin, whereas the transforming growth factor-β (TGF-β) control element is activated by TGF-β [9]. Besides SMMHCs, SMCs also express two non-muscle (NM) MHC isoforms A and B; the NMMHC-B is associated with the synthetic phenotype [6]. α-SMA is expressed in vascular SMCs even at early stages of development, thus represents the most general marker of SMC lineage, whereas desmin and SMMHCs are markers of well-differentiated SMCs [6, 7].

Other cytoskeletal proteins have been less extensively studied. In particular, smoothelin, SM22α, calponin, h-caldesmon, and meta-vinculin [6] serve as late differentiation markers. Smoothelin, an α-SMA-binding protein, exists in two isoforms: smoothelin-A (59 kDa) and smoothelin-B (110 kDa); the B isoform is predominantly expressed in vascular SMCs. Recently, Rensen *et al.* [10] showed that smoothelin-B plays a crucial role in vascular SMC contraction by still undiscovered mechanisms besides being a reliable marker of the contractile phenotype.

16.3
In Vitro **Establishment of SMC Phenotypic Switch**

The notion of SMC heterogeneity has been reinforced by *in vitro* description of morphologically distinct SMC populations in many species, including humans (for review see [11]). Understanding the biological features of the distinct SMC subpopulations is instrumental to developing strategies aimed at controlling SMC accumulation into the intimal thickening.

16.3.1
Identification of SMC Subpopulations in Animal Models

In vitro SMC heterogeneity has been primarily established in rat vessels by identifying cultured SMC populations with two distinct morphologies: spindle-shaped (S) SMCs with the classical "hills and valleys" growth pattern and epithelioid (E) SMCs, in which cells grow as a monolayer and exhibit cobblestone morphology at confluence [12–15]. The S-SMCs were isolated from the media of carotid artery and aorta whereas the E-SMCs were obtained from the experimental intimal thickening induced 15 days after endothelial injury of these vessels.

Several groups using different methods of cell isolation have demonstrated that E-SMCs exist within the normal media, supporting the hypothesis that this particular population is prone to accumulate into the intimal thickening [14, 16]. In particular, by producing clones from the media and intimal thickening, we have demonstrated that S- and E-SMC clones can be recovered from both locations albeit in different proportions, the media predominantly yielding S-SMC clones and the intimal thickening yielding a majority of E-SMC clones [14]. Others have confirmed the production of distinct SMC clones from the media of rat [15, 17, 18] and mouse [19]. According to the age of the rat, S- and E-SMCs have been recovered in variable proportions from the healthy aorta [20–26]. S-SMCs were predominant in fetuses at different developmental stages [25] as well as in newborn rats (4–5 days) [23, 24, 27] whereas E-SMCs were prevalent in old rats (>18 months) [22, 24]. In this respect, several studies have shown that the intimal thickening in response to injury is greater in old rats (>18 months) compared with young adult rats (3–4 months) [28–30]. It is noteworthy, however, that a predominant population of E-SMCs is recovered from the media of 12-day-old rats, an age when sexual maturation occurs [20, 21, 26]. Taken together, these studies provide evidence that (1) the media contains phenotypically heterogeneous SMCs, (2) a variable proportion of SMCs exhibiting an E-phenotype *in vitro* exists within the media throughout the whole lifespan and that this proportion increases with age, and (3) the intimal thickening develops essentially from a distinct medial subpopulation exhibiting an E-phenotype *in vitro*.

Distinct phenotypes similar to those isolated in the rat have been recovered in arteries of dog [31], cow [32], and pig [33], extending the notion of SMC heterogeneity to large animals. In the canine carotid artery, spherical SMCs, similar to the rat E-SMCs were isolated from the abluminal part of the media and were predomi-

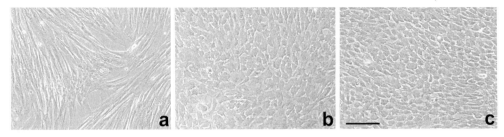

Figure 16.3 Phase-contrast microphotographs showing the morphological features of coronary artery smooth muscle cell subpopulations. (a) Spindle-shaped smooth muscle cells are isolated from the normal media and (b,c) rhomboid smooth muscle cells are respectively derived from the normal media and thickened intima 15 days after stent implantation. Bar = 100 µm. (From [33], with permission of the publisher.)

nant in the intimal thickening produced 14 days after endothelial injury [31]. SMC subpopulations exhibiting S- and E-phenotypes as well as rhomboid (R) phenotype (i.e., flat but more elongated than rat E- SMCs) were isolated from morphologically distinct compartments within the media of bovine pulmonary artery and aorta [32]. Our group has identified two distinct SMC subpopulations from the porcine coronary artery [33]. S- and R-SMCs were isolated from the media. R-SMCs were recovered in higher proportion when SMCs were cultured from the intimal thickening induced after experimental stent implantation compared with the media (Figure 16.3). Hence R-SMCs represent good candidates for the formation of intimal thickening in the porcine coronary artery.

16.3.2
Identification of SMC Subpopulations in Humans

Exporting the concept of SMC heterogeneity from animal models to human has been hampered by the limited availability of human arterial SMCs and difficult experimental standardization. The description of distinct SMC subpopulations is of particular interest in human even though these studies have been sporadically performed [34–45]. The early works by Orekhov *et al.* [34] in the 1980s showed the presence of SMCs with differing morphology in the uninvolved and atherosclerotic intima of human arteries *in situ*. Elongated (i.e., differentiated) SMCs were predominant in unaffected intima, whereas stellate (i.e., de-differentiated) SMCs were predominant in atherosclerotic intima. This SMC diversity was partially retained in primary culture [35]. Björkerud [36] also isolated two different SMC phenotypes from the aortic media according to their different adhesive properties. As observed in animal models, other studies have indicated the presence of distinct SMC subpopulations in healthy human arteries (i.e., media of abdominal aorta [38], pulmonary [40, 42], mammary [43], and intracranial basilar [44] arteries), as well as in diseased vessels (i.e., aortic aneurysm [39] and atherosclerotic carotid artery [39]). The most conclusive work to date demonstrating the presence of SMC subpopulations in human was based on cell cloning from the media of undiseased arteries [41]. Taken together, the findings that SMCs exhibiting a synthetic phenotype can be retrieved from the media of normal arteries as well [34, 36, 40–42, 44], support the hypothesis that expansion of a SMC subset takes place in atherosclerotic lesions.

16.3.3
Biological Features of SMC Subpopulations

16.3.3.1 Proliferative Activity and Apoptosis

SMC proliferation and apoptosis are essential processes in experimental intimal thickening, atherosclerotic, and restenotic lesion formation. Although they stop growing at confluence as a result of cell contact inhibition, animal E- and R-SMCs show a higher proliferative activity than S-SMCs [12, 14, 32, 33]. In humans the corresponding phenotypes exhibit similar proliferative features [37, 38, 40–43]. It is noteworthy that growth in the absence of serum has been observed in rat E-SMCs [13, 14, 20, 22, 26, 46, 47], as well as in cow E- and R-SMCs [32] and chick embryo E-SMCs [48]. Rat E-SMCs produce platelet-derived growth factor (PDGF)-BB, which is a potent SMC mitogen [20, 26], and fail to respond to the growth inhibitory effect of TGF-β [47]. Pig R-SMCs do not display autonomous growth [33]; in this respect they are similar to human [41]. An enhanced susceptibility of rat E-SMCs to apoptosis induced by reactive oxygen species [18], retinoic acid, and antimitotic drugs [49] has also been described. All together, the studies demonstrating the enhanced proliferative and apoptotic activities of E- and R-SMCs in various species fit well with the expected features of participants in intimal thickening.

16.3.3.2 Migratory and Proteolytic Activities

Cell migration, a major event of the intimal thickening, is a complex process that includes the degradation of extracellular matrix components by proteolytic enzymes. In the rat [14, 50], pig [33], and humans [41, 43] E- and R-SMCs exhibit a higher migratory activity than S-SMCs. This has been correlated to high tissue-type plasminogen activator (tPA) activity in rat E-SMCs [51] and high urokinase-type PA (uPA) activity in pig R-SMCs [33]. Likewise, Lau [17] has shown that rat E-SMCs may produce tPA, uPA, and metalloproteinase-2 under particular growth conditions.

16.3.3.3 Cytoskeletal Features

When placed in culture, all SMCs tend to show a de-differentiated phenotype [4, 5]; nevertheless the phenotypic variations of cultured SMCs furnish important information concerning the influence of many factors on their biological features. α-SMA, desmin, and SMMHC isoforms are generally more abundant in S- than in E- or R-SMCs in any species studied (Figure 16.2) [14, 24, 26, 27, 31–33, 41, 50]. Smoothelin, SM22α, calponin, h-caldesmon, and meta-vinculin [6] that serve as late differentiation markers are again more abundant in S-SMCs than in E- or

R-SMCs [31–33, 48]. The level of expression of these proteins (e.g., desmin and SMMHCs) is much higher in larger animals, such as pig and cow, than in rodents [32, 33]. In this respect, SMCs isolated from large animals, such as pig, behave similarly to SMCs derived from human arteries [36, 37, 41]. The data obtained in different species suggest that the degree of differentiation of SMC changes with the phenotype.

16.3.3.4 **Other Features**
It has been reported that phenotypic switch is accompanied by important changes in sarcolemma composition and organization, possibly accounting for impaired contractile properties in the synthetic phenotype [52]. S-SMCs are more responsive than E-SMCs to vasoactive factors such as endothelin-1 [16], angiotensin II, histamine, and norepinephrine [41] in terms of contraction or variation of calcium concentration. It has been suggested that despite a large production of nitric oxide, E-SMCs proved to be less sensitive to its action than S-SMCs [15, 30, 53]. Neylon *et al.* [54] have shown that Ca^{2+}-activated K^+ channels, which are involved in limiting Ca^{2+} influx through voltage-gated Ca^{2+} channels, reduce vascular contractility responsiveness to vasoactive substances in rat proliferating SMCs (similar to E-SMCs), compared with their differentiated counterparts (similar to S-SMCs). The same group has further cloned an intermediate-conductance Ca^{2+}-activated K^+ channel, which is markedly upregulated in proliferating smooth muscle cells [55]. Therefore excitation–contraction coupling, a key feature of muscle cell biology, may influence the phenotypic switch [56].

An interesting correlation has been demonstrated, albeit occasionally, between de-differentiated (and/or highly proliferating) SMC phenotypes and increased LDL uptake [40, 45, 57–59] or decreased high-density lipoprotein (HDL)–binding sites [60]. The role of LDL and HDL processes in atheromatous plaque formation with respect to SMC heterogeneity should be further investigated.

16.3.4
Biomarkers of SMC Subpopulations

The identification of differentially expressed genes and/or proteins among the distinct SMC subpopulations is instrumental to track the phenotypic changes that occur *in vivo* during atherosclerosis (Figure 16.2). By using the technique of 2D-PAGE followed by protein sequencing, we have identified cellular retinol–binding protein-1 (CRBP-1), a protein involved in retinoid metabolism, as a specific marker of rat aortic E-SMCs [61, 62]. After endothelial injury, CRBP-1 is rapidly activated in a subset of medial SMCs located towards the lumen and is expressed in the large majority of intimal SMCs; it disappears when re-endothelialization is complete [62]. During these phases, CRBP-1 is present in replicating SMCs during the initial phase and in apoptotic SMCs during the late phase of intimal thickening formation [62]. In this respect, SMCs cultured from re-endothelialized intimal thickening exhibit exclusively a S-phenotype [13]. Moreover, cultured rat E-SMCs are more sensitive to apoptosis than S-SMCs [18, 49]. These results suggest that

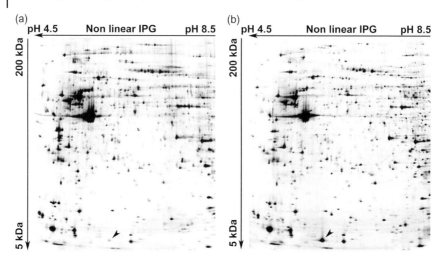

Figure 16.4 Representative silver-stained 2D-PAGE of (a) Spindle-shaped smooth muscle cells and (b) rhomboid smooth muscle cells. Arrowheads point to S100A4. (From [63], with permission of the publisher.)

a predisposed subset of medial SMCs becomes rapidly CRBP-1 positive after injury, undergoes replication during the early phase of IT development and then disappears, allegedly through apoptosis, when re-endothelialization takes place [62]. This also indicates that CRBP-1 is a marker of the E-phenotype *in vitro* and possibly of SMC activation after endothelial injury *in vivo*. It should be noted that CRBP-1 is a marker restricted to rodent models.

More recently, comparison of porcine S- and R-SMCs again by 2D-PAGE followed by tandem mass spectrometry allowed us to identify S100A4 as a marker of the R-SMC population *in vitro* (Figure 16.4) [63]. S100A4 belongs to the S100 Ca^{2+}-binding protein family and exists as an intra- and extracellular protein (for review see [64, 65]). Several observations support a role of S100A4 in invasive growth and metastasis. In our model, S100A4 can be localized in stress fibers of quiescent R-SMCs and is upregulated in migrating R-SMCs [63]. Silencing of S100A4 mRNA decreased R-SMC proliferation. *In vivo*, S100A4 is absent in the media and is upregulated in SMCs of experimentally induced intimal thickening in porcine coronary artery. In humans, it is overexpressed in intimal thickening, atheromatous plaques and restenotic lesions. Therefore S100A4 appears to represent a novel marker of intimal SMC and could play a role during this process.

Cytokeratin 8 and 18, intermediate filament proteins, as well as zonula occludens-2 protein and cingulin, proteins of tight junctions, were thought to be exclusively expressed in epithelial or endothelial cells. They have since been identified as markers of rodent E-SMCs [19, 62, 66] and are expressed in experimental intimal thickening [66] and human atheromatous plaque [67]. Connexin (Cx) 40 and 43, components of gap junction channels, are typical of arterial SMCs (for review see [68, 69]). Cx43 is expressed in healthy arteries and is upregulated in human atherosclerotic plaque and mouse intimal thickening. Cx43 downregulation reduces atherosclerotic plaque development in mice [68, 69]. Cx40 is also expressed in arterial SMCs but only in large animals and humans [70]. In the porcine coronary artery, Cx40 is highly expressed in normal media whereas Cx43 is predominant in stent-induced intimal thickening [71]. This correlates *in vitro* with the high expression of Cx40 in S-SMCs and of Cx43 in R-SMCs [71]. Studies of these proteins could give further insight into the mechanisms of SMC phenotypic modulation (see Section 16.4).

Several other genes have been discovered, mainly in rodents, as specific or at least more abundant in one SMC population compared with the others. Rat E-SMCs overexpress osteopontin [72–74], an extracellular matrix protein involved in bone mineralization. *In vivo*, osteopontin is transiently upregulated in experimental intimal thickening and accumulates in calcified areas of the human atheromatous plaque (for review see [75]). In addition, rat E-SMCs express tropoelastin [26, 66], PDGF-BB [26], cytochrome P450 [76], and peroxisome proliferator-activated receptor γ (PPARγ) [66] whereas S-SMCs overexpress procollagen type I and PDGFα receptor [26, 66]. A study performed in the cow pulmonary artery model has shown that autonomously growing R-SMCs exhibit constitutively activated extracellular signal-regulated kinase and eicosanoid production [77].

16.4
Factors Influencing SMC Phenotypic Switch

The chance to orchestrate SMC switching, thus influencing atherosclerosis formation and evolution, may be instrumental for the development of preventing and therapeutic strategies. Inhibitors of SMC proliferation and/or differentiation factors, for example, heparin, TGF-β, and retinoic acid, and stimulators of SMC proliferation and/or de-differentiation factors, for example, PDGF-BB and fibroblast growth factor-2 (FGF-2), are among the most studied agents employed to modulate SMC phenotype. Some of these show species-specific and phenotype-dependent effects.

Heparin and TGF-β inhibit proliferation and increase differentiation in rat [13, 78] and pig SMCs [33] regardless of their phenotype, while only R-SMCs are affected in the bovine model [32]. The effects on cell morphology are absent or partial depending on the initial phenotype; moreover a complete morphological switch is never achieved. Retinoic acid is effective only in the rat, where it acts mainly on the poorly differentiated phenotype by inducing differentiation and partial morphologic changes; however it decreases proliferation and increases migration of both phenotypes [79]. Recently, we studied the effects of different cytostatic drugs on porcine S- and R-SMCs. Although imatinib, and to a lesser extent curcumin, may represent valid therapeutic options by promoting SMC quiescence and differentiation [80], SMCs regained a contractile phenotype only partially.

In human SMCs only E-SMCs show a strong response to PDGF-BB and FGF-2 in terms of proliferation and to PDGF-BB alone in terms of migration [41]. In pig,

SMCs both increase migration and proliferation; interestingly they induce a switch from the S- to the R-phenotype [33]. The PDGF-driven phenotypic switch is accompanied by an upregulation of S100A4 [63] as well as Cx43 and a loss of Cx40. S- to R-phenotype change induced by PDGF-BB is prevented by a reduction of Cx43 expression (α-SMA expression is upregulated and S100A4 appearance is prevented), suggesting that Cx43 may play a role in the development of atherosclerosis and restenosis [71]. We have also observed that coculture of endothelial cells with porcine S-SMCs induces a transition toward the R-phenotype; in this experiment endothelial cells never reached quiescence, thus mimicking a dysfunctional endothelium [33]. Taken together, these studies offer insights on how microenvironmental factors can influence essential cellular processes (i.e., proliferation, migration, and de-differentiation) and selectively modulate the behavior of distinct SMC subpopulations. It is noteworthy that after implantation into the rat carotid artery S- and E-SMCs keep their distinct features, as defined by the expression level of α-SMA, SMMHCs, and CRBP-1 [81]. Contrasting with the pig studies, this result indicates that rat SMC phenotypes depend more on their intrinsic features rather than on their environment.

16.5
Conclusions and Perspectives

Studies on SMC heterogeneity in different species have led to the identification of particular SMC subpopulations. E- and/or R-SMCs exhibit features that explain their capacity to accumulate in the intimal thickening, thus representing an atheroma-prone phenotype. Nevertheless SMCs retain the capacity to achieve a bidirectional phenotypic switch in response to both physiological (e.g., morphogenesis) and pathological conditions (e.g., atherosclerosis, restenosis, and vascular injury). It is noteworthy that in pathologic conditions a complete modulation from the synthetic toward the contractile phenotype has not yet been achieved.

Some of the animal models have been instrumental in identifying markers that may play a role in SMC behavior, especially in phenotypic switching. By using a proteomic strategy we have identified markers of the E- and/or R-phenotype in both the rat (CRBP-1) and the pig (S100A4) models. S100A4 proved to be relevant to the human situation. Up to now, isolation of human SMC subpopulations has not been exploited enough to clearly identify genetic or proteomic markers specific to the human atheroma-prone phenotype that can offer new insights for the development of strategies tailored to the human situation. Identification of phenotype-specific molecules could lead to the development of strategies aimed at preventing the acquisition of a synthetic phenotype and/or inducing the differentiation of a potentially noxious cell into a less aggressive one.

The ultimate aim of identifying the different SMC phenotypes is to provide a better understanding of their role in both physiological and pathological situations and to develop tools capable of influencing the evolution of atherosclerotic lesions. Targeting atheroma-prone SMCs could prevent SMC accumulation into the intima during early stages of atherosclerosis and/or restenosis. It could inhibit their further expansion in already developed lesions, thereby stabilizing the lesion and eventually leading to plaque regression.

Acknowledgments

The authors acknowledge the support of the Swiss National Science Foundation (grant no. 32-226595), Ernst and Lucie Schmidheiny Foundation, and Carife Foundation.

References

1 Ross, R. (1999) Atherosclerosis: an inflammatory disease. *N. Engl. J. Med.*, **340**, 115–126.

2 Benditt, E.P. and Benditt, J.M. (1973) Evidence for a monoclonal origin of human atherosclerotic plaques. *Proc. Natl Acad. Sci. U. S. A.*, **70**, 1753–1756.

3 Murry, C.E., Gipaya, C.T., Bartosek, T., Benditt, E.P., and Schwartz, S.M. (1997) Monoclonality of smooth muscle cells in human atherosclerosis. *Am. J. Pathol.*, **151**, 697–705.

4 Campbell, G.R. and Campbell, J.H. (1990) The phenotypes of smooth muscle expressed in human atheroma. *Ann. N. Y. Acad. Sci.*, **598**, 143–158.

5 Thyberg, J., Blomgren, K., Hedin, U., and Dryjski, M. (1995) Phenotypic modulation of smooth muscle cells during the formation of neointimal thickenings in the rat carotid artery after balloon injury: an electron-microscopic and stereological study. *Cell Tissue Res.*, **281**, 421–433.

6 Sartore, S., Franch, R., Roelofs, M., and Chiavegato, A. (1999) Molecular and cellular phenotypes and their regulation in smooth muscle. *Rev. Physiol. Biochem. Pharmacol.*, **134**, 235–320.

7 Owens, G.K., Kumar, M.S., and Wamhoff, B.R. (2004) Molecular regulation of vascular smooth muscle cell differentiation in development and disease. *Physiol. Rev.*, **84**, 767–801.

8 Rensen, S.S., Doevendans, P.A., and van Eys, G.J. (2007) Regulation and characteristics of vascular smooth muscle cell phenotypic diversity. *Neth. Heart J.*, **15**, 100–108.

9 Kawai-Kowase, K. and Owens, G.K. (2007) Multiple repressor pathways contribute to phenotypic switching of vascular smooth muscle cells. *Am. J. Physiol. Cell. Physiol.*, **292**, C59–C69.

10 Rensen, S.S., Niessen, P.M., van Deursen, J.M., Janssen, B.J., Heijman, E., Hermeling, E., Meens, M., Lie, N., Gijbels, M.J., Strijkers, G.J., Doevendans, P.A., Hofker, M.H., De Mey, J.G., and van Eys, G.J. (2008) Smoothelin-B deficiency results in reduced arterial contractility, hypertension, and cardiac hypertrophy in mice. *Circulation* **118**, 828–836.

11 Hao, H., Gabbiani, G., and Bochaton-Piallat, M.L. (2003) Arterial smooth muscle cell heterogeneity: implications for atherosclerosis and restenosis development. *Arterioscler. Thromb. Vasc. Biol.*, **23**, 1510–1520.

12 Walker, L.N., Bowen-Pope, D.F., Ross, R., and Reidy, M.A. (1986) Production of platelet-derived growth factor-like molecules by cultured arterial smooth muscle cells accompanies proliferation after arterial injury. *Proc. Natl Acad. Sci. U. S. A.*, **83**, 7311–7315.

13 Orlandi, A., Ehrlich, H.P., Ropraz, P., Spagnoli, L.G., and Gabbiani, G. (1994) Rat aortic smooth muscle cells isolated from different layers and at different times after endothelial denudation show distinct biological features in vitro. *Arterioscler. Thromb.*, **14**, 982–989.

14 Bochaton-Piallat, M.L., Ropraz, P., Gabbiani, F., and Gabbiani, G. (1996) Phenotypic heterogeneity of rat arterial smooth muscle cell clones. Implications for the development of experimental intimal thickening. *Arterioscler. Thromb. Vasc. Biol.*, **16**, 815–820.

15 Yan, Z.Q. and Hansson, G.K. (1998) Overexpression of inducible nitric oxide synthase by neointimal smooth muscle cells. *Circ. Res.*, **82**, 21–29.

16 Villaschi, S., Nicosia, R.F., and Smith, M.R. (1994) Isolation of a morphologically and functionally distinct smooth muscle cell type from the intimal aspect of the normal rat aorta. Evidence for smooth muscle cell heterogeneity. *In Vitro Cell Dev. Biol. Anim.*, **30A**, 589–595.

17 Lau, H.K. (1999) Regulation of proteolytic enzymes and inhibitors in two smooth muscle cell phenotypes. *Cardiovasc. Res.*, **43**, 1049–1059.

18 Li, W.G., Miller, F.J., Jr., Brown, M.R., Chatterjee, P., Aylsworth, G.R., Shao, J., Spector, A.A., Oberley, L.W., and Weintraub, N.L. (2000) Enhanced H2O2-induced cytotoxicity in "epithelioid" smooth muscle cells: implications for neointimal regression. *Arterioscler. Thromb. Vasc. Biol.*, **20**, 1473–1479.

19 Ehler, E., Jat, P.S., Noble, M.D., Citi, S., and Draeger, A. (1995) Vascular smooth muscle cells of H-2Kb-tsA58 transgenic mice. Characterization of cell lines with distinct properties. *Circulation*, **92**, 3289–3296.

20 Seifert, R.A., Schwartz, S.M., and Bowen-Pope, D.F. (1984) Developmentally regulated production of platelet-derived growth factor- like molecules. *Nature*, **311**, 669–671.

21 Gordon, D., Mohai, L.G., and Schwartz, S.M. (1986) Induction of polyploidy in cultures of neonatal rat aortic smooth muscle cells. *Circ. Res.*, **59**, 633–644.

22 McCaffrey, T.A., Nicholson, A.C., Szabo, P.E., Weksler, M.E., and Weksler, B.B. (1988) Aging and arteriosclerosis. The increased proliferation of arterial smooth muscle cells isolated from old rats is associated with increased platelet-derived growth factor-like activity. *J. Exp. Med.*, **167**, 163–174.

23 Hültgardh-Nilsson, A., Krondahl, U., Querol-Ferrer, V., and Ringertz, N.R. (1991) Differences in growth factor response in smooth muscle cells isolated from adult and neonatal rat arteries. *Differentiation*, **47**, 99–105.

24 Bochaton-Piallat, M.L., Gabbiani, F., Ropraz, P., and Gabbiani, G. (1993) Age influences the replicative activity and the differentiation features of cultured rat aortic smooth muscle cell populations and clones. *Arterioscler. Thromb. Vasc. Biol.*, **13**, 1449–1455.

25 Cook, C.L., Weiser, M.C., Schwartz, P.E., Jones, C.L., and Majack, R.A. (1994) Developmentally timed expression of an embryonic growth phenotype in vascular smooth muscle cells. *Circ. Res.*, **74**, 189–196.

26 Lemire, J.M., Covin, C.W., White, S., Giachelli, C.M., and Schwartz, S.M. (1994) Characterization of cloned aortic smooth muscle cells from young rats. *Am. J. Pathol.*, **144**, 1068–1081.

27 Bochaton-Piallat, M.L., Gabbiani, F., Ropraz, P., and Gabbiani, G. (1992) Cultured aortic smooth muscle cells from newborn and adult rats show distinct cytoskeletal features. *Differentiation*, **49**, 175–185.

28 Hariri, R.J., Alonso, D.R., Hajjar, D.P., Coletti, D., and Weksler, M.E. (1986) Aging and arteriosclerosis. I. Development of myointimal hyperplasia after endothelial injury. *J. Exp. Med.*, **164**, 1171–1178.

29 Stemerman, M.B., Weinstein, R., Rowe, J.W., Maciag, T., Fuhro, R., and Gardner, R. (1982) Vascular smooth muscle cell growth kinetics *in vivo* in aged rats. *Proc. Natl Acad. Sci. U. S. A.*, **79**, 3863–3866.

30 Chen, L., Daum, G., Fischer, J.W., Hawkins, S., Bochaton-Piallat, M.L., Gabbiani, G., and Clowes, A.W. (2000) Loss of expression of the β subunit of soluble guanylyl cyclase prevents nitric oxide-mediated inhibition of DNA synthesis in smooth muscle cells of old rats. *Circ. Res.*, **86**, 520–525.

31 Holifield, B., Helgason, T., Jemelka, S., Taylor, A., Navran, S., Allen, J., and Seidel, C. (1996) Differentiated vascular myocytes: are they involved in neointimal formation? *J. Clin. Invest.*, **97**, 814–825.

32 Frid, M.G., Aldashev, A.A., Dempsey, E.C., and Stenmark, K.R. (1997) Smooth muscle cells isolated from discrete compartments of the mature vascular media exhibit unique phenotypes and distinct growth capabilities. *Circ. Res.*, **81**, 940–952.

33 Hao, H., Ropraz, P., Verin, V., Camenzind, E., Geinoz, A., Pepper, M.S., Gabbiani, G., and Bochaton-Piallat, M.L. (2002) Heterogeneity of smooth muscle cell populations cultured from pig coronary artery. *Arterioscler. Thromb. Vasc. Biol.*, **22**, 1093–1099.

34 Orekhov, A.N., Karpova, II, Tertov, V.V., Rudchenko, S.A., Andreeva, E.R., Krushinsky, A.V., and Smirnov, V.N. (1984) Cellular composition of atherosclerotic and uninvolved human aortic subendothelial intima. Light-microscopic study of dissociated aortic cells. *Am. J. Pathol.*, **115**, 17–24.

35 Orekhov, A.N., Krushinsky, A.V., Andreeva, E.R., Repin, V.S., and Smirnov, V.N. (1986) Adult human aortic cells in primary culture: heterogeneity in shape. *Heart Vessels*, **2**, 193–201.

36 Björkerud, S. (1985) Cultivated human arterial smooth muscle displays heterogeneous pattern of growth and phenotypic variation. *Lab. Invest.*, **53**, 303–310.

37 Babaev, V.R., Antonov, A.S., Domogatsky, S.P., and Kazantseva, I.A. (1992) Phenotype related changes of intimal smooth muscle cells from human aorta in primary culture. *Atherosclerosis*, **96**, 189–202.

38 Benzakour, O., Kanthou, C., Kanse, S.M., Scully, M.F., Kakkar, V.V., and Cooper, D.N. (1996) Evidence for cultured human vascular smooth muscle cell heterogeneity: isolation of clonal cells and study of their growth characteristics. *Thromb. Haemost.*, **75**, 854–858.

39 Bonin, L.R., Madden, K., Shera, K., Ihle, J., Matthews, N., Aziz, S., Perez-Reyes, N., McDougall, J.K., and Conroy, S.C. (1999) Generation and characterization of human smooth muscle cell lines derived from atherosclerotic plaque.

Arterioscler. Thromb. Vasc. Biol., **19**, 575–587.

40 Llorente-Cortes, V., Martinez-Gonzalez, J., and Badimon, L. (1999) Differential cholesteryl ester accumulation in two human vascular smooth muscle cell subpopulations exposed to aggregated LDL: effect of PDGF-stimulation and HMG-CoA reductase inhibition. *Atherosclerosis*, **144**, 335–342.

41 Li, S., Fan, Y.S., Chow, L.H., Van Den Diepstraten, C., van Der Veer, E., Sims, S.M., and Pickering, J.G. (2001) Innate diversity of adult human arterial smooth muscle cells: cloning of distinct subtypes from the internal thoracic artery. *Circ. Res.*, **89**, 517–525.

42 Upton, P.D., Wharton, J., Davie, N., Ghatei, M.A., Smith, D.M., and Morrell, N.W. (2001) Differential adrenomedullin release and endothelin receptor expression in distinct subpopulations of human airway smooth-muscle cells. *Am. J. Respir. Cell Mol. Biol.*, **25**, 316–325.

43 Blindt, R., Vogt, F., Lamby, D., Zeiffer, U., Krott, N., Hilger-Eversheim, K., Hanrath, P., vom Dahl, J., and Bosserhoff, A.K. (2002) Characterization of differential gene expression in quiescent and invasive human arterial smooth muscle cells. *J. Vasc. Res.*, **39**, 340–352.

44 Wang, Z., Rao, P.J., Shillcutt, S.D., and Newman, W.H. (2003) Phenotypic diversity of smooth muscle cells isolated from human intracranial basilar artery. *Neurosci. Lett.*, **351**, 1–4.

45 Argmann, C.A., Sawyez, C.G., Li, S., Nong, Z., Hegele, R.A., Pickering, J.G., and Huff, M.W. (2004) Human smooth muscle cell subpopulations differentially accumulate cholesteryl ester when exposed to native and oxidized lipoproteins. *Arterioscler. Thromb. Vasc. Biol.*, **24**, 1290–1296.

46 Schwartz, S.M., Foy, L., Bowen-Pope, D.F., and Ross, R. (1990) Derivation and properties of platelet-derived growth factor-independent rat smooth muscle cells. *Am. J. Pathol.*, **136**, 1417–1428.

47 McCaffrey, T.A. and Falcone, D.J. (1993) Evidence for an age-related dysfunction in the antiproliferative response to transforming growth factor-β?in vascular

smooth muscle cells. *Mol. Biol. Cell*, **4**, 315–322.

48 Topouzis, S. and Majesky, M.W. (1996) Smooth muscle lineage diversity in the chick embryo. Two types of aortic smooth muscle cell differ in growth and receptor-mediated transcriptional responses to transforming growth factor-β. *Dev. Biol.*, **178**, 430–445.

49 Orlandi, A., Francesconi, A., Cocchia, D., Corsini, A., and Spagnoli, L.G. (2001) Phenotypic heterogeneity influences apoptotic susceptibility to retinoic acid and cis-platinum of rat arterial smooth muscle cells *in vitro*: implications for the evolution of experimental intimal thickening. *Arterioscler. Thromb. Vasc. Biol.*, **21**, 1118–1123.

50 Li, Z., Cheng, H., Lederer, W.J., Froehlich, J., and Lakatta, E.G. (1997) Enhanced proliferation and migration and altered cytoskeletal proteins in early passage smooth muscle cells from young and old rat aortic explants. *Exp. Mol. Pathol.*, **64**, 1–11.

51 Bochaton-Piallat, M.-L., Gabbiani, G., and Pepper, M.S. (1998) Plasminogen activator expression in rat arterial smooth muscle cells depends on their phenotype and is modulated by cytokines. *Circ. Res.*, **82**, 1086–1093.

52 Matschke, K., Babiychuk, E.B., Monastyr-skaya, K., and Draeger, A. (2006) Phenotypic conversion leads to structural and functional changes of smooth muscle sarcolemma. *Exp. Cell. Res.*, **312**, 3495–3503.

53 Yan, Z.Q., Sirsjo, A., Bochaton-Piallat, M.L., Gabbiani, G., and Hansson, G.K. (1999) Augmented expression of inducible NO synthase in vascular smooth muscle cells during aging is associated with enhanced NF-κB activation. *Arterioscler. Thromb. Vasc. Biol.*, **19**, 2854–2862.

54 Neylon, C.B., Avdonin, P.V., Dilley, R.J., Larsen, M.A., Tkachuk, V.A., and Bobik, A. (1994) Different electrical responses to vasoactive agonists in morphologically distinct smooth muscle cell types. *Circ. Res.*, **75**, 733–741.

55 Neylon, C.B., Lang, R.J., Fu, Y., Bobik, A., and Reinhart, P.H. (1999) Molecular cloning and characterization of the intermediate-conductance Ca^{2+}-activated K$^+$ channel in vascular smooth muscle: relationship between K(Ca) channel diversity and smooth muscle cell function. *Circ. Res.*, **85**, e33–e43.

56 Wamhoff, B.R., Bowles, D.K., and Owens, G.K. (2006) Excitation-transcrip-tion coupling in arterial smooth muscle. *Circ. Res.*, **98**, 868–878.

57 Campbell, J.H., Reardon, M.F., Campbell, G.R., and Nestel, P.J. (1985) Metabolism of atherogenic lipoproteins by smooth muscle cells of different phenotype in culture. *Arteriosclerosis*, **5**, 318–328.

58 Parlavecchia, M., Skalli, O., and Gabbiani, G. (1989) LDL accumulation in cultured rat aortic smooth muscle cells with different cytoskeletal phenotypes. *J. Vasc. Med. Biol.*, **1**, 308–313.

59 Thyberg, J. (2002) Caveolae and cholesterol distribution in vascular smooth muscle cells of different phenotypes. *J. Histochem. Cytochem.*, **50**, 185–195.

60 Dusserre, E., Bourdillon, M.C., Pulcini, T., and Berthezene, F. (1994) Decrease in high density lipoprotein binding sites is associated with decrease in intracellular cholesterol efflux in dedifferentiated aortic smooth muscle cells. *Biochim. Biophys. Acta*, **1212**, 235–244.

61 Cremona, O., Muda, M., Appel, R.D., Frutiger, S., Hughes, G.J., Hochstrasser, D.F., Geinoz, A., and Gabbiani, G. (1995) Differential protein expression in aortic smooth muscle cells cultured from newborn and aged rats. *Exp. Cell Res.*, **217**, 280–287.

62 Neuville, P., Geinoz, A., Benzonana, G., Redard, M., Gabbiani, F., Ropraz, P., and Gabbiani, G. (1997) Cellular retinol-binding protein-1 is expressed by distinct subsets of rat arterial smooth muscle cell *in vitro* and in vivo. *Am. J. Pathol.*, **150**, 509–521.

63 Brisset, A.C., Hao, H., Camenzind, E., Bacchetta, M., Geinoz, A., Sanchez, J.C., Chaponnier, C., Gabbiani, G., and Bochaton-Piallat, M.L. (2007) Intimal smooth muscle cells of porcine and human coronary artery express S100A4, a marker of the rhomboid phenotype in vitro. *Circ. Res.*, **100**, 1055–1062.

64 Donato, R. (2007) RAGE: a single receptor for several ligands and different cellular responses: the case of certain S100 proteins. *Curr. Mol. Med.*, **7**, 711–724.

65 Heizmann, C.W., Ackermann, G.E., and Galichet, A. (2007) Pathologies involving the S100 proteins and RAGE. *Subcell. Biochem.*, **45**, 93–138.

66 Adams, L.D., Lemire, J.M., and Schwartz, S.M. (1999) A systematic analysis of 40 random genes in cultured vascular smooth muscle subtypes reveals a heterogeneity of gene expression and identifies the tight junction gene zonula occludens 2 as a marker of epithelioid "pup" smooth muscle cells and a participant in carotid neointimal formation. *Arterioscler. Thromb. Vasc. Biol.*, **19**, 2600–2608.

67 Jahn, L., Kreuzer, J., von Hodenberg, E., Kubler, W., Franke, W.W., Allenberg, J., and Izumo, S. (1993) Cytokeratins 8 and 18 in smooth muscle cells. Detection in human coronary artery, peripheral vascular, and vein graft disease and in transplantation-associated arteriosclerosis. *Arterioscler. Thromb.*, **13**, 1631–1639.

68 Haefliger, J.A., Nicod, P., and Meda, P. (2004) Contribution of connexins to the function of the vascular wall. *Cardiovasc. Res.*, **62**, 345–356.

69 Chadjichristos, C.E. and Kwak, B.R. (2007) Connexins: new genes in atherosclerosis. *Ann. Med.*, **39**, 402–411.

70 van Kempen, M.J., ten Velde, I., Wessels, A., Oosthoek, P.W., Gros, D., Jongsma, H.J., Moorman, A.F., and Lamers, W.H. (1995) Differential connexin distribution accommodates cardiac function in different species. *Microsc. Res. Tech.*, **31**, 420–436.

71 Chadjichristos, C.E., Morel, S., Derouette, J.P., Sutter, E., Roth, I., Brisset, A.C., Bochaton-Piallat, M.L., and Kwak, B.R. (2008) Targeting connexin 43 prevents platelet-derived growth factor-BB-induced phenotypic change in porcine coronary artery smooth muscle cells. *Circ. Res.*, **102**, 653–660.

72 Gadeau, A.P., Campan, M., Millet, D., Candresse, T., and Desgranges, C. (1993) Osteopontin overexpression is associated with arterial smooth muscle cell proliferation in vitro. *Arterioscler. Thromb.*, **13**, 120–125.

73 Giachelli, C.M., Bae, N., Almeida, M., Denhardt, D.T., Alpers, C.E., and Schwartz, S.M. (1993) Osteopontin is elevated during neointima formation in rat arteries and is a novel component of human atherosclerotic plaques. *J. Clin. Invest.*, **92**, 1686–1696.

74 Shanahan, C.M., Weissberg, P.L., and Metcalfe, J.C. (1993) Isolation of gene markers of differentiated and proliferating vascular smooth muscle cells. *Circ. Res.*, **73**, 193–204.

75 Giachelli, C.M., Speer, M.Y., Li, X., Rajachar, R.M., and Yang, H. (2005) Regulation of vascular calcification: roles of phosphate and osteopontin. *Circ. Res.*, **96**, 717–722.

76 Giachelli, C.M., Majesky, M.W., and Schwartz, S.M. (1991) Developmentally regulated cytochrome P-450IA1 expression in cultured rat vascular smooth muscle cells. *J. Biol. Chem.*, **266**, 3981–3986.

77 Frid, M.G., Aldashev, A.A., Nemenoff, R.A., Higashito, R., Westcott, J.Y., and Stenmark, K.R. (1999) Subendothelial cells from normal bovine arteries exhibit autonomous growth and constitutively activated intracellular signaling. *Arterioscler. Thromb. Vasc. Biol.*, **19**, 2884–2893.

78 Au, Y.P., Kenagy, R.D., Clowes, M.M., and Clowes, A.W. (1993) Mechanisms of inhibition by heparin of vascular smooth muscle cell proliferation and migration. *Haemostasis*, **23** (Suppl. 1), 177–182.

79 Neuville, P., Yan, Z., Gidlof, A., Pepper, M.S., Hansson, G.K., Gabbiani, G., and Sirsjo, A. (1999) Retinoic acid regulates arterial smooth muscle cell proliferation and phenotypic features *in vivo* and *in vitro* through an RARα-dependent signaling pathway. *Arterioscler. Thromb. Vasc. Biol.*, **19**, 1430–1436.

80 Prunotto, M., Bacchetta, M., Jayaraman, S., Galloni, M., Van Eys, G., Gabbiani, G., and Bochaton-Piallat, M.L. (2007) Cytostatic drugs differentially affect

phenotypic features of porcine coronary artery smooth muscle cell populations. *FEBS Lett.*, **581**, 5847–5851.

81 Bochaton-Piallat, M.L., Clowes, A.W., Clowes, M.M., Fischer, J.W., Redard, M., Gabbiani, F., and Gabbiani, G. (2001) Cultured arterial smooth muscle cells maintain distinct phenotypes when implanted into carotid artery. *Arterioscler. Thromb. Vasc. Biol.*, **21**, 949–954.

17
Platelets: Their Role in Atherogenesis and Thrombosis in Coronary Artery Disease

Matthew T. Harper, Lucy MacCarthy-Morrogh, Matthew L. Jones, Olga Konopatskaya, and Alastair W. Poole

17.1
Introduction

Platelets, the smallest of the blood cells, have an important role in hemostasis: the process of maintaining the integrity of the cardiovascular system. Platelets adhere to sites of vascular damage, aggregate to form a platelet plug, and enhance coagulation to form a fibrin-based clot [1, 2]. Arterial thrombosis, however, is the pathologic formation of a platelet-rich mass, commonly at the site of an atherosclerotic plaque in a damaged artery. Rupture or erosion of the endothelial cell layer of the plaque exposes subendothelial adhesive proteins that result in platelet adhesion and thrombus formation [3, 4]. The importance of platelets in arterial thrombosis is underlined by the benefits of antiplatelet agents, such as aspirin and clopidogrel, in the treatment and prevention of thrombosis in coronary artery disease.

Platelets may also have a key role in atherogenesis in addition to their critical role in thrombus formation. Atherogenesis has a major inflammatory component and platelets both respond to, and secrete, chemokines. Moreover, platelets are able to bind to "inflamed" endothelial cells, and chemokines found in platelets recruit other inflammatory cells to the inflamed vessel wall [5].

In this chapter, we will consider the roles of platelets in atherogenesis and arterial thrombosis. Central to the function of platelets are their dual characteristics as adhesive and secretory cells.

17.2
Platelets

Platelets are small (2–5 μm in diameter), anucleate cells derived from megakaryocytes in the bone marrow. Megakaryocytes are derived from pluripotent stem cells and display the unique property of multiple DNA replications without cell division

Atherosclerosis: Molecular and Cellular Mechanisms. Edited by Sarah Jane George and Jason Johnson
Copyright © 2010 WILEY-VCH Verlag GmbH & Co. KGaA, Weinheim
ISBN: 978-3-527-32448-4

(endomitosis), leading to polyploidization with functional gene amplification. In humans, approximately 1×10^8 platelets are produced daily from megakaryocytes. Each platelet has a lifespan of 7–10 days in the circulation. Circulating platelets are maintained in an unactivated state by inhibitory factors released by the endothelium, such as prostacyclin (prostaglandin I_2; PGI_2) and nitric oxide (NO), but they adhere to subendothelial extracellular matrix (ECM) proteins, such as collagen, and von Willebrand factor (vWF), that are exposed on vessel damage through specific receptors on the platelet surface.

In dynamic flow models, it is clear that platelet interaction with ECM proteins is a coordinated multistep process involving tethering, rolling, and adhesion. The process is similar for the platelet–endothelial cell interaction, even under high shear stress *in vivo*, and involves receptors, including selectins, glycoproteins, and integrins [5]. Ligation and activation of these receptors result in cell adhesion and the relay of receptor-specific signals to both platelet and the adhesive endothelial cell. Activated platelets secrete numerous soluble messengers and proteins from their granules (α-granules, dense granules and lysosomes) and synthesize thromboxane A_2 (TXA_2), which recruits more platelets or inflammatory cells [3, 4]. In addition, the surface of the activated platelet acts as a nidus for accumulation of coagulation factors, which results in the formation of a stable clot that is specifically limited to the site of injury.

17.2.1
Platelet Secretory Organelles

In order to perform the multiple functions that platelets undertake, they release a range of components from their secretory granules: α-granules, dense granules, and lysosomes. The structure of the platelet and secretory granules is shown in Figure 17.1. Secreted granule contents amplify and sustain the initial platelet activation, and recruit and activate other cells such as monocytes. The known contents of specific granules are summarized in Table 17.1.

17.2.1.1 α-Granules
α-Granules are round or oval in shape, 200–500 nm in diameter, and are the most numerous of the platelet organelles (an average 80 per platelet). The interior structure is divided into compartments: submembrane (containing vWF, arranged in tube-like structures), peripheral (less electron-dense), and central (more dense). α-Granules contain procoagulant molecules, fibrinolytic regulators, growth factors, chemokines, immunologic modulators, adhesion molecules, and other proteins. P-selectin in the α-granule membrane is translocated to the plasma membrane during secretion.

The development of α-granules occurs during megakaryocyte maturation. The synthesis of granule content starts in the rough endoplasmic reticulum and proteins are sorted in the *trans*-Golgi network. Some proteins are taken up into platelets from the plasma, via coated pits and the process of receptor-mediated endocytosis. Fibrinogen, for example, is taken up by $\alpha_{IIb}\beta_3$.

A

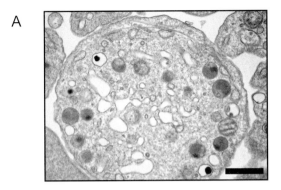

B

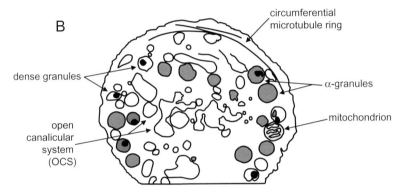

Figure 17.1 Platelet structure. (a) Transmission electron micrograph of a resting platelet. Scale bar represents 0.5 μm. (b) Schematic diagram of the same platelet, indicating some visible organelles.

17.2.1.2 Dense Granules

The smallest platelet granules (mean diameter 150 nm), dense granules are characterized by the presence of an electron-opaque spherical body, which is separated from the enclosing membrane by what appears to be an empty space. Dense granules are the storage sites for adenosine nucleotides (ADP : ATP ratio of 1.5), serotonin (5-HT), magnesium, and calcium. The latter forms complexes with serotonin and pyrophosphate, producing electron opacity. Dense granules are acidic and contain membrane proteins also found in lysosomes, such as CD63 (LAMP3) and LAMP2. Of note also are the small GTP-binding proteins, Ral and Rab27, responsible for exocytosis, as well as P-selectin, which is also present on the dense granule membrane, though at lower levels than in α-granules.

17.2.1.3 Lysosomes

Lysosomes are formed at early stages of megakaryocyte maturation. They contain a variety of digestive enzymes, which are thought be released only when platelets are exposed to a maximum stimulation *in vitro*, although there is very little evidence of lysosomal exocytosis *in vivo* except in one study by Ciferri and colleagues

Table 17.1 Platelet granule contents.

Dense granule – "Pro-aggregating factors"

Nucleotides	Adenine: ATP, ADP
	Guanine: GTP, GDP
Amines	Serotonin (5-HT)
	Histamine
Bivalent cations	Calcium
	Magnesium
	Pyrophosphate

α-Granule – "Adhesion and repairing factors"

Proteoglycans	Platelet-specific: β-TG, PF4
	Serglycin, HRGP
	β-TG Ag molecules: PBP, CTAP-III, NAP-2
Adhesive glycoproteins	Fibronectin, vitronectin, vWF, thrombospondin
Hemostasis factors and cofactors	Fibrinogen, Factor V, VII, XI, XIII
	Kininogens, protein S, plasminogen
Cellular mitogens	PDGF, TGFβ, ECGF, EGF, VEGF/VPF, IGF, IL-Iβ
Protease inhibitors	α_2-Macroglobulin, α_2-antitrypsin, PDCI
	α_2-Antiplasmin, PAI1, TFPI, α_1-PI, PIXI
	PN-2/APP, C1 inhibitor
Miscellaneous	Immunoglobulins: IgG, IgA, IgM
	Albumin, GPIa/multimerin

Lysosome – "Clearing factors"

Acid proteases	Cathepsins D, E
	Carboxypeptidases (A, B)
	Proline carboxypeptidase
	Collagenase
	Acid phosphatase
	Arylsulfatase
	Glycohydrolases: Heparinase, β-*N*-acetyl-glucosaminidase, β-glucuronidase, β-galactosidase, β-glycerophosphatase, α-D-glucosidase, β-D-glucosidase, α-L-fucosidase, β-D-fucosidase, α-L-arabinosidase, β-D-mannosidase

Adapted from Rendu and Brohard-Bohn [6].

[7]. There has been no significant role of lysosomes in platelet function described so far.

17.2.1.4 Granule Heterogeneity

α-Granules may be organized into separate and distinct subsets of granules. Fibrinogen and vWF are differentially localized in resting human platelets, most probably in different granules [8]. VEGF and endostatin also reside in separate α-granules [9], and thrombospondin-1, basic fibroblast growth factor, platelet factor 4, and placental growth factor may also be segregated into discrete compartments. Interestingly, two distinct classes of granules may be differentially secreted, with pro- or anti-angiogenic factors released depending on which receptor is activated [10], which suggests that other α-granule contents may be differentially secreted.

17.3
Role of Platelets in Atheroprogression

17.3.1
Platelets Adhere to Activated Endothelium

Platelets do not normally adhere to the endothelium. Indeed, the endothelium has been recognized for many years as a source of factors that potently inhibit platelet activity, such as PGI_2 and NO, as well as expressing inhibitory adhesion molecules such as PECAM-1. However, under conditions where the endothelial cells are dysfunctional, activated, or inflamed, the vessel wall becomes adhesive to platelets, even in the absence of endothelial denudation. *In vitro*, platelets adhere to a human umbilical vein endothelial cell layer after "stimulation" either with herpes virus infection or interleukin-1 (IL-1) [11, 12].

Platelet–endothelial cell interaction has also been studied *in vivo* in apolipoprotein E–deficient ($Apoe^{-/-}$) mice, where disruption of the endogenous *Apoe* gene causes animals to exhibit five times normal serum plasma cholesterol levels when fed a high-fat diet, leading to spontaneous atherosclerotic lesions. In this model system, platelets adhere to intact endothelium of lesion-prone sites such as the carotid artery bifurcation before either leukocyte invasion or appearance of atherosclerotic lesions [13]. Similarly, in hypercholesteremic rabbits, platelets were observed to bind to predicted sites of atherosclerotic plaque formation before lesions were detected [14].

17.3.1.1 Tethering and Rolling

The first tentative interaction with the endothelium is mediated principally by endothelial cell–expressed P-selectin (CD62P) and platelet-expressed counterreceptors, including P-selectin glycoprotein ligand-1 (PSGL-1) and GPIb (see Figure 17.1). Whereas the initial tethering of platelets on P-selectin is mediated primarily by platelet PSGL-1, in the rolling phase of contact GPIb also plays an important role. GPIb is a component of the multimeric receptor for vWF, GPIb-IX-V, found on inactive platelet surfaces. This interaction leads to some platelet activation [15–17]. Inflammatory stimuli induce translocation of endothelial P-selectin from Weibel–Palade bodies to the plasma membrane. This surface-exposed endothelial P-selectin can mediate platelet rolling in both arterioles and venules in acute inflammatory processes [15, 18]. Similarly E-selectin, also expressed on inflamed endothelial cells, allows a loose contact with platelets [18]. Although P-selectin is also expressed in platelets, it is not required for rolling as both P-selectin and E-selectin null platelets roll on endothelial cell layers to the same degree as wild-type platelets [17]. Since P-selectin surface exposure on platelets requires platelet activation and α-granule secretion, this demonstrates that platelet rolling does not require platelet activation.

17.3.1.2 Stable Adhesion

Stable adhesion may not be achieved by P-selectin and PSGL-1 or GPIb alone, because of the low affinity and rapid reversibility of these interactions. However, intravital microscopy studies have shown that firm adhesion between platelets and inflamed but intact endothelium does occur even under high shear stress *in vivo* [15, 17–20]. This is mediated by platelet integrin $\alpha_{IIb}\beta_3$ (GPIIb-IIIa) and endothelial $\alpha_v\beta_3$, and is dependent on soluble fibrinogen, which bridges the interaction between the two integrins. *In vivo*, firm platelet adhesion to the endothelium can be inhibited by blocking antibodies to $\alpha_{IIb}\beta_3$. Furthermore, platelets defective in, or lacking $\alpha_{IIb}\beta_3$, do not firmly adhere to activated endothelial cells [19].

17.3.2
Consequences of Platelet Adhesion: Endothelial Cell and Platelet Activation, Leukocyte Recruitment, and Monocyte Differentiation

The process of adhesion itself is not a passive event, but adherent platelets are activated and release, or expose on their surface, multiple potent inflammatory and vasoactive substances, causing further endothelial cell activation [5]. Platelets release adhesion molecules, growth factors, chemokines (such as RANTES (CCL5) and platelet factor 4 (PF4, CXCL4)), cytokines (such as IL-1β and CD40 ligand) and coagulation factors, from their three types of granules.

Interleukin-1β (IL-1β), expressed on the surface of activated platelets, leads endothelial cells to secrete IL-6 and IL-8, and enhances endothelial secretion of monocyte chemoattractant protein-1 (MCP-1) [5], increasing monocyte recruitment to the inflamed endothelium. In addition, expression of ICAM-1 and $\alpha_v\beta_3$ integrin on the surface of endothelial cells is also increased, enhancing neutrophil and monocyte adhesion to the endothelium.

Finally, release of multiple chemokines, such as RANTES and CD40 ligand (CD40L), from platelets induces further chemokine release from various cells of the blood vessel wall. Some of these chemokines, including RANTES, further enhance platelet activation and induce monocyte recruitment to the plaque [21]. PF4, highly and specifically expressed in platelets, is secreted from α-granules by

Table 17.2 Effect of platelet-secreted factors on endothelial cells (ECs) and atherogenesis.

Platelet factor	Effect on ECs	Effect on atherogenesis
IL-1β	IL-6 and IL-8 secretion [23]	Neutrophil and monocyte chemoattractants
	NF-κB-dependent MCP-1 secretion [24, 25]	Monocyte recruitment to inflamed tissue and atherosclerosis [21, 26]
	Increased surface expression of ICAM-1 and $\alpha_\omega\beta_3$ [79]	Neutrophils and monocytes adhesion
	MCP-1 secretion	Monocyte adhesion and neointima formation [27]
CD40L	Increase release of IL-8 (CXCL8) and MCP-1 [28]	Monocytes and neutrophil chemoattractants
	Increase expression of adhesion receptors E-selectin, VCAM-1 and ICAM-1 [28]	Adhesion of leukocytes
	Tissue factor expression [29]	Thrombosis
	Expression and release of MMP-9 and uPAR [30]	Destruction and remodeling of inflamed tissue by these matrix degrading enzymes
MMP-2		Matrix degradation [31]
RANTES (CCL5)		Triggers monocyte arrest on inflamed and atherosclerotic endothelium [21]
PF-4 (CXCL4)		Possible neutrophil and monocyte chemoattractant [32]
		Monocyte survival and differentiation into macrophages [33]
		Lipoprotein retention [34].
CXCL-1 (GRO1)		Monocyte arrest [35]

activated platelets. PF4 is a monocyte chemoattractant, causes their differentiation into macrophages, and subsequently induces macrophages to take up greater quantities of oxidized LDL [22]. The actions of these agents, and other major ones released by activated platelets upon interaction with inflamed endothelial cells, are summarized in Table 17.2.

17.3.3
Platelet Interactions with Leukocytes

There is good evidence that platelets play an important role in recruiting other inflammatory blood cells to the site of atherogenesis. They do this in at least two

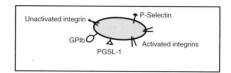

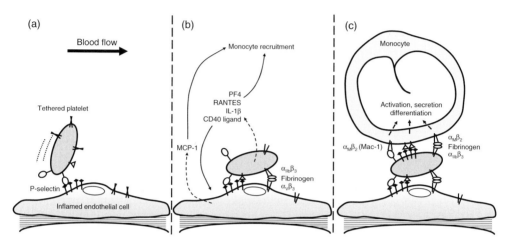

Figure 17.2 Platelets in atherogenesis. (a) Platelets weakly tether to inflamed endothelial cells. (b) Strong adhesion is mediated by platelet integrin $\alpha_{IIb}\beta_3$ and endothelial integrin $\alpha_v\beta_3$ via soluble fibrinogen. Activated platelets release or surface-express numerous cytokines that enhance endothelial activation and recruit monocytes. (c) Monocytes adhere to platelets through several receptors, including monocytic integrin $\alpha_M\beta_2$ (Mac-1).

ways (Figure 17.2). First, platelet-mediated endothelial cell activation recruits leukocytes to inflamed endothelium by releasing multiple chemoattractants. Second, platelets directly bind leukocytes and thus they can act as bridging cells between the endothelium and plaque-forming cells. Leukocytes initially tether via PSGL-1 to P-selectin on the surface of activated platelets [36, 37]. Subsequent firm adhesion is mediated by binding of Mac-1 ($\alpha_M\beta_2$) on monocytes to platelet GPIb, and to $\alpha_{IIb}\beta_3$ via fibrinogen [38]. Other platelet surface receptors involved in the interaction between platelets and the Mac-1 receptor include JAM-3 [39] and ICAM-2 [40]. These interactions are summarized in Figure 17.2. For more details in reviews on this subject, see [5, 41, 42].

There is evidence that this translates into human atherogenesis; however, the data are much more limited. Although there has been some evidence provided that platelet activation is enhanced in coronary artery disease [43], probably more convincing is the finding that the highly platelet-specific PF-4 is found in atherosclerotic plaques [44]. Clearly however, more work is required to follow up the *in vitro* and animal model findings in human patients.

17.4
Role of Platelets in Atherothrombosis

The roles of platelets in thrombosis are much better understood than their suggested roles in atherogenesis. As with atherogenesis, the central platelet functions of adhesion and secretion are essential to thrombus formation at the sites of plaque damage. In addition, activated platelets accelerate coagulation, leading to a platelet-rich fibrin-based clot at the site of vascular damage.

17.4.1
Plaque Rupture Initiates Coagulation and Exposes Platelet Activators

Atherosclerotic plaques are rich in procoagulant proteins and platelet activators that are exposed on plaque rupture. Tissue factor–exposing macrophages in the lipid core initiate the coagulation cascade and thrombin generation [4, 45], and platelets adhere to collagens and become activated, forming a pro-aggregatory and pro-coagulant surface (Figure 17.3). The lipid core also contains a high concentration of lysophosphatidic acid (LPA), which enhances platelet responses to other activators [46–48].

17.4.1.1 Tissue Factor Initiates Coagulation

Plaque rupture brings plasma into contact with tissue factor (TF) and initiates a cascade of serine protease activation. Macrophages are the main source of TF in atherosclerotic plaques, although smooth muscle cells may also provide significant TF activity. Coagulation *in vivo* involves three phases: initiation, amplification, and propagation [49]. During initiation, plasma factor VII is activated by TF. The TF/FVIIa complex activates factors IX and X. FXa activates membrane-bound FV in a Ca^{2+}-dependent manner. Finally, the FXa/FVa ("prothrombinase") complex converts prothrombin (FII) to active thrombin (FIIa), which is essential for stabilization of a growing thrombus by generating fibrin from plasma fibrinogen. This initiation phase of coagulation generates picomolar amounts of thrombin, which then amplifies and propagates coagulation [49]. During the amplification phase, thrombin activates FXI, FVIII, and more FV. FXIa activates more FIX. FVIIIa regulates generation of FXa by FIXa, leading to greater assembly of the prothrombinase complex and more thrombin generation. Thrombin also activates FXIII, a transglutaminase that catalyzes covalent crosslinking of fibrin monomers into a stable mesh. Finally, thrombin is a potent platelet agonist that enhances the pro-coagulant activity of collagen-adhered platelets.

17.4.1.2 Platelets Adhere to Exposed Collagen via Multiple Receptors

Plaque rupture exposes numerous subendothelial matrix proteins to flowing blood. Many of these proteins, in particular collagens, can support platelet adhesion, localizing platelets to the site of vascular injury, and induce activation. Col-

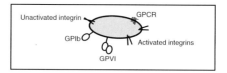

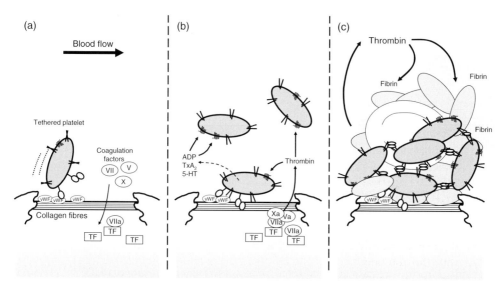

Figure 17.3 Platelets in thrombosis. (a) When the vessel wall is damaged, platelets weakly tether to exposed collagen fibers via von Willebrand factor (vWF) and plasma coagulation factors are activated by exposed tissue factor (TF). (b) Platelets firmly adhere to collagen through multiple receptors (see main text). Activated platelets release several platelet activating agents. Together with thrombin generated by the coagulation cascade, these agents recruit further platelets to the growing thrombus. (c) Activated platelets aggregate via $\alpha_{IIb}\beta_3$ and soluble fibrinogen, and accelerate thrombin generation by exposing a procoagulant surface. Thrombin cleaves fibrinogen into fibrin, forming a fibrin mesh that traps platelets and erythrocytes.

lagens, especially types I and III, have been detected in excised plaques [3]. *In vitro,* adhesion to collagen at shear rates of $>1000\,s^{-1}$ (such as found in arterioles) is almost entirely dependent on vWF, [1]. vWF is likely to be similarly important for platelet adhesion under the higher shear rates of stenotic arteries. vWF-mediated tethering is critically dependent on platelet GPIb, lack of which results in Bernard–Soulier syndrome [50], a congenital bleeding disorder. Experimental disruption of GPIb in mice is associated with a severe bleeding defect. As with platelet adhesion to inflamed endothelial cells, tethering recruits platelets to the site of vascular injury and reduces their velocity but does not lead to their complete arrest, as the GPIb-vWF interaction has a rapid dissociation rate and is unable to mediate stable adhesion under flow [51]. Firm adhesion to the damaged vessel wall requires the participation of other collagen receptors, such as glycoprotein VI (GPVI), integrin $\alpha_2\beta_1$, the fibronectin receptor ($\alpha_5\beta_1$), and integrin $\alpha_{IIb}\beta_3$.

17.4.1.3 GPVI is the Major Signaling Receptor for Collagen Whereas $\alpha_2\beta_1$ Promotes Firm Adhesion on Collagen

The single-transmembrane-domain protein GPVI is a low-affinity collagen receptor. GPVI surface expression and signaling is dependent on its noncovalent association with a disulfide-linked Fc receptor (FcR) γ-chain homodimer [52]. Crosslinking of GPVI induces Src family tyrosine kinase–dependent phosphorylation of the FcRγ ITAM motif. Syk tyrosine kinase is recruited via its SH2 domains to the phosphorylated ITAM, leading to Syk autophosphorylation and also phosphorylation by Src kinases. Downstream signaling from Syk leads to formation of a signal complex around the transmembrane adapter, LAT, and the cytosolic adapters, SLP-76 and Gads. The major effector of this complex is phospholipase Cγ (PLCγ), leading to formation of inositol-1,4,5-trisphosphate (IP_3) and diacylglycerol (DAG), which activate Ca^{2+} release from intracellular Ca^{2+} stores, and protein kinase C (PKC), respectively. In addition, the LAT signal complex activates phosphoinositide 3-kinase γ (PI3Kγ), whose major product, PIP_3, is required for full activation of PLCγ, and regulates multiple additional effectors. Vav is also phosphorylated, leading to activation of the small GTPase, Rac. The result of these multiple signaling pathways are cytoskeletal reorganization, activation of the surface integrins $\alpha_2\beta_1$ and $\alpha_{IIb}\beta_3$, PKC-dependent granule secretion, and loss of plasma membrane asymmetry [2, 52]. These signaling pathways are summarized in Figure 17.4.

The high-affinity collagen receptor $\alpha_2\beta_1$ (CD49b/CD29) reinforces GPVI signaling and is necessary for firm platelet adhesion to collagen under flow *in vitro* [53, 54], making adherent platelets more resistant to shear [51], and may enhance GPVI-dependent signaling. However, mice lacking either α_2 or β_1 do not have altered bleeding times or substantially changed thrombus formation *in vivo* [55], making the role of this integrin uncertain.

17.4.1.4 Activated Platelets Recruit Further Platelets to the Thrombus

Soluble factors, such as ADP and TXA_2 are released by platelets adhered to collagen, and recruit further platelets to the growing thrombus from flowing blood [1]. Platelets express two G protein-coupled receptors (GPCRs) for ADP, namely $P2Y_1$ and $P2Y_{12}$. $P2Y_1$ is coupled by the heterotrimeric G protein, G_q, to activation of PLCβ. As with PLCγ downstream of GPVI, PLCβ activation leads to Ca^{2+} release and increased PKC activity. $P2Y_{12}$ is coupled through G_i to inhibition of adenylyl cyclase, and through the βγ subunits of G_i to PI3K. $P2Y_{12}$ activity is required for thrombus stability and is the target of the commonly used antiplatelet agent clopidogrel [56]. 5-HT is an important vasoconstrictor in addition to its effects on platelets. The 5-HT receptor 5-HT_{2A} is coupled to PLCβ by G_q, in a similar manner to $P2Y_1$.

TXA_2 synthesis from arachidonate-containing phospholipids is catalyzed by the sequential actions of cytosolic phospholipase A_2 ($cPLA_2$), cyclooxygenase (COX),

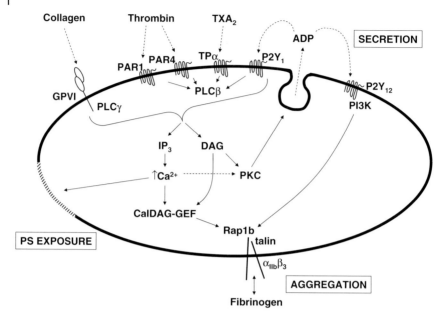

Figure 17.4 Central role of Ca^{2+} and PKC in platelet signaling. Platelets have many receptors, most of which are coupled to phospholipase C (PLC) β or γ, leading to increased $[Ca^{2+}]_i$ and protein kinase C (PKC) activation. Together, this signaling regulates integrin $\alpha_{IIb}\beta_3$ activation, granule secretion and phosphatidylserine (PS) exposure. For more details, see main text.

and thromboxane synthase. $cPLA_2$ activity requires increased cytosolic calcium concentration ($[Ca^{2+}]_i$) and phosphorylation. TPα, the TXA_2 receptor on platelets, is coupled through G_q to PLCβ, and also through $G_{12/13}$ to Rho/ROCK signaling, which is necessary for platelet shape change [57].

In vitro, ADP, 5-HT, and TXA_2 are each capable of inducing platelet aggregation. In addition to these GPCR-dependent signaling pathways, platelets also express at least one ligand-gated cation channel, $P2X_1$, which is activated by secreted ATP. Although stimulation of $P2X_1$ is not sufficient to induce aggregation, this channel may support platelet aggregation by increasing $[Ca^{2+}]_i$, since platelets deficient in $P2X_1$ are protected from thrombosis in small vascular beds [58] and overexpression of $P2X_1$ enhances thrombus formation [59].

17.4.1.5 Thrombin is a Potent Platelet Activator

Thrombin activates protease-activated receptor-1 (PAR1) and PAR4 on human platelets by cleaving a fragment from the N-terminus, revealing a new N-terminal domain that acts a ligand for the receptor [60]. Mouse platelets lack PAR1, but instead express PAR3 and PAR4 [61]. PAR3 has no known signaling role, but may act as a coreceptor to PAR4, and loss of PAR3 protects mice from thrombosis [62]. These GPCRs are coupled to G_q and $G_{12/13}$, leading to a robust increase in $[Ca^{2+}]_i$,

PKC-dependent granule secretion, $\alpha_{IIb}\beta_3$ activation, and loss of plasma membrane asymmetry. Thrombin induces robust aggregation *in vitro*, and platelet activation by thrombin is likely to be central in thrombosis. Although fibrinogen$^{-/-}$ mice form unstable thrombi [63], prothrombin$^{-/-}$ mice have a much greater disruption of hemostasis: half of the embryos die of cardiovascular collapse and almost all the remainder exsanguinate at or just after birth [64, 65]. Simultaneous deletion of the major mouse platelet thrombin receptor (PAR4) and fibrinogen gives a similar hemostatic defect (although not the embryonic lethality) to that seen in the pro-thrombin$^{-/-}$ mice [66].

Furthermore, PAR4$^{-/-}$ mice display much reduced thrombus propagation, despite normal fibrin formation. These data suggest that platelet activation and fibrinogen cleavage are the most important roles of thrombin in hemostasis, and also demonstrate the importance of platelet activation by thrombin in thrombus formation.

17.4.1.6 Inside-Out Activation of $\alpha_{IIb}\beta_3$ is Required for Platelet Aggregation

The common end-point of platelet activation by collagen, thrombin, TXA$_2$, or ADP is activation of integrin $\alpha_{IIb}\beta_3$, a process known as "inside-out signaling". Integrin $\alpha_{IIb}\beta_3$ binds multiple ligands, including vWF, fibronectin and, in particular, fibrinogen. The latter contains multiple $\alpha_{IIb}\beta_3$-binding sites, forming a bridge between two activated platelets and leading to platelet aggregation [51].

The molecular mechanisms of $\alpha_{IIb}\beta_3$ inside-out signaling are partially understood. Rap1b, a small GTPase, is central. Rap1b$^{-/-}$ platelets have deficient aggregation to all major platelet activators [67]. Inactive GDP-bound Rap1b is converted to its active GTP-bound form under regulation by CalDAG-GEF1, itself activated by DAG and a rise in $[Ca^{2+}]_i$. This directly connects PLC activation by various platelet receptors to Rap1b activation. Rap1b regulates interaction between the cytoskeletal protein talin and $\alpha_{IIb}\beta_3$ and activation of the latter. Absence of talin results in impaired platelet aggregation and a bleeding disorder in mice [68, 69]. Kindlin-3 also regulates $\alpha_{IIb}\beta_3$ and its absence leads to severe bleeding [70].

Rap1b activity is maintained by PI3K-dependent signaling downstream of the ADP receptor P2Y$_{12}$, which is required for sustained $\alpha_{IIb}\beta_3$ activation [71, 72]. Inhibition of P2Y$_{12}$ strongly inhibits platelet aggregation and promotes disaggregation [71]. Dense granule secretion is therefore a key factor in stable thrombus development. Since P2Y$_{12}$ is desensitized and internalized following prolonged stimulation [73], continuous P2Y$_{12}$ signaling also requires the receptor's rapid resenitization [74].

17.4.1.7 Activated Platelets Enhance TF-Mediated Thrombin Generation

Collagen and thrombin induce loss of platelet plasma membrane asymmetry. Exposed aminophospholipids, such as phosphatidylserine (PS) in the outer leaflet of platelets promote efficient thrombin generation by providing a surface for assembly of the TF/FVIIa, tenase, and prothrombinase complexes. Loss of plasma membrane asymmetry takes several minutes and requires a sustained rise in $[Ca^{2+}]_i$. If Ca^{2+} is extruded from the cell or sequestered in intracellular storage organelles then membrane asymmetry can be restored. Secreted factors, especially ADP, may be important in sustaining the rise in $[Ca^{2+}]_i$ sufficiently to allow prolonged PS exposure [75].

17.5
Mechanisms of Platelet Granule Secretion

Secretion of platelet granules plays an essential role not only in the late-stage thrombus formation at sties of plaque rupture, but also in the development and progression of the atherosclerotic plaque itself. The highly proinflammatory and pro-thrombotic contents of platelet secretions suggest that exocytosis must be tightly regulated to prevent pathological complications. Although some of the important parts of this regulation are known, it is likely that differential regulation of granule secretion can be achieved by selective activation of specific isoforms of secretory proteins, such as SNAREs, and regulatory proteins, such as PKCs.

17.5.1
Molecular Regulation of Platelet Exocytosis

Exocytosis involves reorganization of the actin structure, the movement of granules into close apposition with the plasma membrane, granule–plasma membrane fusion, and release of the intracellular content. SNARE protein interaction recruits NSF and SNAPs to the 20S fusion complex. ATP hydrolysis by NSF is critical for disassembly of the 20S complex and subsequent bilayer fusion. The SNARE model of docking and fusion of vesicles to their target membranes describes an interaction between a v-SNARE (vesicle membrane proteins) and t-SNARE (target membrane proteins) that determines the specificity of vesicle targeting and docking. Many SNARE and exocytotic core complex proteins have been demonstrated in platelets, including syntaxin 2, syntaxin 4, SNAP-23, SNAP-25, SNAP-29, VAMP2 (vesicle-associated membrane protein 2; synaptobrevin), VAMP3 (cellubrevin), VAMP7, and VAMP8 (endobrevin) [76–85]. Different classes of secretory granules may use discrete N-ethylmaleimide-sensitive fusion (NSF) protein attachment protein receptor (SNARE) proteins to mediate their release in response to different stimuli. SNAP-23 has been shown to regulate lysosome, dense and α-granule release; syntaxin 2 specifically mediates exocytosis from dense granules, but this function also seems to overlap with syntaxin 4 for α-granule and lysosome secretion. Dense granule release from human platelets also requires VAMP3, although VAMP3$^{-/-}$ mice lack a secretion defect, which may indicate functional redundancy between various VAMP isoforms, as well as species specificity in function. A role for VAMP3 and VAMP8 has been shown for α-granule-dependent exocytosis. Recent studies described a strong correlation between single nucleotide polymorphism (SNP) in the human VAMP8 gene and early onset of myocardial infarction. This observation may suggest that altered levels of VAMP8 expression can be associated with increased platelet secretion and prothrombotic ability.

The secretion mechanism in platelets is tightly regulated by members of the Munc protein family (18a, b, c and Munc 13-4), and may also involve a number of Rab GTP-binding proteins, although the function of the latter is not fully established [77, 86].

The important role of PKC in calcium-dependent granule secretion has been known for a long time. Increased $[Ca^{2+}]_i$ is sufficient to induce platelet release reaction but PKC synergizes with Ca^{2+} to amplify secretion [87]. Several PKC targets may be involved. PKC-dependent MARCKS (myriostoylated alanine-rich C-kinase substrate) phosphorylation precedes granule secretion, and pseudosubstrate peptide domain blocks activation of PKCs and phosphorylation of MARCKS. Munc 18c, syntaxin 4, and SNAP-23 are also targets of PKC for phosphorylation. Munc 18c phosphorylation interferes with its interaction with syntaxin 4, and phosphorylation of syntaxin 4 inhibits its binding to SNAP-23 [77]. Moreover, SNAP-23 phosphorylation by PKC parallels the rate of dense and α-granule secretion.

Platelets express multiple PKC isoforms (α, β, δ, θ), although the current understanding of distinct roles for specific PKC isoforms in regulation of secretion remain incomplete, partly owing to the lack of sufficiently isoform-selective inhibitors [88]. The contribution of individual PKC isoforms has been addressed directly in platelets from knockout mouse models. Recent studies revealed a prominent positive role played by PKCα in platelet dense and α-granule exocytosis, as well as dense granule biogenesis. Importantly, PKCα-dependent dense granule release critically determined other physiological platelet responses, such as $\alpha_{IIb}\beta_3$ integrin activation and thrombus formation under flow [89]. A positive regulation for α-granule, but not dense granule, release has been suggested for PKCθ [90]. PKCϵ may also positively regulate granule secretion in mouse platelets [91], but this isoform is not strongly expressed in human platelets.

17.6
Conclusion

Platelets play key roles at the beginning and end of atherosclerosis, by enhancing monocyte recruitment and activation at sites of endothelial inflammation, and by generating potentially fatal thrombi when the subsequent plaque ruptures. There are considerable differences in the actions of platelets in these two events, but also considerable similarities. In both events, platelets bind to an adhesive substrate, become activated, and form a site for further adhesion of platelets, monocytes or coagulation complexes. Central in this process is the capacity of platelets to secrete a vast array of factors that recruit more cells to the site of injury and activate them. Understanding platelet adhesion and secretion, and the signaling pathways that regulate them will therefore be important in reducing the pathologic character of platelets.

References

1 Jackson, S.P., Nesbitt, W., and Kulkarni, S. (2003) Signaling events underlying thrombus formation. *J. Thromb. Haemost.*, **1**, 1602–1612.

2 Sachs, U.J. and Nieswandt, B. (2007) *In vivo* thrombus formation in murine models. *Circ. Res.*, **100** (7), 979–991.

3 Penz, S., Reininger, A., Brandl, R., Goyal, P., Rabie, T., Bernlochner, I., Rother, E., Goetz, C., Engelmann, B., Smethurst, P.A., Ouwehand, W.H., Farndale, R., Nieswandt, B., and Siess, W. (2005) Human atheromatous plaques stimulate thrombus formation by activating platelet glycoprotein VI. *FASEB J.*, **19** (8), 898–909.

4 Rauch, U., Osende, J., Fuster, V., Badimon, J.J., Fayad, Z., and Chesebro, J.H. (2001) Thrombus formation on atherosclerotic plaques: pathogenesis and clinical consequences. *Ann. Intern. Med.*, **134** (3), 224–238.

5 Gawaz, M., Langer, H., and May, A.E. (2005) Platelets in inflammation and atherogenesis. *J. Clin. Invest.*, **115** (12), 3378–3384.

6 Rendu, F. and Brohard-Bohn, B. (2001) The platelet release reaction: granules' constituents, secretion and functions. *Platelets*, **12** (5), 261–273.

7 Ciferri, S., Emiliani, C., Guglielmini, G., Orlacchio, A., Nenci, G.G., and Gresele, P. (2000) Platelets release their lysosomal content *in vivo* in humans upon activation. *Thromb. Haemost.*, **83** (1), 157–164.

8 Sehgal, S. and Storrie, B. (2007) Evidence that differential packaging of the major platelet granule proteins von Willebrand factor and fibrinogen can support their differential release. *J. Thromb. Haemost.*, **5**, 2009–2016.

9 Italiano, J.E., Jr., Richardson, J.L., Patel-Hett, S., Battinelli, E., Zaslavsky, A., Short, S., Ryeom, S., Folkman, J., and Klement, G.L. (2008) Angiogenesis is regulated by a novel mechanism: pro- and antiangiogenic proteins are organized into separate platelet alpha granules and differentially released. *Blood*, **111**, 1227–1233.

10 Ma, L., Perini, R., McKnight, W., Dicay, M., Klein, A., Hollenberg, M.D., and Wallace, J.L. (2005) Proteinase-activated receptors 1 and 4 counter-regulate endostatin and VEGF release from human platelets. *Proc. Natl Acad. Sci. U. S. A.*, **102** (1), 216–220.

11 Bombeli, T., Schwartz, B., and Harlan, J.M. (1998) Adhesion of activated platelets to endothelial cells: evidence for a GPIIbIIIa-dependent bridging mechanism and novel roles for endothelial intercellular adhesion molecule 1 (ICAM-1), alphavbeta3 integrin, and GPIbalpha. *J. Exp. Med.*, **187** (3), 329–339.

12 Etingin, O.R., Silverstein, R., and Hajjar, D.P. (1993) von Willebrand factor mediates platelet adhesion to virally infected endothelial cells. *Proc. Natl Acad. Sci. U. S. A.*, **90** (11), 5153–5156.

13 Massberg, S., Brand, K., Grüner, S., Page, S., Müller, E., Müller, I., Bergmeier, W., Richter, T., Lorenz, M., Konrad, I., Nieswandt, B., and Gawaz, M. (2002) A critical role of platelet adhesion in the initiation of atherosclerotic lesion formation. *J. Exp. Med.*, **196** (7), 887–896.

14 Theilmeier, G., Michiels, C., Spaepen, E., Vreys, I., Collen, D., Vermylen, J., and Hoylaerts, M.F. (2002) Endothelial von Willebrand factor recruits platelets to atherosclerosis-prone sites in response to hypercholesterolemia. *Blood*, **99** (12), 4486–4493.

15 Frenette, P.S., Johnson, R., Hynes, R.O., and Wagner, D.D. (1995) Platelets roll on stimulated endothelium *in vivo*: an interaction mediated by endothelial P-selectin. *Proc. Natl Acad. Sci. U. S. A.*, **92** (16), 7450–7454.

16 Frenette, P.S., Mayados, T., Rayburn, H., Hynes, R.O., and Wagner, D.D. (1996) Susceptibility to infection and altered hematopoiesis in mice deficient in both P- and E-selectins. *Cell*, **84** (4), 563 574.

17 Massberg, S., Enders, G., Leiderer, R., Eisenmenger, S., Vestweber, D., Krombach, F., and Messmer, K. (1998) Platelet-endothelial cell interactions

during ischemia/reperfusion: the role of P-selectin. *Blood*, **92** (2), 507–515.

18 Frenette, P.S., Moyna, C., Hartwell, D.W., Lowe, J.B., Hynes, R.O., and Wagner, D.D. (1998) Platelet-endothelial interactions in inflamed mesenteric venules. *Blood*, **91** (4), 1318–1324.

19 Massberg, S., Enders, G., Matos, F.C., Tomic, L.I., Leiderer, R., Eisenmenger, S., Messmer, K., and Krombach, F. (1999) Fibrinogen deposition at the postischemic vessel wall promotes platelet adhesion during ischemia-reperfusion *in vivo*. *Blood*, **94** (11), 3829–3838.

20 Massberg, S., Grüner, S., Konrad, I., Garcia Arguinzonis, M.I., Eigenthaler, M., Hemler, K., Kersting, J., Schulz, C., Muller, I., Besta, F., Nieswandt, B., Heinzmann, U., Walter, U., and Gawaz, M. (2004) Enhanced *in vivo* platelet adhesion in vasodilator-stimulated phosphoprotein (VASP)-deficient mice. *Blood*, **103** (1), 136–142.

21 von Hundelshausen, P., Weber, K., Huo, Y., Proudfoot, A.E., Nelson, P.J., Ley, K., and Weber, C. (2001) RANTES deposition by platelets triggers monocyte arrest on inflamed and atherosclerotic endothelium. *Circulation*, **103** (13), 1772–1777.

22 Nassar, T., Sachais, B., Akkawi, S., Kowalska, M.A., Bdeir, K., Leitersdorf, E., Hiss, E., Ziporen, L., Aviram, M., Cines, D., Poncz, M., and Higazi, A.A. (2003) Platelet factor 4 enhances the binding of oxidized low-density lipoprotein to vascular wall cells. *J. Biol. Chem.*, **278** (8), 6187–6193.

23 Kaplanski, G., Farnarier, C., Kaplanski, S., Porat, R., Shapiro, L., Bongrand, P., and Dinarello, C.A. (1994) Interleukin-1 induces interleukin-8 secretion from endothelial cells by a juxtacrine mechanism. *Blood*, **84** (12), 4242–4248.

24 Gawaz, M., Brand, K., Dickfeld, T., Pogatsa-Murray, G., Page, S., Bogner, C., Koch, W., Schömig, A., and Neumann, F. (2000) Platelets induce alterations of chemotactic and adhesive properties of endothelial cells mediated through an interleukin-1-dependent mechanism. Implications for atherogenesis. *Atherosclerosis*, **148** (1), 75–85.

25 Gawaz, M., Page, S., Massberg, S., Nothdurfter, C., Weber, M., Fisher, C., Ungerer, M., and Brand, K. (2002) Transient platelet interaction induces MCP-1 production by endothelial cells via I kappa B kinase complex activation. *Thromb. Haemost.*, **88** (2), 307–314.

26 Schober, A., Manka, D., von Hundelshausen, P., Huo, Y., Hanrath, P., Sarembock, I.J., Ley, K., and Weber, C. (2002) Deposition of platelet RANTES triggering monocyte recruitment requires P-selectin and is involved in neointima formation after arterial injury. *Circulation*, **106** (12), 1523–1519.

27 Gawaz, M., Neumann, F., Dickfeld, T., Koch, W., Laugwitz, K.L., Adelsberger, H., Langenbrink, K., Page, S., Neumeier, D., Schömig, A., and Brand, K. (1998) Activated platelets induce monocyte chemotactic protein-1 secretion and surface expression of intercellular adhesion molecule-1 on endothelial cells. *Circulation*, **98** (12), 1164–1171.

28 Henn, V., Slupsky, J., Gräfe, M., Anagnostopoulos, I., Förster, R., Müller-Berghaus, G., and Kroczek, R.A. (1998) CD40 ligand on activated platelets triggers an inflammatory reaction of endothelial cells. *Nature*, **391** (6667), 591–594.

29 Slupsky, J.R., Kalbas, M., Willuweit, A., Henn, V., Kroczek, R.A., and Müller-Berghaus, G. (1998) Activated platelets induce tissue factor expression on human umbilical vein endothelial cells by ligation of CD40. *Thromb. Haemost.*, **80** (6), 1008–1014.

30 May, A.E., Kälsch, T., Massberg, S., Herouy, Y., Schmidt, R., and Gawaz, M. (2002) Engagement of glycoprotein IIb/IIIa (alpha(IIb)beta3) on platelets upregulates CD40L and triggers CD40L-dependent matrix degradation by endothelial cells. *Circulation*, **106** (16), 2111–2117.

31 Sawicki, G., Sanders, E., Murat, J., Miszta-Lane, H., and Radomski, M.W. (1997) Release of gelatinase A during platelet activation mediates aggregation. *Nature*, **386** (6625), 616–619.

32 Deuel, T.F., Senior, R., Chang, D., Griffin, G.L., Heinrikson, R.L., and Kaiser, E.T. (1981) Platelet factor 4 is

chemotactic for neutrophils and mono-cytes. *Proc. Natl Acad. Sci. U. S. A.*, **78** (7), 4584–4587.

33 Scheuerer, B., Ernst, M., Dürrbaum-Landmann, I., Fleischer, J., Grage-Griebenow, E., Brandt, E., Flad, H.D., and Petersen, F. (2000) The CXC-chemokine platelet factor 4 promotes monocyte survival and induces monocyte differentiation into macrophages. *Blood*, **95** (4), 1158–1166.

34 Sachais, B.S., Kuo, A., Nassar, T., Morgan, J., Kariko, K., Williams, K.J., Feldman, M., Aviram, M., Shah, N., Jarett, L., Poncz, M., Cines, D.B., and Higazi, A.A. (2002) Platelet factor 4 binds to low-density lipoprotein receptors and disrupts the endocytic machinery, resulting in retention of low-density lipoprotein on the cell surface. *Blood*, **99** (10), 3613–3622.

35 Smith, D.F., Galkina, E., Ley, K., and Huo, Y. (2005) GRO family chemokines are specialized for monocyte arrest from flow. *Am. J. Physiol. Heart. Circ. Physiol.*, **289** (5), H1976–H1984.

36 Evangelista, V., Manarini, S., Sideri, R., Rotondo, S., Martelli, N., Piccoli, A., Totani, L., Piccardoni, P., Vestweber, D., de Gaetano, G., and Cerletti, C. (1999) Platelet/polymorphonuclear leukocyte interaction: P-selectin triggers protein-tyrosine phosphorylation-dependent CD11b/CD18 adhesion: role of PSGL-1 as a signaling molecule. *Blood*, **93** (3), 876–885.

37 Yang, J., Furie, B.C., and Furie, B. (1999) The biology of P-selectin glycoprotein ligand-1: its role as a selectin counterreceptor in leukocyte-endothelial and leukocyte-platelet interaction. *Thromb. Haemost.*, **81** (1), 1–7.

38 Simon, D.I., Chen, Z., Xu, H., Li, C.Q., Dong, J., McIntire, L.V., Ballantyne, C.M., Zhang, L., Furman, M.I., Berndt, M.C., and López, J.A. (2000) Platelet glycoprotein ibalpha is a counterreceptor for the leukocyte integrin Mac-1 (CD11b/CD18). *J. Exp. Med.*, **192** (2), 193–204.

39 Santoso, S., Sachs, U., Kroll, H., Linder, M., Ruf, A., Preissner, K.T., and Chavakis, T. (2002) The junctional adhesion molecule 3 (JAM-3) on human platelets is a counterreceptor for the leukocyte integrin Mac-1. *J. Exp. Med.*, **196** (5), 679–691.

40 Inoue, O., Suzuki-Inoue, K., Dean, W.L., Frampton, J., and Watson, S.P. (2003) Integrin alpha2beta1 mediates outside-in regulation of platelet spreading on collagen through activation of Src kinases and PLCgamma2. *J. Clin. Invest.*, **94** (3), 1243–1251.

41 May, A.E., Seizer, P., and Gawaz, M. (2008) Platelets: inflammatory firebugs of vascular walls. *Arterioscler. Thromb. Vasc. Biol.*, **28** (3), s5–s10.

42 Lindemann, S., Krämer, B., Seizer, P., and Gawaz, M. (2007) Platelets, inflammation and atherosclerosis. *J. Thromb. Haemost.*, **5** (Suppl. 1), 203–211.

43 Willoughby, S., Holmes, A., and Loscalzo, J. (2002) Platelets and cardiovascular disease. *Eur. J. Cardiovasc. Nurs.*, **1** (4), 273–288.

44 Pitsilos, S., Hunt, J., Mohler, E.R., Prabhakar, A.M., Poncz, M., Dawicki, J., Khalapyan, T.Z., Wolfe, M.L., Fairman, R., Mitchell, M., Carpenter, J., Golden, M.A., Cines, D.B., and Sachais, B.S. (2003) Platelet factor 4 localization in carotid atherosclerotic plaques: correlation with clinical parameters. *Thromb. Haemost.*, **90** (6), 1112–1120.

45 Viles-Gonzalez, J.F. and Badimon, J.J. (2004) Atherothrombosis: the role of tissue factor. *Int. J. Biochem. Cell. Biol.*, **36** (1), 25–30.

46 Siess, W. and Tigyi, G. (2004) Thrombogenic and atherogenic activities of lysophosphatidic acid. *J. Cell. Biochem.*, **92** (6), 1086–1094.

47 Pandey, D., Goyal, P., and Siess, W. (2007) Lysophosphatidic acid stimulation of platelets rapidly induces Ca^{2+}-dependent dephosphorylation of cofilin that is independent of dense granule secretion and aggregation. *Blood Cells Mol. Dis.*, **38** (3), 269–279.

48 Haserück, N., Erl, W., Pandey, D., Tigyi, G., Ohlmann, P., Ravanat, C., Gachet, C., and Siess, W. (2004) The plaque lipid lysophosphatidic acid stimulates platelet activation and platelet-monocyte aggregate formation in whole blood: involvement of P2Y1 and P2Y12 receptors. *Blood*, **103** (7), 2585–2592.

49 Monroe, D.M., and Hoffman, M. (2006) What does it take to make the perfect clot? *Arterioscler. Thromb. Vasc. Biol.*, **26**, 41–84.

50 de la Salle, C., Lanza, F., and Cazenave, J.P. (1995) Biochemical and molecular basis of Bernard-Soulier syndrome: a review. *Nouv. Rev. Fr. Hematol.*, **37** (4), 215–222.

51 Varga-Szabo, D., Pleines, I., and Nieswandt, B. (2008) Cell adhesion mechanisms in platelets. *Arterioscler. Thromb. Vasc. Biol.*, **28** (3), 402–412.

52 Watson, S.P., Auger, J., McCarty, O.J., and Pearce, A.C. (2005) GPVI and integrin alphaIIb beta3 signaling in platelets. *J. Thromb. Haemost.*, **3** (8), 1752–1762.

53 Kuijpers, M.J., Schulte, V., Bergmeier, W., Lindhout, T., Brakebusch, C., Offermanns, S., Fässler, R., Heemskerk, J.W., and Nieswandt, B. (2003) Complementary roles of glycoprotein VI and alpha2beta1 integrin in collagen-induced thrombus formation in flowing whole blood ex vivo. *FASEB J.*, **17** (6), 685–687.

54 Inoue, O., Suzuki-Inoue, K., Dean, W.L., Frampton, J., and Watson, S.P. (2003) Integrin alpha2beta1 mediates outside-in regulation of platelet spreading on collagen through activation of Src kinases and PLCgamma2. *J. Cell Biol.*, **160** (5), 769–780.

55 Grüner, S., Prostredna, M., Schulte, V., Krieg, T., Eckes, B., Brakebusch, C., and Nieswandt, B. (2003) Multiple integrin-ligand interactions synergize in shear-resistant platelet adhesion at sites of arterial injury *in vivo*. *Blood*, **102** (12), 4021–4027.

56 Gachet, C. (2005) The platelet P2 receptors as molecular targets for old and new antiplatelet drugs. *Pharmacol. Ther.*, **108** (2), 180–192.

57 Moers, A., Wettschureck, N., and Offermanns, S. (2004) G13-mediated signaling as a potential target for antiplatelet drugs. *Drug News Perspect.*, **17** (8), 493–498.

58 Hechler, B., Lenain, N., Marchese, P., Vial, C., Heim, V., Freund, M., Cazenave, J.P., Cattaneo, M., Ruggeri, Z.M., Evans, R., and Gachet, C. (2003) A role of the fast ATP-gated P2X1 cation channel in thrombosis of small arteries *in vivo*. *J. Exp. Med.*, **198** (4), 661–667.

59 Oury, C., Kuijpers, M.J., Toth-Zsamboki, E., Bonnefoy, A., Danloy, S., Vreys, I., Feijge, M.A., De Vos, R., Vermylen, J., Heemskerk, J.W., and Hoylaerts, M.F. (2003) Overexpression of the platelet P2X1 ion channel in transgenic mice generates a novel prothrombotic phenotype. *Blood*, **101** (10), 3969–3976.

60 Coughlin, S.R. (2005) Protease-activated receptors in hemostasis, thrombosis and vascular biology. *J. Thromb. Haemost.*, **3** (8), 1800–1814.

61 Kahn, M.L., Zheng, Y., Huang, W., Bigornia, V., Zeng, D., Moff, S., Farese, R.V., Jr., Tam, C., and Coughlin, S.R. (1998) A dual thrombin receptor system for platelet activation. *Nature*, **394** (6694), 690–694.

62 Weiss, E.J., Hamilton, J.R., Lease, K.E., and Coughlin, S.R. (2002) Protection against thrombosis in mice lacking PAR3. *Blood*, **100** (9), 3240–3244.

63 Ni, H., Denis, C.V., Subbarao, S., Degen, J.L., Sato, T.N., Hynes, R.O., and Wagner, D.D. (2000) Persistence of platelet thrombus formation in arterioles of mice lacking both von Willebrand factor and fibrinogen. *J. Clin. Invest.*, **106**, 385–392.

64 Sun, W.Y., Witte, D., Degan, J.L., Colbert, M.C., Burkart, M.C., Holmbäck, K., Xiao, Q., Brugge, T.H., Degen, S.J., (1998) Prothrombin deficiency results in embryonic and neonatal lethality in mice. *Proc. Natl Acad. Sci. U. S. A.*, **95**, 7595–7602.

65 Xue, J., Wu, Q., Westfield, L.A., Tuley, E.A., Lu, D., Zhang, Q., Shim, K., Zheng, X., and Sadler, J.E., (1998) Incomplete embryonic lethality and fatal neonatal hemorrhage caused by prothrombin deficiency in mice. *Proc. Natl Acad. Sci. U. S. A.*, **95**, 7603–7607.

66 Camerer, E., Duong, D.N., Hamilton, J.R., and Coughlin, S.R. (2004) Combined deficiency of protease-activated receptor-4 and fibrinogen recapitulates the hemostatic defect but not the embryonic lethality of prothrombin deficiency. *Blood*, **103**, 152–154.

67 Chrzanowska-Wodnicka, M., Smyth, S., Schoenwaelder, S.M., Fischer, T.H., and

White, G.C., 2nd (2005) Rap1b is required for normal platelet function and hemostasis in mice. *J. Clin. Invest.*, **15** (3), 680–687.

68 Nieswandt, B., Moser, M., Pleines, I., Varga-Szabo, D., Monkley, S., Critchley, D., and Fässler, R. (2007) Loss of talin1 in platelets abrogates integrin activation, platelet aggregation, and thrombus formation *in vitro* and *in vivo*. *J. Exp. Med.*, **204** (13), 3113–3118.

69 Petrich, B.G., Marchese, P., Ruggeri, Z.M., Spiess, S., Weichert, R.A., Ye, F., Tiedt, R., Skoda, R.C., Monkley, S.J., Critchley, D.R., and Ginsberg, M.H. (2007) Talin is required for integrin-mediated platelet function in hemostasis and thrombosis. *J. Exp. Med.*, **204** (13), 3103–3111.

70 Moser, M., Nieswandt, B., Ussar, S., Pozgajova, M., and Fässler, R. (2008) Kindlin-3 is essential for integrin activation and platelet aggregation. *Nat. Med.*, **14** (3), 325–330.

71 Cosemans, J.M., Munnix, I., Wetzker, R., Heller, R., Jackson, S.P., and Heemskerk, J.W. (2006) Continuous signaling via PI3K isoforms beta and gamma is required for platelet ADP receptor function in dynamic thrombus stabilization. *Blood*, **108** (9), 3045–3052.

72 Schoenwaelder, S.M., Ono, A., Sturgeon, S., Chan, S.M., Mangin, P., Maxwell, M.J., Turnbull, S., Mulchandani, M., Anderson, K., Kauffenstein, G., Rewcastle, G.W., Kendall, J., Gachet, C., Salem, H.H., and Jackson, S.P. (2007) Identification of a unique co-operative phosphoinositide 3-kinase signaling mechanism regulating integrin alpha IIb beta 3 adhesive function in platelets. *J. Biol. Chem.*, **282** (39), 28648–28658.

73 Hardy, A.R., Conley, P.B., Luo, J., Benovic, J.L., Poole, A.W., and Mundell, S.J. (2005) P2Y1 and P2Y12 receptors for ADP desensitize by distinct kinase-dependent mechanisms. *Blood*, **105** (9), 3552–3560.

74 Mundell, S.J., Barton, J.F., Mayo-Martin, M.B., Hardy, A.R., and Poole, A.W. (2008) Rapid resensitization of purinergic receptor function in human platelets. *J. Thromb. Haemost.*, **6** (8), 1393–1404.

75 van der Meijden, P.E., Schoenwaelder, S.M., Feijge, M.A., Cosemans, J.M., Munnix, I.C., Wetzker, R., Heller, R., Jackson, S.P., and Heemskerk, J.W. (2008) Dual P2Y 12 receptor signaling in thrombin-stimulated platelets–involvement of phosphoinositide 3-kinase beta but not gamma isoform in Ca2+ mobilization and procoagulant activity. *FEBS J.*, **275** (2), 371–385.

76 Lemons, P.P., Chen, D., Bernstein, A.M., Bennett, M.K., and Whiteheart, S.W. (1997) Regulated secretion in platelets: identification of elements of the platelet exocytosis machinery. *Blood*, **90**, 1490–1500.

77 Reed, G.L., Houng, A., and Fitzgerald, M.L. (1999) Human platelets contain SNARE proteins and a Sec1p homologue that interacts with syntaxin 4 and is phosphorylated after thrombin activation: implications for platelet secretion. *Blood*, **93**, 2617–2626.

78 Polgár, J. and Reed, G. (1999) A critical role for *N*-ethylmaleimide-sensitive fusion protein (NSF) in platelet granule secretion. *Blood*, **94**, 1313–1318.

79 Flaumenhaft, R., Croce, K., Chen, E., Furie, B., and Furie, B.C. (1999) Proteins of the exocytotic core complex mediate platelet alpha-granule secretion. *J. Biol. Chem.*, **274**, 2492–2501.

80 Chen, D., Lemons, P.P., Schraw, T., and Whiteheart, S.W. (2000) Molecular mechanisms of platelet exocytosis: role of SNAP-23 and syntaxin 2 and 4 in lysosome release. *Blood*, **96**, 1782–1788.

81 Lemons, P.P., Chen, D., and Whiteheart, S.W. (2000) Molecular mechanisms of platelet exocytosis: requirements for alpha-granule release. *Biochem. Biophys. Res. Commun.*, **267**, 875–880.

82 Bernstein, A.M. and Whiteheart, S.W. (1999) Identification of a cellubrevin/vesicle associated membrane protein 3 homologue in human platelets. *Biochem. Biophys. Res. Commun.*, **93**, 571–579.

83 Polgár, J., Chung, S.H., and Reed, G.L. (2002) Vesicle-associated membrane protein 3 (VAMP-3) and VAMP-8 are present in human platelets and are required for granule secretion. *Biochem. Biophys. Res. Commun.*, **100**, 1081–1083.

84 Schraw, T.D., Rutledge, T., Crawford, G.L., Bernstein, A.M., Kalen, A.L., Pessin, J.E., and Whiteheart, S.W. (2003) Granule stores from cellubrevin/VAMP-3 null mouse platelets exhibit normal stimulus-induced release. *Blood*, **102**, 1716–1722.

85 Ren, Q., Barber, H.K., Crawford, G.L., Karim, Z.A., Zhao, C., Choi, W., Wang, C.C., Hong, W., and Whiteheart, S.W. (2007) Endobrevin/VAMP-8 is the primary v-SNARE for the platelet release reaction. *Mol. Biol. Cell.*, **18**, 24–33.

86 Shirakawa, R., Higashi, T., Tabuchi, A., Yoshioka, A., Nishioka, H., Fukuda, M., Kita, T., and Horiuchi, H. (2004) Munc13-4 is a GTP-Rab27-binding protein regulating dense core granule secretion in platelets. *J. Biol. Chem.*, **279**, 10730–10737.

87 Walker, T.R. and Watson, S.P. (1993) Synergy between Ca^{2+} and protein kinase C is the major factor in determining the level of secretion from human platelets. *Biochem. J.*, **289**, 277–282.

88 Harper, M.T. and Poole, A.W. (2007) Isoform-specific functions of protein kinase C: the platelet paradigm. *Biochem. Soc. Trans.*, **35** (5), 1005–1008.

89 Konopatskaya, O., Gilio, K., Harper, M.T., Zhao, Y., Cosemans, J.M., Karim, Z.A., Whiteheart, S.W., Molkentin, J.D., Verkade, P., Watson, S.P., Heemskerk, J.W., and Poole, A.W. (2009) Calpha regulates platelet granule secretion and thrombus formation in mice. *J. Clin. Invest.*, **119** (2), 399–407.

90 Hall, K.J., Harper, M.T., Gilio, K., Cosemans, J.M., Heemskerk, J.W., and Poole, A.W. (2008) Genetic analysis of the role of protein kinase Cθ in platelet function and thrombus formation. *PLoS One*, **3**, e3277.

91 Pears, C.J., Thornber, K., Auger, J.M., Hughes, C.E., Grygielska, B., Protty, M.B., Pearce, A.C., and Watson, S.P. (2008) Differential roles of the PKC novel isoforms, PKCdelta and PKCepsilon, in mouse and human platelets. *PLoS ONE*, **3** (11), e3793.

18
Modulators of Monocyte and Macrophage Phenotypes in Atherosclerosis

Jason L. Johnson and Nicholas P. Jenkins

18.1
Introduction

Host defense systems of multicellular organisms comprise two major arms: (i) the innate and (ii) the adaptive parts of the immune system. The innate immune system has a humoral arm consisting of antimicrobial peptides and opsonins and a cellular arm involving specialized cells capable of phagocytosis.

Phagocytic cells of the innate immune system were originally classified as part of the reticuloendothelial system however, over the last 30 years this has been reclassified to the mononuclear phagocyte system (MPS) and includes the committed bone marrow precursors of monocytes, circulating monocytes, and macrophages. Tissue macrophages are a remarkably heterogeneous population, which in addition to the roving scavenger cells also include more specialized cell types, including Kupffer cells in the liver, lung alveolar macrophages, dendritic cells, and osteoclasts.

The primary role of the MPS is the identification and phagocytosis of microorganisms and the scavenging of certain toxic byproducts of metabolism. MPS activation, leading to tissue inflammation, also has an important role in chronic inflammatory diseases, such as granulomatous diseases, rheumatoid arthritis, and atherosclerosis. This chapter will highlight the origins and functional differences between monocyte and macrophage subsets. In addition, we will describe the evidence supporting their potential pro- and anti-inflammatory involvement in atherosclerosis.

18.1.1
Origin of Monocyte and Macrophage Populations

It is accepted that circulating monocytes originate in the bone marrow from hematopoietic stem cells and subsequently from CD34$^+$ myeloid progenitor cells. These progenitor cells are also termed granulocyte/macrophage colony–forming

Atherosclerosis: Molecular and Cellular Mechanisms. Edited by Sarah Jane George and Jason Johnson
Copyright © 2010 WILEY-VCH Verlag GmbH & Co. KGaA, Weinheim
ISBN: 978-3-527-32448-4

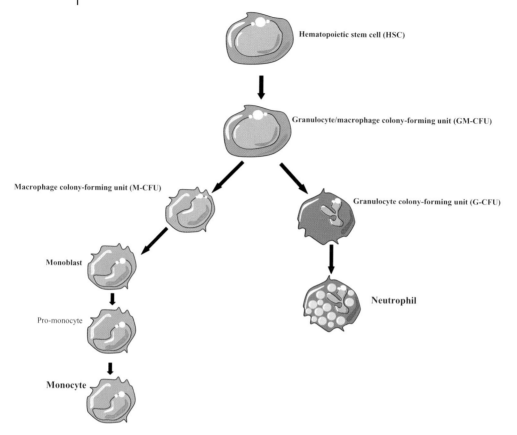

Figure 18.1 Bone marrow evolution of monocytes. Monocytes and neutrophils originate from a common hematopoietic stem cell and subsequently from CD34⁺ myeloid progenitor cells, also termed granulocyte/ macrophage-forming units (GM-CFUs). During development, exposure to either macrophage colony-forming unit (M-CFU) or GM-CFU commits precursor cells to the monocyte or neutrophil lineage, respectively.

units (GM-CFUs) as it is these common precursors which give rise to both circulating monocytes and neutrophils [1]. From this common precursor monocytes develop along a single lineage pathway as depicted in Figure 18.1 to form "mature" monocytes. Emigration from the bone marrow is necessary to maintain the circulating reservoir of monocytes, which in turn replenishes resident macrophage populations and makes monocytes available for recruitment to inflamed tissues.

The factors controlling monocyte emigration from the bone marrow are poorly understood. There is, however, clearly a cytokine dialog between tissues and bone marrow such that during periods of infection increased numbers of monocytes emigrate from the bone marrow into the circulation [2]. Work by Serbina *et al.* has shown that in mice this is at least in part mediated by the binding of circulating monocyte chemoattractant protein-1 (MCP-1) to its ligand chemokine (CC motif) receptor 2 (CCR2) on the surface of monocytes in the bone marrow [3].

18.1.2
Human Monocyte Heterogeneity

The study of circulating monocyte populations and their definition by both morphological characteristics and typical cell surface markers has been a complex area of immunology over the last 20 years. Morphological heterogeneity of monocytes makes them extremely difficult to distinguish from other circulating leukocytes, such as lymphocytes and natural killer (NK) cells [1]. While certain characteristics, such as irregularity of shape with kidney- or oval-shaped nuclei, cytoplasmic vesicles and a high cytoplasm-to-nucleus ratio, may allude to the likelihood of cell type, typical cell surface markers have proven more useful.

Until 30 years ago circulating monocytes were identified as a homogeneous population by the expression of large amounts of the lipopolysaccharide (LPS) receptor CD14 [4, 5]. In the 1980s it became apparent from the work of several groups that monocytes could be divided into at least two functionally separate groups. Ziegler-Heitbrock *et al.* showed that pooled human monocytes, which expressed CD14, could be separated into two subgroups based on their expression of CD16, known to be one of the family of IgG receptors and also known as FcγR-III [5]. Following this discovery, further work has been done to further elucidate both the differing cytokine receptor profiles and the functional properties of these two subsets (Table 18.1 and Figure 18.2).

18.1.2.1 CD14⁺/CD16⁻ Monocytes
The CD14⁺/CD16⁻ monocytes are often called "classical" monocytes and make up 80–90% of circulating monocytes. These monocytes express high levels of the MCP-1 (monocyte chemoattractant protein-1) receptor CCR2 and low levels of the fractalkine receptor CX₃CR1 [16] and MHC class II [17]. *In vitro* stimulation with the bacterial product LPS produces interleukin-10 (IL-10), which was originally known as cytokine synthesis inhibitory factor and explains its predominant role in suppressing the generation of proinflammatory cytokines, such as tumor necrosis factor-α (TNF-α) and IL-1 [18].

18.1.2.2 CD14⁺/CD16⁺ Monocytes
Conversely the CD14⁺/CD16⁺ subset make up only 10–20% of the circulating monocytes and are referred to as "non-classical" monocytes. In contrast to CD14⁺/CD16⁻ monocytes, they express high levels of CX₃CR1 and low levels of CCR2 [16]. Fractalkine (CX₃CL1) is known to be expressed on the surface of activated endothelial cells and has been implicated in the pathogenesis of chronic inflammatory diseases, such as rheumatoid arthritis, atherosclerosis, and some chronic infections, for example, HIV [15]. When stimulated *in vitro* with LPS CD14⁺/CD16⁺ monocytes produce large amounts of proinflammatory cytokines, such as IL-1 and TNF-α [18] and very little IL-10. These monocytes also express far higher levels of

Table 18.1 Differential expression pattern of surface markers and cytokine receptors between monocyte subsets from both human and mouse [6–15].

	Human CD14$^+$/CD16$^-$	Human CD14$^+$/CD16$^+$	Mouse Ly6C^{high}	Mouse Ly6C^{low}
Surface marker				
CD11c (integrin family; cell adhesion/activation)	+	++	−	+
CD14 (LPS receptor)	+	+	ND	ND
CD15 (selectin ligand: cell adhesion/activation)	+	−	ND	ND
CD16 (FcγRIII IgG receptor)	−	+	+	+
CD32 (FcγRII IgG receptor)	++	+	+	+
CD38 (cell adhesion/activation)	+	−	ND	ND
CD62L (L-selectin; homing, attachment and rolling of monocytes)	++	−	+	−
CD64 (FcγRI IgG receptor)	+	−	ND	ND
CD163 (hemoglobin scavenger receptor)	++	−	ND	ND
MHC class II (antigen presentation)	+	++	−	−
Cytokine receptors				
CCR1 (ligands include MCP-2, HCC-1, MIP-1α)	+	−	ND	ND
CCR2 (MCP-1(CCL2) receptor)	+	−	+	−
CCR4 (MDC (CCL4) receptor)	+	−	ND	ND
CCR7 (ligand for CCL19, CCL21 involved in chemotaxis and cell remodeling)	+	−	ND	ND
CX$_3$CR1 (fractalkine (CX$_3$CL1) receptor)	+	++	+	++

ND, not determined.

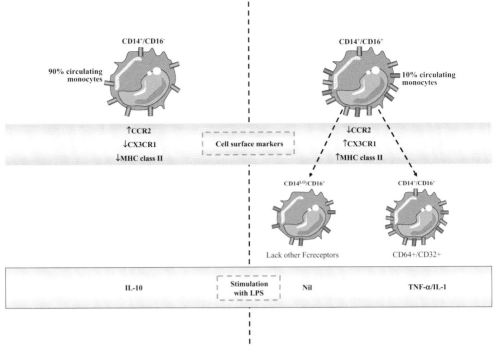

Figure 18.2 Circulating human monocyte subsets. Human circulating monocytes are present as two phenotypes, defined by their CD14/CD16 surface expression profile. CD14⁺/CD16⁻ monocytes, termed "classical" monocytes, express differing levels of cell surface markers compared with "non-classical" CD14⁺/CD16⁺ monocytes. Similarly, *in vitro* stimulation with the bacterial product LPS produces divergent responses between the two subsets with regards to expression of anti-inflammatory (IL-10) and proinflammatory cytokines (TNF-α and IL-1). A further subpopulation of CD14⁺/CD16⁺ monocytes has been appreciated with characteristically weak expression of CD14, termed CD14 low (CD14ᴸᴼ)/CD16⁺ monocytes. They are functionally distinct from the other CD16⁺ cells in that they display poor phagocytic activity and when stimulated *in vitro* with LPS they do not produce TNF-α or IL-1.

MHC class II and as such it can be assumed that they have a much more prominent role in antigen presentation [17]. These monocytes have therefore come to be known as "proinflammatory" monocytes. Interestingly, several studies have shown that in patients with sepsis presenting with either localized tissue infections (i.e., leishmaniasis), or indeed those with other acute inflammatory responses, the CD14⁺/CD16⁺ subpopulation of monocytes is greatly expanded [5, 19, 20].

Over the last 10 years a further subpopulation of CD14⁺/CD16⁺ monocytes has been appreciated with a differing functional profile [21, 22]. Due to a characteristically weak expression of CD14 they have been termed CD14 low (CD14ᴸᴼ) or CD14 diminished (CD14ᴰᴵᴹ)/CD16⁺ monocytes. They are functionally distinct from the other CD16⁺ cells in that they display poor phagocytic activity and when stimulated *in vitro* with LPS they do not produce TNF-α or IL-1. To date, it is unclear what the exact role of this subset is, however similar to other CD16$^+$ cells this population has been seen to expand in critically ill patients with sepsis or following cardiac surgery [19].

18.1.3
Mouse Monocyte Heterogeneity

Understanding of the mononuclear phagocyte system in mice and drawing parallels to human equivalent cell sets has been of obvious importance in enabling us to decipher responses in murine models of human diseases. Although there are similarities between subsets of monocytes in mouse and humans, there are also striking functional differences which we shall discuss in brief below. In addition, although CD14$^+$/CD16$^+$ monocytes predominate in humans, in mice the two populations are more equally distributed in the blood [23].

As in their human counterparts, circulating monocytes in mice may not be morphologically distinct from other leukocytes. Based primarily on their homing capabilities, mouse monocytes have been identified and termed "inflammatory" and "resident" monocytes [16]. These two mouse monocyte populations can be identified by cell surface markers (see Table 18.1), including the following:

i) Macrophage colony–stimulating factor (M-CSF) receptor, CD115
ii) Membrane activating complex-1 (Mac-1), CD11b
iii) F4/80 antigen
iv) Variable expression of Ly6C (Gr1) antigen.

18.1.3.1 CD115$^+$/Ly6C^{high} Monocytes
These monocytes are phenotypically similar to the CD14$^+$/CD16$^-$ subset of human monocytes in that they express high levels of CCR2 and low levels of CX$_3$CR1 [24]. Understandably, they are seen to migrate towards inflammatory tissue expressing the ligand CCL2 (MCP-1). They are, however, functionally similar to human CD14$^+$/CD16$^+$ monocytes, in that they produce high levels of TNF-α and IL-1 during bacterial infection or when there is tissue damage [25]. It has also been shown by Serbina and co-workers that during bacterial infection large numbers of Ly6C^{high} monocytes emigrate from the bone marrow and that this is dependent on signals mediated by the CCR2 receptor [3]. Hence, this subset has been termed "proinflammatory" but because of their cell surface marker profile some confusion has arisen when extrapolating the murine monocyte model to their human orthologs.

18.1.3.2 CD115$^+$/Ly6C^{low} Monocytes
This subpopulation of monocytes is characterized by low levels of CCR2 and high levels of CX$_3$CR1 [24]. In a similar fashion to the above subset, they are phenotypic orthologs of the human CD14$^+$/CD16$^+$ monocyte subset, however functionally resemble the "anti-inflammatory" CD14$^+$/CD16$^-$ type of human monocyte. These

cells have a longer half-life *in vivo* and patrol along the luminal border of the endothelium with extravasation being a rare event in the steady state [26]. While their true function is unknown, it may well be that these monocytes play a role in policing the endothelium and phagocytosing metabolic debris and *de novo* pathogens.

Adding weight to this argument is that following infection with the bacteria *Listeria monocytogenes*, these Ly6C^{low} monocytes show a profound inflammatory response within 1 h of infection. This involves extravasation and the release of multiple inflammatory cytokines. This reaction appears to be transient in that at 8 h post infection it is the Ly6C^{high} monocytes which are the major local cell type and producers of inflammatory cytokines [26].

In summary, useful comparisons can be drawn between mouse and human monocyte subsets. Functionally it would appear that the human CD14$^+$/CD16$^+$ human monocytes correlate with the Ly6C^{high} mouse subsets in their importance in the proinflammatory state. Conversely, the human CD14$^+$/CD16$^-$ human monocytes bear functional resemblance to the mouse Ly6C^{low} monocytes, and it would appear that their role may be more in patrolling tissues and potentially the replenishment of resident macrophage populations. Interestingly, the evidence above indicates that chemokine receptor expression on the monocyte populations appears disparate among species. It is therefore plausible that mouse "inflammatory" monocytes, which have high CCR2 expression, preferentially utilize MCP-1 as their major chemoattractant ligand towards inflamed tissue, while human "inflammatory" monocytes, which express abundant CX$_3$CR1, react primarily to fractalkine for chemotaxis. The opposite would hold true for "anti-inflammatory" monocytes.

18.1.4
Human Macrophage Heterogeneity

Macrophages are a widely studied subset of leukocytes whose broad function is tissue homeostasis and clearance of pathogens, apoptotic cells, and toxic products of metabolism. They have a remarkably dualistic nature, including proinflammatory versus anti-inflammatory and tolerogenic versus immunogenic activity, which we are only beginning to understand. Heterogeneity and plasticity are key hallmarks of macrophage populations. Despite our increasing knowledge of separate monocyte subpopulations it is still unclear whether certain populations of macrophages are either lineage specific and have predetermined roles before the influence of their tissue microenvironment, or harbor the ability to adapt to any role they are stimulated to perform.

Broadly speaking we can divide macrophage populations into two major subsets:

i) **recruited macrophages:** those that are derived from circulating monocytes and extravasate to sites of tissue inflammation, and
ii) **resident macrophages:** those that reside in specific tissues where they develop specialized roles according to the microenvironment.

18.1.4.1 Recruited Macrophages

It has long been recognized that macrophages can be recruited to areas of tissue inflammation from the circulating monocyte pool. It is also clear from the study of chronic inflammatory conditions, such as granuloma formation, rheumatoid arthritis, and atherosclerosis, that more than one functional set of macrophages exist (as reviewed by [1]). The early phase of inflammation is characterized by cytotoxicity and a proinflammatory cascade. Conversely, in the later stages there is an emphasis on tissue repair and regeneration. There also exist situations, such as in granuloma formation, where both functional sets of macrophages coexist at the same time, though they may be spatially distinct [27]. While immunologists have tried to define specific functional subsets of macrophages it is still unclear how much plasticity exists between these types. It remains to be elucidated whether macrophages have a distinct fate once given their cytokine cues, or whether they have an ability to switch phenotype should they encounter a change in stimuli.

Nonetheless, it is accepted that in response to specific cytokine cues macrophages express specialized and polarized functional properties and thus the following nomenclature is generally used to describe these two macrophage phenotypes: (i) classically activated M1 macrophages which harbor proinflammatory functional properties or (ii) alternatively activated M2 macrophages that are associated with an anti-inflammatory role [28] (Figure 18.3). While the nomenclature of M1/M2 macrophages can be useful it is clearly an oversimplification in that it represents the ends of what is certainly a spectrum of cell phenotype and function.

M1 Polarization As illustrated in Figure 18.3, classically activated macrophages can be induced by stimulation with proinflammatory cytokines such as TNF-α, interferon-γ (IFN-γ), LPS, or IL-1 [29, 30]. Once polarized, these macrophages are themselves very efficient producers of TNF-α, as well as reactive oxygen species and other proinflammatory cytokines. Their major role in tissues is promotion of type 1 inflammatory reactions, tissue destruction, killing of intracellular parasites and tumor resistance [31].

M2 Polarization M2 polarized macrophages are also termed generically as "nonclassically" or "alternatively" activated due to their inability to be activated by classical cytokines as described above (see Figure 18.3). M2 macrophages are produced in response to IL-4, IL-13, IL-21, immune complexes, and as expected by anti-inflammatory glucocorticoid or secosteroid hormones [28, 29]. They express more mannose receptors, which recognize and assist in the phagocytosis of bacterial and fungal pathogens [28]. While the exact function of M2 macrophages may vary depending on the local microenvironment and combination of cytokine cues, they are capable of orchestrating inflammatory responses and adaptive Th1 immunity, scavenge debris, and promote angiogenesis, tissue remodeling and repair (as reviewed by [32]).

Attached monocyte

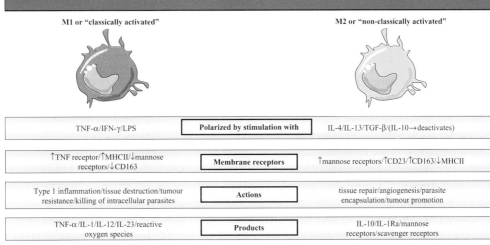

M1 or "classically activated"		M2 or "non-classically activated"
TNF-α/IFN-γ/LPS	**Polarized by stimulation with**	IL-4/IL-13/TGF-β/(IL-10→deactivates)
↑TNF receptor/↑MHCII/↓mannose receptors/↓CD163	**Membrane receptors**	↑mannose receptors/↑CD23/↑CD163/↓MHCII
Type 1 inflammation/tissue destruction/tumour resistance/killing of intracellular parasites	**Actions**	tissue repair/angiogenesis/parasite encapsulation/tumour promotion
TNF-α/IL-1/IL-12/IL-23/reactive oxygen species	**Products**	IL-10/IL-1Ra/mannose receptors/scavenger receptors

Figure 18.3 Polarization of recruited macrophages. After attachment to the endothelium and subsequent transmigration into tissue, monocytes differentiate into macrophages. Differing cytokine stimulation polarizes macrophages towards either a classically activated/M1 or an alternatively activated/M2 phenotype. These two macrophage subsets differ in their cell surface receptor expression and also exhibit divergent biological actions. M1 macrophages are termed inflammatory while M2 macrophages appear to play a role in tissue repair/healing.

Deactivated States It is also recognized that recruited macrophages can enter a deactivated state, often referred to as a subset of the M2 macrophages (M2c). This can be produced by stimulation with cytokines such as IL-10, transforming growth factor-β (TGF-β) or indeed with exposure to glucocorticoids [31]. This is associated with reduced MHC class II expression and release of a profile of anti-inflammatory cytokines [29]. The fate of these quiescent macrophages is as yet unknown.

18.1.4.2 Resident Macrophages

These tissue-specific macrophages originate from the circulating pool of monocytes and become highly specialized in response to their surrounding microenvironment (i.e., the surface and secretory products of surrounding cells and the extracellular matrix in which they reside). These resident macrophage populations can be replenished by three potential mechanisms:

i) mitotic expansion of macrophage populations within tissues,
ii) extravasation and differentiation of circulating monocytes, or

iii) extramedullary proliferation and differentiation of adult bone marrow derived progenitor cells.

These specialist cell types, which include dendritic cells, Langerhans cells, Kupffer cells, osteoclasts, microglia, and alveolar macrophages, are discussed in brief below.

Dendritic Cells   The classification of dendritic cells within a particular lineage of the immune system has proven difficult as they have been shown to arise from either myeloid or lymphoid progenitors [33]. For the purposes of the mononuclear phagocyte system they are included in this discussion and known as myeloid-derived dendritic cells. They are themselves a heterogeneous population of cells with differing profiles of cell surface receptors and varying functional phenotypes depending on the profile of cytokines used to stimulate them [34, 35]. The hallmark of dendritic cell function is in their ability to present antigen to naïve T and B cells in local lymphoid tissue and therefore launching cascades of the adaptive immune system. They have been referred to as key "bridges" between the innate and the adaptive immune system by conveying information between the two [35].

Langerhans Cells   Langerhans cells are a highly specialized type of dendritic cell located in the epidermis and forming the primary immune response to the external environment. Their major role is in antigen presentation, and on contact with pathogens they migrate to regional lymph nodes and initiate an immune response by presentation of processed antigen to T cells by the MHC class II molecule [36, 37]. It has been shown in both mouse and human models that these cell populations are self-maintained, as local tissue proliferation preserves the population and it is rarely replenished from circulating monocyte precursors [36, 38].

Kupffer Cells   Kupffer cells reside in the hepatic sinusoids of most mammalian species and are the largest population of resident tissue macrophages. Their primary role appears to be in the phagocytosis of pathogens and metabolic products from the portal circulation. They also play an important role in bilirubin metabolism, via breakdown of senescent erythrocytes [39]. The maintenance of Kupffer cell populations appears to be via either local proliferation or via replenishment by circulating monocyte precursors depending on prevailing metabolic conditions [40].

Osteoclasts   Osteoclasts are monocyte-derived multinucleated cells whose function is to resorb bone. Their differentiation and function is dependent on several osteoblast-derived factors, such as M-CSF, TGF-β, and the receptor activator of NF-κB ligand (RANKL) [41, 42], as culture of unfractionated monocytes with M-CSF and RANKL stimulates cell differentiation into osteoclast-like macrophages [43].

Microglia   Microglia comprise the majority of tissue-resident macrophages in the central nervous system (CNS) of mammals. As with other tissue macrophages the major role of these cells is the surveillance and endocytosis of pathogens. Similar to other organs populations of macrophages they are found at strategic locations such as the meninges and choroid plexi, as well as perivascularly, where they are more likely to encounter pathogens [44]. They themselves form a functionally heterogeneous population. Using a combination of bone marrow transplantation experiments and *in vivo* labeling it is apparent that these populations can be replenished by either *in situ* proliferation or replenishment from bone marrow–derived precursors [45, 46].

Alveolar Macrophages   As suggested, these macrophages reside in the free air spaces of mammalian lungs. They play a vital role in pathogen phagocytosis and antigen presentation. It has also been shown that they play an important role in the pathogenesis of certain inflammatory lung diseases, such as cryptogenic fibrosing alveolitis, although the exact mechanisms remain to be elucidated [47]. As with CNS-resident macrophages it has been shown that alveolar macrophage populations are derived from circulating precursors, as well as local proliferation [48, 49].

18.2
Role of Monocytes/Macrophages in Atherosclerosis

Monocytes and differentiated monocyte-derived macrophages are found in atherosclerotic plaques at all stages [50]. Accumulation of low-density lipoprotein (LDL) within the vascular intima triggers the recruitment of monocytes from the circulation, which differentiate to macrophages and accumulate cholesteryl esters to become macrophage-derived foam cells [51, 52]. This process results in the generation of intimal lesions termed "fatty streaks," also referred to as type 2 lesions, consisting mainly of foamy macrophages and a few foamy smooth muscle cells (SMCs). Recent work suggests that although monocytes are heterogeneous, the $Ly6C^{high}$ phenotype is selectively recruited into atherosclerotic plaques [53, 54]. With continuing hypercholesterolemia, type 3 (pre-atheroma) and 4 (atheroma) lesions develop, exhibiting extracellular lipid pools of increasing size, respectively [55]. Type 5 lesions (fibro-atheroma) have a pronounced fibrous cap, which grows mainly by recruitment and proliferation of SMCs [56]. Macrophages may stimulate fibrous cap formation by producing growth factors and extracellular proteases that aid migration and proliferation of resident SMCs [57, 58]. However, macrophages are abundant at the shoulder region of thin cap fibro-atheromas (TCFAs), the type 5 lesions that most frequently give rise to fibrous cap rupture [59, 60]. The presence of macrophages in these plaques correlates with reduced levels of interstitial collagen and high levels of extracellular proteases, including several matrix-degrading metalloproteinases (MMPs) and cathepsins (thiol proteases) [61]. There are also fewer SMCs in TCFAs, perhaps because macrophages provoke apoptosis of SMCs [62]. Macrophage apoptosis may also contribute to cellular rarefaction of the plaque cap, which is commonly observed in type 5 lesions. Plaque disruption, in most cases plaque rupture, is followed by mural or occlusive thrombus formation. If these events are not fatal, the thrombus will become organized, leading to type 6 lesions (complicated plaques). Macrophages play an important part in fibrosis during thrombus organization [6]. In summary, therefore, during fibrous cap formation and thrombus reorganization macrophages seem to promote recruitment of SMC-like cells and laying down of extracellular matrix (ECM) but in the shoulder region of plaques vulnerable to rupture macrophages have been proposed to cause matrix destruction and promote death of VSMC. Interestingly, MMPs derived from macrophages also seem to have a role in both fibrous cap formation and rupture [58]. Literature support for this concept is extensive but perhaps best illustrated by our work in early lesions in the brachiocephalic artery of apolipoprotein E (ApoE)-null mice [63]. We showed that knockout of MMP-3 or MMP-9 led to larger, less stable lesions, implying that MMP-3 and MMP-9 contribute to cap formation and stability at this site and stage of disease. On the other hand knockout of MMP-12 led to smaller, more stable lesions, implying a role for this MMP in plaque rupture.

18.3
Monocyte Polarization and Atherosclerosis

As mentioned previously, circulating monocytes in the mouse exist as two major subsets as defined by their differing chemokine receptor expression pattern: $Ly6C^{high}$ (which are also phenotypically $Ly6C/G^{high}CCR2^+CX_3CR1^{low}$) monocytes, which are short-lived and commonly termed inflammatory monocytes, and $Ly6C^{low}$ ($Ly6C/G^{low}CCR2^-CX_3CR1^{high}$) monocytes which persist longer in tissues and are considered to give rise to resident myeloid cells, such as dendritic cells [16]. These populations are similar to the $CD14^+CD16^-$ and $CD14^{low}CD16^+$ subsets observed in humans, which can also be defined by their CX_3CR1 and CCR2 expression [16, 64]. In order to mimic human atherosclerosis, mouse models of atherosclerosis invariably rely on elevating circulating total cholesterol levels. Subsequently, several studies have focused on the effects of hypercholesterolemia on monocyte heterogeneity. Consumption of a high-fat diet increases circulating monocyte numbers in ApoE-deficient mice, which is termed hypercholesterolemia-associated monocytosis [53, 54]. Additionally, it was shown that hypercholesterolemia impairs the conversion of $Ly6C^{high}$ monocytes to the $Ly6C^{low}$ phenotype, resulting in $Ly6C^{high}$ monocytosis. Furthermore, proinflammatory $Ly6C^{high}$ monocytes accumulate more efficiently in murine atherosclerotic lesions than their $Ly6C^{low}$ counterparts, which are more prone to differentiate into dendritic cell-like monocyte/macrophages, as ascertained by their expression of CD11c [54].

Interestingly, conversion of monocytes to a dendritic-like cell is associated with their increased capacity to emigrate from tissues (see review [65]). Therefore, as well as continued monocyte accumulation, the progression of atherosclerotic plaques may also result from reduced monocyte egress from lesions. In support of this notion, Llodra and coworkers [66] demonstrated that monocyte-derived cells egress from lesions undergoing regression, whereas little emigration was observed in progressive plaques. Moreover, pro-atherogenic molecules such as lysophospha-tidic acid (LPA) and platelet-activating factor (PAF) retard the transformation of monocytes into migratory cells and thus facilitate their retention in the intimal space.

Coronary occlusive atherosclerosis/thrombosis gives rise to myocardial infarc-tion (MI), and the subsequent healing of MI requires monocyte/macrophages. Nahrendorf and colleagues [67] demonstrated that this healing process involves two phases of monocyte participation undertaken by two distinct monocyte subsets with divergent properties. Ly6C^{high} monocytes dominate the early phase of healing and exhibit phagocytic, proteolytic, and inflammatory functions involved in clear-ing damaged tissue. Conversely, Ly6C^{low} monocytes govern the later phase, pro-moting healing through myofibroblast accumulation, angiogenesis, and deposition of collagen. Consequently, MI in atherosclerotic mice, which exhibit chronic Ly6C^{high} monocytosis, exhibit an impaired healing response. An additional finding demonstrated that Ly6C^{high} monocytes accumulate preferentially via CX$_3$CR1 while Ly6C^{low} monocytes accumulate via CCR2. Similarly, Tacke and coworkers [54] also found that Ly6C^{low} monocytes, which they term as "non-classical" monocytes, did not require CX$_3$CR1 to enter murine aortic atherosclerotic plaques. However, they also determined that Ly6C^{low} monocytes were not reliant on CCR2, but require CCR5 to accumulate in atherosclerotic lesions. Moreover, it was observed that Ly6C^{high} monocytes, termed "classical" or "inflammatory" monocytes, rely on CCR2, CX$_3$CR1, and CCR5 to enter atherosclerotic plaques.

Thus, multiple chemokine receptors can work in concert to regulate the migra-tion of a single population of monocytes. Furthermore, although CCR2 is funda-mental for monocyte recruitment in most models of inflammation, it would appear that CX$_3$CR1 maybe athero-specific, so it has been postulated that targeting of CX$_3$CR1 may ameliorate CCR2$^+$ (Ly6C^{high}) monocyte recruitment to atheroscle-rotic lesions while sparing their CCR2-dependent responses to other inflammatory stimuli. Indeed, it has been shown that humans with a polymorphism in CX$_3$CR1 exhibit relative protection from cardiovascular disease [68]. A more indepth per-spective of the role of chemokines in atherosclerosis is given in Chapter 2.

18.4
Macrophage Polarization and Atherosclerosis

A simple web search demonstrates that there is a paucity of published material examining canonical macrophage phenotypes in atherosclerosis. However, if you scratch deeper it becomes apparent that what appears a dearth of information is indeed a matter of semantics. As previously mentioned, macrophages have been described as a heterogeneous cell population with a variety of physiological and pathophysiological functions dependent on microenvironmental signals [69, 70]. Atherosclerotic lesions from human subjects and those developed in animal models are heterogeneous entities that contain diverse cell types, creating a complex cytokine milieu. Thus the environment exists for macrophage polariza-tion to occur in atherosclerotic lesions, and if this is the case, can the nomenclature used by immunologists and macrophage biologists to identify the divergent macrophage phenotypes be applied?

To mimic the atherosclerotic environment, Wintergerst and colleagues [71] demonstrated that culturing human monocytes in Teflon-coated bags containing autologous plasma and T lymphocytes for 14 days induced the differentiation of monocytes into macrophages, giving rise to three distinct macrophage subpopulations based on the expression of CD14, CD36, and LDL-receptor. Moreover, Poston and Hussain [72] performed an elegant study over 15 years ago demonstrating immunohistochemical heterogeneity of macrophages in human atherosclerotic lesions. They went on to describe that the properties of the macrophages could be linked to stages of differentiation and to position in the arterial wall, with relation to anatomic site and lesion phenotype. When comparing their findings to that observed in other lymphoid tissues, it was concluded that recently recruited blood-derived macrophages can be distinguished from longer residing cells. Similarly, Komohara and colleagues [73] also reported a specialized subpopulation of macrophages in granulomatous diseases such as tuberculosis, sarcoidosis, and foreign body reactions with relation to hemoglobin scavenger receptor (CD163) expression, a molecule associated with an anti-inflammatory macrophage phenotype, in part due to its induction of IL-10 and heme oxygenase-1 [6]. Interestingly, it was observed that the macrophages encircling granulomas are predominantly CD163[+], while the macrophages within granulomas are negative or only weakly positive. In atherosclerotic lesions, CD163[+] macrophages with no or little lipid accumulation are observed in the outer aspects of the plaque, whereas foam cell macrophages associated with the lipid-rich core in advanced lesions are negative or only weakly positive [73]. These findings suggest that recently recruited macrophages are of an anti-inflammatory "M2" phenotype, while transformation to a foam cell macrophage induces differentiation to a proinflammatory "M1" phenotype.

Accordingly, it has recently been demonstrated that a CD163[high]/human leukocyte antigen (HLA)-DR[low] macrophage subpopulation presides in atherosclerotic plaques in response to intraplaque hemorrhage [74]. Moreover it is implied that the CD163[high] subset are of an anti-inflammatory and antioxidant nature due to their reduced H_2O_2 production and increased IL-10 secretion, therefore defining this macrophage subset as atheroprotective.

As well as differential expression in a monocyte subpopulation [75], CD16 has been proposed as a marker of a functional subset of macrophages able to participate in immune responses [76]. Indeed, in human atherosclerotic lesions CD16[+] macrophages were correlated with CD40/CD40L expression [76]. However, it is unclear whether CD16 expression is upregulated by CD40/CD40L signaling or vice versa. Nonetheless, it is clear that atherosclerotic lesion macrophages are heterogeneous and contain subsets active in different tasks, for example, endocytosis, phagocytosis, cytokine secretion, and inflammatory responses.

It has been shown in monocyte-derived macrophages that CD14 and CD16 expression is differentially expressed between two populations, dependent on their differentiation with divergent growth factors [77]. Differentiation with granulocyte-macrophage colony–stimulating factor (GM-CSF) gave rise to an anti-inflammatory macrophage phenotype expressing genes known to mediate reverse cholesterol transport and emigration from the vessel wall. Conversely, M-CSF stimulated macrophages which lack the atheroprotective genes but instead express proinflammatory genes. Furthermore, it was shown that CD14$^+$ macrophages, which were deemed similar to M-CSF differentiated macrophages, predominate in human atherosclerotic plaques; whereas CD14$^-$ macrophages, similar to GM-CSF-treated macrophages, are abundant in areas devoid of disease. In accordance with this, hypercholesterolemia and incidence of coronary artery disease are associated with an increased number of monocytes expressing both CD14 and CD16 [75, 78]. Furthermore, regulators of cholesterol hemostasis such as peroxisome proliferator–activator receptor-γ (PPARγ) and liver X receptor-α (LXR-α) were differentially expressed between the two populations and both increased with GM-CSF differentiation.

Interestingly, Bouhlel and colleagues recently reported that PPARγ primes human monocytes into differentiation towards an anti-inflammatory (M2) phenotype [79]. In addition, they provide clear evidence that M2 anti-inflammatory macrophages reside in human atherosclerotic lesions, and correlate with PPARγ expression. Furthermore, building on a previous study [80], it is shown that PPARγ-induced M2 macrophages can exert anti-inflammatory effects on proinflammatory M1 macrophages. GM-CSF deficiency also increased lesion size and macrophage accumulation in diet-induced hyperlipidemic mice, in part due to reduced PPARγ expression [81]. Conversely, M-CSF deficiency results in decreased atherosclerosis [82], and macrophage M-CSF expression can be downregulated by PPARγ [83]. Thus it is plausible that divergent macrophage phenotypes in atherosclerotic lesions are growth factor–dependent and regulated via PPARγ.

Anti-inflammatory GM-CSF-differentiated macrophages also exhibit 20-fold increased expression of tissue inhibitor of matrix metalloproteinase-3 (TIMP-3) compared with M-CSF-treated proinflammatory macrophages [77]. TIMP-3 is capable of inhibiting the proteolytic activity of a broad range of MMPs and is present in human atherosclerotic plaques [84]. Increased macrophage MMP expression and unchecked activity is associated with atherosclerotic plaque progression and rupture [85]. Studies in rabbits and humans illustrate divergent MMP and TIMP expression in subsets of macrophages. Although many MMPs are widely expressed in macrophages from atherosclerotic plaques of all stages, MMP-11 is detected exclusively in advanced lesions [86]. Both MMP-7 and MMP-12 are localized in macrophages juxtaposed between the necrotic core and the rupture-prone shoulders of human plaques [87]. While in advanced rabbit plaques MMP-12 is confined to deep-lying foam cell macrophages [88].

We have recently shown that TIMP-3 is downregulated and MMP-14 upregulated in a subpopulation of rabbit and human foam cell macrophages [89]. This phenotype display proinflammatory properties including: increased degradation of, and invasion through synthetic extracellular matrix; heightened proliferative rates; and an increased susceptibility to apoptosis, all factors expected to potentiate plaque development and instability. Moreover, this subset localize to the periphery of the necrotic core where arginase-1 expression, a phenotypic marker of proinflammatory M1 macrophages, is concomitantly reduced [29]. Considering TIMP-3 expression is associated with GM-CSF-differentiated macrophages, also classed as anti-inflammatory M2 macrophages, which are inversely related with atherosclerotic lesion severity [77], it is tempting to speculate that the TIMP-3^{low}/MMP-14^{high} macrophage subset are of the M1 phenotype, however further confirmatory evidence is necessary.

18.5
Concluding Remarks

It is apparent that monocytes, macrophages, and foam cell macrophages display heterogeneity in atherosclerosis (Figure 18.4). What is unclear is the regulation of

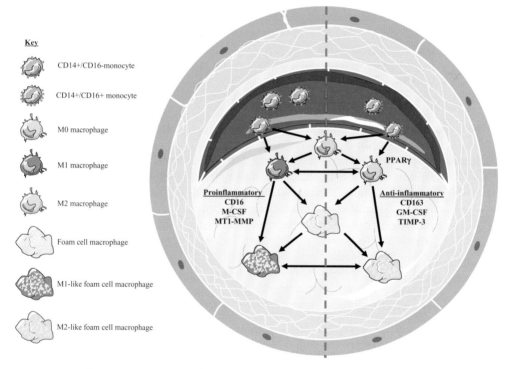

Key

CD14+/CD16-monocyte

CD14+/CD16+ monocyte

M0 macrophage

M1 macrophage

M2 macrophage

Foam cell macrophage

M1-like foam cell macrophage

M2-like foam cell macrophage

Figure 18.4 Monocyte and macrophage subsets in atherosclerosis. Monocytes, macrophages, and foam cell macrophages display heterogeneity in atherosclerosis and it is unclear if these discordant subpopulations are plastic or permanently differentiated. What is clear is that CD14$^+$/CD16$^+$ monocytes, M1 macrophages, and M1-like foam cell macrophages display proinflammatory properties while CD14$^+$/CD16$^-$ monocytes, M2 macrophages, and M2-like foam cell macrophages express anti-inflammatory associated molecules. However as the arrows indicate, it remains unknown, for example, whether proinflammatory monocytes (CD14$^+$/CD16$^+$) give rise solely to proinflammatory M1 macrophages, or under the requisite stimuli also generate anti-inflammatory M2 macrophages, and vice versa for anti-inflammatory monocytes (CD14$^+$/CD16$^-$).

these discordant subpopulations down their linage pathways and if these can be perturbed. Further studies are needed to determine whether the subsets are plastic or differentiate permanently. Defining whether the cytokine/growth factor environment recruits and/or activates individual monocyte subsets or instead polarizes already accumulated/resident subpopulations of specific macrophages is of utmost importance.

References

1 Gordon, S. and Taylor, P.R. (2005) Monocyte and macrophage heterogeneity. *Nat. Rev. Immunol.*, **5** (12), 953–964.

2 Serbina, N.V., Jia, T., Hohl, T.M., and Pamer, E.G. (2008) Monocyte-mediated defense against microbial pathogens. *Annu. Rev. Immunol.*, **26**, 421–452.

3 Serbina, N.V. and Pamer, E.G. (2006) Monocyte emigration from bone marrow during bacterial infection requires signals mediated by chemokine receptor CCR2. *Nat. Immunol.*, **7** (3), 311–317.

4 Ziegler-Heitbrock, H.W., Passlick, B., and Flieger, D. (1988) The monoclonal antimonocyte antibody My4 stains B lymphocytes and two distinct monocyte subsets in human peripheral blood. *Hybridoma*, **7** (6), 521–527.

5 Ziegler-Heitbrock, L. (2007) The CD14+ CD16+ blood monocytes: their role in infection and inflammation. *J. Leukoc. Biol.*, **81** (3), 584–592.

6 Philippidis, P., Mason, J.C., Evans, B.J., Nadra, I., Taylor, K.M., Haskard, D.O., and Landis, R.C. (2004) Hemoglobin scavenger receptor CD163 mediates interleukin-10 release and heme oxygenase-1 synthesis: antiinflammatory monocyte-macrophage responses *in vitro*, in resolving skin blisters *in vivo*, and after cardiopulmonary bypass surgery. *Circ. Res.*, **94** (1), 119–126.

7 Musso, T., Deaglio, S., Franco, L., Calosso, L., Badolato, R., Garbarino, G., Dianzani, U., and Malavasi, F. (2001) CD38 expression and functional activities are up-regulated by IFN-gamma on human monocytes and monocytic cell lines. *J. Leukoc. Biol.*, **69** (4), 605–612.

8 Postigo, A.A., Corbi, A.L., Sanchez-Madrid, F., and de Landazuri, M.O. (1991) Regulated expression and function of CD11c/CD18 integrin on human B lymphocytes. Relation between attachment to fibrinogen and triggering of proliferation through CD11c/CD18. *J. Exp. Med.*, **174** (6), 1313–1322.

9 Gragegriebenow, E., Lorenzen, D., Fetting, R., Flad, H.D., and Ernst, M. (1993) Phenotypical and functional-characterization of Fc-gamma receptor-I (Cd64)-negative monocytes, a minor human monocyte subpopulation with high accessory and antiviral activity. *Eur. J. Immunol.*, **23** (12), 3126–3135.

10 Lo, S.K., Golenbock, D.T., Sass, P.M., Maskati, A., Xu, H., and Silverstein, R.L. (1997) Engagement of the Lewis X antigen (CD15) results in monocyte activation. *Blood*, **89** (1), 307–314.

11 Tsou, C.L., Gladue, R.P., Carroll, L.A., Paradis, T., Boyd, J.G., Nelson, R.T., Neote, K., and Charo, I.F. (1998) Identification of C-C chemokine receptor 1 (CCR1) as the monocyte hemofiltrate C-C chemokine (HCC)-1 receptor. *J. Exp. Med.*, **188** (3), 603–608.

12 Lloyd, C.M., Delaney, T., Nguyen, T., Tian, J., Martinez, C., Coyle, A.J., and Gutierrez-Ramos, J.C. (2000) CC chemokine receptor (CCR)3/eotaxin is followed by CCR4/monocyte-derived chemokine in mediating pulmonary T helper lymphocyte type 2 recruitment after serial antigen challenge in vivo. *J. Exp. Med.*, **191** (2), 265–273.

13 De Martinis, M., Modesti, M., and Ginaldi, L. (2004) Phenotypic and functional changes of circulating monocytes and polymorphonuclear leucocytes from elderly persons. *Immunol. Cell Biol.*, **82** (4), 415–420.

14 Sanchez-Sanchez, N., Riol-Blanco, L., and Rodriguez-Fernandez, J.L. (2006)

The multiple personalities of the chemokine receptor CCR7 in dendritic cells. *J. Immunol.*, **176** (9), 5153–5159.

15 Landsman, L., Bar-On, L., Zernecke, A., Kim, K.W., Krauthgamer, R., Shagdar-suren, E., Lira, S.A., Weissman, I.L., Weber, C., and Jung, S. (2009) CX(3)CR1 is required for monocyte homeostasis and atherogenesis by promoting cell survival. *Blood*, **113** (4), 963–972.

16 Geissmann, F., Jung, S., and Littman, D.R. (2003) Blood monocytes consist of two principal subsets with distinct migratory properties. *Immunity*, **19** (1), 71–82.

17 Ziegler-Heitbrock, H.W., Fingerle, G., Strobel, M., Schraut, W., Stelter, F., Schutt, C., Passlick, B., and Pforte, A. (1993) The novel subset of CD14+/CD16+ blood monocytes exhibits features of tissue macrophages. *Eur. J. Immunol.*, **23** (9), 2053–2058.

18 Belge, K.U., Dayyani, F., Horelt, A., Siedlar, M., Frankenberger, M., Frankenberger, B., Espevik, T., and Ziegler-Heitbrock, L. (2002) The proinflammatory CD14(+)CD16(+)DR(++) monocytes are a major source of TNF. *J. Immunol.*, **168** (7), 3536–3542.

19 Fingerle-Rowson, G., Auers, J., Kreuzer, E., Fraunberger, P., Blumenstein, M., and Ziegler-Heitbrock, L.H.W. (1998) Expansion of CD14(+)CD16(+) mono-cytes in critically ill cardiac surgery patients. *Inflammation*, **22** (4), 367–379.

20 Strauss-Ayali, D., Conrad, S.M., and Mosser, D.M. (2007) Monocyte subpopu-lations and their differentiation patterns during infection. *J. Leukoc. Biol.*, **82** (2), 244–252.

21 Grage-Griebenow, E., Flad, H.D., and Ernst, M. (2001) Heterogeneity of human peripheral blood monocyte subsets. *J. Leukoc. Biol.*, **69** (1), 11–20.

22 Auffray, C., Sieweke, M.H., and Geissmann, F. (2009) Blood monocytes: development, heterogeneity, and relationship with dendritic cells. *Annu. Rev. Immun.*, **27** (1), 669–692.

23 Passlick, B., Flieger, D., and Ziegler-Heitbrock, H.W. (1989) Identification and characterization of a novel monocyte subpopulation in human peripheral blood. *Blood*, **74**, (7), 2527–2534.

24 Palframan, R.T., Jung, S., Cheng, G., Weninger, W., Luo, Y., Dorf, M., Littman, D.R., Rollins, B.J., Zweerink, H., Rot, A., and von Andrian, U.H. (2001) Inflammatory chemokine transport and presentation in HEV: a remote control mechanism for monocyte recruitment to lymph nodes in inflamed tissues. *J. Exp. Med.*, **194** (9), 1361–1373.

25 Dunay, I.R., DaMatta, R.A., Fux, B., Presti, R., Greco, S., Colonna, M., and Sibley, L.D. (2008) Gr1(+) inflammatory monocytes are required for mucosal resistance to the pathogen *Toxoplasma gondii*. *Immunity*, **29** (4), 660–660.

26 Auffray, C., Fogg, D., Garfa, M., Elain, G., Join-Lambert, O., Kayal, S., Sarnacki, S., Cumano, A., Lauvau, G., and Geissmann, F. (2007) Monitoring of blood vessels and tissues by a population of monocytes with patrolling behavior. *Science*, **317** (5838), 666–670.

27 Yoshikai, Y. (2006) Immunological protection against *Mycobacterium tuberculosis* infection. *Crit. Rev. Immunol.*, **26** (6), 515–526.

28 Mantovani, A., Sica, A., and Locati, M. (2007) New vistas on macrophage differentiation and activation. *Eur. J. Immunol.*, **37** (1), 14–16.

29 Gordon, S. (2003) Alternative activation of macrophages. *Nat. Rev. Immunol.*, **3** (1), 23–35.

30 Scotton, C.J., Martinez, F.O., Smelt, M.J., Sironi, M., Locati, M., Mantovani, A., and Sozzani, S. (2005) Transcriptional profiling reveals complex regulation of the monocyte IL-1 beta system by IL-13. *J. Immunol.*, **174** (2), 834–845.

31 Mantovani, A., Sica, A., Sozzani, S., Allavena, P., Vecchi, A., and Locati, M. (2004) The chemokine system in diverse forms of macrophage activation and polarization. *Trends Immunol.*, **25** (12), 677–686.

32 Mantovani, A., Sozzani, S., Locati, M., Allavena, P., and Sica, A. (2002) Macrophage polarization: tumor-associ-ated macrophages as a paradigm for polarized M2 mononuclear phagocytes. *Trends Immunol.*, **23** (11), 549–555.

33 del Hoyo, G.M., Martin, P., Vargas, H.H., Ruiz, S., Arias, C.F., and Ardavin, C. (2002) Characterization of a common

precursor population for dendritic cells. *Nature*, **415** (6875), 1043–1047.

34 Hume, D.A. (2008) Macrophages as APC and the dendritic cell myth. *J. Immunol.*, **181** (9), 5829–5835.

35 Ueno, H., Klechevsky, E., Morita, R., Aspord, C., Cao, T., Matsui, T., Di Pucchio, T., Connolly, J., Fay, J.W., Pascual, V., Palucka, A.K., and Banchereau, J. (2007) Dendritic cell subsets in health and disease. *Immunol. Rev.*, **219**, 118–142.

36 Merad, M., Manz, M.G., Karsunky, H., Wagers, A., Peters, W., Charo, I., Weissman, I.L., Cyster, J.G., and Engleman, E.G. (2002) Langerhans cells renew in the skin throughout life under steady-state conditions. *Nat. Immunol.*, **3** (12), 1135–1141.

37 Schuler, G. and Steinman, R.M (1985) Murine epidermal langerhans cells mature into potent immunostimulatory dendritic cells-invitro. *J. Exp. Med.*, **161** (3), 526–546.

38 Kanitakis, J., Petruzzo, P., and Dubernard, J. (2004) Turnover of epidermal Langerhans' cells. *N. Eng. J. Med.*, **351** (25), 2661–2662.

39 Maines, M.D. (1988) Heme oxygenase–function, multiplicity, regulatory mechanisms, and clinical-applications. *FASEB J.* **2** (10), 2557–2568.

40 Crofton, R.W., Diesselhoffdendulk, M.M.C., and Vanfurth, R. (1978) Origin, kinetics, and characteristics of kupffer cells in normal steady-state. *J. Exp. Med.*, **148** (1), 1–17.

41 Suzumoto, R., Takami, M., and Sasaki, T. (2005) Differentiation and function of osteoclasts cultured on bone and cartilage. *J. Electron. Microsc.*, **54** (6), 529–540.

42 Udagawa, N., Takahashi, N., Akatsu, T., Tanaka, H., Sasaki, T., Nishihara, T., Koga, T., Martin, T.J., and Suda, T. (1990) Origin of osteoclasts–mature monocytes and macrophages are capable of differentiating into osteoclasts under a suitable microenvironment prepared by bone marrow-derived stromal cells. *Proc. Natl Acad. Sci. U. S. A.*, **87** (18), 7260–7264.

43 Arai, F., Miyamoto, T., Ohneda, O., Inada, T., Sudo, T., Brasel, K., Miyata, T.,

Anderson, D.M., and Suda, T. (1999) Commitment and differentiation of osteoclast precursor cells by the sequential expression of c-Fms and receptor activator of nuclear factor kappa B (RANK) receptors. *J. Exp. Med.*, **190** (12), 1741–1754.

44 McMenamin, P.G. (1999) Distribution and phenotype of dendritic cells and resident tissue macrophages in the dura mater, leptomeninges, and choroid plexus of the rat brain as demonstrated in wholemount preparations. *J. Comp. Neurol.*, **405** (4), 553–562.

45 Degroot, C.J.A., Huppes, W., Sminia, T., Kraal, G., and Dijkstra, C.D. (1992) Determination of the origin and nature of brain macrophages and microglial cells in mouse central-nervous-system, using nonradioactive insitu hybridization and immunoperoxidase techniques. *Glia*, **6** (4), 301–309.

46 Lawson, L.J., Perry, V.H., and Gordon, S. (1992) Turnover of resident microglia in the normal adult-mouse brain. *Neuroscience*, **48** (2), 405–415.

47 Furuie, H., Yamasaki, H., Suga, M., and Ando, M. (1997) Altered accessory cell function of alveolar macrophages: a possible mechanism for induction of Th2 secretory profile in idiopathic pulmonary fibrosis. *Eur. Respir. J.*, **10** (4), 787–794.

48 Tarling, J.D., Lin, H., and Hsu, S. (1987) Self-renewal of pulmonary alveolar macrophages–evidence from radiation chimera studies. *J. Leukoc. Biol.*, **42** (5), 443–446.

49 Thomas, E.D., Ramberg, R.E., Sale, G.E., Sparkes, R.S., and Golde, D.W. (1976) Direct evidence for a bone-marrow origin of alveolar macrophage in man. *Science*, **192** (4243), 1016–1018.

50 Ross, R., Masuda, J., Raines, E.W., Gown, A.M., Katsuda, S., Sasahara, M., Malden, L.T., Masuko, H., and Sato, H. (1990) Localization of PDGF-B protein in macrophages in all phases of atherogenesis. *Science*, **248**, 1009–1011.

51 Hansson, G.K. (2005) Mechanisms of disease–Inflammation, atherosclerosis, and coronary artery disease. *N. Engl. J. Med.*, **352** (16), 1685–1695.

52 Nakashima, Y., Fujii, H., Sumiyoshi, S., Wight, T.N., and Sueishi, K. (2007) Early

human atherosclerosis: accumulation of lipid and proteoglycans in intimal thickenings followed by macrophage infiltration. *Arterioscler. Thromb. Vasc. Biol.*, **27** (5), 1159–1165.

53 Swirski, F.K., Libby, P., Aikawa, E., Alcaide, P., Luscinskas, F.W., Weissleder, R., and Pittet, M.J. (2007) Ly-6Chi monocytes dominate hypercholesterolemia-associated monocytosis and give rise to macrophages in atheromata. *J. Clin. Invest.*, **117** (1), 195–205.

54 Tacke, F., Alvarez, D., Kaplan, T.J., Jakubzick, C., Spanbroek, R., Llodra, J., Garin, A., Liu, J., Mack, M., van Rooijen, N., Lira, S.A., Habenicht, A.J., and Randolph, G.J. (2007) Monocyte subsets differentially employ CCR2, CCR5, and CX3CR1 to accumulate within atherosclerotic plaques. *J. Clin. Invest.*, **117** (1), 185–194.

55 Stary, H.C. (2000) Natural history and histological classification of atherosclerotic lesions. *Arterioscler. Thromb. Vasc. Biol.*, **20**, 1177–1178.

56 Hoofnagle, M.H., Thomas, J.A., Wamhoff, B.R., and Owens, G.K. (2006) Origin of neointimal smooth muscle – We've come full circle. *Arterioscler. Thromb. Vasc. Biol.*, **26** (12), 2579–2581.

57 Tedgui, A. and Mallat, Z. (2006) Cytokines in atherosclerosis: pathogenic and regulatory pathways. *Physiol. Rev.*, **86** (2), 515–581.

58 Newby, A.C. (2005) Dual role of matrix metalloproteinases (matrixins) in neointima formation and atherosclerotic plaque rupture. *Physiol. Rev.*, **85** (1), 1–31.

59 Davies, M.J. (1996) Stability and instability: two faces of coronary atherosclerosis – The Paul Dudley White Lecture 1995. *Circulation*, **94** (8), 2013–2020.

60 Virmani, R., Burke, A.P., Farb, A., and Kolodgie, F.D. (2006) Pathology of the vulnerable plaque. *J. Am. Coll. Cardiol.*, **47** (8), C13–C18.

61 Dollery, C.M. and Libby, P. (2006) Atherosclerosis and proteinase activation. *Cardiovasc. Res.*, **69** (3), 625–635.

62 Boyle, J.J., Weissberg, P.L., and Bennett, M.R. (2002) Human macrophage-

induced vascular smooth muscle cell apoptosis requires NO enhancement of fas/fas-L interactions. *Arterioscler. Thromb. Vasc. Biol.*, **22**, 1624–1630.

63 Johnson, J., Carson, K., Williams, H., Karanam, S., Newby, A., Angelini, G., George, S., and Jackson, C. (2005) Plaque rupture after short periods of fat-feeding in the apolipoprotein E knockout mouse: model characterisation, and effects of pravastatin treatment. *Circulation*, **111**, 1422–1430.

64 Weber, C., Belge, K.U., von Hundelshausen, P., Draude, G., Steppich, B., Mack, M., Frankenberger, M., Weber, K.S., and Ziegler-Heitbrock, H.W. (2000) Differential chemokine receptor expression and function in human monocyte subpopulations. *J. Leukoc. Biol.*, **67** (5), 699–704.

65 Randolph, G.J. (2008) Emigration of monocyte-derived cells to lymph nodes during resolution of inflammation and its failure in atherosclerosis. *Curr. Opin. Lipidol.*, **19** (5), 462–468.

66 Llodra, J., Angeli, V., Liu, J.H., Trogan, E., Fisher, E.A., and Randolph, G.J. (2004) Emigration of monocyte-derived cells from atherosclerotic lesions characterizes regressive, but not progressive, plaques. *Proc. Natl Acad.. Sci. U. S. A.*, **101** (32), 11779–11784.

67 Nahrendorf, M., Swirski, F.K., Aikawa, E., Stangenberg, L., Wurdinger, T., Figueiredo, J.-L., Libby, P., Weissleder, R., and Pittet, M.J. (2007) The healing myocardium sequentially mobilizes two monocyte subsets with divergent and complementary functions. *J. Exp. Med.*, **204** (12), 3037–3047.

68 McDermott, D.H., Fong, A.M., Yang, Q., Sechler, J.M., Cupples, L.A., Merrell, M.N., Wilson, P.W.F., D'Agostino, R.B., O'Donnell, C.J., Patel, D.D., and Murphy, P.M. (2003) Chemokine receptor mutant CX3CR1-M280 has impaired adhesive function and correlates with protection from cardiovascular disease in humans. *J. Clin. Invest.*, **111** (8), 1241–1250.

69 Stout, R.D., Jiang, C.C., Matta, B., Tietzel, I., Watkins, S.K., and Suttles, J. (2005) Macrophages sequentially change their functional phenotype in response to

changes in microenvironmental influences. *J. Immunol.*, **175** (1), 342–349.

70 Van Ginderachter, J.A., Movahedi, K., Ghassabeh, G.H., Meerschaut, S., Beschin, A., Raes, G., and De Baetselier, P. (2005) Classical and alternative activation of mononuclear phagocytes: picking the best of both worlds for tumor promotion. *Immunobiology*, **211** (6–8), 487–501.

71 Wintergerst, E.S., Jelk, J., and Asmis, R. (1998) Differential expression of CD14, CD36 and the LDL receptor on human monocyte-derived macrophages – A novel cell culture system to study macrophage differentiation and heterogeneity. *Histochem. Cell Biol.*, **110** (3), 231–241.

72 Poston, R.N. and Hussain, I.F. (1993) The immunohistochemical heterogeneity of atheroma macrophages – comparison with lymphoid-tissues suggests that recently blood-derived macrophages can be distinguished from longer-resident cells. *J. Histochem. Cytochem.*, **41** (10), 1503–1512.

73 Komohara, Y., Hirahara, J., Horikawa, T., Kawamura, K., Kiyota, E., Sakashita, N., Araki, N., and Takeya, M. (2006) AM-3K, an anti-macrophage antibody, recognizes CD163, a molecule associated with an anti-inflammatory macrophage phenotype. *J. Histochem. Cytochem.*, **54** (7), 763–771.

74 Boyle, J.J., Harrington, H.A., Piper, E., Elderfield, K., Stark, J., Landis, R.C., and Haskard, D.O. (2009) Coronary intraplaque hemorrhage evokes a novel atheroprotective macrophage phenotype. *Am. J. Pathol.*, **174** (3), 1097–1108.

75 Rothe, G., Gabriel, H., Kovacs, E., Klucken, J., Stohr, J., Kindermann, W., and Schmitz, G. (1996) Peripheral blood mononuclear phagocyte subpopulations as cellular markers in hypercholestero-lemia. *Arterioscler. Thromb. Vasc. Biol.*, **16** (12), 1437–1447.

76 Hakkinen, T., Karkola, K., and Yla-Herttuala, S. (2000) Macrophages, smooth muscle cells, endothelial cells, and T-cells express CD40 and CD40L in fatty streaks and more advanced human atherosclerotic lesions – Colocalization with epitopes of oxidized low-density lipoprotein, scavenger receptor, and

CD16 (Fc gamma RIII). *Virchows Arch. Int. J. Pathol.*, **437** (4), 396–405.

77 Waldo, S.W., Li, Y., Buono, C., Zhao, B., Billings, E.M., Chang, J., and Kruth, H.S. (2008) Heterogeneity of human macrophages in culture and in athero-sclerotic plaques. *Am. J. Pathol.*, **172** (4), 1112–1126.

78 Schlitt, A., Heine, G.H., Blankenberg, S., Espinola-Klein, C., Dopheide, J.F., Bickel, C., Lackner, K.J., Iz, M., Meyer, J., Darius, H., and Rupprecht, H.J. (2004) CD14+CD16+ monocytes in coronary artery disease and their relationship to serum TNF-alpha levels. *Thromb. Haemost.*, **92** (2), 419–424.

79 Bouhlel, M.A., Derudas, B., Rigamonti, E., Dièvart, R., Brozek, J., Haulon, S., Zawadzki, C., Jude, B., Torpier, G., Marx, N., Staels, B., and Chinetti-Gba-guidi, G. (2007) PPARγ activation primes human monocytes into alternative M2 macrophages with anti-inflammatory properties. *Cell Metab.*, **6** (2), 137–143.

80 Chinetti, G., Fruchart, J.C., and Staels, B. (2003) Peroxisome proliferator-activated receptors: new targets for the pharmaco-logical modulation of macrophage gene expression and function. *Curr. Opin. Lipidol.*, **14** (5), 459–468.

81 Ditiatkovski, M., Toh, B.-H., and Bobik, A. (2006) GM-CSF deficiency reduces macrophage PPAR-γ expression and aggravates atherosclerosis in ApoE-defi-cient mice. *Arterioscler. Thromb. Vasc. Biol.*, **26** (10), 2337–2344.

82 Qiao, J.H., Tripathi, J., Mishra, N.K., Cai, Y., Tripathi, S., Wang, X.P., Imes, S., Fishbein, M.C., Clinton, S.K., Libby, P., Lusis, A.J., and Rajavashisth, T.B. (1997) Role of macrophage colony-stimulating factor in atherosclerosis – studies of osteopetrotic mice. *Am. J. Pathol.*, **150** (5), 1687–1699.

83 Bonfield, T.L., Thomassen, M.J., Farver, C.F., Abraham, S., Koloze, M.T., Zhang, X., Mosser, D.M., and Culver, D.A. (2008) Peroxisome proliferator-activated receptor-γ regulates the expression of alveolar macrophage macrophage colony-stimulating factor. *J. Immunol.*, **181** (1), 235–242.

84 Fabunmi, R.P., Sukhova, G.K., Sugi-yama, S., and Libby, P. (1998) Expression

of tissue inhibitor of metalloproteinases-3 in human atheroma and regulation in lesion-associated cells. *Circ. Res.*, **83**, 270–278.

85 Newby, A.C. (2008) Metalloproteinase expression in monocytes and macrophages and its relationship to atherosclerotic plaque instability. *Arterioscler. Thromb. Vasc. Biol.*, **28** (12), 2108–2114.

86 Schönbeck, U., Mach, F., Sukhova, G.K., Atkinson, E., Levesque, E., Herman, M., Graber, P., Basset, P., and Libby, P. (1999) Expression of stromelysin-3 in atherosclerotic lesions: regulation via CD40-CD40 ligand signaling *in vitro* and *in vivo*. *J. Exp. Med.*, **189**, 843–853.

87 Halpert, I., Sires, U.I., Roby, J.D., PotterPerigo, S., Wight, T.N., Shapiro, S.D., Welgus, H.G., Wickline, S.A., and Parks, W.C. (1996) Matrilysin is expressed by lipid-laden macrophages at sites of potential rupture in atherosclerotic lesions and localizes to areas of versican deposition, a proteoglycan substrate for the enzyme. *Proc. Natl Acad. Sci. U. S. A.*, **93** (18), 9748–9753.

88 Thomas, A.C., Sala-Newby, G.B., Ismail, Y., Johnson, J.L., Pasterkamp, G., and Newby, A.C. (2007) Genomics of foam cells and nonfoamy macrophages from rabbits identifies arginase-1 as a differential regulator of nitric oxide production. *Arterioscler. Thromb. Vasc. Biol.*, **27**, 571–577.

89 Johnson, J.L., Sala-Newby, G.B., Ismail, Y., Aguilera, C.N.M., and Newby, A.C. (2008) Low tissue inhibitor of metalloproteinases 3 and high matrix metalloproteinase 14 levels defines a subpopulation of highly invasive foam-cell macrophages. *Arterioscler. Thromb. Vasc. Biol.*, **28** (9), 1647–1653.

Index

Atherosclerosis: Molecular and Cellular Mechanisms. Edited by Sarah Jane George and Jason Johnson
Copyright © 2010 WILEY-VCH Verlag GmbH & Co. KGaA, Weinheim
ISBN: 978-3-527-32448-4